PREVENTION AND TREATMENT OF AGE-RELATED DISEASES

Prevention and Treatment of Age-related Diseases

Edited by

Suresh I.S. Rattan
Danish Centre for Molecular Gerontology, University of Aarhus, Denmark

and

Moustapha Kassem
University Hospital of Odense, Denmark

Springer

A C.I.P. Catalogue record for this book is available from the Library of Congress.

ISBN-13 978-90-481-7217-7
ISBN-10 1-4020-5058-5 (e-book)
ISBN-13 978-1-4020-5058-9 (e-book)

Published by Springer,
P.O. Box 17, 3300 AA Dordrecht, The Netherlands.
www.springer.com

Printed on acid-free paper

CONTENTS

PREFACE

During the last 40 years, biogerontology – the study of the biological basis of aging – has progressed tremendously, and it has now become an independent and respectable field of study and research. Numerous universities, medical institutes and research centers throughout the world now offer full-fledged courses on the biology of aging. Pharmaceutical, cosmeceutical, and neutriceutical industry's ever increasing interest in aging research and therapy is also highly apparent. Moreover, increased financial support by the national and international financial agencies to biogerontological research has given much impetus to its further development.

Biogerontologists are now in a position to construct general principles of aging and explore various possibilities of intervention using rational approaches. While not giving serious consideration to the claims made by charlatans, it cannot be ignored that several researchers are making genuine attempts to test and develop various means of intervention for the prevention and treatment of age-related diseases and for achieving healthy old age.

This book takes status of the molecular, cellular, hormonal, nutritional and lifestyle strategies being tested and applied for the prevention and treatment of age-related diseases. The book is comprised of inter-dependent chapters written in the form of critical reviews by the leading researchers and practitioners in their respective fields. The format of the articles is in semi-academic style in which research data from various experimental systems is presented while focusing on their applications in human beings with respect to the prevention and treatment of age-related impairments. Although each chapter does provide an authoritative and up-to-date account of a specific topic, a comprehensive list of original research papers and review articles has also been included for those readers who may like to follow the subject at greater depths.

The target readership is the undergraduate and graduate students in the universities, medical- and nursing-colleges, post-graduate students taking up research projects on different aspects of biogerontology, and practicing clinicians. This books could be an important volume for the college, university and state libraries maintaining a good database in biology, medical and biomedical sciences. Furthermore, this book will also be of much interest to pharmaceutical, and nutrition and healthcare industry for an easy access to accurate and reliable information in the field of aging research and intervention.

Suresh I.S. Rattan and Moustapha Kassem
Editors

CHAPTER 1

BIOLOGICAL CAUSES OF AGING AND AGE-RELATED DISEASES

SURESH I.S. RATTAN
Laboratory of Cellular Ageing, Danish Centre for Molecular Gerontology, Department of Molecular Biology, University of Aarhus, Denmark

Abstract: Aging is a progressive accumulation of molecular damage in nucleic acids, proteins and lipids. The inefficiency and failure of maintenance, repair and turnover pathways is the main cause of age-related accumulation of damage, which is also the basis of all age-related diseases. Research in molecular gerontology is aimed at understanding the genetic and epigenetic regulation of molecular mechanisms at the levels of transcription, post-transcriptional processing, post-translational modifications, and interactions among various gene products. Concurrently, several approaches are being tried and tested to modulate aging. The ultimate aim of such studies is to improve the quality of human life in old age and prolong the health-span. Various gerontomodulatory approaches include gene therapy, hormonal supplementation, nutritional modulation and intervention by free radical scavengers and other molecules. A recent approach is that of applying hormesis in aging research and therapy, which is based on the principle of stimulation of maintenance and repair pathways by repeated exposure to mild stress. A combination of molecular, physiological and psychological modulatory approaches can be effective to prevent and/or treat various age-related diseases

Keywords: lifespan, survival, longevity, stress, hormesis, homeostasis, homeodynamics

1. INTRODUCTION

The significant increase in human life expectancy during the last three generations, achieved primarily by reducing birth-related parturient-deaths and infant-deaths, is however not matched by an equivalent improvement in the health-span in old age. As a biosocial issue, aging is the underlying basis of almost all major human diseases, such as atherosclerosis, cancer, cardiovascular defects, cataract, diabetes, dementia, macular degeneration, neurodegeneration, osteoporosis and sarcopenia.

S. Rattan and M. Kassem (eds.), Prevention and Treatment of Age-related Diseases, 1–13.

Although the optimal treatment of each and every disease, irrespective of age, is a social and moral necessity, preventing the onset of age-related diseases by intervening in the basic process of aging is the best solution for improving the quality of human life in old age.

Biogerontology, the study of the biological basis of aging, has so far unveiled mysteries of aging by describing age-related changes in organisms, organs, tissues, cells and macromolecules. The large body of published data clearly shows that aging has many facets. Most significantly, the progression and rate of aging is highly variable in different species, in organisms within a species, in organs and tissues within an organism, in cell types within a tissue, in sub-cellular compartments within a cell type, and in macromolecules within a cell. Thus, there is neither a single way of defining aging, nor is there a single cause. Furthermore, these observations have led most biogerontologists to abandon the notion of aging being genetically programmed and to consider it as being stochastic and individualistic. A combination of genes, environment and chance appear to determine the course and consequences of aging and the duration of survival of an individual (longevity) (Rattan and Clark, 2005).

2. PRINCIPLES OF AGING

Although the descriptive data about aging suggest that there are no universal markers of aging, some general principles can still be derived, which can be useful for future research and intervention.

First, aging is considered as an emergent phenomenon seen primarily in protected environments which allow survival beyond the natural lifespan in the wild. The natural lifespan of a species has also been termed "essential lifespan" (ELS) (Rattan, 2000), or the "warranty period" of a species (Carnes et al., 2003). ELS is defined as the time required to fulfil the Darwinian purpose of life, that is successful reproduction for the continuation of generations. Species undergoing fast maturation and early onset of reproduction with large reproductive potential generally have a short ELS. In contrast, slow maturation, late onset of reproduction, and small reproductive potential of a species is concurrent with its long ELS. For example, the ELS of *Drosophila* is less than a week as compared with that of about 50 years of *Homo sapiens*, even though in protected environments (laboratories and modern societies), a large proportion of populations of both species can and do live for much longer than that. Therefore, the period of extended survival beyond ELS is also the period of aging.

Second, aging is characterized by a progressive accumulation of molecular damage in nucleic acids, proteins and lipids. The inefficiency and failure of maintenance, repair and turnover pathways is the main cause of age-related accumulation of damage. Since homeostasis or homeodynamic ability of a living system is primarily due to its maintenance and repair processes, it is the progressive failure of maintenance and repair mechanisms which is the universal biochemical basis of aging and age-related diseases (Holliday, 1995, 2000).

Third, unlike development, which is a highly programmed and well-coordinated genetic process in the evolutionary life history of an organism, there is no genetic programme which determines the exact duration of survival of an organism. Furthermore, studies on establishing an association between genes and longevity have reported that the genetic heritability of variance in lifespan is less than 35% (Herskind et al., 1996; Finch and Tanzi, 1997; Korpelainen, 2000; Gudmundsson et al., 2000). The evolutionary theories of aging and longevity have developed sophisticated and convincing arguments against the existence of genes that may have evolved specifically to cause aging and to determine the lifespan of an organism (for a detailed analysis of evolutionary arguments, see (Rose, 1991; Kirkwood and Austad, 2000; Gavrilov and Gavrilova, 2001).

3. THE ROLE AND NATURE OF GERONTOGENES

Genes that do influence longevity are those that have evolved in accordance with the life history of a species for assuring ELS. Several lines of evidence support the view that natural survival and longevity of a species is a function of its maintenance and repair capacities. For example, positive correlations between species lifespan and the ability to repair DNA, to defend against reactive oxygen species, to respond and to counteract stress, and to proliferate and facilitate turnover of cells have been reported. In contrast, there is a negative correlation between longevity and the rate of damage accumulation, including mutations, epimutations, macromolecular oxidation and aggregation (Holliday, 1995; Rattan, 1989; Rattan, 1995).

A lack of specific gerontogenes which cause aging does not imply that genes do not or cannot influence survival, longevity and the rate of aging. There is ample evidence from studies performed on yeast, fungi (Jazwinski, 1999), nematodes (Johnson et al. , 2000; Johnson, 2002), insects (Rogina et al., 2000; Tatar et al., 2001), rodents and humans that mutations in certain genes can either prolong or shorten the lifespan, and are the cause of premature aging syndromes (Arking et al., 2002; Kuro-o et al., 1997; Yu et al., 1996; Martin and Oshima, 2000). Interestingly, these genes cover a wide range of biochemical pathways, such as insulin metabolism, kinases and kinase receptors, transcription factors, DNA helicases, membrane glucosidases, GTP-binding protein coupled receptors, and cell cycle arrest pathways with little or no overlap among them (Rattan, 2000; Johnson, 2002; Martin and Oshima, 2000; Warner, 2005).

Additionally, genetic linkage studies for longevity in mice have identified major histocompatibility complex (MHC) regions (Gelman et al., 1998), and quantitative trait loci on chromosomes 7, 10, 11, 12, 16, 18 and 19 (Miller et al., 1998; De Haan et al., 1998) as putative genes for aging. In human centenarians, certain alleles of HLA locus on chromosome 6 (Gelman et al., 1988), regions of chromosome 4 (Puca et al., 2001), different alleles of APO-E and APO-B, and DD genotype of angiotensin converting enzyme (ACE) have been linked to exceptional longevity. Similarly, several other studies have been published reporting an association between human longevity and single nucleotide polymorphisms in a variety of genes, including heat shock response, restoration of the affected organ/tissue to its original disease-free state. What will be the "treatment" of aging and to what original "age-free" stage one would hope to be restored – to day 1, year 1, 10, 30, 50 or what? Considering aging as a disease and then trying to cure that disease is unscientific and misguided. Similarly, although piecemeal replacement of non-functional or half-functional body parts with natural or synthetic parts made of more durable material may provide a temporary solution to the problems of age-related impairments, it does not modulate the underlying aging process as such.

Scientific and rational anti-aging strategies aim to slow down aging, to prevent and/or delay the physiological decline, and to regain lost functional abilities. However, the history of anti-aging research and therapy is replete with fraud, pseudoscience and charlatanism, and has often given a bad name to the whole field (Boia, 2004). Claims for miraculous remedies and promises for extremely long lifespan are prevalent even today. Recently, highly critical analyses of such approaches have been made by biogerontologists with a view to educate and inform people about the science and non-sense of aging-intervention research (Olshansky et al., 2002).

While not giving serious consideration to the claims made by charlatans, it cannot be ignored that several researchers are making genuine attempts to test and develop various means of intervention for the prevention and treatment of age-related diseases, for regaining the functional abilities and for prolonging the lifespan of experimental organisms. Some of the main anti-aging approaches include supplementation with hormones including growth hormone, dehydroepiandrosterone (DHEA), melatonin and estrogen, and nutritional supplementation with synthetic and natural antioxidants in purified form or in extracts prepared from plant and animal sources (Rattan, 2003; Ferrari, 2004). Although some of these approaches have been shown to have some clinical benefits in the treatment of some diseases in the elderly, none of these really modulate the aging process itself (Olshansky et al., 2002). Furthermore, claims for the benefits of intake of high doses of vitamins and various antioxidants and their supposed anti-aging and life-prolonging effects have very little scientific evidence to back them (Le Bourg, 2005).

In contrast to this, nutritional modulation through caloric restriction (CR) has been shown to be an effective anti-aging and longevity extending approach in rodents and monkeys, with possible applications to human beings (Roth et al., 2004). But, this is a highly debatable issue at present both in terms of the practicalities of defining CR and of applying CR in human beings in physiological and evolutionary contexts (Demetrius, 2004).

Some studies have reported an extension of lifespan of experimental animals by gene manipulation. For example, overexpression of superoxide dismutase and catalase genes and of heat shock protein (hsp) genes have resulted in the increase in average lifespan in *Drosophila* and nematodes, respectively (Orr and Sohal, 1994; Yokoyama et al., 2002). Such a gene-therapy approach for gerontomodulation requires redesigning the blueprint for structural and functional units of the body at the level of genes, gene products, macromolecular interactions, molecular-milieu interactions, and so on. Considering how little information and knowledge we have at present about all those interacting variants of genes, molecules, milieu and chance, it is not clear what this approach really means in practical and achievable terms. Similarly, although piecemeal replacement of non-functional or half-functional body parts with natural or synthetic parts made of more durable material may provide a temporary solution to the problems of age-related impairments, it does not modulate the underlying aging process as such.

5.1 Hormesis

In a more realistic and near-future scenario, a promising approach in aging intervention and prevention is based in making use of an organism's intrinsic homeodynamic property of self maintenance and repair. Since aging is characterized by a decrease in the adaptive abilities due to progressive failure of homeodynamics, it has been hypothesized that if cells and organisms are exposed to brief periods of stress so that their stress response-induced gene expression is upregulated and the related pathways of maintenance and repair are stimulated, one should observe anti-aging and longevity-promoting effects. Such a phenomenon in which stimulatory responses to low doses of otherwise harmful conditions improve health and enhance lifespan is known as hormesis.

Although the phenomenon of hormesis has been defined variously in different contexts, for example in toxicology, pharmacology and radiation biology (Calabrese and Baldwin, 2000; Parsons, 2000), hormesis in aging is characterized by the beneficial effects resulting from the cellular responses to mild repeated stress (Rattan, 2001). The paradigm of hormesis is moderate exercise which is well known to have numerous beneficial effects despite it being a generator of free radicals, acids, and other damaging effects (McArdle et al., 2002).

During the last few years, research done in our labs has shown hormetic effects of mild stress. We have demonstrated the hormetic effects of repeated mild stress (RMS) on human cells undergoing aging in culture. Using a mild stress regime of exposing human skin fibroblasts to 41°C for 1 hr twice a week throughout their replicative lifespan *in vitro*, several beneficial and anti-aging effects have been observed (Rattan et al., 2004). It is important to note that whereas several age-related alterations, such as accumulation of oxidized proteins, levels of various hsp, proteasome activities, and stress resistance, were affected by RMS, there was no change in the proliferative behaviour of cells. This has implications in separating the phenomenon of aging from longevity. It appears that the progression of cellular aging *in vitro* as the increased molecular disorder can be slowed down without upsetting the regulatory mechanisms of cell cycle arrest (Rattan et al., 2004; Rattan et al., 2003). Thus the quality of life of the cell in terms of its structural and functional integrity can be improved without pushing these cells in to potentially carcinogenic hyperproliferative mode.

Other chemical, physical and biological treatments can be used to unravel various pathways of maintenance and repair whose sustained activities improve the physiological performance and survival of cells and organisms. Stresses that have been reported to delay aging and prolong longevity in various systems (for example, yeast, *Drosophila*, nematodes, rodents and human cells) include temperature shock, irradiation (UV-, gamma- and X-rays), heavy metals, pro-oxidants, acetaldehyde, alcohols, hypergravity, exercise and CR (Minois, 2000; Hercus et al., 2003; Rattan, 2004). Hormesis-like beneficial effects of chronic but mild undernutrition have been reported for human beings (Raji et al., 1998). For example, it was reported that peripheral blood lymphocytes isolated from people with low body mass index, representing a group with natural intake of restricted food calories, had higher DNA repair capacity and higher levels of DNA polymerase α, which were also maintained during aging (Raji et al., 1998). Intermittent fasting has been reported to have beneficial effects on glucose metabolism and neuronal resistance to injury (Anson et al., 2003).

Although at present there are only a few studies performed which utilize mild stress as a modulator of aging and longevity, hormesis can be a useful experimental approach in biogerontology. However, there are several issues that remain to be resolved before mild stress can be used as a tool to modulate aging and prevent the onset of age-related impairments and pathologies. Some of these issues are: (1) to establish biochemical and molecular criteria for determining the hormetic levels for different stresses; (2) to identify differences and similarities in stress response pathways initiated by different stressors; (3) to quantify the extent of various stress responses; (4) to determine the interactive and pleiotropic effects of various stress response pathways; (5) to adjust the levels of mild stress for age-related changes in the sensitivity to stress; (6) to determine the biological and evolutionary costs of repeated exposure to stress; and (7) to determine the biological significance of relatively small hormetic effects, which may or may not have large beneficial effects during the entire lifespan. Resolution of these issues requires much more research on hormesis than being carried out at present.

The proof of the hormetic principle has now been provided by experiments with a wide variety of biological systems and by using a range of physical, chemical and biological stressors. Two of the main lifestyle interventions, exercise and reduced food intake, both of which bring their beneficial and anti-aging effects through hormesis (McArdle et al., 2002; Singh, 2002; Masoro, 1998, 2000; Yu and Chung, 2001), are being widely recognized and increasingly practiced as an effective means of achieving a healthy old age.

One can also expect the availability of certain nutriceutical and pharmacological hormetic agents to mimic the HS response and CR. For example, bimoclomal, a nontoxic, hydroxylamine derivative with hsp-inducing activity and cytoprotective effects is under Phase II clinical trials (Vigh et al., 1997; Vigh et al., 1998). Celastrol, a quinone methide triterpene which is an active component of certain Chinese medicinal herbs is another hsp-inducing hormetic agent under test for its cytoprotective effects (Westerheide et al., 2004). Curcumin, an Indian yellow spice, has also been shown to have cytoprotective effects through its hormetic action in stimulating the synthesis of hsp (Dunsmore et al., 2001). Similarly, various chemical mimetics of CR, such as 2-deoxy-D-glucose and its analogues (Lane et al., 2002), and resveratrol, which is a polyphenol found in red wine, are being tested for their use as anti-aging hormetic agents (Lamming et al., 2004; Wood et al., 2004).

Another small molecule, N^6-furfuryladenine or kinetin, has been shown to have significant anti-aging (Rattan and Clark, 1994; Rattan, 2002), and anti-thrombotic (Hsiao et al., 2003) effects in human cells. Kinetin is considered to work both as a direct antioxidant (Olsen et al., 1999; Verbeke et al., 2000), and as a hormetic agent by inducing the synthesis of other protective enzymes and hsp (Rattan, 2002; Barciszewski et al., 1999; Holmes-Davis et al., 2001). Although at present the use of kinetin has been limited to being a cosmeceutical ingredient in a range of cosmetics products, its usefulness as a hormetic nutriceutical agent is under investigation.

In the consideration of irradiation as a hormetic agent, epidemiologic studies of the public, medical cohorts, and occupational workers confirm that low doses of radiation are associated with reduced mortality from all causes, decreased cancer mortality, and reduced mutation load observed in aging and cancer (Pollycove and Feinendegen, 2001). Increasing use of low-dose total body irradiation as an immunotherapy for cancer (Safwat, 2000) also has its basis in hormesis, which, in the not-so-distant future, will be developed into a safe and preventive strategy against a variety of age-related diseases. Hormesis through mental challenge and through mind-concentrating meditational techniques (Bierhaus et al., 2003; De Nicolas, 1998; Kyriazis, 2003) may be useful in stimulating inter- and intra-cellular debris-removal processes, and thus preventing the neuronal loss that leads to the onset of age-related neurodegenerative diseases.

Finally, it must be emphasized that the goal of research on aging is not to increase human longevity regardless of the consequences, but to increase active longevity free from disability and functional dependence. Healthy old age is an achievable goal that however requires significantly more research support and efforts in biogerontology.

ACKNOWLEDGEMENTS

Research in the Laboratory of Cellular Ageing is supported by grants from the Danish Medical Research council FSS, from shared cost action under the EU-Biomed & Health Programme, and research grants from Senetek PLC.

REFERENCES

Altomare, K., Greco, V., Bellizzi, D., Berardelli, M., et al. (2003) The allele (A)-110 in the promoter region of the HSP70-1gene is unfavourable to longevity in women. Biogerontology, 4: 215–220.

Anson, R.M., Guo, Z., de Cabo, R., Lyun, T., et al. (2003) Intermittent fasting dissociates beneficial effects of dietaryrestriction on glucose metabolism and neuronal resistance toinjury from calorie restriction. Proc Natl. Acad. Sci. USA, 100: 6216–6220.

Arking, D.E., Krebsova, A., Macek Sr., M., Macek, J., M., et al. (2002) Association of human aging with a functional variant of klotho. Proc. Natl. Acad. Sci. USA, 99: 856–861.

Atzmon, G., Rincon, M., Rabizadeh, P. and Barzilai, N. (2005) Biological evidence for inheritance of exceptional longevity. Mech. Age. Dev., 126: 341–345.

Barciszewski, J., Rattan, S.I.S., Siboska, G. and Clark, B.F.C. (1999) Kinetin – 45 years on. Plant Sci., 148: 37–45.

Bessenyei, B., Marka, M., Urban, L., Zeher, M., et al. (2004) Single nucleotide polymorphisms: aging and diseases. Biogerontology, 5: 291–300.

Bierhaus, A., Wolf, J., Andrassy, M., Rohleder, N., et al. (2003) A mechanism converting psychosocial stress into mononuclear cellactivation. Proc. Natl. Acad. Sci. USA, 100: 1920–1925.

Boia, L. (2004) Forever Young: A Cultural History of Longevity. London: Reaktion Books Ltd.

Calabrese, E.J. and Baldwin, L.A. (2000) Tales of two similar hypotheses: the rise and fall of chemical and radiation hormesis. Hum. Exp. Toxicol., 19: 85–97.

Carnes, B.A., Olshansky, S.J. and Grahn, D. (2003) Biological evidence for limits to the duration of life. Biogerontology, 4: 31–45.

De Haan, G., Gelman, R., Watson, A., Yunis, E., et al. (1998) Aputative gene causes variability in lifespan among gentoypically identiacal mice. Nat. Genet., 19: 114–116.

Demetrius, L. (2004) Calorie restricition, metabolic rate andentropy. J. Gerontol. Biol. Sci., 59A: 902–915.

De Nicolas, A.T. (1998) The biocultural paradigm: the neural connection between science and mysticism. Exp. Gerontol., 33: 169–182.

Dunsmore, K.E., Chen, P.G. and Wong, H.R. (2001) Curcumin, amedicinal herbal compound capable of inducing heat shock response. Crit. Care Med., 29: 2199–2204.

Ferrari, C.K.B. (2004) Functional foods, herbs and neutraceuticals: towards biochemical mechanisms of healthy aging. Biogerontology, 5: 275–289.

Finch, C.E. and Tanzi, R.E. (1997) Genetics of aging. Science, 278: 407–411.

Franceschi, C., Valensin, S., Bonafè, M., Paolisso, G., et al. (2000) The network and the remodeling theories of aging: historical background and new perspectives. Exp. Gerontol., 35: 879–896.

Franceschi, C., Olivieri, F., Marchegiani, F., Cardelli, M., et al. (2005) Genes involved in immune response/inflammation, IGF/insulin pathway and response to oxidative stress play a majorrole in the genetics of human longevity: the lesson of centenarians. Mech. Age. Dev., 126: 351–361.

Gavrilov, L.A. and Gavrilova, N.S. (2001) The reliability theory of aging and longevity. J. Theor. Biol., 213: 527–545.

Gelman, R., Watson, A., Bronson, R. and Yunis, E. (1988) Murine chromosomal regions correlated with longevity. Genetics, 118: 693–704.

Gudmundsson, H., Gudbjartsson, D.F., Kong, A.N.T., Gudbjartsson, H., et al. (2000) Inheritance of human longevity in Iceland. Eur.J. Hum. Genet., 8: 743–749.

Harley, C.B., Vaziri, H., Counter, C.M. and Allsopp, R.C. (1992) The telomere hypothesis of cellular aging. Exp. Gerontol., 27: 375–382.

Harman, D. (1994) Free-radical theory of aging. Increasing thefunctional lifespan. Annal. N.Y. Acad. Sci., 717: 1–15.

Hercus, M.J., Loeschcke, V. and Rattan, S.I.S. (2003) Lifespan extension of Drosophila melanogaster through hormesis by repeated mild heat stress. Biogerontology, 4: 149–156.

Herskind, A.M.M., M., Holm, N.V., Sørensen, T.I.A., Harvald, B., et al. (1996) The heritability of human longevity: a population-based study of 2872 Danish twin pairs born 1870–1900. Hum. Genet., 97: 319–323.

Holliday, R. (1995) Understanding Ageing. Cambridge: Cambridge University Press. 207.

Holliday, R. (1996) The current status of the protein errortheory of aging. Exp. Gerontol., 31: 449–452.

Holliday, R. (2000) Ageing research in the next century. Biogerontology, 1: 97–101.

Holmes-Davis, R., Payne, S.R. and Comai, L. (2001) The effects ofkinetin and hydroxyurea on the expression of the endogeneous and transgenic *Heat Shock Cognate 80* (HSC80). Plant Cell rep., 20: 744–748.

Howitz, K.T., Bitterman, K.J., Cohen, H.Y., Lamming, D.W., et al. (2003) Small molecule activators of sirtuins extend *Saccharomyces cerevisiae* lifespan. Nature, 425: 191–196.

Hsiao, G., Shen, M.Y., Lin, K.H., Chou, C.Y., et al. (2003) Inhibitory activity of kinetin on free radical formation ofactivated platelets in vitro and on thrombus formation in vivo. Eur. J. Pharmacol., 465: 281–287.

Jazwinski, S.M. (1999) Longevity, genes, and aging: a view provided by a genetic model system. Exp. Gerontol., 34: 1–6.

Johnson, T.E., Cypser, J., de Castro, E., de Castro, S., et al. (2000) Gerontogenes mediate health and longevity in nematodes through increasing resistance to environmental toxins andstressors. Exp. Gerontol., 35: 687–694.

Johnson, T.E. (2002) A personal retrospective on the genetics of aging. Biogerontology, 3: 7–12.

Kanungo, M.S. (1994) Genes and Aging. Cambridge: Cambridge University Press. 325.

Kapahi, P., Boulton, M.E. and Kirkwood, T.B.L. (1999) Positive correlation between mammalian life span and cellular resistanceto stress. Free Radic. Biol. Med., 26: 495–500.

Kirkwood, T.B.L. and Austad, S.N. (2000) Why do we age? Nature, 408: 233–238.

Korpelainen, H. (2000) Variation in the heritability and evolvability of human lifespan. Naturwissenchaften, 87: 566–568.

Kowald, A. and Kirkwood, T.B.L. (1996) A network theory of ageing: the interactions of defective mitochondria, aberrantproteins, free radicals and scavengers in the ageing process. Mutat. Res., 316: 209–236.

Kuro-o, M., Matsumura, Y., Aizawa, H., Kawaguchi, H., et al. (1997) Mutation of the mouse *klotho* gene leads to asyndrome resembling ageing. Nature, 390: 45–51.

Kyriazis, M. (2003) Practical applications of chaos theory to the modulation of human ageing: nature prefers chaos to regularity. Biogerontology, 4: 75–90.

Lakowski, B. and Hekimi, S. (1996) Determination of life-span in *Caenorhabditis elegans* by four clock genes. Science, 272: 1010–1013.

Lamming, D.W., Wood, J.G. and Sinclair, D.A. (2004) Small molecules that regulate lifespan: evidence for xenohormesis. Mol. Microbiol., 53: 1003–1009.

Lane, M.A., Ingram, D.K. and Roth, G.S. (2002) The serious searchfor an anti-aging pill. Sci. Amer., 287: 24–29.

Larsen, P.L. (1993) Aging and resistance to oxidative damage in Caenorhabditis elegans. Proc. Natl. Acad. Sci. USA, 90: 8905–8909.

Le Bourg, E. (2005) Antioxidants and aging in human beings., In: Rattan, S.I.S., Editor. Aging Interventions and Therapies., inpress. World Scientific Publishers.: Singapore.

Lithgow, G.J., White, T.M., Melov, S. and Johnson, T.E. (1995) Thermotolerance and extended lifespan conferred by single-genemutations and induced by thermal stress. Proc. Natl. Acad. Sci. USA, 92: 7540–7544.

Luckinbill, L.S. and Foley, P. (2000) Experimental and empirical approaches in the study of aging. Biogerontology, 1: 3–13.

Martin, G.M. (1997) The Werner mutation: does it lead to a "public" or "private" mechanism of aging? Mol. Med., 3: 356–358.

Martin, G.M. and Oshima, J. (2000) Lessons from progeroidsyndromes. Nature, 408: 263–266.

Masoro, E.J. (1998) Hormesis and the antiaging action of dietary restriction. Exp. Gerontol., 33: 61–66.

Masoro, E.J. (2000) Caloric restriction and aging: an update. Exp. Gerontol., 35: 299–305.

Masoro, E.J. and Austad, S.N. (1996) The evolution of the antiaging action of dietary restriction: a hypothesis. J.Gerontol. Biol. Sci., 51A: B387–B391.

McArdle, A., Vasilaki, A. and Jackson, M. (2002) Exercise and skeletal muscle ageing: cellular and molecular mechanisms. Ageing Res. Rev., 1: 79–93.

Miller, R.A., Chrisp, C., Jackson, A.U. and Burke, D. (1998) Marker loci associated with life span in genetically heterogeneous mice. J. Gerontol. Med. Sci., 53A: M257–M263.

Minois, N. (2000) Longevity and aging: beneficial effects of exposure to mild stress. Biogerontology, 1: 15–29.

Morley, A.A. (1995) The somatic mutation theory of ageing. Mutat. Res., 338: 19–23.

Olovnikov, A.M. (1996) Telomeres, telomerases, and aging: origin of the theory. Exp. Gerontol., 31: 443–448.

Olsen, A., Siboska, G.E., Clark, B.F.C. and Rattan, S.I.S. (1999) N^6-furfuryladenine, kinetin, protects against Fentonreaction-mediated oxidative damage to DNA. Biochem. Biophys. Res.Commun., 265: 499–502.

Olshansky, S.J., Hayflick, L. and Carnes, B.A. (2002) No truth tothe fountain of youth. Sci. Amer., 286: 92–95.

Olshansky, S.J., Hayflick, L. and Carnes, B.A. (2002) Position statement on human aging. J. Gerontol. Biol. Sci., 57A: B292–B297.

Orr, W.C. and Sohal, R.S. (1994) Extension of life-span by over expression of superoxide dismutase and catalase in Drosophila melanogaster. Science, 263: 1128–1130.

Parsons, P.A. (2000) Hormesis: an adaptive expectation with emphasis on ionizing radiation. J. Appl. Toxicol., 20: 103–112.

Pollycove, M. and Feinendegen, L.E. (2001) Biologic responses tolow doses of ionizing radiation: detriment versus hormesis. Part2. Dose responses of organisms. J. Nucl. Med., 42: 26N–37N.

Puca, A.A., Daly, M.J., Brewster, S.J., Matsie, T.C., et al. (2001) A genome-wide scan for linkage to human exceptional longevity identifies a locus on chromosome 4. Proc. Natl. Acad.Sci. USA, 98: 10505–10508.

Raji, N.S., Surekha, A. and Subba Rao, K. (1998) Improved DNA-repair parameters in PHA-stimulated peripheral blood lymphocytes of human subjects with low body mass index. Mech.Ageing Dev., 104: 133–148.

Rattan, S.I.S., Eskildsen-Helmond, Y.E.G. and Beedholm, R. (2003) Molecular mechanisms of anti-aging hormetic effects of mild heatstress on human cells. Nonlinear. Biol. Toxicol. Med., 2: 105–116.

Rattan, S.I.S. (2004) Aging intervention, prevention, and therapy through hormesis. J. Gerontol. Biol. Sci., 59A: 705–709.

Rattan, S.I.S., Gonzales-Dosal, R., Nielsen, E.R., Kraft, D.C., et al. (2004) Slowing down aging from within: mechanistic aspectsof anti-aging hormetic effects of mild heat stress on humancells. Acta Biochimica Polonica, 51: 481–492.

Rattan, S.I.S. series editor; Biology of Aging and its Modulation. 5-volume series. Kluwer Academic Publishers: Dordrecht.

Rattan, S.I.S. (1989) DNA damage and repair during cellularaging. Int. Rev. Cytol., 116: 47–88.

Rattan, S.I.S. (1995) Ageing – a biological perspective. Molec. Aspects Med., 16: 439–508.

Rattan, S.I.S. (1995) Gerontogenes: real or virtual? FASEB J., 9: 284–286.

Rattan, S.I.S. (1998) The nature of gerontogenes and vitagenes. Antiaging effects of repeated heat shock on human fibroblasts. Annal. NY Acad. Sci., 854: 54–60.

Rattan, S.I.S. (2000) Ageing, gerontogenes, and hormesis. Ind. J.Exp. Biol., 38: 1–5.

Rattan, S.I.S. (2001) Applying hormesis in aging research and therapy. Hum. Exp. Toxicol., 20: 281–285.

Rattan, S.I.S. (2002) N6-furfuryladenine (kinetin) as a potential anti-aging molecule. J. Anti-aging Med., 5: 113–116.

Rattan, S.I.S., ed. *Modulating Aging and Longevity.* 2003, Kluwer Academic Publ.: Dordrecht, The Netherlands.

Rattan, S.I.S. and Clark, B.F.C. (1994) Kinetin delays the onset of ageing characteristics in human fibroblasts. Biochem. Biophys. Res. Commun., 201: 665–672.

Rattan, S.I.S. and Clark, B.F.C. (2005) Understanding and modulating ageing. IUBMB Life, 57: 297–304.

Rogina, B., Reenan, R.A., Nilsen, S.P. and Helfand, S.L. (2000) Extended life-span conferred by cotransporter gene mutation in *Drosophila.* Science, 290: 2137–2140.

Rose, M.R. (1991) Evolutionary Biology of Aging. New York: Oxford University Press. 220.

Roth, G.S., Mattison, J.A., Ottinger, M.A., Chachich, M.E., et al. (2004) Aging in Rhesus monkeys: relevance to human health interventions. Science, 305: 1423–1426.

Safwat, A. (2000) The role of low-dose total body irradiation in treatment of non-Hodgkins lymphoma: a new look at an old method. Radiother. Oncol., 56: 1–8.

Singh, A.M.F. (2002) Exercise comes of age: rationale and recommendations for geriatric exercise prescription. J. Gerontol. Med. Sci., 57A: M262–M282.

Singh, R., Kølvraa, S., Bross, P., Gregersen, N., et al. (2004) Association between low self-rated health and heterozygosity for -110A-C polymorphism in the promoter region of HSP70-1 in aged Danish twins. Biogerontology, 5: 169–176.

Tan, Q., De Benedictis, G., Yashin, A.I., Bonafe, M., et al. (2001) Measuring the genetic influence in modulating the humanlife span: gene-environment interaction and the sex-specificgenetic effect. Biogerontology, 2: 141–53.

Tatar, M., Kopelman, A., Epstein, D., Tu, M.P., et al. (2001) Amutant *Drosophila* insulin receptor homolog that extendslife-span and impairs neuroendocrine function. Science, 292: 107–110.

Tyner, S.D., Venkatachalam, S., Choi, J., Jones, S., et al. (2002) p53 mutant mice that display early ageing-associated phenotypes. Nature, 415: 45–53.

Verbeke, P., Siboska, G.E., Clark, B.F.C. and Rattan, S.I.S. (2000) Kinetin inhibits protein oxidation and glyoxidation invitro. Biochem. Biophys. Res. Commun., 276: 1265–1267.

Vigh, L., Literati, P.N., Horváth, I., Török, Z., et al. (1997) Bimoclomol: a nontoxic, hydroxylamine derivative with stress protein-inducing activity and cytoprotective effects.Nature Medicine, 3: 1150–1154.

Vigh, L., Maresca, B. and Harwood, J.L. (1998) Does the membrane's physical state control the expression of heat shockand other genes? TIBS, 23: 369–374.

Vijg, J. (2000) Somatic mutations and aging: a re-evaluation. Mutat. Res., 447: 117–135.

Warner, H. (2005) Longevity genes: from primitive organisms tohumans. Mech. Age. Dev., 126: 235–242.

Weindruch, R. (1996) Calorie restriction and aging. Sci. Amer., 274: 32–38.

Westerheide, S.D., Bosman, J.D., Mbadugha, B.N.A., Kawahara, T.L.A., et al. (2004) Celastrols as inducers of the heat shock response and cytoprotection. J. Biol. Chem., 279: 56053–56060.

Wood, J.G., Rogina, B., Lavu, S., Howitz, K.T., et al. (2004) Sirtuin activators mimic caloric restricition and delay ageing inmetazoans. Nature, 430: 686–689.

Yokoyama, K., Fukumoto, K., Murakami, T., Harada, S., et al. (2002) Extended longevity of *Caenorhabditis elegans* byknocking in extra copies of hsp70F, a homolog of mot-2(mortalin)/mthsp70/Grp75. FEBS Lett., 516: 53–57.

Yu, C.-E., Oshima, J., Fu, Y.-H., Wijsman, E.M., et al. (1996) Positional cloning of the Werner's syndrome gene. Science, 272: 258–262.

Yu, B.P. and Chung, H.Y. (2001) Stress resistance by caloric restriction for longevity. Ann. N.Y. Acad. Sci., 928: 39–47.

Yu, B.P. (1999) Approaches to anti-aging intervention: the promises and the uncertainities. Mech. Ageing Dev., 111: 73–87.

CHAPTER 2

IMMUNITY, INFLAMMATION AND INFECTIONS DURING AGING

The susceptibility to infections in elderly individuals

MIRIAM CAPRI[1,2], STEFANO SALVIOLI[1,2], FEDERICA SEVINI[1,2], ELISA CEVENINI[1,2], MICHELA PIERINI[1,2], LAURA CELANI[1,2], LAURA BUCCI[3], RITA OSTAN[1], MARIA SCURTI[1,2], DANIELA MAZZA[1,2], DANIELA MONTI[3] AND CLAUDIO FRANCESCHI[1,2,4]

[1] *Department of Experimental Pathology, Via S.Giacomo, 12, University of Bologna*

[2] *CIG, Interdepartmental Center Galvani, Via S. Giacomo, 12, University of Bologna*

[3] *Department of Experimental Oncology and Pathology, Via Morgagni, 50, University of Firenze*

[4] *INRCA, National Institute for Research on Aging, Via Birarelli 8, Ancona, Italy*

Abstract: The major changes occurring life long in the human immune system are here described. The progressive inflammatory status which is established during aging together with the progressive susceptibility to infectious diseases are discussed in the frame of the genetic variant influence. Finally, the possibility to counteract the susceptibility to infections by coping with or slowing down immunosenescence, using different molecules or strategies, is argued

Keywords: aging, cytokines, immunity, inflammation, macrophage, T lymphocyte, infectious, genes, anti-immunosenescence

1. INTRODUCTION

Historically, immunity (from the Latin word *Immunitas*) meant protection from diseases and, more specifically, infectious diseases. Leucocytes and molecules, such as cytokines[1] and products of the inflammatory response, are the main responsible for immunity. Actually, the term leucocytes means cells belonging to both natural immunity (or innate, or native) and specific immunity (or adaptive, or clonotipycal).

S. Rattan and M. Kassem (eds.), Prevention and Treatment of Age-related Diseases, 15–29.

The former is established by monocytes/macrofages, granulocytes, Natural Killer cells (NK); the latter by lymphocytes (B and T lymphocytes with different biological and phenotypical[2] proprieties). Indeed, these two compartments are completely integrated in a network and the coordinate attack to foreign substances or micro-organisms is called immune response.

During aging, this immune response can be affected and deregulated. The senescence of the immune system (IS), or Immunosenescence is part of the more general phenomenon of body senescence, and the different theories of aging, which have been proposed during the last century, can also apply to the cells of the IS. Among these theories, "the remodelling theory of aging" (Franceschi and Cossarizza, 1995), based on experimental evidences from studies on healthy young, elderly and centenarian subjects, conceptualised the dynamic adaptation of the body to the age-dependent modifications. These modifications are well characterised in the immune system, in both innate and specific compartments, as it will be described in the next paragraphs. Likely, immunosenescence is responsible for a series of age-related phenomena, among which the increased susceptibility to infectious diseases, thus, it is possible to hypothesize that strategies aimed to counteract the aging of IS, will lead to a decrease of the incidence of infectious diseases. This topic will be discussed at the end of the chapter.

1.1 Immunosenescence within an evolutionary perspective

It is important to outline that most of our ancestors, in the hostile environment of thousands or millions years ago, lived until reproduction. Indeed, the average life expectancy until 1800 was about 40 years even in the most economically developed Countries. In fact, only genetic variants (or polymorphisms[3]) favourable for assuring survival until 30–40 years of age have been selected, despite their possible detrimental role in old age (60 years or more). Recently, our species, *H. sapiens sapiens*, was able to drastically change its environment and to improve living conditions (nutrition, heating, hygiene and medication) and thus the IS must serve the soma of individuals living 80–120 years, an enormous amount of time, largely unpredicted by evolution. Therefore, our IS, selected to help the body only until the age of reproduction, has now to cope with an unprecedented exposure to antigenic burden for a period of time of several decades longer than in the recent past. Thus, we can hypothesize that the IS is evolutionary "unfit" to the recently emerged human longevity. Indeed the immune system appears to be very efficient in neutralizing and eliminating agents which provoke acute infections in young bodies, while it is much less capable of mounting effective immune response towards agents which provoke infections in aged bodies. In this case the causal agents are not neutralized properly and they remain in the body of old people provoking chronic infections which can persist for decades, being responsible of a chronic stimulation of the IS.

1.2 "Inflamm-aging"

It is a trivial topic that elderly people are more susceptible to infections than young people; moreover, they need of more time to recover completely and the mortality due to viruses and bacteria almost exclusively concerns elderly people. In fact, the infections are the major cause of death in the elderly (Mocchegiani et al., 2000). Clearly, this could be also due to the concomitance of different diseases (co-morbidities), but, as a general trend, old individuals are more susceptible to common pathogens. Why? To answer this question a great amount of scientific data shows that aging modifies activities and phenotype of the cells, together with the intensity, duration and quality of cellular responses. This aspect is true also for the cells of the IS, which are responsible for the good health status of each one of us. According to many experimental data, it seems that the phenomenon of immunosenescence likely impinges upon both acquired immunity and natural immunity, which are both hyper-stimulated by the life long exposure to antigens (Franceschi et al., 2000d).

As far as natural immunity is concerned, monocytes or macrophages[4] play an important role in the immune network as one of the first line of defence against micro-organisms. Moreover, their ability to produce different types of cytokines is relevant for the enrollment and differentiation of lymphocytes, responsible of the antigen-specific response[5]. Data from literature, based on the analysis of cell activation markers as well as on biological activity assays, indicate that monocytes appear to be more activated in aged subjects. In specific, their production of cytokines, such as Interleukin-1β (IL-1β), Interleukin-6 (IL-6), Interleukin-10 (IL-10), together with some chemokines[6], is up-regulated during aging (Sadeghi et al., 1999; Olivieri et al., 2002; Mariani et al., 2002). These cytokines/chemokines, except for IL-10, are all involved in inflammatory phenomena.

In this respect, our group argued that the chronic exposure to antigens leads to a progressive activation of macrophages and related cells in most organs and tissues of the body. In other words, the continuous antigenic challenge could be responsible for a progressive pro-inflammatory status, which appears to be one of the major characteristics of the aging process. We named this phenomenon *inflamm-aging* (Franceschi et al., 2000b; Franceschi et al., 2000c). The remodelling of the organism occurring with age could be, at least in part, orchestrated by a shift of cytokine production toward a pro-inflammatory profile, together with other endocrine and metabolic alterations (Paolisso et al., 2000).

A contribution to the onset of an inflammatory status could also be provided by other cells of the natural IS, such as NK and granulocytes. NK cells are defined as non-B, non-T lymphocytes and they have a fundamental role against viruses and tumours. We reported an increased number of cells with NK markers (as both absolute number and percentage) and a well preserved MHC non-restricted cytotoxic activity in the elderly and even more in centenarians (Sansoni et al., 1993). This finding has been subsequently confirmed by other authors (Mariani et al., 1999; Miyaji et al., 2000). It has been proposed that the increased number of NK cells can be a compensatory mechanism that counteracts the age-dependent decrease in the functionality of such cells. Interestingly, the same Authors found an increased production of Interferon-gamma (IFN-γ) by NK cells in both middle-aged subjects and centenarians (Miyaji et al., 2000). IFN-γ is another important pro-inflammatory cytokine. In addition it has also been demonstrated that NK cells derived from healthy nonagenarians retain the ability to synthesize some chemokines and are able to up-regulate their production in response to stimulation by IL-12 and IL-2 cytokines (Mariani et al., 2002). It is important to remember that IL-12 and IL-2 are among the most effective inducers of NK activity and play a key role in the initiation and maintenance of immune response. Thus, these data confirm that the aging process could be responsible also for the up-regulation and differentiation of NK cells towards a specific pro-inflammatory profile. Moreover, in unhealthy centenarians, a high number of T lymphocytes expressing NK markers and producing high amount of IFN-γ has been found (Miyaji et al., 2000).

As far as granulocytes are concerned, they are typically involved in the inflammatory response for counteracting a large variety of antigens and pathogens. Their production of cytokines is also affected by aging. Indeed, it has been found that IL-1β and Tumour Necrosis Factor-alpha (TNF-α), another pro-inflammatory cytokine mainly produced by granulocytes, are up-regulated in centenarians (Miyaji et al., 2000).

Moreover, we recently reported that another pro-inflammatory cytokine, i.e. IL-18, increases with age and that centenarians display significantly higher IL-18 serum level compared to people of younger ages (Gangemi et al., 2003). However, higher levels of IL-18-binding protein, i.e. a protein which binds and neutralizes IL-18, is also increased, suggesting that compensatory mechanisms capable of quenching the pro-inflammatory activity of IL-18 likely occur with age.

To this regard it is interesting to remember that high serum levels of TNF-α are considered as a strong predictor of mortality in both 80-years-old people (Bruunsgaard et al., 2003a) and centenarians (Bruunsgaard et al., 2003b).

On the whole, many studies support the general concept that aging, up to the extreme ages, is characterized by a shift in the production of cytokines in favour of the pro-inflammatory ones. It is at present unknown whether the derangement in the regulation of inflammatory reactions is a cause or rather an effect of the aging process as a whole. Nevertheless, an altered inflammatory response can probably be the result of a life long exposure to stressors[7] such as antigens, but also chemical and physical agents that threaten the integrity of the organism (Franceschi et al., 2000c).

The chronic pro-inflammatory status can be in some cases an important cause of damage, by itself or by interacting with other pathological molecular mechanisms, thus contributing to the acceleration of the onset of different diseases, or to their severity. Indeed, it has been demonstrated that a pro-inflammatory status renders the subjects more prone to a variety of infectious and non infectious diseases (cardiovascular diseases, neurodegenerative disorders, osteoporosis[8], sarcopenia[9] and diabetes, among others) (De Martinis et al., 2005).

1.3 Specific immunity: remodelling and filling of the "immunological space"

Specific immune response is both humoral (that is, mediated by antibodies produced by B lymphocytes), and cell-mediated (that is, mediated by T lymphocytes, whose two main subclasses are named CD4+helper[10] and CD8+cytotoxic[11] lymphocytes). Both types of response, humoral and cell-mediated, are modified and remodelled by aging. As far as humoral response, we found that the number of circulating B lymphocytes decreased with age, and concomitantly an increase of the serum level of immunoglobulin classes (IgG[12] and IgA[13] but not IgM[14]) was observed (Paganelli et al., 1992). Tissue-specific autoantibodies[15] were also observed to increase in old people, but not in healthy centenarians (Mariotti et al., 1992).

As far as cell-mediated response, is concerned we and others observed that the major characteristic of immunosenescence appears to be the accumulation of memory and effector antigen-experienced T cells[16], accompanied by a decrease of virgin, antigen-non experienced, T cells (Cossarizza et al., 1996; Fagnoni et al., 1996; Fagnoni et al., 2000; Wack et al., 1998; Pennesi et al., 2001). Thus, the progressive expansion of clones[17] of memory cells, together with the age-related decrease of thymic[18] production of virgin T cells (thymic output), able to recognise and to cope with new antigens, leads to a progressive accumulation of cells less responsive or even inactive towards antigens, and in general to a weakening of the IS responses. We proposed to indicate this phenomenon as the "filling of the immunological space" with memory cells (Franceschi et al., 2000a; Franceschi et al., 2000c).

Moreover, recent data suggest that also T lymphocytes aged subjects display a shift toward the production of pro-inflammatory cytokines (Zanni et al., 2003). CD8+ T lymphocytes (or cytotoxic T lymphocytes) appear to be the most affected by aging; indeed the number and the percentage of this cell subset increase during aging together with the loss of their functionality. In particular, cytotoxic T lymphocytes lose CD28 costimulatory molecules (Fagnoni et al., 1996; Fagnoni et al., 2000) and reduce their antiviral effector function (Effros, 2004). Actually, our recent data demonstrated that a large clonal expansion of peripheral CD8+ T lymphocytes specific for cytomegalovirus[19] (CMV) and Epstein Barr virus[20] (EBV) are common in elderly individuals, thus confirming that immunosenescence is strictly associated to the life long exposure to a wide antigenic load (Vescovini et al., 2004).

Likely, it can be hypothesized that the filling of the immunological space with clonally expanded, virus-specific, T lymphocytes, together with the persistence of an antigenic burden could impair the antigen processing[21] and in particular the activity of the immunoproteasome[22] (Mishto et al., 2003), which is also modulated by different cytokines. The antigen recognition by T lymphocytes can occur when the antigen processing is correctly made, otherwise T cell function is deranged and the susceptibility to infections increases.

Interestingly, longitudinal studies[23], performed on lymphocytes from the same group of old individuals over many years, show that an "immune risk phenotype (IRP)", predictor of mortality, can be determined in very old people. This IRP is described in a recent paper (Pawelec et al., 2004) and it is defined as the concomitant presence of a series of different features, such as the ratio of CD4+ T cells vs CD8+ T cells lower than 1; a poor T-cell proliferative response to mitogens[24]; an increase in CD8+, CD28−, CD57+[25] cells; a low number of B cells; the seropositivity[26] to CMV and EBV.

On the whole, these modifications (chronic inflammatory status, and progressive derangement of lymphocyte activity), likely account for the proneness of old people to infectious diseases. In specific skin, lung, together with other tissues or organs, can be infected when immune system weakens during aging (Laube, 2004; Meyer, 2004; Gavazzi and Krause, 2002). Influenza seems to be the major health problem among elderly people in industrialized Countries. An estimated 90% of the 10,000–40,000 excess death caused annually by flu in the United States occurs in subjects aged more than 65 years (Castle, 2000). Actually, diseases such as emphysema[27], diabetes or chronic renal failure[28] and in general co-morbidities can also increase the risk of infections.

Considering that aging impacts on the capability to produce different levels of cytokines and to mount an immune and inflammatory response, (and to respond to specific antigenic stimuli), and that all these phenomena are characterized by an extensive individual variability, key questions are to be ascertained: 1. whether genetic variants of genes involved in innate immunity, inflammation and specific immunity play a role in immunosenescence; 2. whether a peculiar genetic profile of these genes is correlated to longevity; 3. whether a relationship exist between such a genetic profile and resistance/susceptibility to infectious diseases in the elderly. The last question, which is the most relevant from a clinical and biomedical point of view, is unfortunately difficult to answer at present, owing to the scanty data available. Thus, in the next section we will focus on the available data on the functional genetic variants of pro- and anti-inflammatory cytokine genes in nonagenarians and centenarians. We will argue that these data are consistent with the hypothesis that genetics and antagonistic pleiotropy[29] play a role in immunosenescence, as well as in longevity and resistance/susceptibility to infections in old age.

2. CYTOKINES AND GENES

Pro-inflammatory and anti-inflammatory cytokines have a fundamental role in the regulation of immune response against pathogens all along our life and during aging too (Rink and Kirchner, 2000; Pawelec, 1995). As above mentioned, abnormal increments of pro-inflammatory cytokines are involved in the appearance of some of the most common age-related disease, as well as infections (Mocchegiani et al., 2000).

Interestingly, in a recent study (Naumova et al., 2003) ten families with long living members from Bulgarian population were analysed. The authors found a significant[30] association among longevity, genotype of anti-inflammatory cytokine, and the absence of IRP. Thus, they concluded that a combination of specific genetic variants, together with the absence of IRP, could contribute to successful aging and to maintaining healthy status in the elderly. From a general point of view these results fit the hypothesis we are testing since several years that a genetically determined capability of producing low amounts of pro-inflammatory cytokines and concomitantly high levels of anti-inflammatory cytokines favour human longevity (Franceschi and Bonafè, 2003)

Accordingly, we surmise that genes of immunity, and in particular those neutralizing/counteracting the onset of the chronic pro-inflammatory status which develops with age, could have an important role also for coping with infectious diseases. Indeed, it is well known that some functional, genetic variants can modulate the serum level of the respective cytokine. When we studied the effect of a genetic polymorphism at position −174 in the IL-6 gene promoter[31] (a cytosine to guanine transition, −174 C/G) on IL-6 production in old subjects, we found that male (but not female) subjects with a GG genotype had significantly higher serum levels of IL 6 with respect to subjects with CC and CG genotypes. Accordingly, in male centenarians the frequency of GG subjects was lower than in young people (Olivieri et al., 2002). These data have been further confirmed by other groups (Rea et al., 2003; Ross et al., 2003).

We also studied the IFN-γ cytokine, in particular the polymorphism of +874T/A, where the presence of the +874A allele[32] is known to be associated with low IFN-γ production. 174 Italian centenarians and 248 control subjects were analysed and it was found that the +874A allele was found more frequently in centenarian women than in centenarian men or in control women (Lio et al., 2002a). The presence of this allele, significantly increases the possibility to achieve extended longevity, and fits the hypothesis that an anti-inflammatory cytokine profile could be crucial for successful aging.

These considerations are further confirmed by studies on the anti-inflammatory cytokine IL-10. This cytokine has a genetic polymorphism (−1082 G/A) that has been suggested to be correlated with high production of IL-10, and subjects carrying the −1082GG genotype are found to be more represented in centenarians (Lio et al., 2002b) and to be less affected by age-related diseases such as myocardial infarction and Alzheimer's disease (Lio et al., 2003; Lio et al., 2004). Thus, high serum levels of an anti-inflammatory cytokine such as IL-10 might favour a successful aging. The same considerations apply to another important anti-inflammatory cytokine such as TGF-beta1 (TGF-β1). We observed an increased plasma level of active TGF-β1 in centenarians in comparison to young subjects, active TGF-β1 plasma levels were significantly increased in the elderly group, but no relationship with TGF-β1 gene polymorphisms was observed (Carrieri et al., 2004).

Moreover, recently we analysed the −308G/A polymorphism of TNF-α in old subjects affected or not affected by infectious diseases and we found that the frequency of −308A allele is increased in subjects suffering by infectious disesases in comparison with healthy old controls (Cipriano et al., 2005). This last finding suggests an association between allelic variants of cytokine genes and the susceptibility to infections during aging.

Nevertheless, quite paradoxically, pro-inflammatory characteristics have also been documented in healthy centenarians. In this perspective, chronic inflammatory response, as already mentioned, might represent an attempt of the organism to counteract stressors, including antigens, and to restore homeostasis (Franceschi and Bonafè, 2003). As discussed later in this session, we have argued that a pro-inflammatory status might represent the first, necessary but *per se* insufficient hit to frailty, disease and death (Franceschi and Bonafè, 2003; Cipriano et al., 2005).

Thus, inflamm-aging, despite being an inescapable result of the long lasting exposure to acute and chronic infections and to the consequent life long antigenic burden, by itself is not a sufficient condition to trigger age-related diseases, and we can hypothesize that a second, or more than two-hits, are necessary, including a genetic predisposition to the onset of specific age-related diseases and to strong inflammatory responses (Lio et al., 2004; Carrieri et al., 2004; Cipriano et al., 2005; Franceschi et al., 2000d; Ginaldi et al., 2005).

Moreover, it is likely that not only genetic variants related to cytokines could be useful to counteract the susceptibility to infections, but also other genes related to metabolism could be involved in the protection of the organism during aging, as recently reviewed (Franceschi et al., 2005). In addition, other genes such as Human Leucocytes Antigens (HLA) alleles and haplotypes could be relevant to susceptibility or resistance to infections during aging (Caruso et al., 2001; Caruso et al., 2000).

3. ANTI-IMMUNOSENESCENCE STRATEGIES

In this section, we will discuss the possibility to modulate the age-related immune system reshaping in order to counteract the susceptibility to infections.

As described above, immunosenescence is characterised by the following three main aspects: 1. the shrinkage of the T cell repertoire[33] together with an increase in number and size of clones of memory/effector cells; 2. the exhaustion of virgin T cells; 3. the inflamm-aging status. These three aspects are deeply interconnected, and likely share a common pathogenic origin, that is the continuous exposure to the antigenic load, together with the early involution of the thymus. Thus, a main strategy for delaying immunosenescence should take into consideration the following features:

1. To avoid any extra immunological burdens, and to pay a careful attention to neglected sources of antigenic stimulation, such as chronic sub-clinical infections in the oral cavity, the gastrointestinal tract and uro-genital tract, among others, which probably represent a major source of antigenic stimulation. From this point of view, a systematic search for chronic viral infections in the elderly, and the establishment of safe procedures to eradicate them, would be likely to have a beneficial impact on the escape of infections and the reaching of longevity. On the other hand, sometimes it appears impossible to completely eliminate some infectious diseases such as flu during winter, and good strategies of vaccination should be applied to prevent the increasing of mortality among elderly, as recently confirmed in Great Britain (Armstrong et al., 2004).
2. To avoid the shrinkage of T cell repertoire. Indeed, as described before, immunosenescence is accompanied by an expansion of specific clones, such as

CMV-specific CD8+ T lymphocytes; these are unnecessary cells contributing to the IRP and it is legitimate to wonder whether deleting them would improve matters for the individual. Nowadays, this is still an unanswered question, and strategies to this aim have been used only *in vitro* or in animal model systems.

3. To force the expression of CD28 in order to allow a functional recovery of the cells and to allow homeostatic processes to eliminate cells in excess. The feasibility of this latter approach has been demonstrated *in vitro* using specific cells in which the re-introduction of CD28 has reconstituted the capability to produce IL-2 (Topp et al., 2003). Physical removal of the CD28− T cells might in theory enable the expansion of more functional CD8+ T cells and the expansion of their repertoire.
4. To prevent the accumulation of CD28- effector T cells right from the beginning. Since CMV seems to be the main driving factor for their expansion, early vaccination against CMV should be considered. Application of antiviral agents might also become an option because these are already in use in other contexts. Immunization strategies against CMV should be potentially protective from this point of view, as they should avoid the accumulation of terminally differentiated T cell clones (Bernstein et al., 2002)
5. To rejuvenate the thymus and/or delay its involution. Studies by several groups on this topic are very promising (Andrew and Aspinall, 2001; Nasi et al., 2006). An increased output of newly produced virgin thymic cells would counteract the progressive impoverishment of the T lymphocyte repertoire, but some problematic aspects can be anticipated, owing to the possible concomitant enlargement of the immunological space due to the well documented lack of regulation between the thymic input and the size of the peripheral lymphoid tissue (Andrew and Aspinall, 2001). We hope that our study in progress on IL-7[34] production and the presence of virgin T lymphocytes in the peripheral blood of the oldest old, including centenarians, will contribute to elucidate this question (Nasi et al., 2006).
6. To counteract inflamm-aging and all of its deleterious consequences. Data from recent studies suggest that patients treated with anti-inflammatory drugs for long periods of time are apparently protected from age-related diseases, such as Alzheimer's disease (Berzins et al., 2002; Ferrucci et al., 2002; Franceschi et al., 2001). On the basis of what we discussed above, it is reasonable that anti-inflammatory drugs could be also useful to counteract the age-dependent decrease in the capability to cope with infections.
7. To provide old subjects with a correct dietary intake. Indeed, it is important to underline that elderly individuals often have an unbalanced diet, which can cause malnutrition, frailty and weakening of the IS. Thus, it is fundamental to prevent malnutrition and sometimes to add minerals or vitamins to the diet. It was shown that the dietary supplementation with the recommended daily intakes of zinc for one or two months decreases the incidence of infections and increases the rate of survival to further infections in the elderly (Mocchegiani et al., 2000).

4. CONCLUSIONS

In conclusion, we can try to answer the question whether genes or genetic variants involved in immune response have an influence on the susceptibility to infections in the elderly. All the data here reported suggest that a "pro-inflammatory risk" with genetic bases exists. In addition, it is hypothesized that other genes (HLA genes, genes involved in stress response and energy metabolism) could also be relevant for susceptibility or resistance to infections, even if direct evidences are not yet available.

By the way, as discussed, these genes are mostly the same that have been claimed to be associated to human longevity (Franceschi et al., 2005). Indeed, it is conceivable that an allele variant of a gene having a protective effect against age-associated diseases can promote longevity. As stated at the beginning of this Chapter and discussed all along it, the immune function is of primary importance for survival, but likely our IS has been selected only to fit for survival at young ages, but not later on. Thus, one of the most important goal of the next years for biomedicine will be to increase the fitness of our IS with pharmacological or genetic strategies in order to allow it to work in optimal conditions even after 100 years of life.

NOTES

1. Cytokines: hormone-like proteins produced by many different cell types. They mediate inflammatory and immune reactions and affect the behaviour of other cells.
2. Phenotypical: related to all the physical characteristics (morphology, physiology, biochemical features) that result from genetic code.
3. Polymorphism: Natural variation in a gene, DNA sequence, that have no adverse effects on the individual and occurs with fairly high frequency in the general population. The most common polymorphisms are the so-called Single Nucleotide Polymorphisms" (SNPs), whose position on the sequence is indicated with "+" or "−" (e.g. −174; +874) basing on the fact that they are upstream or downstream of the transcription starting point.
4. Macrophages: resident large phagocytic cells derived from circulating monocytes.
5. Antigen-specific response: immune response specifically developed against different microbes and macromolecules.
6. Chemokines: family of structurally related glycoproteins with chemotactic and leukocyte activation activity.
7. Stressors: any chemical, physical, or biological entity that can induce adverse effects on cells or organisms.
8. Osteoporosis: a pathological condition in which there is a decrease in bone mass and bone density and an increased risk and/or incidence of bone fracture.
9. Sarcopenia: loss of muscle mass and function that, generally, comes with aging. This condition strongly influences muscle strength and mobility; it is a factor involved in the occurrence of frailty, falls and fractures in the elderly.
10. CD4+ T helper cells: cells that carry the CD4 co-receptor protein and they are involved in the activation of monocytes and lymphocytes by secreting different types of cytokines.
11. CD8+ T cytotoxic cells: cells that carry the co-receptor protein CD8 and they are involved in the killing of infected cells and tumour cells.
12. IgG: the most abundant immunoglobulin in the blood. It is responsible for the elimination of extracellular bacteria and toxins.
13. IgA: immunoglobulin that represents about 15 to 20% of immunoglobulins in the blood although it is primarily secreted across the mucosae. It is responsible for mucosal immunity.

14. IgM: it is the first antibody that is produced to the exposure to an antigen and it is important for the elimination of extracellular bacteria and toxins.
15. Autoantibodies: antibodies produced against the body's own tissues. They are created by the immune system when it fails to recognize between "self" (the body's normal constituents) and "non-self" (foreign pathogens) and starts to attack its own cells, tissues, and/or organs, leading to the so-called "auto immune diseases".
16. Memory and effector antigen-experienced T cells: two T lymphocyte subpopulations in two different phases of their life, after antigen activation.
17. Clones: a population of cells derived from a single progenitor cell.
18. Thymic production: secreted by thymus, a primary lymphoid organ lying in the thoracic cavity, above and behind the heart.
19. Cytomegalovirus (CMV): member of the herpesvirus family, associated with persistent, latent and recurrent infection. It is usually not very harmful to healthy people.
20. Epstein Barr virus (EBV): human herpesvirus that usually causes an asymptomatic infection. It is the causative agent of infective mononucleosis and has been linked to the development of several cancers, particularly lymphomas in immunosuppressed persons.
21. Antigen processing: sequence of events that convert antigen protein into peptides which mount molecules of Major Hystocompatibility Complex (MHC, important molecules responsible for graft rejection).
22. Immunoproteasome: multimeric proteolitic complex inside the cytokine activated cells.
23. Longitudinal studies: research design where subjects are assessed at several different times in their lives in order to monitor the occurance of risk factors and the health status.
24. Mitogens: molecules able to induce cell division.
25. CD57+: cell-surface glycoprotein principally expressed on different types of cells such as NK, monocytes, some subsets of T and B cells.
26. CMV and EBV seropositivity: presence of antibodies to CMV and EBV in the blood detected by appropriate laboratory tests.
27. Enphysema: a lung pathology featuring the loss of lung elasticity and an abnormal accumulation of air in lung alveoli (tiny air sacs).
28. Chronic renal failure: a pathological condition featuring a slow and progressive deterioration of kidney function. It is also called kidney failure and is usually irreversible.
29. Pleiotropy: the phenomenon whereby a single gene affects several unrelated aspects of the phenotype of an organism.
30. Significant: a possible outcome of a significance test; it is performed to determine statistically if an observed value differs enough from a hypothesized value of a parameter. The choice of the "statistically significant" value is somewhat arbitrary but by convention levels of .05 and .01 are most commonly used.
31. Promoter: short sequence of DNA to which specific enzymes bind in order to initiate transcription of a gene
32. Allele: one of several alternative forms of a gene or DNA sequence at a specific chromosomal locus. At each autosomal locus an individual possesses two alleles, one inherited from the father and one from the mother.
33. Cell repertoire: all the lymphocytes which recognize different antigens in the organism.
34. IL-7: cytokine involved in signalling between cells of the immune system with a specific role for lymphocyte maturation.

REFERENCES

Andrew, D. and Aspinall, R. (2001) Il-7 and not stem cell factor reverses both the increase in apoptosis and the decline in thymopoiesis seen in aged mice. J. Immunol., Feb 1;166(3): 1524–30.

Armstrong, B.G., Mangtani, P., Fletcher, A., Kovats, S., McMichael, A., Pattenden, S. and Wilkinson, P. (2004) Effect of influenza vaccination on excess deaths occurring during periods of high circulation of influenza: cohort study in elderly people. BMJ, Sep 18;329(7467): 660.

Bernstein, D.I., Schleiss, M.R., Berencsi, K., Gonczol, E., Dickey, M., Khoury, P., Cadoz, M., Meric, C., Zahradnik, J., Duliege, A.M. and Plotkin, S. (2002) Effect of previous or simultaneous immunization with canarypox expressing cytomegalovirus (CMV) glycoprotein B (gB) on response to subunit gB vaccine plus MF59 in healthy CMV-seronegative adults. J. Infect. Dis., Mar 1;185(5): 686–90.

Berzins, S.P., Uldrich, A.P., Sutherland, J.S., Gill, J., Miller, J.F., Godfrey, D.I. and Boyd, R.L. (2002) Thymic regeneration: teaching an old immune system new tricks. Trends Mol. Med., Oct;8(10): 469–76.

Bruunsgaard, H., Ladelund, S., Pedersen, A.N., Schroll, M., Jorgensen, T. and Pedersen, B.K. (2003a) Predicting death from tumour necrosis factor-alpha and interleukin-6 in 80-year-old people. Clin. Exp. Immunol. 132: 24–31.

Bruunsgaard, H., Andersen-Ranberg, K., Hjelmborg, J.B., Pedersen, B.K. and Jeune, B. (2003b) Elevated levels of tumor necrosis factor alpha and mortality in centenarians. Am. J. Med., 115: 278–83.

Carrieri, G., Marzi, E., Olivieri, F., Marchegiani, F., Cavallone, L., Cardelli, M., Giovagnetti, S., Stecconi, R., Molendini, C., Trapassi, C., De Benedictis, G., Kletsas, D. and Franceschi, C. (2004) The G/C915 polymorphism of transforming growth factor beta1 is associated with human longevity: a study in Italian centenarians. Aging Cell. 3: 443–8.

Caruso, C., Candore, G., Colonna Romano, G., Lio, D., Bonafè, M., Valensin, S. and Franceschi, C. (2000) HLA, aging, and longevity: a critical reappraisal. Hum. Immunol., Sep;61(9): 942–9.

Caruso, C., Candore, G., Romano, G.C., Lio, D., Bonafè, M., Valensin, S. and Franceschi, C. (2001) Immunogenetics of longevity. Is major histocompatibility complex polymorphism relevant to the control of human longevity? A review of literature data. Mech. Aging Dev., Apr 30;122(5): 445–62.

Castle, S.C. (2000) Clinical relevance of age-related immune dysfunction. Clin. Infect. Dis., Aug;31(2): 578–85.

Cipriano, C., Caruso, C., Lio, D., Giacconi, R., Malavolta, M., Muti, E., Gasparini, N., Franceschi, C. and Mocchegiani, E. (2005) The -308G/A polymorphism of TNF-alpha influences immunological parameters in old subjects affected by infectious diseases. Int. J. Immunogenet., Feb;32(1): 13–8.

Cossarizza, C., Ortolani, R., Paganelli, D., Barbieri, D., Monti, P., Sansoni, U., Fagiolo, G., Castellani, F., Bersani, M., Londei and Franceschi, C. (1996) CD45 isoforms expression on CD4+ and CD8+ T cells throughout life, from newborns to centenarians: implications for T cell memory. Mech. Aging Dev., 86(3): 173–195.

De Martinis, M., Franceschi, C., Monti, D. and Ginaldi, L. (2005) Inflamm-aging and life long antigenic load as major determinants of aging rate and longevity. FEBS Lett., Apr 11;579(10): 2035–9.

Effros, R.B. (2004) T cell replicative senescence: pleiotropic effects on human aging. Ann. N. Y. Acad. Sci., 1019: 123–126.

Fagnoni, F.F., Vescovini, R., Mazzola, M., Bologna, G., Nigro, E., Lavagetto, G., Franceschi, C., Passeri, M. and Sansoni, P. (1996) Expansion of cytotoxic CD8+CD28− T cells in healthy aging people, including centenarians. Immunology, 88(4): 501–507.

Fagnoni, F.F., Vescovini, R., Passeri, G., Bologna, G., Pedrazzoni, M., Lavagetto, G., Casti, A., Franceschi, C., Passeri, M. and Sansoni, P. (2000) Shortage of circulating naive CD8(+) T cells provides new insights on immunodeficiency in aging. *Blood*, 95(9): 2860–2868.

Ferrucci, L., Penninx, B.W., Volpato, S., Harris, T.B., Bandeen-Roche, K., Balfour, J., Leveille, S.G., Fried, L.P. and Md, J.M. (2002) Change in muscle strength explains accelerated decline of physical function in older women with high interleukin-6 serum levels. J. Am. Geriatr. Soc., Dec;50(12): 1947–54.

Franceschi, C. and Cossarizza, A. (1995) Introduction: the reshaping of the immune system with age. Int. Rev. Immunol., 12: 1–4.

Franceschi, C., Monti, D., Sansoni, P. and Cossarizza, A. (1995) The immunology of exceptional individuals: the lesson of centenarians. Immunol. Today, 16: 12–16.

Franceschi, C., Bonafè, M. and Valensin, S. (2000a) Human immunosenescence: the prevailing of innate immunity, the failing of clonotypic immunity, and the filling of immunological space. Vaccine, 18: 1717–1720.

Franceschi, C., Valensin, S., Bonafè, M., Paolisso, G., Yashin, A.I., Monti, D. and De Benedictis, G. (2000b) The network and the remodeling theories of aging: historical background and new perspectives. Exp. Gerontol., 35: 879–896.

Franceschi, C., Bonafè, M., Valensin, S., Olivieri, F., De Luca, M., Ottavini, E. and De Benedictis, G. (2000c) Inflamm-aging. An evolutionary perspective on immunosenescence. Ann. N. Y. Acad. Sci., 908: 244–54.

Franceschi, C., Bonafè, M., Valensin, S., Olivieri, F., De Luca, M., Ottaviani, E. and De Benedictis, G. Inflamm-aging. (2000d) An evolutionary perspective on immunosenescence. Ann. N. Y. Acad. Sci., 908: 244–54.

Franceschi, C., Valensin, S., Lescai, F., Olivieri, F., Licastro, F., Grimaldi, L.M., Monti, D., De Benedictis, G. and Bonafè, M. (2001) Neuroinflammation and the genetics of Alzheimer's disease: the search for a pro-inflammatory phenotype. Aging, Jun;13(3): 163–70.

Franceschi, C. and Bonafè, M. (2003) Centenarians as a model for healthy aging. Biochem. Soc. Trans., 31, 457–461.

Franceschi, C., Olivieri, F., Marchegiani, F., Cardelli, M., Cavallone, L., Capri, M., Salvioli, S., Valensin, S., De Benedictis, G., Di Iorio, A., Caruso, C., Paolisso, G. and Monti, D. (2005) Genes involved in immune response/inflammation, IGF1/insulin pathway and response to oxidative stress play a major role in the genetics of human longevity: the lesson of centenarians. Mech. Aging Dev., Feb;126(2): 351–61.

Gangemi, S., Basile, G., Merendino, R.A., Minciullo, P.L., Novick, D., Rubinstein, M., Dinarello, C.A., Lo Balbo, C., Franceschi, C., Basili, S., D'Urbano, E., Davi, G., Nicita-Mauro, V. and Romano, M. (2003) Increased circulating Interleukin-18 levels in centenarians with no signs of vascular disease: another paradox of longevity? Exp. Gerontol., Jun;38(6): 669–72.

Gavazzi, G. and Krause, K.H. (2002) Aging and infection. Lancet Infect. Dis., Nov;2(11): 659–66.

Ginaldi, L., De Martinis, M., Monti, D. and Franceschi, C. (2005) Chronic antigenic load and apoptosis in immunosenescence. Trends Immunol., 26(2): 79–84.

Laube, S. (2004) Skin infections and aging. Aging Res. Rev., Jan;3(1): 69–89.

Lio, D., Scola, L., Crivello, A., Bonafè, M., Franceschi, C., Olivieri, F., Colonna-Romano, G., Candore, G. and Caruso, C. (2002a) Allele frequencies of +874T– > A single nucleotide polymorphism at the first intron of interferon-gamma gene in a group of Italian centenarians. Exp. Gerontol., Jan–Mar;37(2–3): 315–9.

Lio, D., Scola, L., Crivello, A., Colonna-Romano, G., Candore, G., Bonafè, M., Cavallone, L., Franceschi, C. and Caruso, C. (2002b) Gender-specific association between -1082 IL-10 promoter polymorphism and longevity. Genes Immun., 3: 30–3.

Lio, D., Licastro, F., Scola, L., Chiappelli, M., Grimaldi, L.M., Crivello, A., Colonna-Romano, G., Candore, G., Franceschi, C. and Caruso, C. (2003) Interleukin-10 promoter polymorphism in sporadic Alzheimer's disease. Genes Immun., 4: 234–8.

Lio, D., Candore, G., Crivello, A., Scola, L., Colonna-Romano, G., Cavallone, L., Hoffmann, E., Caruso, M., Licastro, F., Caldarera, C.M., Branzi, A., Franceschi, C. and Caruso, C. (2004) Opposite effects of interleukin 10 common gene polymorphisms in cardiovascular diseases and in successful aging: genetic background of male centenarians is protective against coronary heart disease. J. Med. Genet., Oct;41(10): 790–4.

Mariani, E., Ravaglia, G., Forti, P., Meneghetti, A., Tarozzi, A., Maioli, F., Boschi, F., Fratelli, L., Pizzoferrato, A., Piras, F. and Facchini, A. (1999) Vitamin D, thyroid hormones and muscle mass influence natural killer (NK) innate immunity in healthy nonagenarians and centenarians. Clin. Exp. Immunol., Apr;116(1): 19–27.

Mariani, E., Meneghetti, A., Neri, S., Ravaglia, G., Forti, P., Cattini, L. and Facchini, A. (2002) Chemokine production by natural killer cells from nonagenarians. Eur. J. Immunol., Jun;32(6): 1524–9.

Mariani, E., Pulsatelli, L., Neri, S., Dolzani, P., Meneghetti, A., Silvestri, T., Ravaglia, G., Forti, P., Cattini, L. and Facchini, A. (2002) RANTES and MIP-1alpha production by T lymphocytes, monocytes and NK cells from nonagenarian subjects. Exp. Gerontol., 37: 219–226.

Mariotti, S., Sansoni, P., Barbesino, G., Caturegli, P., Monti, D., Cossarizza, A., Giacomelli, T., Passeri, G., Fagiolo, U., Pinchera, A. and Franceschi, C. (1992) Thyroid and other organ-specific autoantibodies in healthy centenarians. Lancet, 339(8808): 1506–1508.

Meyer, K.C. (2004) Lung infections and aging. Aging Res. Rev., Jan;3(1): 55–67.

Mishto, M., Santoro, A., Bellavista, E., Bonafè, M., Monti, D. and Franceschi, C. (2003) Immunoproteasomes and immunosenescence. Aging Res. Rev., Oct;2(4): 419–32.

Miyaji, C., Watanabe, H., Toma, H., Akisaka, M., Tomiyama, K., Sato, Y. and Abo, T. (2000) Functional alteration of granulocytes, NK cells, and natural killer T cells in centenarians. Hum. Immunol., Sep;61(9): 908–16.

Mocchegiani, E., Muzzioli, M. and Giacconi, R. (2000) Zinc and immunoresistance to infection in aging: new biological tools. Trends Pharmacol. Sci., Jun;21(6): 205–8.

Nasi, M., Troiano, L., Lugli, E., Pinti, M., Ferraresi, R., Monterastelli, E., Mussi, C., Salvioli, G., Franceschi, C. and Cossarizza, A. (2006) Thymic output and functionality of IL-7/IL-7 receptor system in centenarians: implications for the neolymphogenesis at the extreme limit of human life Aging Cell, 5: 167–175.

Naumova, E., Mihaylova, A., Ivanova, M., Michailova, S., Penkova, K. and Baltadjieva, D. (2003) Immunological markers contributing to successful aging in Bulgarians. Exp. Gerontol., 39: 637–644.

Olivieri, F., Bonafè, M., Cavallone, L., Giovagnetti, S., Marchegiani, F., Cardelli, M., Mugianesi, E., Giampieri, C., Moresi, R., Stecconi, R., Lisa, R. and Franceschi, C. (2002) -174 C/G locus affects in vitro/vivo IL-6 production during aging. Exp. Gerontol., 37, 309–314.

Paganelli, R., Quinti, I., Fagiolo, U., Cossarizza, A., Ortolani, C., Guerra, E., Sansoni, P., Pucillo, L.P., Scala, E., Cozzi, E., et al. (1992) Changes in circulating B cells and immunoglobulin classes and subclasses in a healthy aged population. Clin. Exp. Immunol., Nov;90(2): 351–4.

Paolisso, G., Barbieri, M., Bonafè, M. and Franceschi, C. (2000) Metabolic age modelling: the lesson from centenarians. Eur. J. Clin. Invest., 30: 888–894.

Pawelec, G., Adibzadeh, M., Pohla, H. and Schaudt, K. (1995) Immunosenescence: aging of the immune system. Immunol. Today, Sep;16(9): 420–2.

Pawelec, G., Akbar, A., Caruso, C., Effros, R., Grubeck-Loebenstein, B. and Wikby, A. (2004) Is immunosenescence infectious? Trends Immunol., 25(8): 406–410.

Pennesi, G., Morellini, M., Lulli, P., Cappellacci, S., Brioli, G., Franceschi, C. and Trabace, S. (2001) TCR VB repertoire in an italian longeval population including centenarians. J. Amer. Aging Assoc., 24: 63–70.

Rea, I.M., Ross, O.A., Armstrong, M., McNerlan, S., Alexander, D.H., Curran, M.D. and Middleton, D. (2003) Interleukin-6-gene C/G 174 polymorphism in nonagenarian and octogenarian subjects in the BELFAST study. Reciprocal effects on IL-6, soluble IL-6 receptor and for IL-10 in serum and monocyte supernatants. Mech. Aging Develop., 124(4): 555–561.

Rink, L. and Kirchner, H. (2000) Zinc-altered immune function and cytokine production. J. Nutr., May; 130(5S Suppl): 1407S–11S.

Ross, O.A., Curran, M.D., Meenagh, A., Williams, F., Barnett, Y.A., Middleton, D. and Rea, I.M. (2003) Study of age-association with cytokine gene polymorphisms in an aged Irish population. Mech. Aging Develop., 124(2): 199–206.

Sadeghi, H.M., Schnelle, J.F., Thoma, J.K., Nishanian, P. and Fahey, J.L. (1999) Phenotypic and functional characteristics of circulating monocytes of elderly persons. Exp. Gerontol., 3D: 959–970.

Sansoni, P., Cossarizza, A., Brianti, V., Fagnoni, F., Snelli, G., Monti, D., Marcato, A., Passeri, G., Ortolani, C., Forti, E., Fagiolo, U., Passeri, M. and Franceschi, C. (1993) Lymphocyte subsets and natural killer cell activity in healthy old people and centenarians. Blood, Nov., 1;82(9): 2767–73.

Topp, M.S., Riddell, S.R., Akatsuka, Y., Jensen, M.C., Blattman, J.N. and Greenberg, P.D. (2003) Restoration of CD28 expression in CD28 – CD8+ memory effector T cells reconstitutes antigen-induced IL-2 production. J. Exp. Med., Sep 15;198(6): 947–55.

Vescovini, R., Telera, A., Fagnoni, F.F., Biasimi, C., Medici, M.C., Valcavi, P., Di Pede, P., Lucchini, G., Zanlari, L., Passeri, G., Zanni, F., Chezzi, C., Franceschi, C. and Sansoni, P. (2004) Different contribution of EBV and CMV infections in very long-term carriers to age-related alterations of CD8(+) T cells. Exp. Gerontol., 39(8): 1233–1243.

Wack, A., Cossarizza, A., Heltai, S., Barbieri, D., D'Addato, S., Franceschi, C., Dellabona, P. and Castrati, G. (1998) Age-related modifications of the human alphabeta T cell repertoire due to different clonal expansions in the CD4+ and CD8+ subsets. Int. Immunol., 10(9): 1281–1288.

Zanni, F., Vescovini, R., Biasini, C., Fagnoni, F., Zanlari, L., Telera, A., Di Pede, P., Passeri, G., Pedrazzoni, M., Passeri, M., Franceschi, C. and Sansoni, P. (2003) Marked increase with age of type 1 cytokines within memory and effector/cytotoxic CD8+ T cells in humans: a contribution to understand the relationship between inflammation and immunosenescence. Exp. Gerontol., Sep;38(9): 981–7.

CHAPTER 3

PROGRESS AND DEVELOPMENT IN PARKINSON DISEASE THERAPY

CARSTEN R. BJARKAM[1] AND JENS C. SØRENSEN[2]

[1] *Department of Neurobiology, Institute of Anatomy, University of Aarhus, Aarhus, Denmark*
[2] *Department of Neurosurgery, University Hospital of Aarhus, Aarhus, Denmark*

Abstract: Parkinson disease (PD) is a common neurodegenerative disorder affecting 1% of the population aged seventy or more. The causes of PD remain obscure, but basic and clinical research has led to a deep insight into PD pathophysiology, identifying several points of intervention for emerging therapeutic strategies enabling modulation of neural circuits and replacement of lost neurons, neurotransmitters, and neurotrophic factors.

In this chapter we aim, accordingly, to present a overview of the current knowledge on PD pathophysiology and demonstrate how this knowledge provides targets for current and future pharmacological and surgical treatment strategies towards PD

Keywords: Basal ganglia circuitry; Current & future interventions; Neuroprotection; Pharmacological treatment; Surgical treatment

1. PARKINSON DISEASE

The major symptoms of Parkinson disease (PD), comprising rigidity of the limbs, resting tremor, impaired ability to initiate and execute movements (akinesia and bradykinesia) and postural imbalance, were described in 1817 by James Parkinson (Parkinson, 1817). A century later severe loss of the neural cell bodies in the substantia nigra was implicated in the pathogenesis of PD (Tretiakoff, 1919; Freeman, 1925; Greenfield and Bosanquet, 1953) and dopamine was identified as the principal neurotransmitter in the nigrostriatal system (Carlsson et al., 1958; Carlsson, 1959; Bertler and Rosengren, 1959). The main pathological findings of PD are thus a massive loss of more than 80% of the dopaminergic neurons in the brainstem substantia nigra pars compacta and the presence of intraneuronal spherical eosinophilic cytoplasmatic protein aggregates (Lewy

S. Rattan and M. Kassem (eds.), Prevention and Treatment of Age-related Diseases, 31–48.

bodies) in the remaining dopaminergic neurons, resulting in a severe depletion of striatal dopamine (Bernheimer et al., 1973; Kish et al., 1988; Forno, 1996; Dauer and Przedborski, 2003).

Although PD pathology and pathophysiology (see below) are well described, the cause of idiopathic PD remains obscure (Marsden, 1994; Przedborski, 2005) Numerous theories have been proposed, but never consistently proven. They involve environmental factors (Tanner et al., 1997; Tanner, 1989; Semchuk et al., 1992; Rybecki et al., 1993; Tuchsen and Jensen, 2000), oxidative stress (Fahn and Cohen, 1992; Haas et al., 1995; Jenner and Olanow, 1996; Offen et al., 1999), excitotoxicity (Rodriguez et al., 1998), mitochondrial dysfunction (Haas et al., 1995; Duvoisin, 1999; Nakagawa-Hattori et al., 1992; Wooten et al., 1997), infectious agents (Von Economo, 1917; Duvoisin and Yahr, 1965; Nisipeanu et al., 1997), and hereditary causes leading to abnormous protein aggregation and mitochondrial dysfunction (Przedborski, 2005; Krüger, 2004). It should be noted that all these probable etiological causes points towards PD as a multifactorial disease caused by the convergence of multiple external and internal pathogenic factors (Przedborski, 2005).

Parkinson disease is a relatively common neurological disorder, affecting approximately 100–250/100000 of the general population with approximately 11–19/100000 new cases evolving every year (Tanner et al., 1997; Tanner and Ben-Shlomo, 1999; Lindgren et al., 2005). PD is strongly correlated to age, as the prevalence increases steadily throughout the last decades of life, illustrated by a prevalence of 47/100000 for individuals aged 40–49 years, 254/100000 for individuals aged 60–69 years, and 832/100000 for individuals aged 70–79 years (Mutch et al., 1986). Since the population of people over 50 years of age is increasing, the occurrence of PD will continue to grow. It is, therefore, evident that treatments interfering with the causal mechanisms and symptoms of PD will not only be beneficial for the patients, but also have substantial socioeconomic importance (Lindgren et al., 2005).

2. THE BASAL GANGLIA CIRCUITRY (FIGURE 1A)

Normal motor function depends on adequate activity in the basal ganglia which comprises the striatum (Str), the globus pallidus pars externa (GPe), the globus pallidus pars interna (GPi), the subthalamic nucleus (STN), and the brainstem substantia nigra (SN) (Albin et al., 1989; Alexander, 1994; Chesselet and Delfs, 1996).

Under normal conditions (Figure 1A) neural information passes from the cerebral cortex to the striatum (Parent and Hazrati, 1995a) and then through the basal ganglia to the GPi which is considered the main output region of the basal ganglia (Albin et al., 1989; Alexander, 1994; Alexander and Crutcher, 1990; DeLong, 1990; Parent and Hazrati, 1995b). The neurons in the GPi are mainly GABAergic (Smith et al., 1987) and have an inhibitory influence on the ventral anterior and ventrolateral thalamic nuclei (VA-VL). The VA-VL project to the motor cortex and the inhibitory

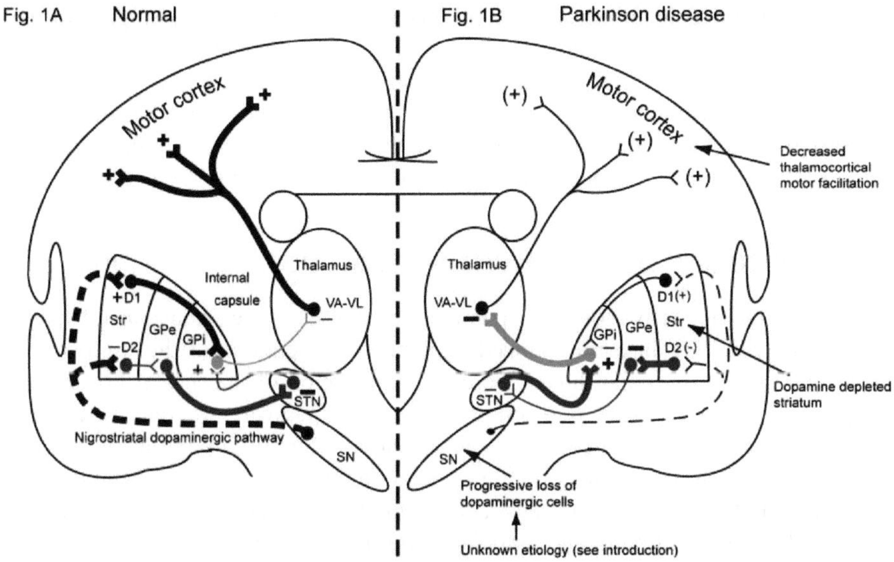

Figure 1. Schematic drawing illustrating the basal ganglia circuitry in health (1A) and PD (1B). Excitatory input is marked by +, whereas inhibitory input is marked by −. Thick lines demarcate facilitated pathways and thin lines demarcate depressed pathways. See text for abbreviations. (1A) Striatum modulates GPi activity via a direct and an indirect pathway. The direct pathway (thin dark line) projects directly to the GPi, whereas the indirect pathway (thin dark gray lines) projects from the Str to the GPe and then to the STN before it reaches GPi. As D1 and D2 receptors act differently on the two pathways, the net result of a proper dopaminergic input to the Str is a reduced inhibitory output from the GPi (thin light gray pathway) and thus a general facilitation of thalamocortical motor activity. (1B) Striatal dopamine depletion leads to an augmented GPi inhibitory output due to reduced activity in the direct pathway and an increased output from the STN to the GPi, causing depression of the thalamocortical motor circuit and thus PD symptoms

input (thin light gray line) from the GPi is, therefore, believed to restrain thalamocortical motor activity. Dopaminergic neurons in the brainstem substantia nigra (SN) supply the striatum with dopamine via the nigrostriatal pathway (fat stippled line), acting on GABAergic neurons expressing either D1 or D2 receptors. GABAergic neurons expressing D1 receptors are stimulated by dopamine, while GABAergic neurons expressing D2 receptors are inhibited. The GABAergic neurons in the striatum influence the activity in the GPi by a **direct pathway** from the striatum to the GPi (fat dark line), and by an **indirect pathway** (thin dark gray lines) from the striatum to the GPi via GABAergic neurons in the GPe and glutaminergic neurons in the STN, the latter having an excitatory influence on the Gpi (Albin et al., 1989; Alexander, 1994; DeLong, 1990; Smith and Parent, 1988; Parent and Hazrati, 1995a) The dopaminergic influence on the D1 receptors of the striatal GABAergic neurons therefore result in an increased inhibition of the GPi output neurons by the direct pathway. The D2 receptors are likewise activated by dopamine, but this activation results in an inhibition of the GABAergic neurons projecting from the striatum to the GPe, and thus in an increased inhibitory influence of the GABAergic neurons projecting from the GPe to the STN which diminishes the excitatory output from the STN to the GPi, resulting in a decreased inhibitory output from the GPi. The net result of a normal dopaminergic influence on the direct and indirect pathways is, accordingly, a reduced inhibitory output from the GPi acting on the thalamocortical motor circuit and thus a general facilitation of motor function (Figure 1A).

3. PARKINSON DISEASE PATHOPHYSIOLOGY (FIGURE 1B)

PD is related to a massive loss of dopaminergic neurons in the brainstem substantia nigra (SN), resulting in a pronounced depletion of dopamine in the nigrostriatal pathway (stippled line) and thus a decreased stimulation of the striatal D1 and D2 receptors. This leads to decreased activity in the direct pathway, whereas the indirect pathway is facilitated. The net result is an increased inhibitory output from the GPi, resulting in decreased thalamocortical motor facilitation and the motor symptoms of PD (Albin et al., 1989; DeLong, 1990; Bjarkam et al., 2001). A similar disturbance in parallel cognitive, limbic and associative cortico-basal ganglia-thalamo-cortical loops is probably responsible for the common occurrence of cognitive decline and psychiatric co-morbidity in PD (Herrero et al., 2002).

Although this model of PD pathophysiology represents a simplification of the complex anatomy and function of the basal ganglia, it has clearly depicted several targets for intervention where different pharmacological and surgical strategies for the treatment of PD may interact and counterbalance the disturbed basal ganglia circuitry (Figures 2 & 3).

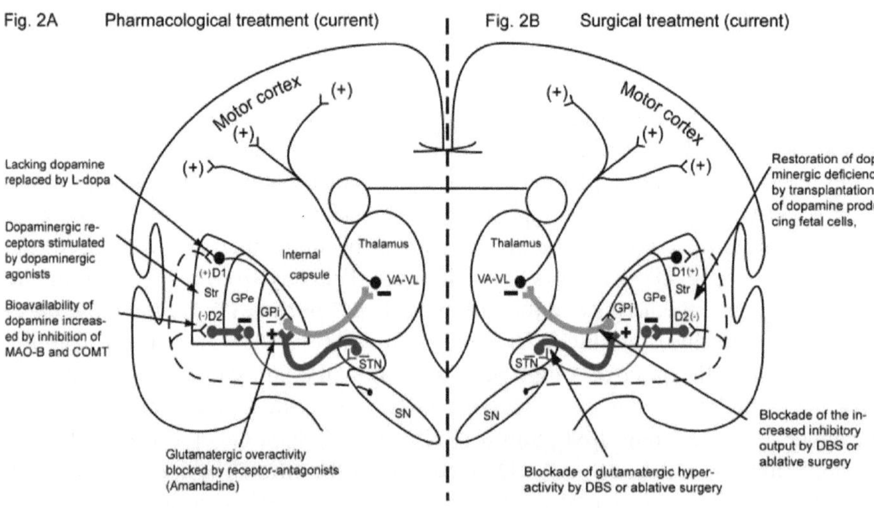

Figure 2. Schematic drawing illustrating where current pharmacological (2A) and surgical (2B) treatment strategies of PD are thought to influence the diseased basal ganglia circuitry

4. PHARMACOLOGICAL TREATMENT OF PARKINSON DISEASE (FIGURE 2A)

Pharmacological treatment of PD is generally the first treatment strategy instituted in the initial phases of PD. In contrast to surgical treatment pharmacological treatment is relatively easy to use and affordable. General problems connected with pharmacological treatments are inadequate drug passage across the blood-brain barrier necessitating use of high drug dosages or drug precursors. Diffuse side effects of the drugs used, on other organs and CNS areas than the basal ganglia likewise, complicate pharmacological treatment strategies.

4.1 L-dopa

The most efficient treatment of PD is replacement of striatal dopamine with a precursor to dopamine, levodopa (L-dopa) (Agid et al., 1999; Rascol et al., 2002; Mercuri and Bernardi, 2005). L-dopa is in contrast to dopamine able to pass the blood-brain-barrier, where after it is converted to dopamine in the striatum. L-dopa is metabolized in the gastrointestinal tract and the liver by the enzyme dopa decarboxylase and is therefore routinely administered together with a peripheral dopa decarboxylase inhibitor (carbidopa or benserazide), which increase the amount of L-dopa available to the CNS.

Long-term use of L-dopa is, however, limited, by the development of motor complications such as dyskinesias, severe motor fluctuations from mobility to immobility (on-off periods) and progressive shortening of the duration of the improved motor response after a dose of L-dopa (Olanow and Stocchi, 2004; Stocchi and Olanow, 2004). It has been speculated that these motor complications are caused by a variable amount of dopamine in the striatal synapses due to the intermittent dosage of L-dopa. Some centers have therefore tried to secure a more constant supply of L-dopa (continuos dopamine stimulation) by oral slow release preparations, i.v.- or intraduodenal drug delivery systems (Stocchi and Olanow, 2004). Slow release preparations of L-dopa are, however, hampered by variable absorption of L-dopa in the gastro-intestinal tract because the drug only is absorbed in the upper part of the small intestine and is further dependent on regular gastric emptying. Slow release preparations have therefore not yet proven them self more useful than commonly used L-dopa preparations. Constant delivery of L-dopa by i.v. and intraduodenal drug delivery systems has proven an effective PD therapy with less motor complications than conventional oral L-dopa treatment (Nyholm and Aquilonius, 2004). These treatments are, however, due to the administration procedure more cumbersome for many patients, difficult to manage for doctors and caregivers, and associated with side effects at the infusion site such as granuloma formation and infections (Stocchi and Olanow, 2004). Transdermal slow release preparations of L-dopa or dopamine agonists may prove to be a future solution to these problems and several promising studies dealing with this drug application method are currently under way (Sudo et al., 2002; Kankkunen et al., 2002; Lewitt and Nyholm, 2004).

4.2 Dopaminergic agonists

Another way to avoid motor complications associated with L-dopa treatment is to use dopaminergic agonists (pergolide, bromocriptine, ropinirole and pramipexole), which acts directly on the dopamine-depleted receptors in the striatum. These drugs are often used as monotherapy in the early phases of PD and then later used as adjunct therapy to L-dopa in the more progressive stages of PD (Jenner, 2003a).

4.3 Monoamine oxidase-B- (MAO-B) inhibitors and Catechol-o-methyltransferase-(COMT) inhibitors

MAO-B inhibitors (selegiline, rasagiline and lazebemide) and COMT inhibitors (entcapone and tolcapone) inhibit two parallel breakdown pathways of dopamine and may therefore secure a larger and more constant level of striatal dopamine in the early stages of PD. These drugs may be used alone in the initial stages of PD as a way to postpone L-dopa treatment and may as adjunct therapy to L-dopa reduce the needed daily L-dopa dose and daily off-time, while on time and motor scores are improved (Rascol et al., 2005; Clarke, 2004).

4.4 Glutaminergic receptor antagonists

Glutamatergic receptor antagonists like amantadine may reduce the postulated excessive glutamatergic output from the subthalamic nucleus and hereby improve PD symptoms. Amantadine has been shown to reduce dyskinesias in advanced PD but this beneficial effect lasted only for 4–9 months where after all patients where withdrawn from treatment due to lack of effect (Clarke, 2004; Thomas et al., 2004). The efficacy of dopaminergic agonists, MAO-B & COMT inhibitors and glutaminergic receptor antagonists is less than that of L-dopa which remains the golden standard for symptomatic pharmacologic PD treatment. Their initial or combined use with L-dopa may, however, avoid or delay the occurrence of drug-induced dyskinesias and neuropsychiatric adverse effects, which often complicate medical treatment of PD (Lang and Lozano, 1998).

5. SURGICAL TREATMENT OF PARKINSON DISEASE (FIGURE 2B)

The efficiency of L-dopa declines over time in a majority of patients where motor-fluctuations and L-dopa induced dyskinesias become frequent (Lang and Lozano, 1998). This has, together with refined stereotaxic neurosurgical procedures and the development of sophisticated brain scanners led to a resurgence of surgical methods for the treatment of PD. Thus, it is estimated that 5–10% of the PD patients will be eligible for surgical procedures encompassing ablative techniques, deep brain stimulation (DBS) and neural transplantation (Hammerstad and Hogarth, 2001; Björklund et al., 2003). Surgical treatment strategies enables generally local and restorative (neural transplantation) interventions in the CNS whereby the diffuse adverse effects seen with pharmacological treatments may be avoided. All neurosurgical procedures, however, carry the risk of causing a potentially life-threatening hemorrhage or introducing infectious agents into the brain. These complications are fortunately rare in most published materials (Hammerstad and Hogarth, 2001), but underscore that the use of surgery must be based on a careful patient selection and pathophysiological models depicting reliable points of intervention. These procedures should only be performed in centers with high neurosurgical standards, enabling meticulous evaluation of inclusion criteria, surgical procedures, and short- and long-term post surgical outcome.

5.1 Neural transplantation

Transplantation of embryonic dopaminergic cells from the midbrain of electively aborted human fetuses to the dopamine depleted striatum in PD has successfully resulted in long term survival of the transplanted tissue, functional reinnervation of the dopamine-depleted striatum and restoration of basal dopamine levels (Hammerstad and Hogarth, 2001; Björklund et al., 2003; Kordower et al., 1995; Piccini et al., 1999; Piccini et al., 2000). Two recently published double-blind studies using this technique have, however, not been able to demonstrate significant efficacy and were both complicated by the occurrence of off-medication dyskinesias (Björklund et al., 2003; Freed et al., 2003; Olanow et al., 2003). The use of this therapeutic strategy is, furthermore, hampered by ethical and practical considerations, e.g. the need of large amounts of fetal dopaminergic nervous tissue (3–4 fetuses for each side of the brain) with an estimated 10% survival rate of the transplanted tissue and lacking consensus regarding the implantation technique (Björklund et al., 2003).

The promising results obtained from neural transplantation in animals and some human trials have increased the efforts to find new ways of generating cells that produce neurotransmitters or neurotrophic substances[60,66]. One way to overcome the problems regarding the use of human fetuses is to use genetically modified stemcells, xenograft material or immortalized cell-lines, which can be reproduced in vitro and harvested when they appear in a sufficient number (Björklund et al., 2003; Martinez-Serrano and Björklund, 1997; Lindvall et al., 2004; Langston, 2005). The cells inserted into the brain may likewise be encapsulated in semipermeable capsules, which protects the genetically modified cells from the host immune system and at the same time allow the neurotrophic factor or dopamine produced by the encapsulated cells to diffuse into the surrounding brain tissue (Yasuhara et al., 2005).

5.2 Ablative techniques

Disruption of the proposed hyperactive STN or of the increased inhibitory output from GPi by localized stereotaxic lesions, aided by advanced brain imaging, microelectrode recordings and prelesional stimulation of the target area (macrostimulation), has been shown to be an effective and reasonably safe procedure (Hammerstad and Hogarth, 2001; Walter and Vitek, 2004; Alvarez et al., 2005). These interventions alleviate contralateral L-dopa induced dyskinesias and improve rigidity, tremor, and to a lesser extent akinesia. However, subthalamotomy may result in severe general chorea that may persist for months and misplaced lesions outside STN and GPi may cause irreversible neurological, cognitive and neuropsychiatric adverse effects (Hammerstad and Hogarth, 2001; Walter and Vitek, 2004; Alvarez et al., 2005).

5.3 Deep brain stimulation

The hyperactive pathways from GPi and STN may also be modulated by implanted electrodes, which block the neural activity in their vicinity with a high-frequency electrical current (Bjarkam et al., 2001; Hammerstad and Hogarth, 2001; Alvarez et al., 2005). This procedure, named deep brain stimulation (DBS), has proven very efficient in the treatment of PD complicated by motor fluctuations and L-dopa induced dyskinesias (The Deep-Brain Stimulation for Parkinson's Disease Study Group, 2001). The therapist can choose between several stimulation leads along the electrode and modify stimulation parameters during and after implantation. Deep brain stimulation hereby represents a more flexible method for basal ganglia circuitry modulation than ablative surgery and can be used bilaterally without the same occurrence of neuropsychiatric and cognitive side effects (Bjarkam et al., 2001; Hammerstad and Hogarth, 2001), although a few cases of severe mood changes and slight deficits in language abilities have been noted postoperatively (Anderson and Mullins, 2003). The most common side effects are, however, related to the surgical implantation (hemorrhage, infections or seizures), or due to the influence of the electrical current on neighboring corticobulbar projections which may result in dyskinesias, dysarthria, diplopia and paraesthesias (The Deep-Brain Stimulation for Parkinson's Disease Study Group, 2001). The latter complications can be diminished by reducing the intensity of the stimulation or moving the electrode, although this may lead to a reduced anti-parkinsonian effect.

6. FUTURE STRATEGIES IN PD (FIGURE 3)

Future treatment strategies apart from optimization of the strategies presented above will probably focus on prevention/neuroprotection in PD, development of vaccines towards neurodegenerative mechanisms in PD, use of gene-therapy, and identification of new pharmacological intervention points such as the recent development of adenosine A_{2A}-receptor antagonists.

6.1 Prevention/Neuroprotection in PD

PD is initiated by a severe nigral loss of dopaminergic neurons. It is therefore obvious that a prevention of this cell loss by neuroprotective treatment strategies could prevent the development or progression of PD. Such strategies in PD

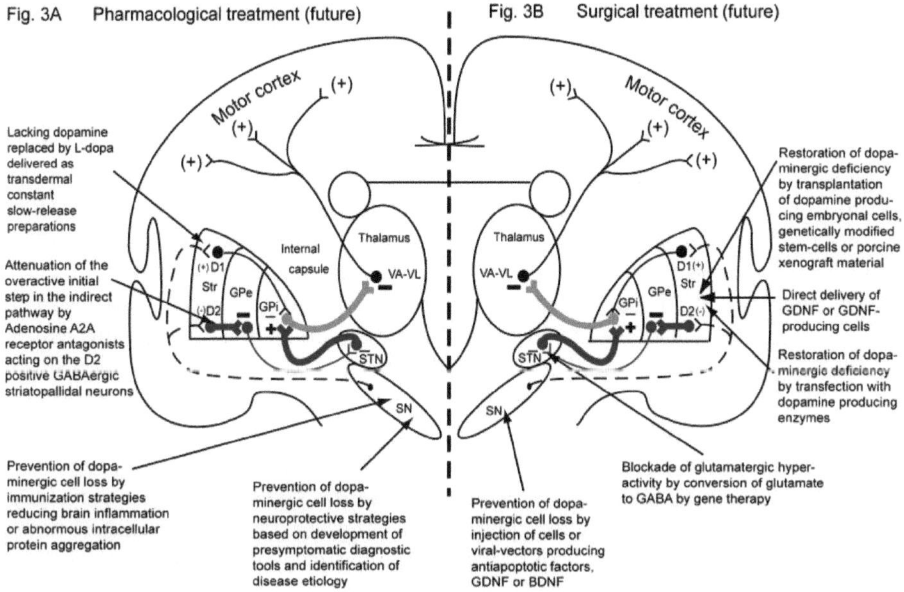

Figure 3. Schematic drawing illustrating where future pharmacological (3A) and surgical (3B) treatment strategies of PD are thought to influence the diseased basal ganglia circuitry

are faced with several problems. Firstly, although several theories have been proposed involving impaired protein degradation and aggregation of insoluble α-synuclein, oxidant stress, mitochondrial dysfunction, excitotoxicity, and inflammatory processes, the cause of the nigral cell loss is unknown (Przedborski, 2005; Nomoto, 2003). A definite target for PD neuroprotection is therefore not available. Secondly, PD symptoms first become clinically apparent when the majority of dopaminergic cells in the substantia nigra are lost. Development of efficient preclinical PD markers enabling cheap and reliable screening of risk populations (persons aged more than 60, or early PD debut among first degree relatives) would therefore be desirable/necessary to provide dopaminergic cells enough for neuroprotective treatment strategies to work (Storch et al., 2004). Finally, several neuroprotective compounds interfering with the proposed mechanisms to nigral cell death have proven effective in "symptomatic" animal models of PD (MPTP or 6-hydroxydopamine intoxication) but failed to do so in humans, indicating that we still lack adequate animal models for the development and cause of human PD.

6.2 Vaccination strategies in PD

Two recent studies have revealed that immunization strategies may prime the immune system to express antibodies or T lymphocytes to interfere with causal mechanisms of PD development/progression in animal models of PD (Benner et al.,

2004; Masliah et al., 2005). The first study demonstrated that i.v. transfer of copolymer-1-immune cells to MPTP-intoxicated mice led to nigral T-cell accumulation, suppression of microglial activation and increased local expression of astrocyte associated GDNF, resulting in significant protection of the nigralstriatal neurons towards the initiated MPTP-intoxication (Benner et al., 2004). The second study showed that active immunization with human alfa-synuclein may prevent neurodegeneration due to abnormous protein accumulation in neuronal cell-bodies and synapses in transgenic mice overexpressing human alfa-synuclein (Masliah et al., 2005). Although both studies are promising they have only shown effect in animal models of PD based on nigral cell loss due to MPTP-intoxication or over expression of alfa-synuclein. These causes represent pathogenetic PD models, which may differ grossly from the actual and still unknown pathogenetic mechanism of human PD. Successful transfer of these vaccine strategies to humans is therefore critically dependent of correct identification of the true pathogenesis of PD, or proper selection of PD patients displaying the PD pathogenesis the actual vaccine is developed against. Another caveat against vaccine strategies is that they may result in an overt immunogenic response in the diseased brain tissue causing more damage than the actual disease process. A clinical trial of vaccine treatment in Alzheimers disease has, thus, been aborted due to the development of aseptic meningoencephalitis in 17 of the 300 participating patients (Schenk, 2002).

6.3 Gene therapy in PD

Several studies have during the past years used gene therapy to modify nigral cell loss and disturbed basal ganglia circuitry in PD animal models. These studies are generally based on viral vectors, as nonviral gene transfer in general is less effective in the CNS (Burton et al., 2003). It has thus been shown that transfection with vectors expressing anti-apoptotic factors (Crocker et al., 2001; Mochizuki et al., 2001) or neurotrophic substances (GDNF or Gli1) (Yasuhara et al., 2005; Bensadoun et al., 2000; Kordower et al., 2000; Kirik et al., 2000; Palfi et al., 2002; Wang et al., 2002; McGrath et al., 2002; Azzouz et al., 2004; Suwelack et al., 2004; Zheng et al., 2005) may be beneficial in neurotoxic (6-hydroxydopamine or MPTP) animal models of PD. The promising effect of GDNF on the dopamine depleted striatum has subsequently led to a phase I study concerning direct intraputaminal delivery of GDNF in five Parkinson patients with promising results and few side effects (Gill et al., 2003). Interestingly, it has been shown that gene therapy using combined transfection with anti-apoptotic and GDNF expressing vectors has a greater (synergistic) effect on the mentioned animal models of PD than transfection with either an anti-apoptotic or GDNF expressing vector alone (Natsume et al., 2001; Eberhardt et al., 2000). Gene therapy may also directly alter the disturbed basal ganglia circuitry by converting the supposed hyperactive excitatory glutamatergic cells in the STN to express GABAergic inhibitory responses, after subthalamic injection of viral vectors expressing glutamic acid decarboxylase that converts glutamate to GABA. The resulting genotypic shift revert a parkinsonian behavioral phenotype in rats (Luo et al., 2002) and has led to the initiation of a phase I clinical trial (During et al., 2001). Another way to increase the amount of striatal dopamine in neurotoxic animal models of PD is to transfect the dopamine depleted striatum with viral vectors that contains genes necessary for the synthesis of dopamine. Thus, simultaneous transfection with viral vector systems expressing tyrosine hydroxylase, GTP cyclohydrolase and aromatic amino acid decarboxylase or combinations hereof have consistently resulted in functional recovery of animals with neurotoxic PD lesions (During et al., 1994; During et al., 1998; Azzouz et al., 2002; Kirik et al., 2002; Muramatsu et al., 2002; Shen et al., 2000; Sun et al., 2003; Carlsson et al., 2005). Finally, has one study demonstrated that intracerebral transfection with a lentiviral vector expressing human alfa-synuclein may reduce the formation of alfa synuclein inclusions and subsequent neurodegeneration in a transgenic mouse model of alfa-synuclein aggregation (Hashimoto et al., 2004).

Gene therapy is, however, not without problems and adverse effects. Thus, acute phase reactions against the viral vector may lead to multisystem organ failure (Chiocca, 2003). The viral vectors may likewise lead to mutagenic conversion and abnormal oncogenic growth of the transfected cells (Hacein-Bey-Abina et al., 2003), while lacking control of the transfected cells may cause unwanted excess production of dopamine or aberrant sprouting responses, which may result in unwanted dyskinesias (Burton et al., 2003; Chiocca, 2003; Hsich et al., 2002).

6.4 Future pharmacological interventions

A continuous effort is ongoing to develop new drugs analogs, dose regimes, and drug combinations based on the established types of PD-medications which hopefully will lead to optimized administration, better anti-parkinsonian effect and minimal adverse effects. An example of such efforts is the development of patches which allow a constant slow transdermal delivery of L-dopa or dopaminergic agonists (Sudo et al., 2002; Kankkunen et al., 2002; Lewitt and Nyholm, 2004; Güldenpfennig et al., 2005). New points of pharmacologic intervention occurs more rarely, but during the last decade has the therapeutic potential of adenosine A_{2A} receptor antagonism shown considerable promise, and pharmacologic agents acting by this mechanism may very well be a integrated part of future PD treatment regimens (Xu et al., 2005). A_{2A} receptor antagonists are thought to exert a dual action on the GABAergic neurons projecting from the striatum to the GPe, by increasing their response to D2 receptor mediated inhibition and diminish their release of GABA. Both effects will consequently attenuate the overactive indirect pathway in PD and thereby result in a better balance in the disturbed basal ganglia circuitry (Xu et al., 2005; Kase, 2003). A_{2A} receptor antagonists have accordingly improved motor deficits in rodent and primate models of PD (Grondin et al., 1999; Shiozaki et al., 1999; Koga et al., 2000; Kanda et al., 1998; Kanda et al., 2000; Jenner, 2003b) and have in several preclinical tests revealed a potential to potentate and prolong the effect of simultaneous given L-dopa and reducing off time, while exhibiting a low side effect profile in PD patients (Bara-Jimenez et al., 2003; Hauser et al., 2003; Stacy, M.A. and the US-005 & US-006 Investigator Group 2004). A_{2A} receptor antagonists have, however, not yet been tested in large patient groups where they may reveal side effects occurring in the cardiovascular, renal, pulmonary and gastrointestinal systems which contains many A_{2A} receptors and were A_{2A} receptor agonism e.g. the opposite of A_{2A} receptor antagonism has been reported beneficial (Stacy, M.A. and the US-005 & US-006 Investigator Group 2004).

7. CONCLUSION

The current review over progress and development in PD therapy illustrates clearly, that a good deal of progress has been made in the elucidation and treatment of PD. Further improvements will undoubtedly occur with the implementation of gene and immune therapy, new drugs and surgical methods. In depth knowledge of the exact pathogenesis of PD and the possibility to identify PD patients before symptoms and dopaminergic cell loss occur are, however, necessary before relevant and effective strategies that prevent PD, or offer sufficient neuroprotection against PD may be developed. Such strategies await future progress in basic and clinical PD research.

REFERENCES

Agid, Y., Ahlskog, E., Albanese, A., Calne, D., Chase, T., De Yebenes, J., Factor, S., Fahn, S., Gershanik, O., Goetz, C., Koller, W., Kurth, M., Lang, A., Lewitt, P., Marsden, D., Melamed, E., Michel, P.P., Mizuno, Y., Obeso, J., Oertel, W., Olanow, W., Poewe, W., Pollak, P., Przedzorski, S., Quinn, N., Raisman-Vozari, R., Rajput, A., Stocchi, F. and Tolosa, E. (1999) Levodopa in the treatment of Parkinson's disease: a consensus meeting. Mov Disord, 14: 911–913.

Albin, R.L., Young, A.B. and Penney, B. (1989) The functional anatomy of basal ganglia disorders. Trends Neurosci, 12: 366–375.

Alexander, G.E. and Crutcher, M.D. (1990) Functional architecture of basal ganglia circuits: neural substrates of parallel processing. Trends Neurosci, 13: 266–271.

Alexander, G.E. (1994) Basal ganglia-thalamocortical circuits: Their role in control of movements. J Clin Neurophysiol, 11: 420–431.

Alvarez, L., Macias, R., Lopez, G., Alvarez, E., Pavon, N., Rodriguez-Oroz, M.C., Juncos, J.L., Maragoto, C., Guridi, J., Litvan, I., Tolosa, E.S., Koller, W., Vitek, J., DeLong, M.R. and Obeso, J.A. (2005) Bilateral subthalamotomy in Parkinson's disease: initial and long-term response. Brain, 128: 570–583.

Anderson, K.E. and Mullins, J. (2003) Behavioral changes associated with deep brain stimulation surgery for Parkinson's disease. Curr Neurol Neurosci Rep, 3: 306–313.

Azzouz, M., Martin-Rendon, E., Barber, R.D., Mitrophanous, K.A., Carter, E.E., Rohll, J.B., Kingsman, S.M., Kingsman, A.J. and Mazarakis, N.D. (2002) Multicistronic lentiviral vector-mediated striatal gene transfer of aromatic L-amino acid decarboxylase, tyrosine hydroxylase, and GTP cyclohydrolase I induces sustained transgene expression, dopamine production, and functional improvement in a rat model of Parkinson's disease. J Neurosci, 22: 10302–10312.

Azzouz, M., Ralph, S., Wong, L.-F., Day, D., Askham, Z., Barber, R.D., Mitrophanous, K.A., Kingsman, S.M. and Mazarakis, N.D. (2004) Neuroprotection in a rat Parkinson model by GDNF gene therapy using EIAV vector. NeuroReport, 15(6): 985–990.

Bara-Jimenez, W., Sherzai, A., Dimitrova, T., Favit, A., Bibbiani, F., Gillespie, M., Morris, M.J., Mouradian, M.M. and Chase, T.N. (2003) Adenosine A(2A) receptor antagonist treatment of Parkinson's disease. Neurology, 61: 293–296.

Benner, E.J., Mosley, R.L., Destache, C.J., Lewis, T.B., Jackson-Lewis, V., Gorantla, S., Nemachek, C., Green, S.R., Przedborski, S. and Gendelman, H.E. (2004) Therapeutic immunization protects dopaminergic neurons in a mouse model of Parkinson's disease. PNAS, 101(25): 9435–9440.

Bensadoun, J.C., Deglon, N., Tseng, J.L., Ridet, J.L., Zurn, A.D. and Aebischer, P. (2000) Lentiviral vectors as a gene delivery system in the mouse midbrain: cellular and behavioral improvements in a 6-OHDA model of Parkinson's disease using GDNF. Exp Neurol, 164: 15–24.

Bernheimer, H., Birkmeyer, W., Hornykiewicz, O., Jellinger, K. and Seitelberger, F. (1973) Brain dopamine and the syndromes of Parkinson and Huntington. J Neurol Sci, 20: 415–455.

Bertler, A. and Rosengren, E. (1959) Occurence and distribution of catecholamines in brain. Acta Physiologica Scandinavica, 47: 350–361.

Björklund, A., Dunnett, S.B., Brundin, P., Stoessl, A.J., Freed, C.R., Breeze, R.E., Levivier, M., Peschanski, M., Studer, L. and Barker, R. (2003) Neural transplantation for the treatment of Parkinson's disease. Lancet Neurol, 2: 437–445.

Bjarkam, C.R., Sørensen, J.C., Sunde, N.Å., Geneser, F.A. and Østergaard, K. (2001) New strategies for the treatment of Parkinson's disease hold considerable promise for the future management of neurodegenerative disorders. Biogerontology, 2: 193–207.

Burton, E.A., Glorioso, J.C. and Fink, D.J. (2003) Gene therapy progress and prospects: Parkinson's disease. Gene Therapy, 10: 1721–1727.

Carlsson, A., Lindqvist, M., Magnuson, T. and Waldeck, B. (1958) On the presence of 3-hydroxythyramin in the brain. Science, 127: 471–471.

Carlsson, T., Winkler, C., Burger, C., Muzyczka, N., Mandel, R.J., Cenci, A., Björklund, A. and Kirik, D. (2005) Reversal of dyskinesias in an animal model of Parkinson's disease by continuous L-DOPA delivery using rAAV vectors. Brain, 128: 559–569.

Carlsson, A. (1959) The occurrence, distribution and physiological role of catecholamines in the nervous system. Pharmacological Reviews, 11: 490–493.

Chesselet, M.-F. and Delfs, J.M. (1996) Basal ganglia and movement disorders: an update. Trends Neurosci, 19: 417–422.

Chiocca, E.A. (2003) Gene therapy: a primer for neurosurgeons. Neurosurgery, 53: 364–373.

Clarke, C.E. (2004) Neuroprotection and pharmacotherapy for motor symptoms in Parkinson's disease. Lancet Neurology, 3: 466–474.

Crocker, S.J., Wigle, N., Liston, P., Thompson, C.S., Lee, C.J., Xu, D., Roy, S., Nicholson, D.W., Park, D.S., MacKenzie, A., Korneluk, R.G. and Robertson, G.S. (2001) NAIP protects the nigrostriatal dopamine pathway in an intrastriatal 6-OHDA rat model of Parkinson's disease. Eur J Neurosci, 14: 391–400.

Dauer, W. and Przedborski, S. (2003) Parkinson's disease: mechanisms and models. Neuron, 39: 889–909.

DeLong, M.R. (1990) Primate models of movement disorders of basal ganglia origin. Trends Neurosci, 13: 281–285.

During, M.J., Naegele, J.R., O'Malley, K.L. and Geller, A.I. (1994) Long-term behavioral recovery in parkinsonian rats by an HSV vector expressing tyrosine hydroxylase. Science, 266: 1399–1403.

During, M.J., Samulski, R.J., Elsworth, J.D., Kaplitt, M.G., Leone, P., Xiao, X., LI, J., Freese, A., Taylor, J.R., Roth, R.H., Sladek, J.R., Jr. O'Malley, K.L. and Redmond, D.E., Jr. (1998) In vivo expression of therapeutic human genes for dopamine production in the caudates of MPTP-treated monkeys using an AAV vector. Gene Ther, 5: 820–827.

During, M.J., Kaplitt, M.G., Stern, M.B. and Eidelberg, D. (2001) Subthalamic GAD gene transfer in Parkinson disease patients who are candidates for deep brain stimulation. Hum Gene Ther, 12: 1589–91.

Duvoisin, R.C. and Yahr, M.D. (1965) Encephalities and Parkinsonism. Arch Neurol, 12: 227.

Duvoisin, R.C. (1999) Genetic and environmental factors in Parkinson's disease In: Stern GM (ed) Parkinson's disease: Advances in Neurology Lippincott Williams & Wilkins, Philadelphia, 80: 161–163.

Eberhardt, O., Coelln, R.V., Kugler, S., Lindenau, J., Rathke-Hartlieb, S., Gerhardt, E., Haid, S., Isenmann, S., Gravel, C., Srinivasan, A., Bahr, M., Weller, M., Dichgans, J. and Schulz, J.B.

(2000) Protection by synergistic effects of adeno-virus-mediated X-chromosome-linked inhibitor of apoptosis and glial cell line-derived neurotrophic factor gene transfer in the 1-methyl-4-phenyl-1,2,3,6-tetrahydropyridine model of Parkinson's disease. J Neurosci, 20: 9126–9134.

Fahn, S. and Cohen, G. (1992) The oxidant strees hypothesis in Parkinson's disease: evidence supporting it. Ann Neurol, 32: 805–812.

Forno, L.S. (1996) Neuropathology of Parkinson's disease. J Neuropathol Exp Neurol, 55: 259–272.

Freed, C.R., Greene, P.E., Breeze, R.E., Tsai, W.Y., DuMouchel, W., Kao, R., Dillon, S., Winfield, H., Culver, S., Trojanowski, J.Q., Eidelberg, D. and Fahn, S. (2003) Transplantation of embryonic dopamine neurons for severe Parkinson's disease. N Engl J Med, 344: 710–719.

Freeman, W. (1925) The pathology of paralysis agitans. Ann Clin Med, 4: 106–16.

Gill, S.S., Patel, N.K., Hotton, G.R., O'Sullivan, K., McCarter, R., Bunnage, M., Brooks, D.J., Svendsen, C.N. and Heywood, P. (2003) Direct brain infusion of glial cell line-derived neurotrophic factor in Parkinson disease. Nature Medicine, 9: 589–595.

Güldenpfennig, W., Poole, K.H., Sommerville, K.W. and Boroojerdi, B. (2005) Safety, tolerability, and efficacy of continuous transdermal dopaminergic stimulation with rotigotine patch in early stage idiopathic Parkinson disease. Clin Neuropharmacol, 28: 106–110.

Greenfield, J.G. and Bosanquet, F.D. (1953) The brain-stem lesions in parkinsonism. J Neurol Neurosurg Psychiatry, 16: 213–26.

Grondin, R., Bedard, P.J., Hadj Tahar, A., Gregoire, L., Mori, A. and Kase, H. (1999) Antiparkinson effect of a new selective adenosine A_{2A} receptor antagonist in MPTP-treated monkeys. Neurology, 52: 1673–1677.

Haas, R.H., Nasirian, F., Nakano, K., Ward, D., Pay, M., Hill, R. and Shults, C.W. (1995) Low platelet mitochondrial complex I and complex II/III activity in early untreated Parkinson's disease. Ann Neurol, 37: 714–22.

Hacein-Bey-Abina, S., von Kalle, C., Schmidt, M., Le Deist, F., Wulffraat, N., McIntyre, E., Radford, I., Villeval, J.L., Fraser, C.C., Cavazzana-Calvo, M. and Fischer, A. (2003) A serious adverse event after successful gene therapy for X-linked severe combined immunodeficiency. N Engl J Med, 348: 255–256.

Hammerstad, J. and Hogarth, P. (2001) Parkinson's disease: Surgical options. Current Neurology and Neuroscience Reports, 1: 313–319.

Hashimoto, M., Rockenstein, E., Mante, M., Crews, L., Bar-On, P., Gage, F.H., Marr, R. and Masliah, E. (2004) An antiaggregation gene therapy strategy for Lewy body disease utilizing β-synuclein lentivirus in a transgenic model. Gene Therapy, 11: 1713–1723.

Hauser, R.A., Hubble, J.P. and Truong, D.D. (2003) Randomized trial of the adenosine A(2A) receptor antagonist istradefylline in advanced PD. Neurology, 61: 297–303.

Herrero, M.-T., Barcia, C. and Navarro, J.M. (2002) Functional anatomy of thalamus and basal ganglia. Child's Nerv System, 18: 386–404.

Hsich, G., Sena-Esteves, M. and Breakefield, X.O. (2002) Critical issues in gene therapy for neurologic disease. Hum Gene Ther, 13: 579–604.

Jenner, P. and Olanow, C.W. (1996) Oxidative stress and the pathogenesis of Parkinson's disease. Neurology, 47: 161–170.

Jenner, P. (2003a) Dopamine agonists, receptor selectivity and dyskinesia induction in Parkinson's disease. Curr Opin Neurol, 16(suppl 1): S3–S7.

Jenner, P. (2003b) A_{2A} antagonists as novel non-dopaminergic therapy for motor dysfunction in PD. Neurology, 61: S32–S38.

Kanda, T., Jackson, M.J., Smith, L.A., Pearce, R.K., Nakamura, J., Kase, H., Kuwana, Y. and Jenner, P. (1998) Adenosine A_{2A} antagonist: a novel antiparkinsonian agent that does not provoke dyskinesia in parkinsonian monkeys. Ann Neurol, 43: 507–513.

Kanda, T., Jackson, M.J., Smith, L.A., Pearce, R.K., Nakamura, J., Kase, H., Kuwana, Y. and Jenner, P. (2000) Combined use of the adenosine A(2A) antagonist KW-6002 with L-DOPA or with selective D1 or D2 dopamine agonists increases antiparkinsonian activity but not dyskinesia in MPTP-treated monkeys. Exp Neurol, 162: 321–327.

Kankkunen, T., Huupponen, I., Lahtinen, K., Sundell, M., Ekman, K., Kontturi, K. and Hirvonen, J. (2002) Improved stability and release control of levodopa and metaraminol using ion-exchange fibers and transdermal iontophoresis. Eur J Pharm Sci, 16(4–5): 273–280.

Kase, H. (2003) Industry forum: Progress in pursuit of therapeutic A_{2A} antagonist. Neurology, 61(suppl 6): S97–S100.

Kirik, D., Rosenblad, C., Bjorklund, A. and Mandel, R.J. (2000) Long-term rAAV-mediated gene transfer of GDNF in the rat Parkinson's model: intrastriatal but not intranigral transduction promotes functional regeneration in the lesioned nigrostriatal system. J Neurosci, 20: 4686–4700.

Kirik, D., Georgievska, B., Burger, C., Winkler, C., Muzyczka, N., Mandel, R.J. and Bjorklund, A. (2002) Reversal of motor impairments in parkinsonian rats by continuous intrastriatal delivery of L-dopa using rAAV-mediated gene transfer. Proc Natl Acad Sci USA, 99: 4708–4713.

Kish, S.J., Shannak, H.K. and Hornykiewicz, O. (1988) Uneven pattern of dopamine loss in the striatum of patients with Parkinson's disease-pathophysiologic and clinical implications. N Engl J Med, 318: 876–880.

Koga, K., Kurokawa, M., Ochi, M., Nakamura, J. and Kuwana, Y. (2000) Adenosine A(2A) antagonists KF17837 and KW-6002 potentiate rotation induced by dopaminergic drug in hemi-Parkinsonian rats. Eur J Pharmacol, 408: 249–255.

Kordower, J.H., Freeman, T.B., Snow, B.J., Vingerhoets, F.J., Mufson, E.J., Sanberg, P.R., Hauser, R.A., Smith, D.A., Nauert, G.M., Perl, D.P. and Olanow, C.W. (1995) Neuropathological evidence of graft survival and striatal reinnervation after the transplantation of fetal mesencephalic tissue in a patient with Parkinson's disease. N Engl J Med, 332(17): 1118–1124.

Kordower, J.H., Emborg, M.E., Bloch, J., Ma, S.Y., Chu, Y., Leventhal, L., McBride, J., Chen, E.Y., Palfi, S., Roitberg, B.Z., Brown, W.D., Holden, J.E., Pyzalski, R., Taylor, M.D., Carvey, P., Ling, Z., Trono, D., Hantraye, P., Deglon, N. and Aebischer, P. (2000) Neurodegeneration prevented by lentiviral vector delivery of GDNF in primate models of Parkinson's disease. Science, 290: 767–772.

Krüger, R. (2004) Genes in familial parkinsonism and their role in sporadic Parkinson's disease. J Neurol, 251(suppl 6): VI/2–VI/6.

Lang, A.E. and Lozano, A.M. (1998) Parkinson's disease: Second of two parts. N Engl J Med, 339: 1130–1143.

Langston, J.W. (2005) The promise of stem cells in Parkinson disease. J Clin Invest, 115: 23–25.

Lewitt, P.A. and Nyholm, D. (2004) New developments in levodopa therapy. Neurology, 62(suppl 1): S9–S16.

Lindgren, P., von Campenhausen, S., Spottke, E., Siebert, U. and Dodel, R. (2005) Cost of Parkinson's disease in Europe. Eur J Neurol, 12(suppl 1): 68–73.

Lindvall, O., Kokaia, Z. and Martinez-Serrano, A. (2004) Stem cell therapy for human neurodegenerative disorders – how to make it work. Nature Med, 10(suppl 10): S42–S50.

Luo, J., Kaplitt, M.G., Fitzsimons, H.L., Zuzga, D.S., Liu, Y., Oshinsky, M.L. and During, M.J. (2002) Subthalamic GAD gene therapy in a Parkinson's disease rat model. Science, 298: 425–429.

Marsden, C.D. (1994) Parkinson's disease. J Neurol Neurosurg Psychiatr, 57: 672–681.

Martinez-Serrano, A., Björklund, A. (1997) Immortalized neural progenitor cells for CNS gene transfer and repair. Trends Neurosci, 20: 530–538.

Masliah, E., Rockenstein, E., Adame, A., Alford, M., Crews, L., Hashimoto, M., Seubert, P., Lee, M., Goldstein, J., Chilcote, T., Games, D. and Schenk, D. (2005) Effects of α-synuclein immunization in a mouse model of Parkinson's disease. Neuron, 46: 857–868.

McGrath, J., Lintz, E., Hoffer, B.J., Gerhardt, G.A., Quintero, E.M. and Granholm, A.C. (2002) Adeno-associated viral delivery of GDNF promotes recovery of dopaminergic phenotype following a unilateral 6-hydroxydopamine lesion. Cell Transplant, 11: 215–227.

Mercuri, N.B. and Bernardi, G. (2005) The 'magic' of L-dopa: why is it the gold standard Parkinson's disease therapy. Trends in Pharmacol Sci, 26(7): 341–344.

Mochizuki, H., Hayakawa, H., Migita, M., Shibata, M., Tanaka, R., Suzuki, A., Shimo-Nakanishi, Y., Urabe, T., Yamada, M., Tamayose, K., Shimada, T., Miura, M. and Mizuno, Y. (2001) An AAV-derived Apaf-1 dominant negative inhibitor prevents MPTP toxicity as antiapoptotic gene therapy for Parkinson's disease. Proc Natl Acad Sci USA, 98: 10918–10923.

Muramatsu, S., Fujimoto, K.I., Ikeguchi, K., Shizuma, N., Kawasaki, K., Ono, F., Shen, Y., Wang, L., Mizukami, H., Kume, A., Matsumura, M., Nagatsu, I., Urano, F., Ichinose, H., Nagatsu, T., Terao, K., Nakano, I. and Ozawa, K. (2002) Behavioral recovery in a primate model of Parkinson's disease by triple transduction of striatal cells with adeno-associated viral vectors expressing dopamine-synthesizing enzymes. Hum Gene Ther, 13: 345–354.

Mutch, W.J., Dingwall-Fordyce, I., Downie, A.W., Paterson, J.G. and Roy, S.K. (1986) Parkinson's disease in a Scottish city. Br Med J, 292: 534–536.

Nakagawa-Hattori, Y., Yoshino, H. and Kondo, T. (1992) Is Parkinson's disease a mitochondrial disorder?. J Neurol Sci, 107: 29–33.

Natsume, A., Mata, M., Goss, J., Huang, S., Wolfe, D., Oligino, T., Glorios, J. and Fink, D.J. (2001) Bcl-2 and GDNF delivered by HSV-mediated gene transfer act additively to protect dopaminergic neurons from 6-OHDA-induced degeneration. Exp Neurol, 169: 231–238.

Nisipeanu, P., Paleacu, D. and Korczyn, A.D. (1997) Infectious and postinfectious parkinsonism. In: Watts RL and Koller WC (eds) Movement disorders, neurologic principles and practice, pp 307–313. New York.

Nomoto, M. (2003) Clinical pharmacology and neuroprotection in Parkinson's disease. Parkinsonism & Related Disorders, 9: S55–S58.

Nyholm, D. and Aquilonius, S.-M. (2004) Levodopa infusion therapy in Parkinson disease. Clin Neuropharmacol, 27(5): 245–256.

Offen, D., Hochman, A., Gorodin, S., Ziv, I., Shirvan, A., Barzilai, A. and Melamed, E. (1999) Oxidative stress and neuroprotection in Parkinson's disease: Implications from studies on dopamine-induced apoptosis. In: Stern GM (ed) Parkinson's disease: Advances in Neurology, Lippincott Williams & Wilkins, Philadelphia, 80: 265–269.

Olanow, C.W. and Stocchi, F. (2004) COMT inhibitors in Parkinson's disease. Neurology, 62(suppl 1): S72–S81.

Olanow, C.W., Goetz, C.G., Kordower, J.H., Stoessl, A.J., Sossi, V., Brin, M.F., Shannon, K.M., Nauert, G.M., Perl, D.P., Godbold, J. and Freeman, T.B. (2003) A double-blind controlled trial of bilateral fetal nigral transplantation in Parkinson's disease. Ann Neurol, 54: 403–414.

Palfi, S., Leventhal, L., Chu, Y., Ma, S.Y., Emborg, M., Bakay, R., Deglon, N., Hantraye, P., Aebischer, P. and Kordower, J.H. (2002) Lentivirally delivered glial cell line-derived neurotrophic factor increases the number of striatal dopaminergic neurons in primate models of nigrostriatal degeneration. J Neurosci, 22: 4942–4954.

Parent, A. and Hazrati, L.-N. (1995a) Functional anatomy of the basal ganglia. I. The cortico-basal ganglia-thalamo-cortical loop. Brain Res Rev, 20: 91–127.

Parent, A. and Hazrati, L.-N. (1995b) Functional anatomy of the basal ganglia. II. The place of subthalamic nucleus and external pallidum in basal ganglia circuitry. Brain Res Rev, 20: 128–154.

Parkinson, J. (1817) An essay on the shaking palsy. Whittingham and Rowland, London.

Piccini, P., Brooks, D.J., Bjorklund, A., Gunn, R.N., Grasby, P.M., Rimoldi, O., Brundin, P., Hagell, P., Rehncrona, S., Widner, H. and Lindvall, O. (1999) Dopamine release from nigral transplants visualized in vivo in a Parkinson's patient. Nat Neurosci, 2(12): 1137–1140.

Piccini, P., Lindvall, O., Bjorklund, A., Brundin, P., Hagell, P., Ceravolo, R., Oertel, W., Quinn, N., Samuel, M., Rehncrona, S., Widner, H. and Brooks, D.J. (2000) Delayed recovery of movement-related cortical function in Parkinson's disease after striatal dopaminergic grafts. Ann Neurol, 48(5): 689–695.

Przedborski, S. (2005) Pathogenesis of nigral cell death in Parkinson's disease. Parkinsonism & Related Disorders, 11: S3–S7.

Rascol, O., Goetz, C., Koller, W., Poewe, W. and Sampaio, C. (2002) Treatment interventions for Parkinson's disease: an evidence based assessment. Lancet, 359: 1589–1598.

Rascol, O., Brooks, D.J., Melamed, E., Oertel, W., Poewe, W., Stocchi, F., Tolasa, E. and the Largo study group (2005) Rasagiline as an adjunct to levodopa in patients with Parkinson's disease and motor fluctuations (LARGO, Lasting effect in Adjunct theraphy with Rasagiline Given Once daily study): a randomized, double-blind, parallel-group trial. Lancet, 365: 947–954.

Rodriguez, M.C., Obeso, J.A. and Olanow, C.W. (1998) Subthalamic nucleus-mediated excitotoxicity in Parkinson's disease: A target for neuroprotection. Ann Neurol, 44(Suppl 1): S175–S188.

Rybecki, B.A., Johnson, C.C., Uman, J. and Gorell, J.M. (1993) Parkinson's disease mortality and the industrial use of heavy metals in Michigan. Mov Disord, 8: 87–92.

Schenk, D. (2002) Amyloid-β immunotherapy for Alzheimer's disease: the end of the beginning. Nat Rev Neurosci, 3: 824–828.

Semchuk, K., Love, E.J. and Lee, R.G. (1992) Parkinson's disease and exposure to agricultural work and pesticide chemicals. Neurology, 42: 1328–1335.

Shen, Y., Muramatsu, S.I., Ikeguchi, K., Fujimoto, K.I., Fan, D.S., Ogawa, M., Mizukami, H., Urabe, M., Kume, A., Nagatsu, I., Urano, F., Suzuki, T., Ichinose, H., Nagatsu, T., Monahan, J., Nakano, I. and Ozawa, K. (2000) Triple transduction with adeno-associated viral vectors expressing tyrosine hydroxlase, aromatic-L-amino-acid decarboxylase, and GTP cyclohydrolase I for gene therapy of Parkinson's disease. Hum Gene Ther, 11: 1509–1519.

Shiozaki, S., Ichikawa, S., Nakamura, J., Kitamura, S., Yamada, K. and Kuwana, Y. (1999) Actions of adenosine A(2A) antagonist KW-6002 on drug-induced catalepsy and hypokinesia caused by reserpine or MPTP. Psychopharmacology, 147: 90–95.

Smith, Y. and Parent, A. (1988) Neurons of the subthalamic nucleus in primates display glutamate but not GABA immunoreactivity. Brain Res, 453: 353–356.

Smith, Y., Parent, A., Séguéla, P. and Descarries, L. (1987) Distribution of GABA immunoreactive neurons in the basal ganglia of the squirrel monkey (Saimiri sciureus). J Comp Neurol, 259: 50–65.

Stacy, M.A. and the US-005 & US-006 Investigator Group (2004) Istradefylline (KW-6002) as adjunctive therapy in patients with advanced Parkinson's disease: a positive safety profile with supporting efficacy. Mov Disord, 19(S9): S215–S216 (P605).

Stocchi, F. and Olanow, C.W. (2004) Continuous dopaminergic stimulation in early and advanced Parkinson's disease. Neurology, 62(suppl 1): S56–S63.

Storch, A., Hofer, A., Krüger, R., Schulz, J.B., Winkler, J. and Gerlach, M. (2004) New developments in diagnosis and treatment of Parkinson's disease – From basic science to clinical applications. J Neurol, 251(suppl 6): VI/33–VI/38.

Sudo, J., Iwase, H., Higashiyama, K., Kakuno, K., Miyasaka, F., Meguro, T. and Takayama, K. (2002) Elevation of plasma levels of L-dopa in transdermal administration of L-dopa-butylester in rats. Drug Dev Ind Pharm, 28: 59–65.

Sun, M., Zhang, G.R., Kong, L., Holmes, C., Wang, X., Zhang, W., Goldstein, D.S. and Geller, A.I. (2003) Correction of a rat model of Parkinson's disease by coexpression of tyrosine hydroxylase and aromatic amino acid decarboxylase from a helper virus-free herpes simplex virus type 1 vector. Hum Gene Ther, 14: 415–424.

Suwelack, D., Hurtado-Lorenzo, A., Millan, E., Gonzalez-Nicolini, V., Wawrowsky, K., Lowenstein, P.R. and Castro, M.G. (2004) Neuronal expression of the transcription factor Gli1 using the Tα1α-tubulin promoter is neuroprotective in an experimental model of Parkinson's Disease. Gene Therapy, 11: 1742–1752.

Tanner, C.M. (1989) The role of environmental toxins in the etiology of Parkinson's disease. Trends Neurosci, 12: 49–54.

Tanner, C.M. and Ben-Shlomo, Y. (1999) Epidemiology of Parkinson's disease. In: Stern GM (ed) Parkinson's disease: Advances in Neurology, 80:265–269 Lippincott Williams & Wilkins, Philadelphia.

Tanner, C.M., Hubble, J.P. and Chan, P. (1997) Epidemiology and genetics of Parkinson's disease. In: Watts RL and Koller WC (eds) Movement disorders, neurologic principles and practice, pp 137–152. New York.

The Deep-Brain Stimulation for Parkinson's Disease Study Group (2001) Deep brain stimulation of the subthalamic nucleus or the pars interna of the globus pallidus in Parkinson's disease. N Engl J Med, 345: 956–963.

Thomas, A., Iacono, D., Luciano, A., Armellino, K., Di Lorio, A. and Onofrj, M. (2004) Duration of amantadine benefit on dyskinesia of severe Parkinson's disease. J Neurol Neurosurg Psychiatry, 75: 141–143.

Tretiakoff, C. (1919) Contribution a l'etude de l'anatomie pathologique du locus niger de Soemmering avec quelques deductions relatives a la pathogenie des troubles du tonus musculaires et de la maladie de Parkinson. Thesis. University of Paris.
Tuchsen, F. and Jensen, A.A. (2000) Agricultural work and the risk of Parkinson's disease in Denmark, 1981–1993. Scand J Work Environ Health, 26: 359–62.
Von Economo, C. (1917) Encephalitis lethargica. Wien Klin Wochnschr, 30: 581.
Walter, B.L. and Vitek, J.L. (2004) Surgical treatment for Parkinson's disease. Lancet Neurol, 3: 719–728.
Wang, L., Muramatsu, S., Lu, Y., Ikeguchi, K., Fujimoto, K., Okada, T., Mizukami, H., Hanazono, Y., Kume, A., Urano, F., Ichinose, H., Nagatsu, T., Nakano, I. and Ozawa, K. (2002) Delayed delivery of AAV-GDNF prevents nigral neurodegeneration and promotes functional recovery in a rat model of Parkinson's disease. Gene Therapy, 9: 381–389.
Wooten, G.F., Currie, L.J., Bennett, J.P., Harrison, M.B., Trugman, J.M. and Parker, W.D., Jr. (1997) Maternal inheritance in Parkinson's disease. Ann Neurol, 41: 265–268.
Xu, K., Bastia, E. and Schwarzschild, M. (2005) Therapeutic potential of adenosine A2A receptor antagonists in Parkinson's disease. Pharmacology & Therapeutics, 105: 267–310.
Yasuhara, T., Shingo, T., Muraoka, K., Kobayashi, K., Takeuchi, A., Yano, A., Wenji, Y., Kameda, M., Matsui, T., Miyoshi, Y. and Date, I. (2005) Early transplantation of an encapsulated glial cell line-derived neurotrophic factor-producing cell demonstrating strong neuroprotective effects in a rat model of Parkinson's disease. J Neurosurg, 102: 80–89.
Zheng, J.-S., Tang, L.-L., Zheng, S.-S., Zhan, R.-Y., Zhou, Y.-Q., Goudreau, J., Kaufman, D. and Chen, A.F. (2005) Delayed gene therapy of glial cell line-derived neurotrophic factor is efficacious in a rat model of Parkinson's disease. Mol Brain Res, 134: 155–161.

CHAPTER 4

UNDERSTANDING AND TREATING ALZHEIMER'S DISEASE

UMESH KUMAR, ALEXANDER ROLAND AND STEPHEN A. BURBIDGE
Neurology and GI Centre of Excellence for Drug Discovery, GlaxoSmithKline, New Frontiers Science Park, Harlow, Essex, CM19 5AW, United Kingdom Email: Umesh_2_Kumar@gsk.com

Abstract: Alzheimer's disease (AD) remains one of the most disabling health conditions in elderly population worldwide. The socio-economic burden of the disease is likely to increase due to increasing life expectancy. Increasing understanding of AD pathogenesis suggests heterogeneous nature of this disease, with number of underlying mechanisms operating simultaneously, contributing to the ultimate phenotype. Neuropathological hallmarks of AD include senile plaques and neurofibrillary tangles, neuronal atrophy and cortical neurodegeneration. There is currently no cure for AD and the available treatments can provide only a degree of symptomatic benefit to patients with mild-to-moderate AD. In this review, we focus on the current understanding of AD, available symptomatic treatments and potential disease modifying opportunities being pursued in the pharmaceutical industry as well as in academia

Keywords: Aging, neurodegenerative diseases, dementia

1. ALZHEIMER'S DISEASE

Alzheimer's disease (AD) is a progressive neurodegenerative disease that accounts for most cases of dementia seen in the elderly population (Ferri et al., 2005) Clinical manifestations of the disease start with minor lapses in episodic memory. As the disease progresses problems with general cognitive functions such as intellectual abilities, memory, executive functions and speech become more common. The cognitive deficit leads to severe personality changes characterised by agitation, depression and social withdrawal. Over a period of years the condition worsens, resulting in complete immobility, with patients becoming totally dependent on their caregivers for social care.

S. Rattan and M. Kassem (eds.), Prevention and Treatment of Age-related Diseases, 49–70.

In the absence of a proven biological marker, the diagnosis of AD remains based on the clinical judgment that the patient's cognitive function has declined from the past level of ability. An internationally agreed criterion for clinical diagnosis of AD includes a detailed history, functional measures of decline such as instrumental activity of daily living scales, mental status tests, Clinical Dementia Rating (CDR), Disability Assessment for Dementia (DAD), neuropsychological evaluation, neurological and psychiatric examinations, blood tests, and brain imaging. The accuracy of diagnosis of probable AD is now more than 90% based on autopsy confirmation.

The risk factors for AD include age, genetic polymorphism, Down's syndrome, abnormal protein processing, neurotransmitter deficit, oxidative stress, head trauma, and environmental toxins e.g. heavy metals. The interaction over time of these genetic and nongenetic risk factors with the biology of aging brain leads to development of the AD process.

It is estimated that around 24 million people worldwide are suffering from AD. The figure is expected to increase significantly over next 50 years due to increasing life expectancy. Every year 1% of the people over the age of 65 years and 6–8% over the age of 85 are diagnosed with AD in the developed world. The disability weight for dementia is higher than for any other health condition apart from spinal cord injury and terminal cancer. In the United Kingdom half of all the elderly people with cognitive impairment live in institutions a at a cost of £4.6 billion every year. Increase in prevalence of AD with age suggests that every person is likely to develop AD should they live long enough (Ferri et al., 2005; Kukull and Ganguli, 2000; Hebert et al., 2003).

1.1 Neuropathological features of AD

Neuropathological hallmarks of AD include senile plaques and neurofibrillary tangles, neuronal atrophy and cortical neurodegeneration (Dickson, 1997). The senile plaques are extracellular proteinaceous deposits of amyloid-beta (Abeta) peptides. The senile plaques are considered to evolve over a long period of time and their fibrillar nature is due to aggregated 40–42 amino acid long Abeta peptides. Besides Abeta peptides plaques contain several other components. Dystrophic neurites, activated microglia and reactive astrocytes are all seen near the plaques. Aggregated amyloid fibrils and inflammatory mediators secreted by microglial and astrocytic cells contribute to neuronal dystrophy. Neurofibrillary tangles consist of paired helical filaments which are composed of hyperphosphorylated microtubule associated protein tau (Grundke-Iqbal et al., 1986). Presence of both plaques and tangles is used as a definitive criterion for diagnosis of AD.

Neuronal death seen in brain is another pathologic hallmark of AD. Certain populations of neurons tend to be lost selectively, and it has been proposed that the loss of synaptic density is likely to have a more immediate relationship to dementia in AD than Abeta accumulation.

1.2 AD pathogenesis

The pathophysiologic abnormalities at the anatomical, cellular, and molecular levels support the view that a variety of mechanisms may contribute to AD. The clinical phenotype of AD could be a cumulative effect of all these events. There is no compelling evidence that these mechanisms are mutually exclusive, however, over last 10 years a dominant mechanism has been proposed by the 'amyloid hypothesis'.

According to another school of thought tau associated pathology is the underlying cause of AD pathogenesis. Both these hypotheses are described below. The role of amyloid or tau as a primary cause of neurodegeneration has been debated by two rival groups (Baptists and tauists) over several years, however, the controversy has now been settled due to some recent observations suggesting a link between the two hypotheses but several questions still remain unanswered (Mudher and Lovestone, 2002).

1.2.1 Amyloid hypothesis

Amyloid hypothesis proposes accumulation of Abeta peptide in the brain as the primary influence driving AD pathogenesis (Hardy and Higgins, 1992). Abeta peptide is produced by proteolytic cleavage of a membrane bound amyloid precursor protein (APP) by two proteases called β and γ-secretases. Under certain circumstances Abeta production is enhanced by changes in activities of both β and γ secretases which leads to a cascade of events including neurofibrillary tangles and cell death. The strongest evidence in favour of this hypothesis is provided by familial cases of AD. Autosomal dominant mutations in the genes for APP, Presenilin-1 (PS1) and Presenilin-2 (PS2) cause early onset familial AD (FAD) by directly increasing synthesis of the toxic Abeta42 peptide. Transgenic mice over expressing Abeta display pathological features of AD such as age specific deposition of Abeta in brain. The other neuropathological characteristics of AD, such as, astrocytosis, neuritic dystrophy, and microgliosis are also seen in these animals, however, no neurofibrillary tangles or neurodegeneration is observed. Further genetic evidence is provided by chromosome-21 (C-21) trisomy seen in patients with Down's syndrome. C-21 harbours APP gene and one extra copy of APP on C-21 results in over production of Abeta. Patients with Down's syndrome develop all the characteristic signs of AD earlier in their lives suggesting that the Abeta accumulation is sufficient to cause symptoms (Tanzi and Bertram, 2001; Phinney et al., 2003). Although the majority of AD cases appear to be sporadic, the strong association of Abeta42 with FAD makes a compelling argument for involvement of Abeta in the etiology of all forms of AD. According to the Amyloid hypothesis, neurofibrillary tangles develop due to imbalance between Abeta production and Abeta clearance. High levels of Abeta disrupt neuronal metabolic and ionic homeostasis and cause aberrant activation of kinases and/or inhibition of phosphatases. These alterations in kinase and phosphatase activities ultimately lead to hyperphosphorylation of tau and formation of neurofibrillary tangles (Oddo et al., 2003).

1.2.2 Tau hypothesis

According to this hypothesis, neurofibrillary tangles formed primarily from abnormal aggregations of a microtubule-associated protein tau, interfere with nerve cell functioning by impairing axonal transport. The distribution of neurofibrillary tangles spreads as the severity of the AD increases. During the early stages of the disease, neurofibrillary tangles occur predominantly in the entorhinal region. Subsequently, neurofibrillary tangles appear in the hippocampus and nearby regions of the cortex and finally throughout the cortex. These regions possess a concentration of neurons that receive cholinergic input, and also show the greatest degree of degeneration (Mandelkow and Mandelkow, 1998; Goedert, 1996). Decreased levels of acetylcholine and other markers of cholinergic function are characteristically found and have been associated with the deficits in learning and memory seen in AD. Neurotransmitters, including serotonin, glutamate, norepinephrine, and somatostatin, are also decreased, and these changes may contribute to the behavioral abnormalities seen in AD.

In addition to the mechanisms described above, other mechanisms of AD pathogenesis include inflammation, oxidative stress, cerebrovascular stress, hypercholesterolemia, metabolic stress, active cell death and lack of neurotrophic support. A variety of genetic and environmental abnormalities can also contribute to AD, e.g. association of apolipoprotein E4 (apoE4) in familial and sporadic AD. In conclusion AD is a heterogeneous disease with number of underlying mechanisms operating simultaneously, contributing to the ultimate phenotype.

2. TREATMENT

AD is one of the most disabling health conditions with serious socio-economic consequences. With the increasing number of AD patients world-wide and their high dependency, the social and economic impact of this disease is likely to increase exponentially. At present the direct and indirect cost of care of AD patients runs into billions of dollars in the developed world annually. There is currently no cure for AD but the marketed acetylcholinesterase inhibitors (AChEIs), donepezil, galantamine and rivastigmine, can provide a degree of symptomatic benefit to patients with mild-to-moderate AD, however, the clinical efficacy of these agents has come under increased scrutiny in recent years. Memantine, an NMDA antagonist, is approved for the symptomatic treatment of moderate to severe AD (Kumar, 2005). Symptomatic treatments and potential disease modifying opportunities are described below.

3. SYMPTOMATIC TREATMENTS

3.1 Cholinesterase inhibitors

The gradual neuronal loss occurring in AD results in learning and memory impairment; this is thought to be largely due to deteriorating cholinergic neurotransmission. Acetylcholine plays an important part in cognition and therefore maintaining its levels by reducing its degradation provides cognitive benefit. A number of inhibitors have been developed that can inhibit acetylcholinesterase, an enzyme that degrades acetylcholine. Short-term clinical trials (3–6 months) with several different inhibitors have shown cognitive benefit to patients with mild to moderate AD. There is evidence to suggest that such inhibitors alter the course of the underlying disease process; however, it has controversially been reported that acetylcholinesterase inhibitor treatment may delay institutionalization (Geldmacher et al., 2003; Courtney et al., 2004). Four acetylcholinesterase inhibitors (Tacrine, Donepezil, Rivastigmine and Galantamine) have been approved by the FDA (Schneider and Tariot, 2003; Ibach and Haen, 2004).

3.1.1 Tacrine

Tacrine was the first acetylcholinesterase inhibitor approved for AD. Tacrine, given twice daily, was efficacious at a high dose but its clinical utility was limited by its unfavourable side effect profile. In addition to the gastrointestinal adverse effects associated with acetylcholinesterase inhibition, signs of liver damage were frequently observed in tacrine-treated patients. Due primarily to its hepatotoxicity, tacrine is now rarely used for AD (Ibach and Haen, 2004).

3.1.2 Donepezil

Donepezil, a noncompetitive reversible inhibitor of acetylcholinesterase, is given at doses of 5 or 10 mg a day. A number of randomized placebo controlled clinical trials have been conducted, in which patients with mild to moderate AD were treated with donepezil, for period ranging from 3 months to 3 year. The outcome measures used in company-sponsored pivotal trials were: the cognitive portion of the Alzheimer's Disease Assessment Scale (ADAS-cog); Clinician's Interview-Based Impression of Change with caregiver input (CIBIC+); Mini-Mental State Examination (MMSE); Clinical Dementia Rating sum-of-boxes (CDR-sb); and patient-rated quality of life (QoL). In these studies, 12 or 24 weeks of donepezil treatment resulted in statistically significant benefits *versus* placebo with respect to each of the cognitive (ADAS-cog, MMSE) and global (CIBIC+, CDR-sb) endpoints; however, donepezil consistently failed to improve patient-rated quality of life in these trials (Rogers et al., 1998a,b). The main adverse events associated with donepezil treatment are mild gastrointestinal symptoms (Ibach and Haen, 2004).

One-year placebo-controlled studies have suggested that the benefits of donepezil treatment may be maintained over the longer term. In comparison to placebo, significant benefits were reported with respect to cognition (MMSE), global function (Gottfries-Bråne-Steen scale, GBS; Global Deterioration Scale, GDS; CDR-sb), caregiver time, and activities of daily living (Progressive Deterioration Scale, PDS; AD Functional Assessment and Change Scale, ADFACS) – although statistical significance was not apparent at all timepoints. Behaviour, as measured using the Neuropsychiatric Inventory (NPI) was not significantly improved (Winblad et al., 2001; Mohs et al., 2001; Wimo et al., 2004). Based on the results of one of these studies, it has been claimed that donepezil treatment delays clinically evident functional decline by a median of 5 months (Mohs et al., 2001).

In addition to the one-year studies described above, a number of open-label extensions of placebo-controlled trials have also been conducted, as extensions to placebo-controlled trials, with controversial results. One such study, in which patients received active treatment for up to almost five years, reported that mean ADAS-cog and CDR-sb scores showed evidence of clinical improvement within the first 6–9 months before gradually deteriorating (Rogers et al., 2000). Although the decline was reportedly less than would have been expected in untreated patients, this should be interpreted cautiously given the historical nature of the comparison. Most controversial of all has been the claim, based on another open-label extension study, that long-term treatment with effective doses of donepezil delays permanent nursing home placement by an estimated 17 months (Geldmacher et al., 2003). This claim was not supported, however, by a recent, publicly-funded, long-term placebo-controlled study, which found no delay in institutionalization, no improvement in caregiver time, and no delay in disability progression within a three-year study period (Courtney et al., 2004).

3.1.3 Rivastigmine

Rivastigmine, a reversible cholinesterase inhibitor, inhibits both acetylcholinesterase and butyrylcholinesterase. In pivotal, 26-week, placebo-controlled trials, rivastigmine (6–12 mg/day) demonstrated statistically significant effects on cognition (ADAS-cog; MMSE), global function (CIBIC+; GDS), and activities of daily living (Progressive Deterioration Scale, PDS) in patients with mild to moderate AD.

Open-label extension data generated from a further 26 weeks of treatment suggest that the cognitive benefit of rivastigmine may be maintained over a period of a year (Farlow et al., 2000). Although the efficacy of rivastigmine is similar to that of donepezil, the former appears to be less well tolerated than the latter (Wilkinson et al., 2002).

3.1.4 Galantamine

Galantamine is a reversible, competitive and selective inhibitor of acetylcholinesterase that allosterically modulates nicotinic acetylcholine receptors. The recommended dose range is 16–24 mg/day. In pivotal, placebo-controlled 13-, 21- and 26-week trials with galantamine (Raskind et al., 2000; Tariot et al., 2000; Wilcock et al., 2000; Rockwood et al., 2001; Sano et al., 2003), significant treatment benefits were apparent with respect to cognition (ADAS-cog), global function (CIBIC+) and caregiver time. Mixed results were reported with respect to behaviour (NPI) and activities of daily living (Alzheimer's Disease Cooperative Study Activities of Daily Living Inventory, ADCS-ADL; Disability Assessment for Dementia, DAD).

As with the other cholinesterase inhibitors, data from uncontrolled, open-label extension studies are suggestive of a continued treatment benefit over the long term

(Raskind et al., 2004). Interestingly, data from a small, long-term comparative study with donepezil have suggested that galantamine may have superior efficacy versus donepezil, but between-group differences were not statistically significant in the overall population (Wilcock et al., 2003).

In some patients galantamine shows mild adverse effects typical of cholinomimetic agents. Although galantamine 16, 24 and 36 mg/day demonstrated significant improvement in cognition and global function, the drug was less well tolerated at the highest dose (Raskind et al., 2004).

3.2 NMDA receptor antagonist

Glutamate is the primary excitatory amino acid in human brain. Under physiological conditions glutamate activates number of receptors including N-methyl-D-aspartate (NMDA) receptor. Activation of NMDA receptor is associated with learning and memory formation. In AD and other pathological conditions excessive activation of NMDA receptor by glutamate may result in neurodegeneration. NMDA receptor antagonists have therapeutic potential in several central nervous system disorders, including neuroprotective treatment in chronic neurodegenerative diseases, and symptomatic treatment in other neurologic diseases. There is considerable evidence suggesting an excitotoxic component in AD pathogenesis. Neurochemical studies of AD brain show degeneration of glutamatergic pathways and decreased expression of NMDA receptor in hippocampal and cortical regions (Palmer, 2001). Targeting the glutamatergic system may help in reducing neurodegeneration and improving cognition.

3.2.1 Memantine

It is a low to moderate affinity, uncompetitive N-methyl-D-aspartate receptor antagonist. Under physiological conditions Memantine allows normal glutamatergic neurotransmission but under pathological conditions it inhibits excitotoxicity (Parsons et al., 1999). A number of clinical studies indicates Memantine to be safe, well tolerated and effective as a symptomatic treatment for moderate to severe AD. In 24- and 28-week pivotal trials, conducted with 20 mg/day memantine as monotherapy or as an add-on to cholinesterase inhibitor treatment, the drug was superior to placebo on cognition (Severe Impairment Battery, SIB; but not MMSE) and activities of daily living (ADCS-ADL, modified for severe patients). Mixed results were seen with respect to global function (CIBIC+; Functional Assessment Staging scale, FAST) and behaviour (NPI) (Reisberg et al., 2003; Tariot et al., 2004).

3.3 Nicotinic receptor agonists

Nicotinic receptors (NRs) belong to the group of polymeric receptors of the cell membrane and are key elements of cholinergic transmission. Numerous subtypes of NRs exist with the alpha4 beta2 and alpha7 types being encountered most frequently. Alpha 7 NRs have been proposed to exert a direct or indirect action on the mechanism of Abeta toxicity. Nicotine has been reported to protect against Abeta-induced neuronal toxicity and death in rat cortical neurons. This neuroprotection can be blocked by dihydro-beta-erythroidine, an alpha4beta2 nicotinic receptor antagonist. Furthermore, incubation with cytisine, a selective alpha4beta2 nicotinic receptor agonist, can inhibit Abeta cytotoxicity. Deficiencies in NRs seem to play a role in AD (Bourin et al., 2003).

Clinical studies suggest that nicotine may provide cognitive benefit, however (van Duijn and Hofman, 1991), its long-term use may induce desensitization of nicotinic receptors (Marks et al., 1987). Allosteric modulation of NR can circumvent desensitisation. This allosteric interaction amplifies the actions of ACh at post- and presynaptic NR. Allosteric modulation of NR could therefore produce significant therapeutic benefit in AD (Maelicke, 2000).

A number of other receptors like 5-Hydroxytryptamine6 (5-HT6) (Reavill and Rogers, 2001), 5-HT4 (Lezoualc'h and Robert, 2003), histamine-H3 (Bongers et al., 2004) and γ-aminobutyric acid (GABA) (Maubach, 2003) are thought to play a role in learning and memory. Modulators of each of these receptor types have reached clinical development for AD, although proof of concept has yet to be demonstrated in patients.

4. DISEASE MODIFYING TREATMENT

According to the amyloid hypothesis Abeta is central to the pathophysiology of AD. High levels of amyloid peptides, especially Abeta42, initiate aggregation and plaque formation in the areas of brain associated with learning and memory. Therapeutic strategies that lower Abeta formation, prevent aggregation, dissolve plaques or promote clearance from the brain should prove beneficial.

4.1 Inhibition of amyloid formation

Abeta is produced by two sequential cleavages of amyloid precursor protein (APP) by two proteases, called β- and γ-secretase. β-secretase first cleaves APP in the extracellular domain to release a large APP fragment called APP–β and generates a membrane bound carboxy terminal fragment. γ-secretase cleaves the membrane bound fragment within the transmembrane domain to release Abeta peptide. Both these proteases are excellent targets for disease modification.

4.1.1 β-secretase inhibitors

Beta secretase, a membrane-anchored aspartyl protease, initiates the cleavage of APP at the beginning of Abeta peptide. β-secretase knock out mice lack Abeta and are phenotypically normal, suggesting that therapeutic inhibition of β-secretase may be free of mechanism-based side effects. Beta-secretase null mice overexpressing human APP are rescued from Abeta-dependent hippocampal memory deficit which correlates with a reduction of amyloid peptides (Ohno et al., 2004). Due to the potential for disease modification in AD, a number of groups have been trying to develop β-secretase inhibitors. Other aspartyl proteases such as Renin and HIV-1 have provided a rich background for the rational design of potent and selective inhibitors. The elucidation of the crystal structure of β-secretase complexed with inhibitors has further helped in designing of several inhibitors. The β-secretase active site is more open and less hydrophobic than that of other aspartyl proteases (Hong et al., 2002). Several peptide based β-secretase inhibitors have been described to date, however, all are relatively large molecules and are not drug-like (Hussain, 2004).

4.1.2 γ-secretase inhibitors

Gamma-secretase is a membrane protein complex with aspartyl protease activity that cleaves APP in its transmembrane domain to release Abeta and the APP intracellular domain (AICD). The identity of γ-secretase complex has been controversial. Identification of PS1 and PS2 as the possible active component of complex was established by genetic linkage studies in familial AD (FAD). Cleavage of APP by mutant presenilin results in the overproduction of amyloidogenic Abeta42. These mutations account for the majority of the cases of the FAD (Hardy, 2003). A number of co-factors have been identified such as nicastrin (Nct), a single transmembrane protein, the peptides presenilin enhancer protein-2 (PEN-2) and anterior pharynx defective protein-1 (APH-1). APH-1 stabilizes the presenilin holoprotein in the complex, whereas PEN-2 is required for endoproteolytic processing of presenilin and conferring γ-secretase activity to the complex. Nct undergoes a major conformational change during the assembly of the γ-secretase complex. The conformational change is directly associated with γ-secretase function (De Strooper, 2003). Recently, various components of γ-secretase complex when co-expressed in yeast that lacks endogenous γ-secretase activity resulted in reconstitution of γ-secretase activity. This work finally confirmed presenilin (PS), nicastrin (Nct), APH-1 and PEN-2 as essential components of γ-secretase complex (Edbauer et al., 2003). Since several paralogs and alternatively spliced variants of Presenilin and Aph-1 have been identified, γ-secretase may cleave several other membrane proteins, for example, Notch and Erb4, a receptor tyrosine kinase that regulates cell cycle (Kopan and Ilagan, 2004).

Gamma-secretase is an interesting but complex drug target that challenges classical thinking about proteolytic processing. The complete inhibition of γ–secretase activity is likely to result in serious side effects. Elan Pharmaceuticals reported a novel class of compounds that reduce Abeta production by functionally inhibiting γ-secretase. Oral administration of one of these compounds, N-[N-(3,5-difluorophenacetyl)-L-alanyl]-S-phenylglycine t-butyl ester (DAPT), to mice transgenic for human APP(V717F) reduces brain levels of amyloid in a dose-dependent manner within 3h (Dovey et al., 2001). Development of such novel functional γ-secretase inhibitors will enable a clinical examination of the Abeta hypothesis. Lilly are known to have progressed one gamma-secretase inhibitor into clinical trials (Siemers et al., 2005).

Recently retrospective epidemiological studies reported that patients on long-term non-steroidal anti-inflammatory drugs have reduced risk of AD (int'Veld et al., 2001). When tested in vitro and in vivo for their abeta lowering activity, 8 out of 13 NSAIDS and the enantiomers of flurbiprofen were found to be effective. Importantly, flurbiprofen and its enantiomers selectively lower Abeta42 levels in broken cell γ-secretase assays indicating that these compounds directly target the γ-secretase complex (Eriksen et al., 2003).

4.1.3 Rho-Rock pathway inhibitors

Recently, the Rho-Rock pathway has been shown to regulate amyloid precursor protein processing in vitro and a subset of NSAIDs that inhibit Rho activity reduce Abeta42. A selective Rock inhibitors (Y-27632) has also been shown to lower brain levels in a transgenic mouse model of AD (Zhou et al., 2003). The Rho-Rock pathway is a novel therapeutic target for AD.

4.2 Inhibition of Abeta aggregation

The aggregation of soluble Abeta peptide is viewed as a critical event in the pathophysiology of AD, preventing, altering, or reversing aggregation may be of therapeutic value.

4.2.1 Metal chelators

Binding of redox active transition metal ions like Cu^{2+} and Zn^{2+} to Abeta is thought to mediate its reversible aggregation and resistance to proteases. These metal ions are elevated in neocortex of AD patients especially in plaques. Chelating agents can inhibit the binding of these ions to Abeta, therefore these agents have potential therapeutic value. Clioquinol, a bioavailable Cu^{2+}/Zn^{2+} chelator, has been tested for its anti-aggregation activity both in vitro and in vivo. It inhibited and reversed Cu^{2+}/Zn^{2+} mediated aggregation of synthetic Abeta in vitro and solubilized Abeta in deposits in post-mortem AD brain samples. In a transgenic mouse model (APP2576) of AD, the oral administration of clioquinol for nine weeks was associated with significantly lower levels of aggregated Abeta accompanied by increased levels of soluble Abeta (Cherny et al., 2001).

In a clinical trial, 10 patients were given Clioquinol at 20 mg/day dose and 10 more given the same drug at 80 mg/day for 21 days each. Cerebrospinal fluid (CSF) investigations revealed a decrease in Tau protein and growth-associated protein (GAP43). These proteins are increased in AD and considered stable markers. The levels of CSF-Tau protein correlated positively and significantly with the serum levels of copper and also with the serum copper/zinc ratio. Clinical assessment showed slight improvement after 3 weeks treatment with clioquinol in this open study (Regland et al., 2001). In another randomised phase II trial, treatment with metal protein attenuating compound (MPAC, clioquinol) showed equivocal cognitive benefit in treated patients compared to placebo controls. Plasma levels of Abeta 42 decreased in clioquinol group and increased in placebo group (Ritchie et al., 2003). The results support targeting the interactions of Cu^{2+} and Zn^{2+} with Abeta as a novel therapeutic approach for the prevention and treatment of AD.

4.2.2 β-sheet breaker peptides

Several neurodegenerative diseases and systemic amyloidosis are thought to arise from the misfolding and aggregation of an underlying protein. In AD, blockade of the early steps involving the pathological conversion of the soluble peptide into the abnormally folded oligomeric intermediate precursor of the amyloid fibrils is an attractive therapeutic strategy. β-sheet breaker peptides are small synthetic peptides that are homologous to regions involved in the aggregation. In transgenic mouse model of AD (PDAPP) administration one such peptide (iAb5p) subcutaneously blocked amyloid aggregation (Soto et al., 2000). The central hydrophobic domain of amyloid peptide interacts with glycosaminoglycan (GAG) and the interaction is involved in aggregation and deposition. GAG mimetics have been developed and tested in vivo for their anti-aggregation activity. One such mimetic, Alzhemed, is in phase III clinical trials. It is yet unclear whether the inhibition of the defective folding of A-beta peptide is beneficial for the treatment of AD.

4.3 Improved clearance of Abeta

Abnormal accumulation of Abeta in brain is the driving factor for AD pathogenesis therefore its clearance is likely to be of primary therapeutic benefit.

4.3.1 Immunisation

Active immunisation with aggregated Abeta or passive immunisation with peripherally administered anti-amyloid antibodies reduce amyloid associated pathology and cognitive decline in transgenic mouse model of AD. A number of mechanisms have been proposed for antibody-mediated clearance of amyloid from brain. One of the proposed mechanisms by which antibodies may reduce brain amyloid is that the antibodies cross the blood brain barrier (BBB), bind to amyloid plaques and activate microglial phagocytosis of immune complexes. Another proposed mechanism is that there is a dynamic equilibrium of Abeta between brain and periphery and the antibodies in periphery can act as sink, capturing Abeta in the blood stream and indirectly reducing the Abeta burden in the brain by driving the clearance of peptide from brain to plasma. Another proposed mechanism is that anti-Abeta antibodies directed against specific epitopes might protect against neurotoxicity by inhibiting aggregation of Abeta and by disaggregating already established aggregates or plaques (Morgan and Gitter, 2004). Two different anti-amyloid monoclonal antibody therapies are currently being examined in clinical trials (Pangalos et al., 2005).

Active vaccination with Abeta 42 (AN-1792) has been investigated in clinical trials for possible treatment of AD. In phase I, a single dose was found to be safe and consequently, a phase II trial of AN1792 was initiated. The trial was terminated after a small percentage of patients developed signs of meningoencephalitis (Orgogozo et al., 2003). A post-mortem study of one of the patients, who died due to unrelated causes, revealed presence of activated T-lymphocytes suggesting the adverse effects seen in some patients might be due to the cellular immune response rather than antibody response (Nicoll et al., 2003). In a cohort of 30 patients from AN-1792 trial a significant correlation was found between antibody response and cognitive decline. Data on the whole trial population have only recently been published (Gilman et al., 2005; Fox et al., 2005). Despite premature study termination, patient monitoring continued for up to 1 year after final dosing. A comparison between placebo-treated patients and those vaccinated patients who generated a predefined antibody response failed to reveal any significant effect of treatment on standard measures of cognition (ADAS-cog; MMSE), global function (Clinician's Global Impression of Change, CGIC; CDR), or activities of daily living (ADL). A significant treatment benefit was, however, observed with respect to a composite neuropsychological test battery. Interestingly, cerebrospinal fluid levels of tau (but not Abeta) appeared to be reduced in antibody responders versus controls. Counter-intuitively, serial MRI measurements revealed an increase in brain volume in antibody responders, possibly due to plaque disruption and associated cerebral fluid shifts.

Vaccination using smaller fragments of Abeta conjugated to T helper epitopes and various routes of administration are still being pursued in pre-clinical models to study safety and efficacy of this approach (Robinson et al., 2004).

Compounds unrelated to antibodies have been used recently to test the peripheral sink mechanism. Peripheral treatment with gelsolin or ganglioside (GM-1) reduced the level of Abeta in the brain of transgenic mouse model of AD suggesting that sequesteration of plasma Abeta could reduce or prevent brain amyloidosis. Gelsolin, a secretory protein and GM-1, a ganglioside, are known to bind with Abeta with high affinity (Chauhan et al., 1999; Choo-Smith and Surewicz, 1997). Future studies with high affinity Abeta binding small molecules may provide further validation of this approach and amyloid hypothesis.

4.3.2 Neprilysin

The steady-state level of Abeta represents a balance between its biosynthesis from the APP and its catabolism by a variety of proteolytic enzymes like neprilysin (NEP), endothelin-converting enzyme, insulin-degrading enzyme, angiotensin-converting enzyme and plasmin. Neprilysin (NEP) is a major Abeta peptide-degrading enzyme as shown by higher Abeta peptide levels in hippocampus, cortex, thalamus/striatum, and cerebellum of an NEP knockout mouse and by reduction in amyloid load in APP transgenic mice treated with viral vector expressing NEP (Marr et al., 2004). Expression of Neprilysin is down regulated with age and correlates negatively with amyloid deposition in APP transgenic mice. Therapeutic strategies aimed at promoting Abeta degradation may provide a novel approach to treat AD (Carson and Turner, 2002).

4.3.3 Insulin degrading enzyme (IDE)

Epidemiological evidence indicates that insulin resistance in type II diabetes is associated with an increased relative risk for AD. Both genetic linkage and allelic association in the IDE region of chromosome 10 have been reported in families with late-onset AD. This may link diabetes with AD (Ertekin-Taner et al., 2004). Naturally occurring IDE missense mutations that result in partial loss of function have been shown to associate with increased levels of insulin and Abeta in plasma (Farris et al., 2004). Insulin resistance promotes amyloidosis in APP transgenic mice that corresponds with increased γ-secretase activities and decreased insulin degrading enzyme (IDE) activities. Apparent interrelationship of insulin resistance to brain amyloidosis is due to a functional decrease in insulin receptor (IR)-mediated signal transduction in the brain. Decreased signal transduction positively correlate with the generation of brain C-terminal fragment of APP, an index of γ-secretase activity, in the brain of insulin-resistant transgenic mice (Ho et al., 2004).

4.3.4 Receptor for Advanced Glycation End products (RAGE)

Blood brain barrier (BBB) regulates entry of plasma derived Abeta into central nervous system and clearance of brain derived Abeta into periphery through several receptors or carrier proteins such as low density lipoprotein related protein-1 (LRP-1), megalin, cubulin and receptor for advanced glycation end products (RAGE) (Zlokovic, 2004). Alterations in the permeability of the BBB may lead to accumulation of Abeta in brain. RAGE is upregulated in AD brain vasculature (Yan et al., 1996) and increases in transgenic mouse model of AD with age (Kawarabayashi et al., 2001). RAGE mediated transport of circulating Abeta across BBB is an important factor in the pathogenesis of cerebrovascular amyloidosis as shown by lack of Abeta deposition in transgenic mice treated with soluble RAGE (Deane et al., 2003). RAGE knock out mice are viable suggesting that blockade of RAGE with immunotherapeutic or small molecule inhibitor may be an important therapeutic opportunity for developing treatment for AD (Sakaguchi et al., 2003).

4.4 Tau phosphorylation inhibitors

Tau is a microtubule associated peptide that is involved in axonal transport. This transport involves repeated phosphorylation and dephosphorylation of tau. Neurofibrillary tangle formation may be due to an imbalance of this process. Glycogen synthase-3 (GSK-3) and cyclin dependent kinase-5 (Cdk-5) are current targets to reduce tau phosphorylation. Certain mood stabilizers such as lithium and valproate may have complex neuroprotective effects including inhibition of GSK-3. Lithium was recently found to reduce amyloid in mouse model of AD. Valproate will be studied in a multicenter clinical trial in patients with AD (Phiel et al., 2003; Tariot et al., 2002).

5. OTHER THERAPIES

5.1 Cholesterol lowering therapies

A number of epidemiological studies suggest that high cholesterol levels increase the risk of AD significantly, however, there are others which did not report a link (Simons et al., 2001). Numerous laboratory studies implicate cholesterol in the process of Abeta production and accumulation. Changes in APP processing by cholesterol may explain how ApoE4 allele increases risk of developing AD (Frears et al., 1999). Cholesterol is present in the dense cores of senile plaques both in humans and transgenic mice suggesting that cholesterol plays an important role in the formation and/or progression of senile plaques (Mori et al., 2001). Cholesterol rich diet increases intracellular Abeta and levels of Abeta strongly correlate with the levels of cholesterol in plasma and CNS (Shie et al., 2002). Free cholesterol in neurofibrillary tangle-bearing neurons is higher than those of adjacent tangle-free neurons (Distl et al., 2001). Genetic heterogeneity in ApoE allele is associated higher risk of AD. People expressing ApoE4 have higher circulating levels of cholesterol and are at greater risk than people with ApoE2 or ApoE3. ApoE4 accelerate amyloid deposition and promotes Abeta aggregation in cholesterol rich lipid rafts (Kawarabayashi et al., 2004). It is now believed that cholesterol-lowering therapies will be of value as disease modifying agents. Epidemiological studies have shown that statins are associated with decreased risk of developing AD (Crisby et al., 2002). These observations require both preclinical and clinical validation. The former involves testing statins in one or more animal models of AD in order to establish relative efficacy and disease features affected by treatment. The latter requires prospective, randomized, placebo controlled trials to evaluate the effect of statin treatment on cognitive and AD biomarker outcomes. High doses of simvastatin show a strong and reversible reduction of cerebral Abeta42 and Abeta40 levels in the cerebrospinal fluid and brain homogenate of transgenic and guinea pig models (Fassbender et al., 2001). In most of the clinical trials, statins have shown no effect on Abeta levels in plasma or cerebrospinal fluid. In several randomized, placebo-controlled, double-blind clinical trials, statins such as simvastatin or atorvastatin did not alter cerebrospinal fluid levels of Abeta40 and Abeta42 (Hoglund et al., 2004). However, in a double-blind, randomized, placebo-controlled study, lovastatin reduced serum Abeta levels compared to the baseline (Friedhoff et al., 2001). Future controlled clinical trials may help in explaining the contradiction seen in epidemiological and most of the clinical studies. In a recently-reported, small, 1-year clinical trial, atorvastatin treatment significantly improved cognition *versus* placebo at 6 months, but not at 12 months (Sparks et al., 2005). Data from larger studies are awaited.

5.2 Anti-inflammatory therapies

Inflammation clearly occurs in AD brain with full complexity of local inflammatory responses (Akiyama et al., 2000). Microglia, the predominant immune cells in the brain are consistently associated with senile plaques in AD brain and may play a pivotal role in neuroinflammation. There is a considerable body of evidence indicating that the microglia activated by β-amyloid, neurofibrillary tangles or degenerating neurons are the primary source of pro-inflammatory cytokines IL1, IL6, TNF α; chemokines such as MIP1 α, MCP-1, IL5, IL8; superoxide free radicals and neurotoxic substances. Activated microglia help clear Abeta deposits and thus prevent their harmful effects. Nevertheless, chronic activation of microglia may contribute to neurodegeneration. Patients who show Abeta deposition and neurofibrillary tangle, but limited inflammation, have no history of dementia. Transgenic mice overexpressing various inflammatory mediators show AD like pathology as well as cognitive deficit. The animals show decreased acetylcholine production, neurodegeneration, learning deficit and memory impairment in dose and age related manner. Based on observations from neuropathology in AD and animal experimentation, inflammation has been considered a therapeutic target for AD. Inflammation is not a primary event in AD and cannot be considered to have a causal role, however, it may add to the progression of the disease. Epidemiological evidence suggests that anti-inflammatory therapies may reduce the risk of developing AD (Moore and O'Banion, 2000). However, clinical trial data are discouraging for patients with established AD. In a randomised controlled trial rofecoxib or naproxen showed no effect on cognitive decline. One early trial with indomethacin saw some benefit; subsequent trials with rofecoxib, celecoxib, diclofenac, hydroxychloroquine, naproxen and prednisolone have not shown significant benefit (Rogers et al., 1993). Inflammatory pathways contributing to AD pathology never occur in isolation but act concurrently. Chronic activation of inflammatory responses may necessitate therapies that target more than one pathway simultaneously to achieve clinical benefit. This may explain why clinical trials with anti-inflammatory drugs have not shown any beneficial effect.

5.3 Antioxidants

Free-radical oxidative stress, particularly of neuronal lipids, proteins and DNA, is extensive in those AD brain areas in which Abeta is abundant. Abeta-induced oxidative stress leads to neurodegeneration in AD brain. Abeta leads to neuronal lipid peroxidation, protein oxidation and DNA oxidation by means that are inhibited by free-radical antioxidants. Catecholamines involved with oxidation (monoamine oxidase) are abundant in AD brain where as antioxidant enzymes like superoxide dismutase, catalase, glutathione peroxidase and gluatthione reductase are reduced. Therefore, the risk of Alzheimer disease might be reduced by intake of antioxidants that counteract the detrimental effects of oxidative stress (Butterfield et al., 2001).

5.3.1 Vitamins and Selegiline

In a population-based, prospective cohort study conducted in the Netherlands, high dietary intake of vitamin C and vitamin E was associated with lower risk of Alzheimer disease (Engelhart et al., 2002). In a cross-sectional and prospective study of dementia, use of vitamin E and vitamin C supplements in combination was associated with reduced prevalence and incidence of AD. One study found that vitamin E from food, but not other antioxidants, may be associated with a reduced risk of AD. Unexpectedly, this association was observed only among individuals without the APOE epsilon 4 allele. However, another study found that the intake of carotenes and vitamin C, or vitamin E in supplemental or dietary (nonsupplemental) form or in both forms, was not related to a decreased risk of AD. In a double-blind, placebo-controlled, randomized, multicenter trial in patients with AD of moderate severity, a selective monoamine oxidase inhibitor selegiline (10 mg a day) or alpha-tocopherol (vitamin E) 2000 IU a day slowed the progression of disease. A meta analysis of the published trials on treatment with selegiline showed little evidence of improvement with selegiline in the short term in cognition and activities of daily living, which was clinically insignificant. Flavonoids, powerful antioxidants present in wine, tea, fruits and vegetables show inverse correlation with the risk of dementia (Wilcock et al., 2002).

5.3.2 Ginkgo biloba

Ginkgo biloba (GbE) extracts have played a crucial role in Chinese herbal medicine for many centuries. Previous studies have suggested the clinical efficacy of GbE in patients with dementia, cerebral insufficiency, or related cognitive decline. However, the effectiveness of GbE in AD is controversial (Solomon et al., 2002). GbE preparations are approved in many European countries for the treatment of dementia syndromes including AD.

5.4 Neurotrophic factors

5.4.1 Nerve growth factor (NGF)

Transgenic mice expressing anti-NGF factor antibodies show AD like neurodegenerative phenotype which includes plaques, neuronal loss, cholinergic deficit, and tau hyperphosphorylation, associated with neurofibrillary pathology suggesting a direct link between NGF signaling and abnormal processing of amyloid precursor protein (Capsoni et al., 2002). NGF therapy might reduce degeneration of cholinergic neurons. In a short term clinical trial, intracerebroventricular infusion of NGF in AD patients resulted in slight cognitive benefit. Long term therapy may provide clear benefit but association of the intraventricular route of administration with negative side effects appear to outweigh the positive effects (Eriksdotter et al., 1998). Due to lack of brain penetration of NGF, orally bioavailable NGF synthesis stimulators, like idebenone and propentofylline, have been tested in pre-clinical models and found to restore age associated NGF loss. The results suggest that the use of NGF synthesis stimulators may provide a novel therapeutic approach to cholinergic dysfunction (Yamada et al., 1997).

5.4.2 Brain-derived neurotrophic factor (BDNF)

BDNF is a prosurvival factor induced by cortical neurons that is necessary for survival of cholinergic neurons of the basal forebrain, hippocampus and cortex (Bimonte-Nelson et al., 2003). The reduction of BDNF early on in AD could weaken synaptic encoding strength of hippocampus and make it vulnerable to degeneration. A single nucleotide polymorphism, val66met, in the BDNF gene has been associated with poor episodic memory and abnormal hippocampal activation in a cohort of 641 human subjects (Egan et al., 2003).

In a clinical trial, the neurotrophic mixture cerebrolysin was reported to be well tolerated and resulted in significant improvements in the global score and activities of daily living in patients with AD (Panisset et al., 2002). In pre-clinical studies, Neotrofin, a purine derivative, was found to stimulate neuritogenesis, the production of neurotrophic factors and to have memory enhancing properties (Holmes et al., 2003). In a phase I, randomized, double blind, placebo-controlled clinical trial, neotrofin was apparently safe and well tolerated in healthy elderly volunteers (Grundman et al., 2003). Other approaches like intraparenchymal administration, tissue transplantation and use of viral vectors to deliver neurotrophic factors are underway.

5.4.3 Estrogen

Estrogen may have cholinergic, neurotrophic and neuroprotective effects and may enhance cognitive function (Fillit et al., 1986). Observational studies have suggested that postmenopausal hormone treatment may improve cognitive function, but data from randomized clinical trials have been sparse and inconclusive. Recently, in a randomised controlled clinical trial of postmenopausal women, estrogen plus progestin did not improve cognitive function but increased risk of clinically meaningful cognitive decline (Rapp et al., 2003). Raloxifene, a selective estrogen receptor modulator (SERM), that produces both estrogen-agonistic effects on bone and lipid metabolism and estrogen-antagonistic effects on uterine endometrium and breast tissue, have been tested for its safety and efficacy in AD patients. In a randomized double blind osteoporosis treatment trial Raloxifene showed no cognitive benefit after 12 months treatment (Scott et al., 1999).

6. CONCLUSION

AD is a progressive neurodegenerative disease that accounts for most cases of dementia seen in the elderly. The socio-economic burden of the disease is likely to increase due to increasing life expectancy. Early clinical diagnosis and timely treatment of AD patients can maintain patient's quality of life and prevent high costs associated with it. There is no definitive cure for AD and the currently available symptomatic treatments show limited efficacy.

It is clear that there are number of strategies to try and intervene in the process of AD. A number of mechanistic targets for AD have been validated by using *in vitro* and *in vivo* systems and several approaches of disease modification are being pursued in the pharmaceutical industry as well as in academia. Recently, active and passive immunisation have been successful in clearing the amyloid peptide from brain of transgenic mouse model of AD. However, in a clinical trial, serious adverse effects of active vaccination resulted in early termination. Other therapeutic opportunities are provided by physiological responses observed in patients such as oxidative stress and neuroinflammation. Epidemiological and clinical trial studies with antioxidants and anti-inflammatory agents have been contradictory. Further controlled trials are needed to address these issues. In the absence of any disease modifying therapy, symptomatic treatment targeting the cholinergic system is the only current option for treatment of AD. A combination of multiple agents is likely to be the option for treatment of AD in the future.

REFERENCES

Akiyama, H., Barger, S., Barnum, S., et al. (2000) Inflammation and Alzheimer's disease. Neurobiol Aging, 21: 383–421.

Bimonte-Nelson, H.A., Hunter, C.L., Nelson, M.E. and Granholm, A.C. (2003) Frontal cortex BDNF levels correlate with working memory in an animal model of Down syndrome. Behav Brain Res., 139: 47–57.

Bongers, G., Leurs, R., Robertson, J., Raber, J. (2004) Role of H3-receptor-mediated signaling in anxiety and cognition in wild-type and Apoe–/–mice. Neuropsychopharmacology, 29: 441–449.

Bourin, M., Ripoll, N., Dailly, E. (2003) Nicotinic receptors and Alzheimer's disease. Curr Med Res Opin., 19: 169–177.

Butterfield, D.A., Drake, J., Pocernich, C., Castegna, A. (2001) Evidence of oxidative damage in Alzheimer's disease brain: central role for amyloid beta-peptide. Trends Mol Med., 7: 548–554.

Capsoni, S., Giannotta, S., Cattaneo, A. (2002) Beta-amyloid plaques in a model for sporadic Alzheimer's disease based on transgenic anti-nerve growth factor antibodies. Mol Cell Neurosci., 21: 15–28.

Carson, J.A. and Turner, A.J. (2002) Beta-amyloid catabolism: roles for neprilysin (NEP) and other metallopeptidases? J Neurochem., 81: 1–8.

Chauhan, V.P., Ray, I., Chauhan, A., Wisniewski, H.M. (1999) Binding of gelsolin, a secretory protein, to amyloid beta-protein. Biochem Biophys Res Commun., 258: 241–246.

Cherny, R.A., Atwood, C.S., Xilinas, M.E., et al. (2001) Treatment with a copper-zinc chelator markedly and rapidly inhibits beta-amyloid accumulation in Alzheimer's disease transgenic mice. Neuron, 30: 665–676.

Choo-Smith, L.P. and Surewicz, W.K. (1997) The interaction between Alzheimer amyloid beta(1-40) peptide and ganglioside GM1-containing membranes. FEBS Lett., 402: 95–98.

Courtney, C., Farrell, D., Gray, R., et al. (2004) Long-term donepezil treatment in 565 patients with Alzheimer's disease (AD2000): randomised double-blind trial. Lancet, 363: 2105–2115.

Courtney, C., Farrell, D., Gray, R., et al. (2004) Long-term donepezil treatment in 565 patients with Alzheimer's disease (AD2000): randomised double-blind trail. Lancet, 363: 2105–2115.

Crisby, M., Carlson, L.A. and Winblad, B. (2002) Statins in the prevention and treatment of Alzheimer disease. Alzheimer Dis Assoc Disord., 16: 131–136.

Deane, R., Yan, S.D., Submamaryan, R.K., et al. (2003) RAGE mediates amyloid-beta peptide transport across the blood-brain barrier and accumulation in brain. Nat Med., 9: 907–913.

De Strooper, B. (2003) Aph-1, Pen-2, and Nicastrin with Presenilin generate an active gamma-Secretase complex. Neuron, 38: 9–12.

Dickson, D.W. (1997) The pathogenesis of senile plaques. J Neuropathol Exp Neurology, 56: 321–339.

Distl, R., Meske, V., Ohm, T.G. (2001) Tangle-bearing neurons contain more free cholesterol than adjacent tangle-free neurons. Acta Neuropathol (Berl)., 101: 547–554.

Dovey, H.F., John, V., Anderson, J.P., et al. (2001) Functional gamma-secretase inhibitors reduce beta-amyloid peptide levels in brain. J Neurochem., 76: 173–181.

Edbauer, D., Winkler, E., Regula, J.T., et al. (2003) Reconstitution of gamma-secretase activity. Nat Cell Biol., 5: 486–488.

Egan, M.F., Kojima, M., Callicott, J.H., et al. (2003) The BDNF val66met polymorphism affects activity-dependent secretion of BDNF and human memory and hippocampal function. Cell, 112: 257–269.

Engelhart, M.J., Geerlings, M.I., Ruitenberg, A., et al. (2002) Dietary intake of antioxidants and risk of Alzheimer disease. JAMA., 287: 3223–3229.

Eriksdotter Jonhagen, M., Nordberg, A., Amberla, K., et al. (1998) Intracerebroventricular infusion of nerve growth factor in three patients with Alzheimer's disease. Dement Geriatr Cogn Disord., 9: 246–257.

Eriksen, J.L., Sagi, S.A., Smith, T.E., et al. (2003) NSAIDs and enantiomers of flurbiprofen target gamma-secretase and lower Abeta 42 in vivo. J Clin Invest., 112: 440–449.

Ertekin-Taner, N., Allen, M., Fadale, D., et al. (2004) Genetic variants in a haplotype block spanning IDE are significantly associated with plasma Abeta42 levels and risk for Alzheimer disease. Hum Mutat., 23: 334–342.

Farlow, M., Anand, R., Messina, Jr. J., et al. (2000) A 52-Week Study of the Efficacy of Rivastigmine in Patients with Mild to Moderately Severe Alzheimer's Disease. European Neurology, 44: 236–241.

Farris, W., Mansourian, S., Leissring, M.A., et al. (2004) Partial loss-of-function mutations in insulin-degrading enzyme that induce diabetes also impair degradation of amyloid beta-protein. Am J Pathol., 164: 1425–1434.

Fassbender, K., Simons, M., Bergmann, C., et al. (2001) Simvastatin strongly reduces levels of Alzheimer's disease beta -amyloid peptides Abeta 42 and Abeta 40 in vitro and in vivo. PNAS USA., 98: 5856–5861.

Ferri, C.P., Prince, M., Brayne, C., et al. (2005) Global prevalence of dementia: a Delphi consensus study. Lancet, 366: 2112–2117.

Fillit, H., Weinreb, H., Cholst, I., et al. (1986) Observations in a preliminary open trial of estradiol therapy for senile dementia-Alzheimer's type. Psycho-neuroendocrinology, 11: 337–345.

Fox, N.C., Black, R.S., Gilman, S., et al. (2005) Effects of Abeta immunization (AN1792) on MRI measures of cerebral volume in Alzheimer disease. Neurology, 64: 1563–1572.

Frears, E.R., Stephens, D.J., Walters, C.E., et al. (1999) The role of cholesterol in the biosynthesis of beta-amyloid. Neuroreport, 10: 1699–1705.

Friedhoff, L.T., Cullen, E.I., Geoghagen, N.S. and Buxbaum, J.D. (2001). Treatment with controlled-release lovastatin decreases serum concentrations of human beta-amyloid (A beta) peptide. Int J Neuropsychopharmacol., 4: 127–130.

Geldmacher, D.S., Provenzano, G., McRae, T., et al. (2003) Donepezil is associated with delayed nursing home placement in patients with Alzheimer's disease. Journal of the American Geriatrics Society 51: 937–944.

Gilman, S., Koller, M., Black, R.S., et al. (2005) Clinical effects of Abeta immunization (AN1792) in patients with AD in an interrupted trial. Neurology, 64: 1553–1562.

Goedert, M. (1996) Tau protein and the neurofibrillary pathology of Alzheimer's disease. Ann, N. Y. Acad Sci., 777: 121–131.

Grundke-Iqbal, I., Iqbal, K., Tung, Y.C., et al. (1986) Abnormal phosphorylation of the microtubule associated protein t (tau) in Alzheimer cytoskeletoal pathology. PNAS USA., 83: 4913–4917.

Grundman, M., Capparelli, E. and Kim, H.T. (2003) A multicenter, randomized, placebo controlled, multiple-dose, safety and pharmacokinetic study of AIT-082 (Neotrofin) in mild Alzheimer's disease patients. Life Sci., 73: 539–553.

Hardy, J. (2003) Alzheimer's disease: genetic evidence point to a single pathogenesis. Ann. Neurol., 54: 143–144.

Hardy, J.A., Higgins, G.A. (1992) Alzheimer's disease: the amyloid cascade hypothesis. Science., 256: 184–185.

Hebert, L.E., Scherr, P.A., Bienias, J.L., et al. (2003) Alzheimer disease in the US population: prevalence estimates using the 2000 census. Arch Neurol., 60: 1119–1122.

Ho, L., Qin, W., Pompl, P.N., et al. (2004) Diet-induced insulin resistance promotes amyloidosis in a transgenic mouse model of Alzheimer's disease. FASEB J., 18: 902–904.

Hoglund, K., Wiklund, O., Vanderstichele, H., et al. (2004) Plasma levels of beta-amyloid(1-40), beta-amyloid(1-42), and total beta-amyloid remain unaffected in adult patients with hypercholesterolemia after treatment with statins. Arch Neurol., 61: 333–337.

Holmes, M., Maysinger, D., Foerster, A., et al. (2003) Neotrofin, a novel purine that induces NGF-dependent nociceptive nerve sprouting but not hyperalgesia in adult rat skin. Mol Cell Neurosci., 24: 568–580.

Hong, L., Turner., R.T., Koelsch, G., et al. (2002) Crystal structure of memapsin 2 (b-secretase) in complex with an inhibitor OM00-3. Biochemistry, 41: 10963–10967.

Hussain, I. (2004) The potential for BACE1 inhibitors in the treatment of Alzheimer's disease. IDrugs, 7: 653–658.

Ibach, B., Haen, E. (2004) Acetylcholinesterase inhibition in Alzheimer's Disease. Curr Pharm Des., 10: 231–251.

in t' Veld, B.A., Ruitenberg, A., Hofman, A., et al. (2001) Nonsteroidal antiinflammatory drugs and the risk of Alzheimer's disease. N Engl J Med., 345: 1515–1521.

Kawarabayashi, T., Younkin, L.H., Saido, T.C., et al. (2001) Age-dependent changes in brain , CSF and plasma amyloid (beta) protein in the Tg2576 transgenic mouse model of Alzheimer's disease. J Neurosci., 21: 372–381.

Kawarabayashi, T., Shoji, M., Younkin, L.H., et al. (2004) Dimeric amyloid beta protein rapidly accumulates in lipid rafts followed by apolipoprotein E and phosphorylated tau accumulation in the Tg2576 mouse model of Alzheimer's disease. J Neurosci., 24: 3801–3809.

Kopan, R. and Ilagan, M. (2004) γ-Secretase: proteosome of the membrane? Nature Reviews Molecular Cell Biology, 5: 499–504.

Kukull, W.A. and Ganguli, M. (2000) Epidemiology of dementia: concepts and overview. Neurol Clin., 18: 923–950.

Kumar, U. (2005) Alzheimer's Disease: Current and Future treatments. In: Aging Interventions and Therapies (Ed.: Rattan, S.) Pages 329–354, World Scientific, Singapore.

Lezoualc'h, F. and Robert, S.J. (2003) The serotonin 5-HT4 receptor and the amyloid precursor protein processing. Exp Gerontol., 38: 159–166.

Maelicke, A. (2000) Allosteric modulation of nicotinic receptors as a treatment strategy for Alzheimer's disease. Dement Geriatr Cogn Disord., 11 Suppl 1: 11–18.

Mandelkow, E.M., Mandelkow, E. (1998) Tau in Alzheimer's disease. Trends Cell Biol., 8: 425–427.

Marks, M.J., Stitzel, J.A. and Collins, A.C. (1987) Influence of kinetics of nicotine administration on tolerance development and receptor levels. Pharmacol Biochem Behav., 27: 505–512.

Marr, R.A., Guan, H., Rockenstein, E., et al. (2004) Neprilysin regulates amyloid Beta peptide levels. J Mol Neurosci., 22: 5–11.

Maubach, K. (2003) GABA(A) receptor subtype selective cognition enhancers. Curr Drug Targets CNS Neurol Disord., 2: 233–239.

Mohs, R.C., Doody, R.S., Morris, J.C., et al. (2001) A 1-year, placebo-controlled preservation of function survival study of donepezil in AD patients. Neurology, 57: 481–488.

Moore, A.H., O'Banion, M.K. (2000) Neuroinflammation and anti-inflammatory therapy for Alzheimer's disease. Adv Drug Deliv Rev., 54: 1627–1656.

Morgan, D., Gitter, B.D. (2004) Evidence supporting a role for anti-Abeta antibodies in the treatment of Alzheimer's disease. Neurobiol Aging., 25: 605–608.

Mori, T., Paris, D., Town, T., et al. (2001) Cholesterol accumulates in senile plaques of Alzheimer disease patients and in transgenic APP(SW) mice. J Neuropathol Exp Neurol., 60: 778–785.

Mudher, A., Lovestone, S. (2002) Alzheimer's disease-do tauists and baptists finally shake hands? Trends Neurosci., 25: 22–26.

Nicoll, J.A., Wilkinson, D., Holmes, C., et al. (2003) Neuropathology of human Alzheimer disease after immunization with amyloid-beta peptide: a case report. Nat Med., 9: 448–452.

Oddo, S., Caccamo, A., Kitazawa, M., et al. (2003) Amyloid deposition precedes tangle formation in a triple transgenic model of Alzheimer's disease. Neurobiol Aging., 24: 1063–1070.

Ohno, M., Sametsky, E., Younkin, N., et al. (2004) BACE1 deficiency rescues memory deficits and cholinergic dysfunction in a mouse model of Alzheimer's disease. Neuron, 41: 27–33.

Orgogozo, J.M., Gilman, S., Dartigues, J.F., et al. (2003) Subacute meningoencephalitis in a subset of patients with AD after Abeta42 immunization. Neurology, 61: 46–54.
Palmer, G.C. (2001) Neuroprotection by NMDA receptor antagonists in a variety of neuropathologies. Curr Drug Targets, 2: 241–271.
Pangalos, M.N., Jacobsen, S.J. and Reinhart, P.H. (2005) Disease modifying strategies for the treatment of Alzheimer's disease targeted at modulating levels of the beta-amyloid peptide. Biochemical Society Transactions, 33: 553–558.
Panisset, M., Gauthier, S., Moessler, H. and Windisch, M. (2002) Cerebrolysin in Alzheimer's disease: a randomized, double-blind, placebo-controlled trial with a neurotrophic agent. J Neural Transm., 109: 1089–1104.
Parsons, C.G., Danysz, W., Quack, G. (1999) Memantine is a clinically well tolerated N-methyl-D-aspartate (NMDA) receptor antagonist–a review of preclinical data. Neuropharmacology, 38: 735–767.
Phiel, C.J., Wilson, C.A., Lee, V.M. and Klein, P.S. (2003) GSK-3alpha regulates production of Alzheimer's disease amyloid-beta peptides. Nature, 423: 435–439.
Phinney, A.L., Horne, P., Yang, J., et al. (2003) Mouse models of Alzheimer's disease: the long and filamentous road. Neurol Res., 25: 590–600.
Rapp, S.R., Espeland, M.A., Shumaker, S.A., et al. (2003) Effect of estrogen plus progestin on global cognitive function in postmenopausal women: the Women's Health Initiative Memory Study: a randomized controlled trial. JAMA., 289: 2663–2672.
Raskind, M.A., Peskind, E.R., Wessel, T. and Yuan, W. (2000) Galantamine in AD: A 6-month randomized, placebo-controlled trial with a 6-month extension. Neurology, 54: 2261–2268.
Raskind, M.A., Peskind, E.R., Truyen, L., et al. (2004) The cognitive benefits of galantamine are sustained for at least 36 months: a long-term extension trial. Archives of neurology, 61: 252–256.
Raskind, M.A., Peskind, E.R., Truyen, L.B., et al. (2004) The cognitive benefits of galantamine are sustained for at least 36 months: a long-term extension trial. Arch Neurol., 61: 252–256.
Reavill, C. and Rogers, D.C. (2001) The therapeutic potential of 5-HT6 receptor antagonists. Curr Opin Investig Drugs., 2: 104–109.
Regland, B., Lehmann, W., Abedini, I., et al. (2001) Treatment of Alzheimer's disease with clioquinol. Dement Geriatr Cogn Disord., 12: 408–414.
Reisberg, B., Doody, R., Stoffler, A., et al. (2003) Memantine in moderate-to-severe Alzheimer's disease. N Engl J Med., 348: 1333–1341.
Ritchie, C.W., Bush, A.I., Mackinnon, A., et al. (2003) Metal-protein attenuation with iodochlorhydroxyquin (clioquinol) targeting Abeta amyloid deposition and toxicity in Alzheimer disease: a pilot phase 2 clinical trial. Arch Neurol., 60: 1685–1691.
Robinson, S.R., Bishop, G.M., Lee, H.G. and Munch, G. (2004) Lessons from the AN 1792 Alzheimer vaccine: lest we forget. Neurobiol Aging, 25: 609–615.
Rockwood, K., Mintzer, J., Truyen, L., et al. (2001) Effects of a flexible galantamine dose in Alzheimer's disease: a randomised, controlled trial. Journal of neurology, neurosurgery, and psychiatry, 71: 589–595.
Rogers, J., Kirby, L.C., Hempelman, S.R., et al. (1993) Clinical trial of indomethacin in Alzheimer's disease. Neurology, 43: 1609–1611.
Rogers, S.L., Farlow, M.R., Doody, R.S., et al. (1998a) A 24-week, double-blind, placebo-controlled trial of donepezil in patients with Alzheimer's disease. Neurology. 50: 136–145.
Rogers, S.L., Doody, R.S., Mohs, R.C., et al. (1998b) Donepezil improves cognition and global function in Alzheimer disease: a 15-week, double-blind, placebo-controlled study. Donepezil Study Group. Archives of Internal Medicine, 158: 1021–1031.
Rogers, S.L., Doody, R.S., Pratt, R.D., et al. (2000) Long-term efficacy and safety of donepezil in the treatment of Alzheimer's disease: final analysis of a US multicentre open-label Study. European neuropsychopharmacology, 10: 195–203.
Sakaguchi, T., Yan, S.F., Yan, S.D., et al. (2003) Central role of RAGE-dependent neointimal expansion in arterial restenosis. J Clin Invest., 111: 959–972.
Sano, M., Wilcock, G.K., van Baelen, B., et al. (2003) The effects of galantamine treatment on caregiver time in Alzheimer's disease. International Journal of Geriatric Psychiatry, 18: 942–950.

Schneider, L.S. and Tariot, P.N. (2003) Cognitive enhancers and treatments for Alzheimer's disease. In Tasman, A., Kay, J. and Lieberman, J.A. (eds.) Psychiatry, 2nd edition John Wiley and Sons, London.

Scott, J.A., Da Camara, C.C. and Early, J.E. (1999) Raloxifene: a selective estrogen receptor modulator. Am Fam Physician, 60: 1131–1139.

Shie, F.S., Jin, L.W., Cook, D.G., et al. (2002) Diet-induced hypercholesterolemia enhances brain A beta accumulation in transgenic mice. Neuroreport, 13: 455–459.

Siemers, E., Skinner, M., Dean, R.A., et al. (2005) Safety, Tolerability, and Changes in Amyloid beta Concentrations After Administration of a gamma-Secretase Inhibitor in Volunteers. Clinical Neuropharmacology, 28: 126–132.

Simons, M., Keller, P., Dichgans, J. and Schulz, J.B. (2001) Cholesterol and Alzheimer's disease: is there a link? Neurology, 57: 1089–1093.

Solomon, P.R., Adams, F., Silver, A., et al. (2002) Ginkgo for memory enhancement: a randomized controlled trial. JAMA., 288: 835–840.

Soto, C., Saborio, G.P., Permanne, B. (2000) Inhibiting the conversion of soluble amyloid-beta peptide into abnormally folded amyloidogenic intermediates: relevance for Alzheimer's disease therapy. Acta Neurol Scand Suppl., 176: 90–95.

Sparks, D.L., Sabbagh, M.N., Connor, D.J., et al. (2005) Atorvastatin therapy lowers circulating cholesterol but not free radical activity in advance of identifiable clinical benefit in the treatment of mild-to-moderate AD. Current Alzheimer Research, 2: 343–353.

Tanzi, R.E. and Bertram, L. (2001) New frontiers in Alzheimer's disease genetics. Neuron, 32: 181–184.

Tariot, P.N., Solomon, P.R., Morris, J.C., et al. (2000) A 5-month, randomized, placebo-controlled trial of galantamine in AD. Neurology, 54: 2269–2276.

Tariot, P.N., Loy, R., Ryan, J.M., et al. (2002) Mood stabilizers in Alzheimer's disease: symptomatic and neuroprotective rationales. Adv Drug Deliv Rev., 54: 1567–1577.

Tariot, P.N., Farlow, M.R., Grossberg, G.T., et al. (2004) Memantine treatment in patients with moderate to severe Alzheimer disease already receiving donepezil: a randomized controlled trial. JAMA., 291: 317–324.

van Duijn, C.M. and Hofman, A. (1991) Relation between nicotine intake and Alzheimer's disease. BMJ., 302: 1491–1494.

Wilcock, G.K., Lilienfeld, S. and Gaens, E. (2000) Efficacy and safety of galantamine in patients with mild to moderate Alzheimer's disease: multicentre randomised controlled trial. BMJ, 321: 1445–1449.

Wilcock, G.K., Birks, J., Whitehead, A., Evans, S.J. (2002) The effect of selegiline in the treatment of people with Alzheimer's disease: a meta-analysis of published trials. Int J Geriatr Psychiatry., 17: 175–183.

Wilcock, G., Howe, I., Coles, H., et al. (2003) A long-term comparison of galantamine and donepezil in the treatment of Alzheimer's disease. Drugs & aging, 20: 777–789.

Wilkinson, D.G., Passmore, A.P., Bullock, R., et al. (2002) A multinational, randomised, 12-week, comparative study of donepezil and rivastigmine in patients with mild to moderate Alzheimer's disease. International journal of clinical practice, 56: 441–446.

Wimo, A., Winblad, B., Shah, S.N., et al. (2004) Impact of donepezil treatment for Alzheimer's disease on caregiver time. Current medical research and opinion, 20: 1221–1225.

Winblad, B., Engedal, K., Soininen, H., et al. (2001) A 1-year, randomized, placebo-controlled study of donepezil in patients with mild to moderate AD. Neurology, 57: 489–495.

Yamada, K., Nitta, A., Hasegawa, T., et al. (1997) Orally active NGF synthesis stimulators: potential therapeutic agents in Alzheimer's disease. Behav Brain Res., 83: 117–122.

Yan, S.D., Chen, X., Fu, J., et al. (1996) RAGE and amyloid-beta peptide neurotoxicity in Alzheimer's disease. Nature, 382: 685–691.

Zhou, Y., Su, Y., Li, B., et al. (2003) Nonsteroidal anti-inflammatory drugs can lower amyloidogenic Abeta42 by inhibiting Rho. Science, 302: 1215–1217.

Zlokovic, B.V. (2004) Clearing amyloid through the blood-brain barrier. J Neurochem., 89: 807–811.

CHAPTER 5

SLOWING DOWN AGE-RELATED MUSCLE LOSS AND SARCOPENIA

P. NOIREZ[1,2] AND G. BUTLER-BROWNE[1]
[1]*Inserm U787, Université Pierre et Marie Curie, Paris 6, Institut de Myologie*
[2]*Ufr Staps, Université René Descartes, Paris 5*

Abstract: The maintenance of posture is the result of an equilibrium between the actions of the muscle groups on either side of the joints. A failure in this process therefore stems from a disequilibrium between the muscle groups of one or several joints, originating from muscular weakness, which could even cause a person to fall. These well known mechanical characteristics have guided research towards our current knowledge of the molecular mechanisms involved in muscular contraction and help us understand how muscle is affected by aging

Keywords: sarcopenia, fraility, energy, aging

1. WHAT IS A SKELETAL MUSCLE

1.1 Muscle fibres

The muscle fibres are giant cells that can measure several tens of centimetres in length. Although they have numerous nuclei, the fibre size/muscle nucleus ratio remains relatively constant. The number of myonuclei seems to play a mechanistic role in the change in muscle size (Allen et al., 1999; Kadi, 2000) and nuclear domains have a constant size in heathy muscle.

Muscle fibre contraction corresponds to a series of actin-myosin cross-bridge formations, which causes the muscles to shorten. The greater the number of actin-myosin cross-bridges formed, the greater the force developed by the fibre will be.

S. Rattan and M. Kassem (eds.), Prevention and Treatment of Age-related Diseases, 71–85.

1.2 Each muscle is unique

Most correlative morphological and functional studies on human muscle have been performed on the large thigh muscle, the vastus lateralis. The results from these studies have provided us with golden standards for human muscle morphology and function. It has become increasingly evident that each human muscle is unique with respect to its muscle fibre composition, fibre diameter and function (Stal et al., 1994). The smallest natural unit of muscular contraction is called the motor unit: it corresponds to a set of muscle fibres which are innervated by the same motoneuron. Human skeletal muscle display three main types of motor unit: the IIx motor units which are fast and fatigable, the IIa motor units which are fast but resistant to fatigue, and the I motor units which are slowly contracting and resistant to fatigue. In general, the slow motor units are the smallest. Motor unit recruitment varies according to physical effort such that an increased production of force requires not only the recruitment of motor units from the smallest to the largest, but also increasingly smaller time lapses between recruitment. In humans, all the fibres that make up a motor unit have identical characteristics. The muscle fibres of slow motor units are termed type I fibres, contain slow twitch MyHC. Human fast motor units are composed of type II fibres and subgroups thereof (IIA and IIX). They contain different fast isoforms of MyHC, are fast contracting and, depending on the subtype, show various degrees of resistance to fatigue. Some small hand muscles like the interossei have a mixed composition of fibre types and are of large diameters, whereas the lumbricale muscles are almost exclusively composed of type I fibres (Stal et al., 1994; Soukup et al., 2003). The muscle fibre composition in the trapezius muscle differs in the different parts of the muscle and there are obvious differences related to gender (Lindman et al., 1991). Some facial muscles have small sized fibres which contain mainly fast myosin isoforms (Stal et al., 1994) whereas the masticatory muscles are very complex and have fibres which contain mixtures of different myosin isoforms, some of which are not present in limb muscles such as alpha cardiac and fetal myosin (Butler-Browne et al., 1988; Pedrosa-Domellof et al., 1992). These observations suggest that the muscles may also behave differently upon aging and to some extent this is what has been observed. The age related changes in the masseter, a jaw closing muscle, and the lateral pterygoid, a jaw stabilizing muscle, are opposite to those reported for limb and trunk muscles. On the contrary, changes in the anterior and posterior bellies of the digastricus, a jaw opening muscle, resemble those of limb and trunk muscles (Thornell et al., 2003). The individual variability seems also to be large and there is still no consensus on the effects of aging on the vastus lateralis. Some studies have reported an increase in the relative percentage of type I fibres, others a decrease, and a further subset observe no change in fiber proportions (Thornell et al., 2003). Therefore, the heterogeneity and individual variability in the structure and function of the different human muscles should be kept in mind when discussing the different aspects of sarcopenia and its prevention.

1.3 Muscular lesion

Muscles are continually undergoing adaptation to different function needs as well as injury. Injury caused by elongation or contusion of the muscle, represents over 90% of muscle injuries. This type of injury occurs when excessive force is applied to the muscle resulting in over-stretching. More often than not, these lesions are located near the neuromuscular junction of superficial muscles working on two joints, such as the femoris rectus of the quadriceps. A slight lesion corresponds to the tearing of a few muscle fibres, which results in slight discomfort (the twinge scenario) with little or no loss of force or restriction of movement. A moderate lesion corresponds to more significant damage with a decrease in force production. With a severe lesion, the tearing of the muscle affects the whole or part of the muscle, leading to total loss of muscle function (Jarvinen et al., 2005). Luckily, striated skeletal muscle has an incredible capacity for regenerating itself. Even in the absence of severe tearing, the muscle can also suffer a relative degree of damage or remodelling after a mere session of physical exercise (Yu et al., 2004). Even those who practice sport at a high level are not exempt from these micro-lesions of the muscle. They are particularly frequent in physical or sports activities that require the production of maximum force or eccentric muscle contractions. Aching muscle pain (or DOMS syndrome – Delayed Onset Muscle Soreness), representing a pain peak 48 hours after exercise (Cheung et al., 2003), is the soreness that may result either from the degeneration and regeneration phenomena taking place in the damaged muscle or to the remodelling (Yu et al., 2004).

1.4 Muscle Satellite Cells

There exists a particularly interesting cell population situated on the edge of the muscle fibres wich are called satellite or myosatellite cells: these cells are quiescent myoblasts that reside adjacent to the muscle fibre sarcolemma and beneath the basement lamina. Myoblast is a term designating a myogenic cell that is fully determined with respect to its myogenic phenotype. Early during development, multinucleated myotubes are formed by proliferating myoblasts, which withdraw from the cell cycle and fuse with one another. Myoblasts continue to be added to these myotubes allowing them to expand in both length and girth to become mature muscle fibres (Edom et al., 1994). Thus during development and postnatal growth, nuclei are added to the muscle fibres by the fusion of myoblasts to the parent fibre. The identification of satellite cells in 1961 (Mauro, 1961) led to a rapid progress in our understanding of the early events involved in skeletal muscle regeneration. If the quiescent state of satellite cells were a delicate equilibrium between electrical activity, growth factors and extra-cellular matrix composition, disequilibrium of the environment would trigger activation and proliferation of satellite cells. Following a muscle trauma, the satellite cells proliferate and either form new muscle fibres or repair damaged fibres via a process equivalent to muscle histogenesis (Bischoff and Heintz, 1994). In recent years, the importance of satellite cells has been emphasized by the discovery that their proliferation is evoked not only by acute muscle injury but also by muscle overuse and increased muscle tension. A number of factors are involved in this regulation of satellite cell activation (Hawke and Garry, 2001).

2. WHAT HAPPENS TO MUSCLE AS WE AGE

Muscle aging is associated with a decrease in maximum produced force. Maximum force, which increases up to the age of thirty, then decreases by an average 15% per decade as of the age of fifty and by an average 30% after the age of seventy. This decrease in force appears to be greater in the leg muscles than in the arm muscles. Endurance is also reduced in elderly subjects. On the other hand, a decrease of 1% per year is observed in the maximum level of oxygen uptake (V02max) as of the age of thirty (Le Page et al., 2002).

This decrease in force can be at least explained, in part, by a decrease in muscle mass (sarcopenia). Muscle mass decrease by between 35 and 40% between the ages of twenty and eighty, representing 1.9 kg per decade in men and 1.1 kg per decade in women. Moreover, this age-related loss of muscle mass appears preferentially to affect the lower part of the body. This muscular atrophy results from both a loss of individual muscle fibres as well as from a decrease in fibre diameter estimated at 1.4% per year after the age of fifty (Le Page et al., 2002).

The density of the skeletal muscle also decreases with age. Muscle atrophy is accompanied by an increase in the amount of non-contractile tissue: intramuscular fat and conjunctive tissue (Lexell et al., 1988). Communication between the muscle fibres and the blood vessels is less efficient: there are fewer blood capillaries in the muscle and this leads to reduced oxygen uptake, which partially explains the decrease in V02max (Hepple et al., 1997). This could also induce an oxidative stress on the muscle fibres. A decrease in muscle oxidative capacity is also observed, and this contributes to the decreased V02max and increased fatigability (Degens, 1998). Fibrosis can also develop in the muscle over the years. Increased fibrosis not only hinders communication between the muscle fibres and the blood vessels, but also causes stiffening of the muscle, thereby contributing to alterations in muscle function (Gosselin et al., 1994). Moreover, the regenerative capacity of muscle tissue also appears to alter with age (Vignaud et al., 2003).

All these modifications observed in the process of muscle aging are the result of a combination of intrinsic factors (related to the functioning of the muscle cell) and extrinsic factors (such as decreased hormonal status and neuromuscular activity), which we will now describe in more detail.

3. INTRINSIC FACTORS

3.1 Excitation-contraction coupling

Contraction of skeletal muscle cells is controlled by nerve cell or motoneuron activity. The arrival of a nerve signal, or action potential, at the level of the junction between the neuron and the muscle triggers the discharge of a neurotransmitter, acetylcholine, by the nerve cell. The neurotransmitter then binds to its receptor located on the muscle cell membrane and induces the formation of an electric current across the membrane. Excitation-contraction coupling is defined as the biological phenomenon that transforms an order arriving in the form of an electrical signal into a mechanical event: contraction of the muscle cell. This phenomenon is made possible by the presence in certain parts of the cell of calcium reservoirs termed sarcoplasmic reticulum (SR). The SR is bound by its own membrane, which is linked to the cell membrane by binding molecules (one located on the cell membrane, the dihydropyridine receptor (DHPR), and the other on the reservoir membrane, the Ryanodine receptors (RyR). These two binding molecules constitute channels through which the calcium passes and whose opening is controlled by the electric current. Thus, SR discharges its calcium inside the muscle cell when the channels open under the effect of the current (Ryan and Ohlendieck, 2004).

In humans, the speed of contraction and the force developed by the muscles both deteriorate with age. Similar results have been obtained in mice. This loss of force could be explained by excitation-contraction decoupling. In effect, it has been shown that the number of calcium – channels diminishes with age (Delbono, 2003). It was therefore assumed that if for the same electric current fewer channels opened, this should limit the amount of calcium entering the cell and thus lead to a weaker contraction. However, experiments carried out on isolated human muscle cells moderate this theory. The experimental results obtained in vitro on muscle fibres from different subjects in which the reservoirs had been rendered inactive show a drop in developed force in the fibres of elderly subjects compared to that of young subjects (Frontera et al., 2000). Moreover DHPR expression seem to be preserved during the aging process of human skeletal muscle fibres (Ryan et al., 2003). This would indicate that excitalion-contraction decoupling is not the limiting factor in the loss of developed force with age. The number and the force of the actin-myosin crossbridges appear to be the preponderant factors.

However, the question of alterations in the neural control of the expression of muscle genes such as myosin or actin, which also depends on the quantity of calcium discharged by the reservoirs, remains unanswered.

In conclusion, excitation-contraction decoupling due to the reduction in the number of calcium channels in the calcium reservoir membranes is not the direct cause of the loss of muscular force observed in elderly people. Nevertheless, it cannot be excluded that this reduction may modify the expression of the genes encoding myosin, for example, which would lead to a modification in the actin-myosin cross-bridges. The cause-and- effect relationship should be explored in more detail in the forth coming years. It has however been shown that the myosin molecule is susceptible to post-translational modification such as glycation. In addition it has been shown that glutathione can reverse these modifications (Ramamurthy et al., 2003). It could therefore be imagined that physical activity can maintain the number of functional receptors and thus maintain sufficient expression of the muscle genes, thereby making it possible to maintain a high level of force production.

3.2 Mitochondria, oxidative stress and aging

Mitochondria are cell structures that produce energy that is vital to the cells; moreover, they participate in the cascade of cell signalling events. The number of mitochondria varies according to muscle fibre type. Type I fibres have the greatest number, followed by Type lla, and finally Type llx fibres. In addition to this heterogeneous number of mitochondria in muscle cells, it is interesting to note that regular physical activity increases the number of mitochondria in the cells. As previously discussed, the main effects of age on skeletal muscle are sarcopenia and cell death. These two events could be linked to dysfunction of the mitochondria. In effect, these structures responsible for cell respiration can, in certain cases, form reactive oxygen species (ROS) that are toxic for the cells. ROS production increases drastically during aging (Fulle et al., 2004). Free radicals cause severe damage if they are not promptly eliminated by the action of anti-oxidant agents. However, some of these toxic molecules may escape and bind to the mitochondrial DNA causing punctual mutations of the DNA molecule. These mutations could trigger a cascade of events leading to cell death by apoptosis: formation of chemically unstable molecules, induction of mutations on the DNA, formation of mutated enzymes, alteration of the respiratory activity of the mitochondria, which triggers either the accumulation of other unstable molecules (and thus other mutations) or cell death by apoptosis (Kujoth et al., 2005). This is a lengthy process.

Although this is an interesting theory, it is nevertheless controversial. Many questions still remain unanswered. It is undeniable that cells accumulate mutations with age, but not all these mutations induce modifications in mitochondrial activity. Moreover, the induced modifications are not always bad for the cells. We can add to this argument by saying that the mutations that trigger cell death disappear and that the muscle cells reformed by satellite cells no longer present these mutations. This leads us to discuss the advantages of regular exercise in respect of changes in mitochondria in the skeletal muscles of elderly people. The first experiments carried out on patients suffering from mitochondrial myopathies type pathologies are encouraging (Chabi et al., 2005). It is already well know that physical activity improves endurance capacities in healthy subjects, but the same also appears to be true for myopathic patients. The working hypothesis currently put forward by researchers is that, by allowing satellite cells to renew the mitochondria or to strengthen the existing muscle fibres, exercise diminishes the chances of mitochondrial DNA mutations to accumulate.

3.3 Satellite cells and Telomeres

When a muscular lesion occurs, the satellite cells are rapidly activated, proliferate and then fuse either with the damaged fibres in order to repair them, or among themselves in order to form new fibres. One part of the activated satellite cells does not differentiate and renews the stock of quiescent satellite cells. The satellite cells are involved in maintaining the fibre size/muscle nuclei ratio.

The reduction in the number of satellite cells with age could therefore be one of the factors that could explain the loss of muscle mass linked to aging and alterations in the regenerative capacity. Modification, with age, in the capacity of satellite cells to proliferate or fuse could be another factor limiting the action of repairing these cells and of maintaining muscle mass during the aging process.

How the pool of satellite cells evolves during normal aging in human skeletal muscle is still controversial. Using EM, human satellite cells represent 15% of all the myonuclei at birth, 6–10% at two years of age, and 4% in the adult (Tome and Fardeau, 1986; Schmalbruch and Hellhammer, 1976). For older subjects, this value varies between 0.6 and 3.4% in different studies (Thornell et al., 2003).

In our own studies, we have observed values around 5% for the young biceps brachii and masseter, a value which is in close agreement with previous studies which were carried out on the trapezius muscle of young female subjects (Kadi and Thornell, 2000). The proportion of satellite cells we found in corresponding muscles in aged persons (mean age: 74 ± 4.25 years) were relatively low; 1.44% in the biceps brachii and 1,77% in the masseter (Renault et al., 2002). We have also examined in the same way the number of satellite cells in the vastus lateralis of four subjects with a mean age of 88 years. Values obtained were 1.49%, 1.33%, 1.07% and 1.67% giving a mean value of 1.39% (unpublished data). This suggests that there is a significant decrease in the satellite cell number between young and old adults for three different muscles. Further analysis is needed to find out if there is a progressive decrease in satellite cells number during adulthood or whether at some critical time there is a sudden decrease due to altered trophic enviroment in the aged muscle. To obtain this knowledge it will be necessary to carry out a transversal analysis.

It has previously been described in birds and rodents that the satellite cell populations isolated in vitro from fast or slow muscle fibres expressed myosin heavy chain isoforms that reflected the phenotype of the muscle from which they were isolated (Dusterhoft and Pette, 1993; Feldman and Stockdale, 1991; Rosenblatt et al., 1996). In our laboratory, we have shown both by clonal (Edom et al., 1994) and by single fibre (Bonavaud et al., 2001) analyses that all of the myogenic satellite cells when differentiated in culture co-express both fast and slow myosin heavy chains. This suggests that human satellite cells are not lineage restricted, and that the regulation of the program they can express is open and will depend on external factors such as innervation (Edom et al., 1994). One should keep in mind that although human muscle contains in general mixed fibres, the ratio of which is specific for each muscle, there are no specific fast and slow satellite cell lineages in human skeletal muscle. Since human satellite cells upon differentiation are not oriented towards a precise fibre type programme this will allow them to participate in the growth and repair of any fibre in their vicinity regardless of its programme of differentiation (Mouly et al., 2005).

In order to provide sufficient nuclei to repair damaged muscle fibres following activation the satellite cells undergo successive cycles of cell division; Proliferation is therefore one of the key steps involved in muscle regeneration. However it has been well established that human diploid cells are limited in their proliferation capacity. During their life span human cells will gradually replicate more slowly until they reach a non replicative state called replicative senescence. We have studied the number of divisions that human satellite cells can make when they are isolated from donors of different ages. Previous studies on skin fibroblasts have shown that there is a gradual decline in proliferative capacity with increasing donor age. When we carried out a similar study on human satellite cells isolated from donors of increasing age, we did not observe a regular loss of proliferative capacity with donor age. Instead, we have found that there was a rapid loss of proliferative capacity during the first two decades of life (from about 55–60 divisions at birth to about 20 divisions at 20 years of age. Satellite cells isolated from adult muscle independent of age were always able to make between 15–20 divisions (Decary et al., 1997; Renault et al., 2000). The fact that the proliferative potential does not change in adult skeletal muscle would suggest that during normal healthy aging the ability to regenerate skeletal muscle is maintained throughout life even into old age. We can however predict that the situation will be different if proliferation of the satellite cells were to be highly solicited as has been observed in muscular dystrophies (Decary et al., 2000).

One mechanism, which has been suggested to control this limited proliferation, or mitotic clock, is the shortening of the telomeric sequences. Telomeres are specialized DNA fragments located at the end of all eucaryotic chromosomes. In mammals, they consist of short repeated non coding DNA sequences, (TTAGGG)n, which in human are 5–20 kb in length (Harley et al., 1990). During DNA replication, DNA polymerase is unable to copy the 3 '92 terminal segment of each DNA strand. This results in chromosome shortening at each round of cell division (Olovnikov, 1973). In somatic cells, telomere length decreases regularly with cell division. In vivo, a decrease in the length of telomeric DNA with aging has been demonstrated in many human mitotic tissues (Klapper et al., 2001). In a series of studies carried out on three different human muscles, quadriceps (Decary et al., 1997), masseter and biceps (Renault et al., 2002) we found that there is only a very small decrease in the length of the telomeric DNA in skeletal muscle with increasing donor age. However a dramatic decrease in telomeric DNA length was observed in the muscles of children with muscular dystrophy (Decary et al., 2000). Our results would confirm previous observations that skeletal muscle is a very stable tissue and that during the lifetime there is a low turnover of the myonuclei. The results that we have obtained so far seem to point to the fact that number and quality of satellite cells and hence regenerative capacity are not a limiting factor during healthy aging. Limitations would only arise if these factors were to be oversolicited during the lifetime of an individual by sore chronic disease or if the quality of the satellite cell would become modified by a decrease in trophic factors which accompanies aging (Mouly et al., 2005).

Consequently, alternative hypothesis have been proposed based on a defect in the activation of the satellite cells due to changes in their environment caused by age-related changes in the body, such as modification of the hormone status, reduction in certain local factors, or changes in neuromuscular activity.

4. EXTRINSIC FACTORS

4.1 Hormones and growth factors

The human body is a collection of tissues having different activities coordinated over time in such a way as to ensure that the body can feed itself reproduce and react to changes in its environment. The two main coordinators are the nervous system and the endocrinal system.

The endocrinal system includes all the hormones (signaling molecules) and the organs that secrete them. Aging, and in particular, muscular aging, is related to alterations in the secretion of certain hormones such as thyroid hormone, dihydroapiandrosterone (DHEA), growth hormone, and insulin-like growth factor (IGF1). In women, it has been observed that the effects of aging intensify at the age of the menopause when the ovarian cells no longer secrete any progesterone. In men, blood testosterone levels fall by 50% between the ages of twenty and eighty. Experiments have shown that there is a correlation between loss of muscle mass, loss of muscular force and the reduction in sex hormone levels (Shavlakadze and Grounds, 2003).

As previously discussed, loss of muscle mass appears to be greater in the lower limbs. This phenomenon could be explained by the fact that the muscles of the upper body have more testosterone receptors that those of the lower part of the body (Kadi, 2000). Substitution treatments reduce these muscular alterations.

Similar results have been obtained with growth hormone. Presently, the molecule that seems to be the most important in the muscle aging mechanism is IGF1 (Shavlakadze and Grounds, 2003). This factor is secreted by different cell types such as liver cells, skeletal muscle cells and heart cells. When it is over-expressed through genetic manipulation in mice, it increases adult muscle mass by 15% in young adults and maintains muscle mass in elderly adults (Musaro et al., 2001). Its expression level is known to decrease with age but to be increased by exercise. Myostatin, known as a negative muscle mass regulator, would also be a candidat for treatment, however, its expression does not seem to be influenced by aging (Haddad and Adams, 2005).

4.2 Neuromuscular activity

4.2.1 Innervation

The principal function of certain muscles is posture maintenance. These muscles are recruited permanently as long as there is no change in posture and are thus very regularly stimulated by their motoneurons. Other muscles are mainly involved in the production of movement. They are stimulated by their motoneurons only when it is necessary to modify the position of the muscle. Similarly, in any given muscle, slow fibres are more often recruited than fast fibres since movements requiring great force are less frequent in everydaylife than those requiring low levels of force (lower activation threshold for slow motor units than for fast motorunits).

Muscle fibres are controlled by motoneurons with different morphological and electrical characteristics. Today, we know that the relationship between the motoneuron and the muscle cell is much closer than a mere excitation-contraction event. Motoneuron activity thus enables the formation and maintenance of the biochemical composition of the muscle cell. It has also recently been demonstrated that these two cells communicate via growth factor type signaling molecules (Shavlakadze and Grounds, 2003). For reasons not yet well understood, skeletal muscle innervation is modified with age. Fast motor units disappear and are replaced (or not) by slow motor units. We observe a change in the fibre-type composition of the muscles of elderly people towards a slower phenotype. The process of sarcopenia and loss of developed force observed in the muscles of elderly people could thus be due to alterations in skeletal muscle innervation. This phenomenon would lead to the excitation-contraction decoupling mentioned above, which would modify the expression of the muscle genes.

Moreover, this would result in modifications in the communication via signaling molecules. In effect, IGF-I secretion by the muscle cells could be modified, resulting in the initiation of the vicious circle of events that leads to age-related muscle loss. Recently, it was shown that induced overexpression of IGF-1 in spinal cord motoneurones of aging mices prevents muscle fibres specific force decline (Payne et al., 2006). IGF1 is not the only factor secreted by the muscle cells that allows maintenance of the motoneurons, other molecules such as neurotrophines or IGF2 are currently being studied. In the next few years, it is expected other growth factors will be added to these known molecules, making it possible to develop efficient anti-aging therapies in the not too distant future.

4.3 Increased and decreased muscle activity

Physical training is efficient in elderly subjects, their muscles retaining the capability to adapt to functional demand. The effects of force training are characterized by an increase in force production and by muscular hypertrophy. Endurance training improves muscle performance and VO2max (Beere et al., 1999). The anti-oxidizing defense capacity and the oxidative power of the mitochondria also increase (Meijer et al., 2002). Force training (three times a week for ten years) makes it possible to maintain the maximum level of isometric force in elderly subjects aged at a level corresponding to a sedentary young person. Improvements in force production as a result of training can be achieved even in subjects over the age of eighty. The percentage of force gain is similar to that obtained by subjects aged around sixty or by young adults (Le Page et al., 2002).

As the level of activity diminishes with age, it is important to distinguish changes specific to reduced activity from those due to aging. Studies have been carried out on models of diminished muscle activity such as prolonged bed-rest, immobilization or microgravity. The results show that muscular atrophy is accompanied by reduction in muscle fibre size, force production and muscular work capacity as well as alterations in locomotor coordination (Bloomfield, 1997). These effects look very similar to those observed in aging.

5. PREVENT AND/OR TREATMENT

Skeletal muscles are the organs that enable us to maintain posture and movement. As individuals age they frequently become less active and this will progressively lead to muscle atrophy and frailty. The mechanisms that would allow us to explain how muscles age, why we lose both mass and force are still not well understood. From the current results it is not possible to ascertain the exact role of intrinsic factors in the aging process. On the other hand, changes in certain extrinsic factors, such as the secretion of certain hormones and neuromuscular inactivity, appear to be involved in this process.

It has been suggested by several authors (Le Page et al., 2002) that age related muscle loss can be reversed by exercise. However we do not know if the oxidative stress liberated by exercise could be damaging to the muscle especially in elderly individuals in the lack of a certain adaptation to regular exercise. It should be noted that during aging there is a gradual increase in the proinflammatory state which could increase the incidence of muscle injury following exercise (Fulle et al., 2004). However, as stated earlier regular exercise will increase the anti-oxidant response in the skeletal muscles (Meijer et al., 2002) and this is accompanied by an increase in muscle produced interleukin-6 which is thought to counterbalance this pro-inflammatory state (Petersen and Pedersen, 2005). How much exercise is required to maintain muscle force and mass? It is not always easy to formulate an adequate standard exercise protocol for each individual. It is not necessarily the role of the doctor to determine how much exercise a healthy individual should undertake in order to stay healthy. This falls into the domain of preventive medecine to maintain a good quality of life for our aging population. One could imagine however, that the doctor could prescribe a series of regular exercises which are adapted to the health status of the patient, then this would be followed by a specialist in physical education. Nevertheless we could ask the question is this really his role and could not these roles be inverted.

As stated previously the loss of muscle strength and mass occurs rather early, between 30 and 40 years of age, therefore it is important that the idea that regular exercise should become an integral part of the general life style just as brushing ones teeth. It is surprising in our modern day culture that the majority of the population prefers to participate in sport by proxy from their arm chair rather than carrying out some sort of physical exercise themselves.

6. CONCLUSION

At the present time, hormone treatment still remains premature: however, maintaining regular neuromuscular activity could delay the effects of muscular aging; it is nevertheless important to take into account the personal capacities of each individual as well any eventual pathological states before establishing exercise and physical activity programmes.

In elderly people, physical training improves skeletal muscle performance (Le Page et al., 2002), oxydant defense capacity (Meijer et al., 2002), arterial compliance (Tanaka et al., 2000; Monahan et al., 2001), cardiac function during acute exercise (Stratton et al., 1994), maximal oxygen consumption (Beere et al., 1999) and prevents vascular endothelial dysfunction, probably by limiting oxidative stress (Taddei et al., 2000). In addition, exercise training in cardiovascular disease limits the incidence of coronary events (Abete et al., 2001), improves functional capacity to exercise and reduces coronary stenosis in patients with coronary heart disease (Hakim et al., 1999; Gielen et al., 2001).

Moreover, level of physical activity has a direct impact on the level of cognitive activity. Recent studies have shown that improving physical fitness leads to better performances in tasks assessing a diversity of cognitive domains (Renaud and Bherer, 2005).

It thus seems that physical training could improve not only the health of the elderly individual but also serve to enhance and maintain cognitive vitality in older adults. However the current way of life is making us increasingly less active. In order to preserve independence during aging, it would be advisable to encourage our contemporaries to indulge in regular exercise and physical activity.

REFERENCES

Abete, P., Ferrara, N., Cacciatore, F., Sagnelli, E., Manzi, M., Carnovale, V., Calabrese, C., de Santis, D., Testa, G., Longobardi, G., Napoli, C. and Rengo, F. (2001) High level of physical activity preserves the cardioprotective effect of preinfarction angina in elderly patients. J.Am.Coll.Cardiol., 38: 1357–1365.

Allen, D.L., Roy, R.R. and Edgerton, V.R. (1999) Myonuclear domains in muscle adaptation and disease. Muscle Nerve., 22: 1350–1360.

Beere, P.A., Russell, S.D., Morey, M.C., Kitzman, D.W. and Higginbotham, M.B. (1999) Aerobic exercise training can reverse age-related peripheral circulatory changes in healthy older men. Circulation., 100: 1085–1094.

Bischoff, R. and Heintz, C. (1994) Enhancement of skeletal muscle regeneration. Dev.Dyn., 201: 41–54.

Bloomfield, S.A. (1997) Changes in musculoskeletal structure and function with prolonged bed rest. Med.Sci.Sports Exerc., 29: 197–206.

Bonavaud, S., Agbulut, O., Nizard, R., D'honneur, G., Mouly, V. and Butler-Browne, G. (2001) A discrepancy resolved: Human satellite cells are not preprogrammed to fast and slow lineages. Neuromuscul.Disord., 11: 747–752.

Butler-Browne, G.S., Eriksson, P.O., Laurent, C. and Thornell, L.E. (1988) Adult human masseter muscle fibers express myosin isozymes characteristic of development. Muscle Nerve., 11: 610–620.

Chabi, B., Adhihetty, P.J., Ljubicic, V. and Hood, D.A. (2005) How is mitochondrial biogenesis affected in mitochondrial disease? Med.Sci.Sports Exerc., 37: 2102–2110.

Cheung, K., Hume, P. and Maxwell, L. (2003) Delayed onset muscle soreness : Treatment strategies and performance factors. Sports Med., 33: 145–164.

Decary, S., Mouly, V., Hamida, C.B., Sautet, A., Barbet, J.P. and Butler-Browne, G.S. (1997) Replicative potential and telomere length in human skeletal muscle: Implications for satellite cell-mediated gene therapy. Hum.Gene Ther., 8: 1429–1438.

Decary, S., Hamida, C.B., Mouly, V., Barbet, J.P., Hentati, F. and Butler-Browne, G.S. (2000) Shorter telomeres in dystrophic muscle consistent with extensive regeneration in young children. Neuromuscul.Disord., 10: 113–120.

Degens, H. (1998) Age-related changes in the microcirculation of skeletal muscle. Adv.Exp.Med.Biol., 454: 343–348.

Delbono, O. (2003) Neural control of aging skeletal muscle. Aging Cell., 2: 21–29.

Dusterhoft, S. and Pette, D. (1993) Satellite cells from slow rat muscle express slow myosin under appropriate culture conditions. Differentiation., 53: 25–33.

Edom, F., Mouly, V., Barbet, J. P., Fiszman, M. Y. and Butler-Browne, G.S. (1994) Clones of human satellite cells can express in vitro both fast and slow myosin heavy chains. Dev.Biol., 164: 219–229.

Feldman, J.L. and Stockdale, F.E. (1991) Skeletal muscle satellite cell diversity: Satellite cells form fibers of different types in cell culture. Dev.Biol., 143: 320–334.

Frontera, W.R., Suh, D., Krivickas, L.S., Hughes, V.A., Goldstein, R. and Roubenoff, R. (2000) Skeletal muscle fiber quality in older men and women. Am.J.Physiol.Cell.Physiol., 279: C611–8.

Fulle, S., Protasi, F., Di Tano, G., Pietrangelo, T., Beltramin, A., Boncompagni, S., Vecchiet, L. and Fano, G. (2004) The contribution of reactive oxygen species to sarcopenia and muscle ageing. Exp.Gerontol., 39: 17–24.

Gielen, S., Schuler, G. and Hambrecht, R. (2001) Exercise training in coronary artery disease and coronary vasomotion. Circulation., 103: E1–E6

Gosselin, L.E., Martinez, D.A., Vailas, A.C. and Sieck, G.C. (1994) Passive length-force properties of senescent diaphragm: Relationship with collagen characteristics. J.Appl.Physiol., 76: 2680–2685.

Haddad, F. and Adams, G.R. (2005) Aging sensitive cellular and molecular mechanisms associated with skeletal muscle hypertrophy. J.Appl.Physiol.

Hakim, A.A., Curb, J.D., Petrovitch, H., Rodriguez, B.L., Yano, K., Ross, G.W., White, L.R. and Abbott, R.D. (1999) Effects of walking on coronary heart disease in elderly men: The honolulu heart program. Circulation., 100: 9–13.

Harley, C.B., Futcher, A.B. and Greider, C.W. (1990) Telomeres shorten during ageing of human fibroblasts. Nature., 345: 458–460.

Hawke, T.J. and Garry, D.J. (2001) Myogenic satellite cells: Physiology to molecular biology. J.Appl.Physiol., 91: 534–551.

Hepple, R.T., Mackinnon, S.L., Goodman, J.M., Thomas, S.G. and Plyley, M.J. (1997) Resistance and aerobic training in older men: Effects on VO2peak and the capillary supply to skeletal muscle. J.Appl.Physiol., 82: 1305–1310.

Jarvinen, T.A., Jarvinen, T.L., Kaariainen, M., Kalimo, H. and Jarvinen, M. (2005) Muscle injuries: Biology and treatment. Am.J.Sports Med., 33: 745–764.

Kadi, F. (2000) Adaptation of human skeletal muscle to training and anabolic steroids. Acta Physiol.Scand.Suppl., 646: 1–52.

Kadi, F. and Thornell, L.E. (2000) Concomitant increases in myonuclear and satellite cell content in female trapezius muscle following strength training. Histochem.Cell Biol., 113: 99–103.

Klapper, W., Parwaresch, R. and Krupp, G. (2001) Telomere biology in human aging and aging syndromes. Mech.Ageing Dev., 122: 695–712.

Kujoth, G.C., Hiona, A., Pugh, T.D., Someya, S., Panzer, K., Wohlgemuth, S. E., Hofer, T., Seo, A.Y., Sullivan, R., Jobling, W. A., Morrow, J. D., Van Remmen, H., Sedivy, J. M., Yamasoba, T., Tanokura, M., Weindruch, R., Leeuwenburgh, C. and Prolla, T.A. (2005) Mitochondrial DNA mutations, oxidative stress, and apoptosis in mammalian aging. Science., 309: 481–484.

Le Page, C., Riou, B. and Besse, S. (2002) Vieillissement du muscle squelettique : Effet de l'exercice physique. Age & Nutrition., 13: 162–177.

Lexell, J., Taylor, C.C. and Sjostrom, M. (1988) What is the cause of the ageing atrophy? total number, size and proportion of different fiber types studied in whole vastus lateralis muscle from 15- to 83-year-old men. J.Neurol.Sci., 84: 275–294.

Lindman, R., Eriksson, A. and Thornell, L.E. (1991) Fiber type composition of the human female trapezius muscle: Enzyme-histochemical characteristics. Am.J.Anat., 190: 385–392.

Mauro, A. (1961) Satellite cell of skeletal muscle fibers. J.Biophys.Biochem.Cytol., 9: 493–495.

Meijer, E.P., Goris, A.H., van Dongen, J.L., Bast, A. and Westerterp, K.R. (2002) Exercise-induced oxidative stress in older adults as a function of habitual activity level. J.Am.Geriatr.Soc., 50: 349–353.

Monahan, K.D., Tanaka, H., Dinenno, F.A. and Seals, D.R. (2001) Central arterial compliance is associated with age- and habitual exercise-related differences in cardiovagal baroreflex sensitivity. Circulation., 104: 1627–1632.

Mouly, V., Aamiri, A., Bigot, A., Cooper, R.N., Di Donna, S., Furling, D., Gidaro, T., Jacquemin, V., Mamchaoui, K., Negroni, E., Perie, S., Renault, V., Silva-Barbosa, S.D. and Butler-Browne, G.S. (2005) The mitotic clock in skeletal muscle regeneration, disease and cell mediated gene therapy. Acta Physiol.Scand., 184: 3–15.

Musaro, A., McCullagh, K., Paul, A., Houghton, L., Dobrowolny, G., Molinaro, M., Barton, E.R., Sweeney, H.L. and Rosenthal, N. (2001) Localized igf-1 transgene expression sustains hypertrophy and regeneration in senescent skeletal muscle. Nat.Genet., 27: 195–200.

Olovnikov, A.M. (1973) A theory of marginotomy. the incomplete copying of template margin in enzymic synthesis of polynucleotides and biological significance of the phenomenon. J.Theor.Biol., 41: 181–190.

Payne, A.M., Zheng, Z., Messi, M.L., Milligan, C.E., Gonzalez, E. and Delbono, O. (2006) Motor neurone targeting of IGF-1 prevents specific force decline in ageing mouse muscle. J.Physiol., 570: 283–294.

Pedrosa-Domellof, F., Eriksson, P.O., Butler-Browne, G.S. and Thornell, L.E. (1992) Expression of alpha-cardiac myosin heavy chain in mammalian skeletal muscle. Experientia., 48: 491–494.

Petersen, A.M. and Pedersen, B.K. (2005) The anti-inflammatory effect of exercise. J.Appl.Physiol., 98: 1154–1162.

Ramamurthy, B., Jones, A.D. and Larsson, L. (2003) Glutathione reverses early effects of glycation on myosin function. Am.J.Physiol.Cell.Physiol., 285: C419–24.

Renaud, M. and Bherer, L. (2005) Impact on physical fitness on cognitive aging. Psychol.Neuropsychiatr.Vieil., 3: 199–206.

Renault, V., Piron-Hamelin, G., Forestier, C., DiDonna, S., Decary, S., Hentati, F., Saillant, G., Butler-Browne, G.S. and Mouly, V. (2000) Skeletal muscle regeneration and the mitotic clock. Exp.Gerontol., 35: 711–719.

Renault, V., Thornell, L.E., Eriksson, P.O., Butler-Browne, G. and Mouly, V. (2002) Regenerative potential of human skeletal muscle during aging. Aging Cell., 1: 132–139.

Rosenblatt, J.D., Parry, D.J. and Partridge, T.A. (1996) Phenotype of adult mouse muscle myoblasts reflects their fiber type of origin. Differentiation., 60: 39–45.

Ryan, M. and Ohlendieck, K. (2004) Excitation-contraction uncoupling and sarcopenia. Basic Appl Myol., 14(3): 141–154.

Ryan, M., Butler-Browne, G., Erzen, I., Mouly, V., Thornell, L.E., Wernig, A. and Ohlendieck, K. (2003) Persistent expression of the alpha1S-dihydropyridine receptor in aged human skeletal muscle: Implications for the excitation-contraction uncoupling hypothesis of sarcopenia. Int.J.Mol.Med., 11: 425–434.

Schmalbruch, H. and Hellhammer, U. (1976) The number of satellite cells in normal human muscle. Anat.Rec., 185: 279–287.

Shavlakadze, T. and Grounds, M.D. (2003) Therapeutic interventions for age-related muscle wasting importance of innervation and exercice for preventing. In Modulating Aging and Longevity (Rattan, S. I. S., ed.), Kluwer Academic Publishers, The Netherlands, 1–28.

Soukup, T., Pedrosa-Domellof, F. and Thornell, L.E. (2003) Intrafusal fiber type composition of muscle spindles in the first human lumbrical muscle. Acta Neuropathol.(Berl)., 105: 18–24.

Stal, P., Eriksson, P.O., Schiaffino, S., Butler-Browne, G.S. and Thornell, L.E. (1994) Differences in myosin composition between human oro-facial, masticatory and limb muscles: Enzyme-, immunohisto- and biochemical studies. J.Muscle Res.Cell.Motil., 15: 517–534.

Stratton, J.R., Levy, W.C., Cerqueira, M.D., Schwartz, R.S. and Abrass, I.B. (1994) Cardiovascular responses to exercise. effects of aging and exercise training in healthy men. Circulation., 89: 1648–1655.

Taddei, S., Galetta, F., Virdis, A., Ghiadoni, L., Salvetti, G., Franzoni, F., Giusti, C. and Salvetti, A. (2000) Physical activity prevents age-related impairment in nitric oxide availability in elderly athletes. Circulation., 101: 2896–2901.

Tanaka, H., Dinenno, F.A., Monahan, K.D., Clevenger, C.M., DeSouza, C.A. and Seals, D.R. (2000) Aging, habitual exercise, and dynamic arterial compliance. Circulation., 102: 1270–1275.

Thornell, L.E., Lindstrom, M., Renault, V., Mouly, V. and Butler-Browne, G.S. (2003) Satellite cells and training in the elderly. Scand.J.Med.Sci.Sports., 13: 48–55.

Tome, F.M. and Fardeau, M. (1986) Nuclear changes in muscle disorders. Methods Achiev.Exp.Pathol., 12: 261–296.

Vignaud, A., Noirez, P., Besse, S., Rieu, M., Barritault, D. and Ferry, A. (2003) Recovery of slow skeletal muscle after injury in the senescent rat. Exp.Gerontol., 38: 529–537.

Yu, J.G., Carlsson, L. and Thornell, L.E. (2004) Evidence for myofibril remodeling as opposed to myofibril damage in human muscles with DOMS: An ultrastructural and immunoelectron microscopic study. Histochem.Cell Biol., 121: 219–227.

CHAPTER 6

PATHOPHYSIOLOGY, PREVENTION AND TREATMENT OF AGE-RELATED OSTEOPOROSIS IN WOMEN

MOUSTAPHA KASSEM AND KIM BRIXEN

Department of Endocrinology and Metabolism, Odense University Hospital and University of Southern Denmark, Kloevervaenget 6, DK-5000 C, Denmark

Abstract: One of the cardinal manifestations of old age in humans is bone loss leading to fragility of the skeleton and increased risk of fractures, a disease known as osteoporosis. It is estimated that approximately 45% of all women will suffer at least one osteoporotic fracture during their lifetime. Genetic, environmental, nutritional, biomechanical and hormonal factors determine the integrity of the skeleton and age-related bone loss and thus the risk for developing osteoporosis. Several pharmacological agents that are capable for decreasing the risk of fractures are currently available and have proven their efficacy in randomized clinical studies. Among these are the anti-catabolic drugs e.g., calcium, vitamin-D, estrogen, raloxifen, and bisphosphonates (e.g., etidronate, alendronate, risedronate, ibandronate, and pamidronate), anabolic drugs e.g., parathyroid hormone (1–34) and strontium ranelate which has both anti-catabolic and anabolic effects. Also, evidence suggests that individualized advice on lifestyle modification, e.g., increased physical exercise, cessation of smoking, fall prevention and use of hip protectors, should be offered to most patients

Keywords: aging; osteoporosis; bone; bone remodeling; pathophysiology; endocrinology; hormones; bone loss; osteoblasts; osteoclasts light-emitting diode

1. INTRODUCTION AND DEFINITION AND SCOPE OF THE PROBLEM

Osteoporosis is "a disease characterized by low bone mass and deterioration of bone architecture leading to decreased bone strength and increased risk of fractures". Clinically, osteoporosis is defined by the presence of low bone mass phenotype defined as a T-score < -2.5 as measured by dual energy X-ray absorptiometry (DEXA) scan or the presence of one or more vertebral compression fractures as result of no or low-energy trauma. However, osteoporosis is also implicated in most

S. Rattan and M. Kassem (eds.), Prevention and Treatment of Age-related Diseases, 87–104.

low-energy fractures occurring in the elderly population (Riggs and Melton III, 1986). It is estimated that approximately 40–47% of all women at the age of 45–50 years will suffer at least one osteoporotic fracture during their remainder life time (Riggs and Melton III, 1986). Such fractures often have considerable consequences for the patient due to increased morbidity and pain, loss of independence, reduced life expectancy (following hip and vertebral fractures), and reduced health related quality of life. It also imposes enormous costs on the society in terms of hospital treatment, rehabilitation, and nursing home care. The annual costs of osteoporotic fractures and their sequels are estimated to exceed $14.000 million $ in the U.S. alone. The number of osteoporotic fractures is expected to rise due to demographic changes of increasing the number of elderly persons. Thus, it is projected that the number of hip fractures will increase 4–5 folds during the next 40–50 years as a consequence of the increasing population aged 65 years or above. Even more importantly, this increase will be most pronounced in the developing countries.

2. PATHOPHYSIOLOGY OF AGE-RELATED BONE LOSS AND OSTEOPOROSIS

2.1 Patterns of age-related bone loss and bone fractures

Bone mass of the whole skeleton or of a particular region of interest can be measured by a number of different technologies e.g., single photon absorptiometry (SPA), dual photon absorptiometry (DPA), dual energy X-ray absorptiometry (DEXA), and single or dual energy quantitative computer tomography (QCT). Bone mass is usually expressed as area bone mineral density (BMD) and bone mineral content (BMC). Measurements of BMD and BMC employing DEXA machines have become widely used in clinical assessment of fracture risk and the diagnosis of osteoporosis. In addition, extensive studies of large cohorts of men and women using DEXA machines have also provided important insights into the patterns of bone loss during the human life span.

The highest bone mass achieved during the life span of an individual is known as *peak bone mass* and usually reached during the 3rd–4th decade of life (Gilsanz et al., 1997; Lu et al., 1996). Bone loss starts shortly thereafter at some skeletal sites (lumbar spine and proximal femur) and a decade later at other skeletal sites (Matkovic et al., 1994). As shown in Figure 1, two patterns of bone loss are recognized. A continuous, slow, age-related bone loss is observed in both men and women and results in an overall bone loss of 20–25% of both cortical (the outer dense envelop of most bones) and trabecular bone (located internal to the cortical bone at the end of long bones and in the vertebrae and other short or irregular bones). In the perimenopausal period in women, a rapid phase of bone loss is observed during a period of 5–10 years around menopause. This phase leads to bone loss up to 14%. A decade after the menopause, the rapid phase of bone loss terminates and merges with the slow but progressive aged-related bone loss. The rate of bone loss varies between skeletal sites and is generally most pronounced in the spine being

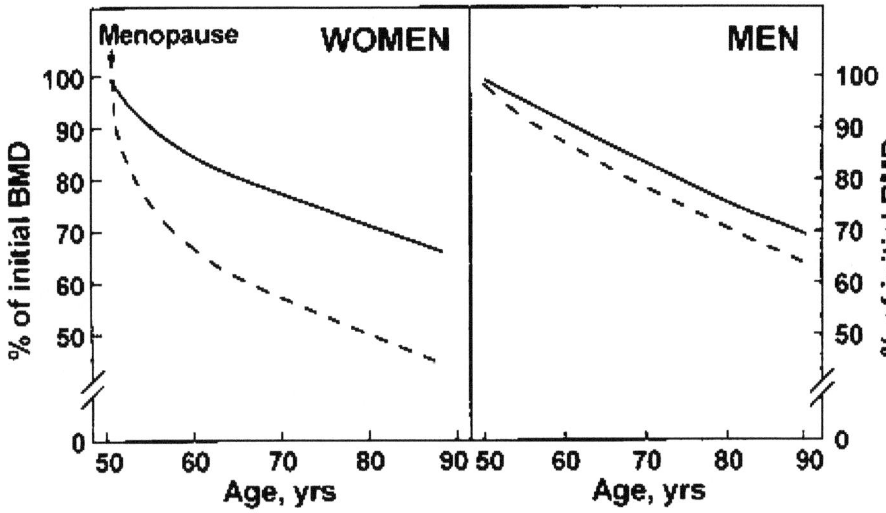

Figure 1. Schematic representation of changes in bone mass over life in cancellous (broken line) and cortical (solid line) bone in women (left panel) and men (right panel) from age 50 onward. In men only one phase of continuous bone loss is observed but in women two phase are recognized: a perimenopausal accelerated phase of bone loss and a late slow phase. Note also that the accelerated phase, but not the slow phase, involves disproportionate loss of cancellous bone (Riggs et al., 1998)

rich in the metabolically active trabecular bone. It is least pronounced in the hip and other sites rich in cortical bone. In addition, to age-related decrease in bone mass, significant changes do also occur in what is known as "bone quality" that includes several parameters e.g., the 3-dimensional structure of bone, the material quality of bone as tissue, the presence of micro-fractures (Mosekilde et al., 1987). Age-related changes in these factors contribute to the deterioration of the mechanical strength of the skeleton (Mosekilde et al., 1987; Ebbesen et al., 1999). Currently, no-invasive methods that measure the bone quality factors are being developed for clinical or epidemiological studies.

The age-related patterns of bone loss are associated with age-related increase in bone fractures. After 50 years of age the fracture risk increases exponentially in both sexes. However, the increase in fracture risk takes place approximately 10 years later in males compared with females. The first fracture type to increase after the menopause is the forearm fracture (Figure 2) which often is related to falls during forward movement, where the energy of the fall is conveyed to the stretched forearm. Hip fractures often occur in elderly people during falls on the side when standing or walking slowly (Cummings and Nevitt 1989).

Osteoporotic bone loss and fractures can thus be perceived as the end result of several pathophysiological mechanisms underlying : 1) low peak bone mass, 2) age-related bone loss, 3) post-menopausal bone loss, or a combination of these factors.

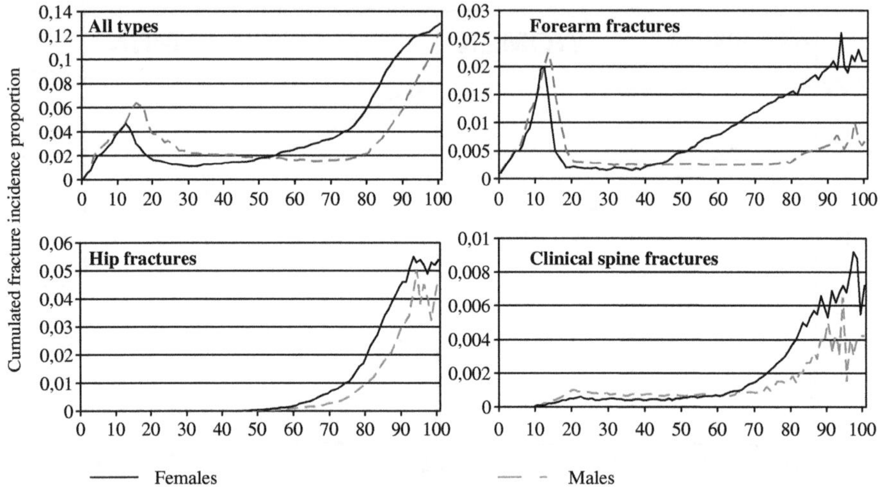

Figure 2. Fracture risk in Denmark (population 5 million) 1995–99 according to age and sex. Based on patients admitted to Danish Hospitals (Danish Hospital Central Register). (Kindly provided by Drs. P. Vestergaard and Leif Mosekilde, unpublished data)

2.2 Why do we lose bone mass as we age?

Our current understanding of the cellular mechanisms responsible for age-related bone loss are based on quantitative studies of bone cell activities in bone biopsies obtained from iliac crest or vertebral bodies of aging human population and by employing histomorphometric techniques (Frost, 2001; Parfitt, 1991; Frost). Bone as a tissue, is composed of bone matrix and bone cells. Bone matrix is built up of type I collagen (90%) and the remaining 10% is composed of a large number of non-collagenous proteins (e.g., osteocalcin, osteonectin, bone sialoproteins and various proteoglycans). Non-collagenous proteins participate in the process of matrix maturation, mineralization and may regulate the functional activity of bone cells. Two main types of bone cells have been identified. Osteoblasts (bone forming cells) and osteoclasts (bone resorbing cells). These cells together with their precursor cells and associated cells (e.g., endothelial cells, nerve cells) are organized in specialized units called bone multicellular units (BMU) that perform bone remodeling activities. Bone remodeling is a bone regenerative process taking place in the adult skeleton aiming at maintaining the integrity of the skeleton by removing old bone of high mineral density and high prevalence of fatigue microfractures and replacing it with young bone of low mineral density and better mechanical properties. This process is important for the biomechanical competence of the skeleton and it also supports the role of the skeleton as an active participant in the divalent ion homeostasis. Bone remodeling consists of a specific sequence of cellular events with a defined temporal sequence occurring at the same anatomical location (Figure 3). It is the same sequence in both trabecular and cortical bone. The remodeling sequence is termed ARF sequence. "**A**" refers to the **a**ttraction of osteoclast precursors to specific bone sites where remodeling will take place. These sites are determined by specific mechanical needs or mechanical signals, the nature of which is not known. This is followed by **a**ctivation to the osteoclast precursor cells to fuse and form functional multinucleated osteoclasts. "**R**" indicates the resorptive phase, where osteoclasts remove a certain thickness of mineralized bone tissue which can be measured histomorphometrically and known as erosion depth. This phase usually lasts 4–6 weeks. "**F**" refers to the formative phase where osteoblasts are recruited from stem cells and precursor cells in the bone marrow. They recreate the amount of bone matrix removed by the osteoclasts and secure a proper mineralization of the newly formed osteoid tissue. The amount of new bone formed can also be measured histomorphometrically and known as mean wall thickness. The duration of the formative phase is usually 3–4 months.

Based on understanding of bone remodeling dynamics maintenance of stable bone mass depends on: i) the balance between the osteoclastic activity indicated by the *erosion depth* and osteoblastic activity indicated by the *mean wall thickness*, and ii) the number of remodeling cycles initiated in unit time per unit bone volume (termed *the activation frequency*). In the young adult, there is a balance between the amount of bone removed by osteoclasts and the amount of bone formed by osteoblast and bone mass is unchanged. Both the erosion depth (Eriksen et al., 1984) and the mean wall thickness (Eriksen et al., 1984) decrease with increasing age. However, in perimenopausal women estrogen deficiency is

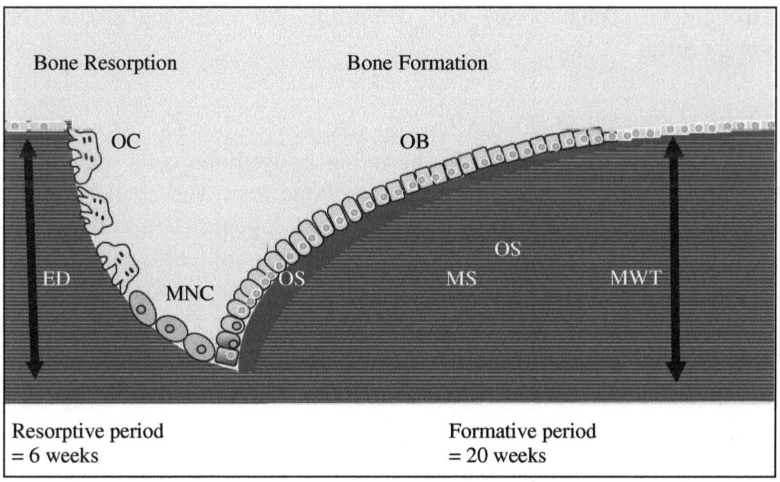

Figure 3. Trabecular bone remodeling following the A-R-F sequence (activation of osteoclasts, resorption by osteoclasts (OC) and mononuclear cells (MNC) and formation by osteoblasts (OB). ED = erosion depth, MWT = mean wall thickness. OS = osteoid (unmineralized bone) surface, MS = mineralized surface

associated with hyperactive osteoclasts and increased bone resorption compared to bone formation (Eriksen et al., 1999). On the other hand, age-related decreased mean wall thickness and impaired osteoblast functions have been observed in several histomorphometric studies in the elderly (Cohen-Solal et al., 1991; Eriksen et al., 1990).

In addition to age-related decrease in bone mass caused by imbalance of bone resorption and bone formation, aging is associated with architectural deterioration of the skeleton as outlined above. These changes are also caused by age-related changes in bone remodeling dynamics. An age-related increase in the activation frequency (turnover) or in resorption depth will by itself threaten the integrity of the 3-dimensional trabecular network (Mosekilde, 1990). During bone resorption, deep osteoclastic lacunae may hit thin trabecular structures leading to trabecular perforations. Concomitant remodeling processes on the opposite sides of thicker trabeculae may have the same consequence. The thinning of trabecular structures with age due to the imbalance between bone resorption and bone formation may also increase the risk of perforations. The consequence of this process is a progressive loss of trabecular elements, deterioration of bones three-dimensional structure and a loss of mechanical strength with age. Complex calculations from trabecular density and intertrabecular distances suggest that age-related trabecular perforations and structural changes contribute more to the age-related decrease in bone strength compared with age-related decrease in bone mass.

The above-mentioned changes in bone cells behaviour are caused by two universal factors present in the whole aging population: intrinsic age-related changes in bone cell functions and age-related changes in the endocrine system. These universal factors interact with individual-related characteristics (e.g., genetics, environmental, behavioural) and determine the individual's risk for developing osteoporosis.

2.2.1 Age-related changes in bone cells

Similar to other cellular compartments in the aging body, bone cells undergo a multitude of age-related changes that contribute to bone loss. The available data suggest that decreased cell proliferation capacity of osteogenic stem cells is the rate limiting factor for bone formation with age (Stenderup et al., 2003). The aging microenvironment may also contribute to the age-related decreased bone formation since sera obtained from old persons (a surrogate for the aging microenvironment of bone) exerted inhibitory effects on osteoblast differentiation of osteoprogenitor cells compared to sera obtained from young persons (Kassem et al. Bone, 2006, in press).

2.2.2 Age-related changes in the endocrine system

Aging is associated with several changes in the endocrine system which in turn affects different organs in the body including the skeleton. Some of the best studied endocrine systems with respect to their impact on bone are: sex steroids,

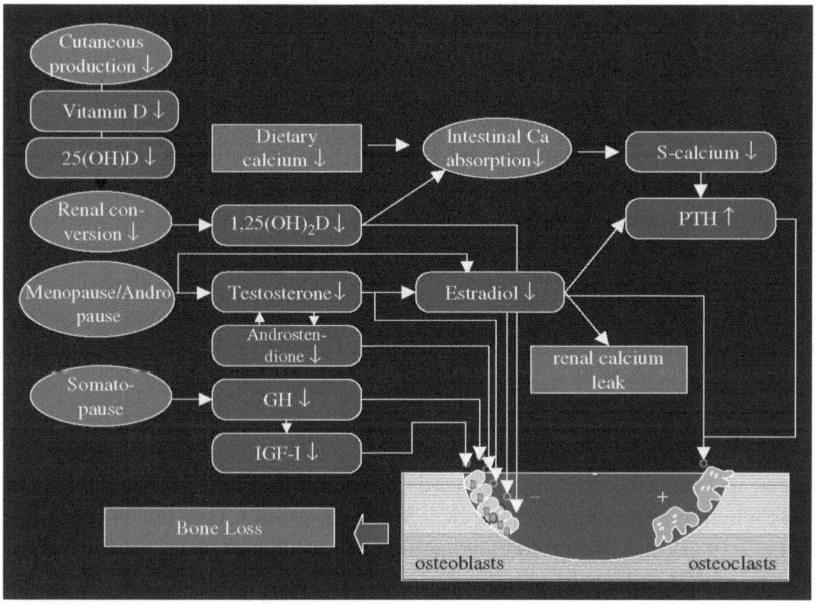

Figure 4. Age-changes in the endocrine system and its contribution to the observed age-related bone loss.25(OH)D = 25-hydroxyvitamin D, 1,25(OH)2D = 1,25-dihydroxyvitamin D. PTH = parathyroid hormone. GH = growth hormone, IGF = insulin-like growth factor. Ca = calcium. All the changes in the endocrine system lead finally to increase (+) in osteoclastic bone resorption and inhibition (−) of osteoblastic bone formation leading to remodelling imbalance and bone loss

parathyroid hormone and growth hormone (GH)/insulin-like growth factor (IGF) system (Figure 4).

A. Sex steroids In women, aging is associated with marked changes in serum levels of estrogen but not androgens. Total estradiol (E_1) decreases from 221 pmol/l in young women to 133 pmol/l in elderly women and estrone (E_2) from 338 pmol/l in young to 78 pmol/l in elderly women while a slight drop in testosterone (T) levels decrease from 1.4 in young to 1.1 nmol/l in elderly women (Khosla et al., 1998).

Estrogen deficiency and bone loss in women

The rapid decrease of estrogen metabolites in the postmenopausal period leads to increased bone turnover, osteoclast activity (Eriksen et al., 1999) and consequently increased bone resorption compared to bone formation leading to bone loss. The molecular basis of increased osteoclastic activity resulting from E deficiency has recently been a topic of intensive investigation. E deficiency has been shown to increase the production of osteoclast-activating cytokines (IL-1, TNF-α, IL-6) and E treatment led to the inhibition of their production (Pacifici, 1996). Also, E is capable for induction of apoptosis in osteoclasts and shortening of osteoclast life span

(Hughes et al., 1996). The direct effects of E on osteoblastic cell functions are less clear.

B. Parathyroid hormone Age-related secondary hyperparathyroidism is caused by age-related impaired mechanisms of calcium conservation. With increasing age, intestinal calcium absorption is impaired because of decreased production of 1,25-dihydroxyvitamine D (Slovik et al., 1981). Also, an age-related increased urinary calcium excretion (urinary calcium leak) has been reported (Heshmati et al., 1998). Recently, Riggs et al., (Riggs et al., 1998) have suggested that the age-related secondary hyperparathyroidism and impaired mechanisms of calcium conservation and homeostasis are caused by the effects of E deficiency on intestine and kidneys.

C. Growth hormone and insulin-like growth factors (IGF) Serum levels of GH reach its peak in late puberty, and afterwards a pronounced age-related decline in serum levels which can be explained by decreased secretion rate (Finkelstein et al., 1972; Ho et al., 1987) and increased clearance rate (Iranmanesh et al., 1991). Serum concentrations of IGF-I largely parallel serum GH with a peak at puberty and a decrease with ageing. Serum IGF-I, but not IGF-II correlates closely to 24-hour integrated GH secretion (Florini et al., 1985). Similarly, serum levels of IGF-I and not IGF-II decrease with age in both men and women (Florini et al., 1985; Copeland et al., 1990; Bennett et al., 1984). The age-related decline in GH and IGF-I parallels the age-related decline in bone mass suggesting that changes in serum GH and IGF-I are responsible for the age-related bone-loss. However, administration of GH to healthy elderly persons was unable to restore and only increased bone mass slightly (Rudman et al., 1990). Therefore, it seems unlikely that GH and IGF are major factors contributing to the skeletal phenotype of senescence except in subgroup of osteoporotic patients with abnormally low levels of the hormones.

2.2.3 Genetic, environmental and individual risk factors

Peak bone mass and the rate of bone loss are affected by a multitude of factors including genetic, behavioral and dietary. They are also affected by diseases and medications received by the persons throughout their life history.

Several studies have shown that part of the varaitions of bone mass of adult skeleton can be explained by polymorphic traits in a number of key extracellular matrix components (collagen type I), hormones receptors (vitamin D receptors, ER, AR, PTH/PTHrp receptors), cytokines (OPG, RANKL, TGF-β). However, the relative contributions of each of these polymorphic traits to age-related bone loss need to be determined (Nguyen et al., 2000; Ralston, 2002).

Smoking, large alcohol intake, exercise levels, decreased in muscle strength due to aging or specific neuromuscular disorders, diet and diseases affecting the skeleton (e.g. hyperthyroidism, anorexia nervosa, chronic exposure to glucocorticoids) are some of a long list of factors that are capable of affecting bone mass and skeletal integrity. These factors can interact with the universal mechanisms of age-related bone loss described above and determine the individual risk for developing osteoporosis.

3. MANAGEMENT OPTIONS FOR AGE-RELATED OSTEOPOROSIS

The ideal pharmacological therapy of osteoporosis should reduce the number of patients with new fractures significantly. This should be documented in one or more randomized, double blind, placebo-controlled trials. In some cases, trials demonstrating non-inferiority comparing with documented efficacious therapy may be acceptable. Also, the mode of action should be known, the frequency of adverse effect should be low, and serious side effects should not occur. Finally, the drug should be affordable (i.e., the cost-efficacy ratio should be favorable) and easy to administer to ensure long-term persistence with therapy.

Three different approaches to therapy can be employed: anti-catabolic, anabolic, or a combination of these. Estrogen, selective estrogen modifiers (SERMs), bisphosphonates, and calcitonin are mainly anti-catabolic drugs while parathyroid hormone receptor agonists (PTH(1–34), PTH(1–84), and PTH-related protein (PTHrp)) are anabolic agents, and strontium ranelate seems to have both anti-catabolic and anabolic effects (Table 1). Anti-catabolic drugs decrease bone resorption and bone remodeling that reduces the remodeling space (*i.e.*, the number of active resorption sites), increases the mean age of the bone tissue, and its degree of mineralization. These changes increase bone strength. The anabolic drugs increase bone remodeling and may initially lead to an apparent decrease in bone mass due to decreased degree of mineralization and expansion of the remodelling space, however, with time bone dimensions, cortical thickness, and the number of trabecular elements increase leading to increasing bone strength.

3.1 Calcium and vitamin-D

Calcium and vitamin-D (ergo- or cholecalciferol) may partly overcome the age-related decrease in calcium absorption thereby lowering serum PTH and thus bone turnover. Moreover, vitamin-D insufficiency is prevalent in the elderly as well as institutionalized persons. Finally, vitamin-D improves muscle function and decreases the risk of falling.

In most studies on pharmacological therapy of osteoporosis, calcium and vitamin-D have been administered to both the active and placebo groups. Two studies, however, have investigated the effect of calcium alone on the occurrence of fractures. In these both of these, calcium supplementation (1000 or 1200 mg/day) decreased the occurrence of vertebral fractures significantly in elderly patients with a low calcium intake.

The effect of vitamin-D alone administered as cod liver oil was investigated in Norwegian study and the effect of oral vitamin-D 10 μg/day was investigated in a Dutch study, however, both studies did not demonstrate any effects of treatment

Table 1. Pharmacological options in prevention and treatment of osteoporosis grouped according to mode of action

	Mechanism of action		
	Anticatabolic	Combined	Anabolic
Currently in use	Bisphosphonates Estrogens SERMs* Calcium Vitamin-D	Strontium ranelat	PTH(1–34)*
Under development	New bisphosphonates Vitamin-K Anti-RANK-L antibodies Osteoprotegerin Integrin antagonists Chloride channel antagonists		PTH(1–84)* PTHrp* Growth hormone IGF-I/IGFBP-3*

*SERMs = selective estrogen receptor modifiers; PTH = parathyroid hormone; PTHrp = parathyroid hormone-related peptide. IGF-I/IGFBP-3 = insulin-like growth factor-I – IGF-binding protein-3-complex.

on bone mass. In a Finnish study, however, annual vitamin-D injections (150,000–300,000 IU/year) decreased the incidence of peripheral fractures by 25% in an open, quasi-randomized study comprising elderly subjects. Similarly, a decrease in the occurrence of peripheral fractures (RR = 0.67 (0.48–0.93)) was found in an English study where a dose of 100,000 IU was administered orally every 4 months. The contrasting results can be explained by differences in vitamin-D or calcium intake, sun-exposure, or the bioavailability of the vitamin-D preparations used for injections.

The effect of combined calcium and vitamin-D has been studied in several studies. In a French study of 3,270 elderly females in nursing home, vitamin-D (800 IE/day) plus calcium (1200 mg/day) reduced the number of hip fractures by 43% and peripheral fractures overall by 32%. In a similar, but smaller study in 583 institutionalized subjects the same investigators, a reduction in risk of hip fracture of the same magnitude (RR = 0.59 (0.33–1.04)) was found, however this did not reach significance due to the small size of the study. In a Danish population-based study comprising 9,605 women and men aged 65 years or above that were block-randomized to calcium plus vitamin-D, fall prevention, calcium plus vitamin-D plus fall prevention, or no intervention, calcium (1000 mg/day) and vitamin-D (400 mg/day) decreased the risk of fractures (0.84 (0.72–0.98)). The potential effect of treatment effect may be higher since only half the participants were compliant with treatment.

The active (i.e., hydroxylated) vitamin-D metabolites (calcitriol and 1-alpha-hydroxy-vitamin-D) have been tested in small studies of sub-optimal design. Unlike newer treatments, hydroxylated vitamin-D metabolites require individual dosing and careful biochemical monitoring. In a single-blind study of three years duration comprising 622 post-menopausal women, significantly less vertebral and peripheral fractures were seen after calcitriol compared with placebo. No effect was seen in a similar but smaller study of two years duration comprising 50 patients. Also, studies on the effect of 1-alpha-hydroxy-vitamin-D have yielded conflicting results.

3.2 Hormone Replacement Therapy (HRT)

Estradiol decreases bone turnover and the number of active resorption lacunae and thereby the bone loss (see above).

In an early study, transdermal estradiol and oral medroxyprogesterone acetate decreased the incidence of new vertebral fractures in 75 women with manifest osteoporosis. In a study comprising 464 post-menopausal women randomized to HRT (2 mg estradiol and 1 mg cypoterone acetate), vitamin-D (300 IU/day), HRT plus vitamin-D, or placebo for 5 years, only HRT significantly reduced the risk of non-vertebral fractures (RR = 0.29 (0.10–0.90)), while no significant effect could be demonstrated with vitamin-D (RR = 0.47 (0.20–1.14)), or combined HRT plus vitamin-D (RR = 0.44 (0.17–1.15)). These results were corroborated in an open study comprising 2,016 post-menopausal women randomized to five years of HRT or no treatment. In this study, HRT tended to decrease the risk of non-vertebral fractures (RR = 0.73 (0.50–1.05)) and significantly reduced the incidence of forearm fractures (RR = 0.45 (0.22–0.90)) in intention-to-treat analysis. In the per-protocol analysis, both all non-vertebral fractures (RR = 0.61 (0.39–0.97)) and the risk of forearm fractures (RR = 0.24 (0.09–0.69)) were significantly reduced.

In the estrogen-progestagen-arm of the WHI (Women Health Initiative) study, 16.608 non-hysterectomized women were randomized to HRT or placebo. The risk of hip fractures were significantly reduced (RR = 0.66 (0.45–0.98). Similar results were recently published from the estrogen-only-arm of this study (hysterectomized women) showing a reduction in "all fractures" (0.70 (0.63–0.79)). However, the most important side effect of HRT as evidenced by the WHI study, is an increased risk of breast cancer (in the estrogen-progestage(n arm), thrombo-embolic events (in both the estrogen-progestagen and the estrogen-only arms), and cardio-vascular events. Most studies have employed estradiol 2 mg/day or equivalent, however, 1 mg estradiol per day has almost the same effect on BMD while side effects may be less frequent.

3.3 Selective estrogen receptor modifiers(SERM) (Ettinger et al., 1999)

Selective estrogen receptor modulators (SERMs) bind to the estrogen receptor, however, not at the ligand-pouch, and change the receptor conformation. This alters the affinity of a number of tissue-specific transcription factors (co-activators and co-repressors) leading to estrogen agonistic effects in some tissues, e.g. bone, and antagonistic effects in other tissues, *e.g.*, breast. This group of compounds comprises tamoxifen, raloxifen, and several other drugs under development. Only raloxifen is currently approved for prevention and treatment of osteoporosis.

In a study comprising 7,705 post-menopausal women with osteoporosis, participants were randomized to 60 mg/day or 120 mg/day of raloxifene or placebo. The risk of vertebral fracture was reduced (RR = 0.7 (0.5–0.8)) following the approved dosage of 60 mg/day. The risk of non-vertebral fracture, however, was not significantly altered by treatment. Similarly, tamoxifen reduces the risk of fractures, although, the increased risk of ovarian cancer precludes its use outside oncology.

Raloxifen has an estrogen agonistic effect on the cardio-vascular system and reduces serum levels of total and LDL cholesterol. In contrast to estrogen, however, it does not increase HDL-cholesterol. While ongoing studies are in the process of assessing the effects of raloxifen on cardiovascular events, a *post-hoc* analysis from previous trials suggest that the event rate is reduced by raloxifene treatment. The effect of raloxifen on the breast and endometrium is estrogen-antagonistic. Thus, treatment causes no breast tenderness and decreases the incidence of estrogen-receptor-positive breast cancer with 76 %. Main side effects of raloxifen therapy is the risk of thrombo-embolism (RR = 2.17 (0.83–5.70)).

3.4 Bisphosphonates

Bisphophonates are synthetic analogues of pyrophosphate that inhibit the osteoclasts and bone resorption. The aminobisphosphonates inhibit the enzyme farnesyl diphosphate synthase and thereby the achoring of a number of intracellular enzymes to the cytoskeleton leading to osteoclastic apoptosis. Other bisphosphonates are metabolized within the osteoclasts to cytotoxic ATP-analogues. In both cases osteoclastic activity and bone resorption as well as bone turnover are decreased.

Several bisphosphonates (etidronate, alendronate, risedronate, ibandronate, and pamidronate) have been demonstrated to decrease the occurrence of vertebral fractures by approximately 50%. In contrast, only alendronate and risedronate have been demonstrated to decrease the incidence of peripheral fractures.

When taken orally, the most prevalent side effects are abdominal pain, nausea, dyspepsia and heart-burn, however, in the many of the placebo-controlled trials the frequency of these side effects have been similar in the placebo and bisphosphonate groups. Erosion or ulceration in esophagus may occur in rare cases during treatment with aminobisphosphonates. Etidronate in high dosages (16–160 times those used in osteoporosis) may inhibit the mineralization of bone; however, this side effect has not been observed with the other compounds. With intravenous administration, flu-like symptoms and low-grad fever may be seen for 1–2 days in a minority of the patients. This has no clinical importance and may be prevented using acetaminophen.

3.5 Strontium ranelate

The divalent cat-ions of stable strontium isotopes may be administered orally as strontium ranelate. Strontium is incorporated in bone and seems to posses dual modes of action; it stimulates bone formation and decreases bone resorption. These effects seem to be mediated by the calcium-sensing receptor. *In vitro* strontium has affinity to this receptor and displays calcimimetic effects. The detailed mechanism of action, however, remains unknown.

In patients with osteoporosis, strontium ranelate (1–2 g/day) increases biochemical markers of bone formation and reduces markers of bone resorption. During treatment, strontium ranelate increases BMD 14.4 percent at the lumbar spine and 8.3 percent at the femoral neck after 3 years. These results, however, should be interpreted in light of the stronger x-ray attenuation (higher atomic mass) of strontium compared with calcium. Thus, approximately 50% of the increase in BMD seems to be to be attributable to the physical properties of strontium within bone.

In a study comprising 1,649 postmenopausal women with manifest osteoporosis the effect of strontium ranelate (2 g/day) for three years was compared with placebo. All participants received calcium and vitamin-D before and during the study. Strontium ranelate reduced the incidence of vertebral fractures significantly (RR 0.59 (0.48 to 0.73)). Similar results were found in a study comprising 5,091 postmenopausal women with osteoporosis where the relative risk of vertebral fractures was reduced by 39–45%. In this study, the occurrence of non-vertebral fractures was also significantly reduced by 16 % (RR = 0.84 (0.702–0.995)) and in a subgroup (n = 1977) with high-risk of hip-fractures (age 74+ years and a femoral neck T-score ≤ -3) the risk of these fractures was reduced significantly by 36%.

Strontium ranelate has few side effects. Diarrhea may be seen initially, but often subsides with time. A small but significant incidence of thrombo-embolic diseases was seen, however, the physiologic basis for this remains unknown. Treatment may increase serum levels of creatine kinase but does not lead to clinical events.

3.6 Parathyroid hormone receptor agonists

A number of drugs activating the PTH-receptor have been approved for treatment of osteoporosis (PTH(1–34)) or are under development (PTH(1–84) and PTHrp (PTH related paptide). Binding of PTH(34) to the receptor activates adenylate cyclase and a number of phospholipases (A, C, and D) and increases intracellular levels of cAMP and calcium. Intermittent treatment with PTH (1–34) increases the number of osteoblasts and bone formation by activation of pre-existing osteoblasts, increased differentiation of *lining cells*, and reduced osteoblast apoptosis. In addition to its effects on bone mass, PTH improves bone structural integrity, bone diameter, and bone strength.

The clinical effect of PTH(1–34) was documented in a study comprising 1,637 post-menopausal women with manifest post-menopausal osteoporosis. Participants were randomized to treatment with PTH(1–34) at a dosage of 20 or 40 μg/day or placebo. All participants received supplementation with calcium (1000 mg/day) and vitamin-D (400–1200 IE/day). The mean duration of treatment was 18 (maximum 24) months. In these patients, the approved dosage of 20 μg/day increased BMD in the lumbar spine and hip by 9.7 % and 2.6 %, respectively, after 18 months.

In comparison with placebo, PTH(1–34) 20 μg/day significantly reduced the incidence of patients with new vertebral fractures by 65 (45–78) %. In absolute terms, 14% of the participants in the placebo group compared with 5% the PTH(1–34) group experienced a new vertebral fracture during the study. Similarly, the incidence of patients with new fractures in the appendicular skeleton was 9.7% in the placebo group and 6.3 in the group receiving PTH(1–34). Fracture protection was evident only after approximately 12 months of therapy. Duration of therapy is restricted to 18–24 months. Following termination of therapy, BMD of the lumbar spine is reduced by approximately 2–3 % after 2½ years, however, fracture prevention extends beyond termination of treatment.

The effect of PTH(1–84) followed by alendronate has been investigated in an open study. Following treatment with PTH(1–84) (50, 75, or 100 μg/day) or placebo for one year, 75 patients were treated with alendronate (10 mg/day) for one additional year. PTH(1–84) increased BMD of the lumbar spine by 7.1 % while the sequential treatment increased BMD by 13.4 % in total.

The most frequent adverse effects during treatment with PTH(1–34) are nausea, headache, dizziness, and leg cramps. Serum levels of calcium, uric acid, and magnesium may be increased and urinary excretion of calcium is increased. In rats, high dosages (8 to 10 times the human dosage) and long duration of therapy increases the occurrence of osteosarcoma, but such an effect has not been seen in human studies. At present, however, PTH(1–34) should not be used in children, pregnant or lactating women, patients with Paget's disease of bone, malignant disease, or patients who have previous received radiation therapy to the skeleton.

3.7 Future treatment options

An array of new anti-catabolic drugs is under development. First, very potent bisphosphonates such as ibandronate and zolendronate may allow once-a-month or once-a-year administration. This may improve compliance considerably; but the anti-fracture-efficacy of these compounds remains to be documented. High dosages of vitamin-K may have positive effects on bone health. Moreover, advances in molecular biology have identified an array of potential target for new drugs such as integrins, osteoprotegerin, RANK-L, and osteoclast-specific chloride channels that may decrease osteoclastic activity via new mechanisms of action. New anabolic drugs under development include PTH(1–84) and PTH-rp. With better knowledge of the molecular pathways mediating the effect of PTH, it is hoped that non-peptide drugs with similar effect allowing oral administration may be developed. Also, growth hormone and recombinant IGF-I/IGFBP-3-complex may be beneficial.

Table 2. Prevention and treatment of post-menopausal osteoporosis. Advise on lifestyle and fall prevention should be individualized according to age and concommitant diseases. Additional calcium and vitamin-D should be adminstered along with specific treatments, while the latter drugs should only rarely be used simultaneously

Target group	Prevention		
	Primary	Secondary	Tertiary
	Population	Osteoporosis	Manifest osteoporosis
Life style modification*	+	+	+
Fall prevention**		+	+
Hip protector**		+	+
Calcium + vitamin-D	+	+	+
HRT***		(+)	(+)
Bisphosphonates		+	+
SERMs		+	+
Strontium ranelate		+	+
PTH(1–34)			+

* Increased physical exercise, cessation of smoking, and reduced alcohol intake should be considered.
** Fall prevention and hip protectors should be considered in patients with increased risk of falling.
*** HRT should be used for a short period only in the lovest possible dosage in patients with substantial climacteric symptoms.

3.8 Current recommendations

A summary on current recommendations on prevention and treatment is shown in Table 2. Primary prevention, i.e., measures directed at the general population without individual risk assessment, may include life-style advice (diet, cessation of smoking, and exercise). In countries of high latitudes where the food is not fortified by addition of vitamin-D and the prevalence of vitamin-D insufficiency in the elderly population is high, supplementation with vitamin-D and calcium seem appropriate above the age of 65 years.

In patients with osteoporosis as determined by dual energy absorptiometry (DEXA), secondary prevention with anti-catabolic agents (bisphonates, SERMs) or strontium ranelate should be considered. HRT may be used in women with climacteric symptoms; however, duration of therapy should be limited. These treatments should be accompanied by calcium and vitamin-D supplementation. Combination of anti-catabolic agents is usually not recommended.

The same options should be considered as tertiary prevention (i.e., in patients with osteoporotic fractures of the spine). In these patients, however, anabolic therapy with PTH(1–34) for 18 months followed by a bisphosphonates may be discussed.

In all cases, the patients symptoms (i.e., prevalent vertebral fractures and fracture history), risk factors for new fractures, and bone mineral density should be balanced against potential side effects. Also, the patient's preference regarding e.g., administration should be considered. Finally, cost-efficacy and national rules on reimbursement should be considered.

REFERENCES

Bennett, A.E., Wahner, H.W., Riggs, B.L., Hintz, R.L. (1984) Insulin-like growth factors I and II: aging and bone density in women. J.Clin.Endocrinol.Metab 59: 701–704.

Beral, V. (2003) Breast cancer and hormone-replacement therapy in the Million Women Study. Lancet 362: 419–27.

Brixen, K.T., et al. (2004) Teriparatide (biosynthetic human parathyroid hormone 1–34): a new paradigm in the treatment of osteoporosis. Basic Clin.Pharmacol.Toxicol. 94: 260–70.

Cohen-Solal, M.E., Shih, M.S., Lundy, M.W., Parfitt, A.M. (1991) A new method for measuring cancellous bone erosion depth: application to the cellular mechanisms of bone loss in postmenopausal osteoporosis. J.Bone Miner.Res. 6: 1331–1338.

Copeland, K.C., Colletti, R.B., Devlin, J.T., McAuliffe, T.L. (1990) The relationship between insulin-like growth factor-I, adiposity, and aging. Metabolism 39: 584–587.

Cummings, S.R., Nevitt, M.C. (1989) A hypothesis: the causes of hip fractures. Journal of Gerontology 44: M107–M111.

Cummings, S.R., et al. (1998) Effect of alendronate on risk of fracture in women with low bone density but without vertebral fractures: results from the Fracture Intervention Trial. JAMA 280: 2077–82.

Ebbesen, E.N., Thomsen, J.S., Beck-Nielsen, H., Nepper-Rasmussen, H.J., Mosekilde, L. (1999) Age- and gender-related differences in vertebral bone mass, density, and strength. Journal of Bone and Mineral Research: the Official Journal of the American Society for Bone and Mineral Research 14: 1394–1403.

Eriksen, E.F., Melsen, F., Mosekilde, L. (1984) Reconstruction of the resorptive site in iliac trabecular bone: a kinetic model for bone resorption in 20 normal individuals. Metabolic Bone Disease & Related Research 5: 235–242.

Eriksen, E.F., Gundersen, H.J., Melsen, F., Mosekilde, L. (1984) Reconstruction of the formative site in iliac trabecular bone in 20 normal individuals employing a kinetic model for matrix and mineral apposition. Metabolic Bone Disease & Related Research 5: 243–252.

Eriksen, E.F., Hodgson, S.F., Eastell, R., Cedel, S.L., O'Fallon, W.M., Riggs, B.L. (1990) Cancellous bone remodeling in type I (postmenopausal) osteoporosis: quantitative assessment of rates of formation, resorption, and bone loss at tissue and cellular levels. J.Bone Miner.Res. 5: 311–319.

Eriksen, E.F., Langdahl, B., Vesterby, A., Rungby, J., Kassem, M. (1999) Hormone replacement therapy prevents osteoclastic hyperactivity: A histomorphometric study in early postmenopausal women. J.Bone Miner.Res. 14: 1217–1221.

Ettinger, B., et al. (1999) Reduction of vertebral fracture risk in postmenopausal women with osteoporosis treated with raloxifene: results from a 3-year randomized clinical trial. Multiple Outcomes of Raloxifene Evaluation (MORE) Investigators. JAMA 282: 637–45.

Finkelstein, J.W., Roffwarg, H.P., Boyar, R.M., Kream, J., Hellman, L. (1972) Age-related change in the twenty-four-hour spontaneous secretion of growth hormone. J.Clin.Endocrinol.Metab 35: 665–670.

Florini, J.R., Prinz, P.N., Vitiello, M.V., Hintz, R.L. (1985) Somatomedin-C levels in healthy young and old men: relationship to peak and 24-hour integrated levels of growth hormone. J.Gerontol. 40: 2–7.

Frost, H.M., Vilanueva, A.R., Jett, S., Eyring, E. Tetracycline-based analysis of bone remodelling in osteopetrosis. Clin.Orthop. 65: 203–217.

Frost, H.M. (2001) The Utah paradigm of skeletal physiology: what is it? Veterinary and Comparative Orthopaedics and Traumatology 14: 179–184.

Gilsanz, V., Kovanlikaya, A., Costin, G., Roe, T.F., Sayre, J., Kaufman, F. (1997) Differential effect of gender on the sizes of the bones in the axial and appendicular skeletons. J.Clin.Endocrinol.Metab 82: 1603–1607.

Heshmati, H.M., Khosla, S., Burritt, M.F., O'Fallon, W.M., Riggs, B.L. (1998) A defect in renal calcium conservation may contribute to the pathogenesis of postmenopausal osteoporosis. J Clin Endocrinol Metab. 83: 1916–20.

Ho, K.Y., Evans, W.S., Blizzard, R.M., Veldhuis, J.D., Merriam, G.R., Samojlik, E., Furlanetto, R., Rogol, A.D., Kaiser, D.L., Thorner, M.O. (1987) Effects of sex and age on the 24-hour profile of growth hormone secretion in man: importance of endogenous estradiol concentrations. J.Clin.Endocrinol.Metab 64: 51–58.

Hughes, D.E., Dai, A., Tiffee, J.C., Li, H.H., Mundy, G.R., Boyce, B.F. (1996) Estrogen promotes apoptosis of murine osteoclasts mediated by TGF-beta. Nat.Med. 2: 1132–1136.

Iranmanesh, A., Lizarralde, G., Veldhuis, J.D. (1991) Age and relative adiposity are specific negative determinants of the frequency and amplitude of growth hormone (GH) secretory bursts and the half-life of endogenous GH in healthy men. J.Clin.Endocrinol.Metab 73: 1081–1088.

Khosla, S., Melton, L.J., III, Atkinson, E.J., O'Fallon, W.M., Klee, G.G., Riggs, B.L. (1998) Relationship of serum sex steroid levels and bone turnover markers with bone mineral density in men and women: a key role for bioavailable estrogen. The Journal of Clinical Endocrinology and Metabolism 83: 2266–2274.

Lu, P.W., Cowell, C.T., LLoyd-Jones, S.A., Briody, J.N., Howman-Giles, R. (1996) Volumetric bone mineral density in normal subjects, aged 5–27 years. J.Clin.Endocrinol.Metab 81: 1586–1590.

Matkovic, V., Jelic, T., Wardlaw, G.M., Ilich, J.Z., Goel, P.K., Wright, J.K., Andon, M.B., Smith, K.T., Heaney, R.P. (1994) Timing of peak bone mass in Caucasian females and its implication for the prevention of osteoporosis. Inference from a cross-sectional model. J.Clin.Invest 93: 799–808.

McClung, M.R., et al. (2001) Effect of risedronate on the risk of hip fracture in elderly women. Hip Intervention Program Study Group. N.Engl.J.Med. 344: 333–40.

Meunier, P.J., et al. (2004) The effects of strontium ranelate on the risk of vertebral fracture in women with postmenopausal osteoporosis. N.Engl.J.Med. 350: 459–68.

Mosekilde, L., Mosekilde, L., Danielsen, C.C. (1987) Biomechanical competence of vertebral trabecular bone in relation to ash density and age in normal individuals. Bone 8: 79–85.

Mosekilde, L. (1990) Consequences of the remodelling process for vertebral trabecular bone structure: a scanning electron microscopy study (uncoupling of unloaded structures). Bone Miner. 10: 13–35.

Mosekilde, L. (2005) Vitamin D and the elderly. Clin.Endocrinol. 62: 265–81.

Neer, R.M., et al. (2001) Effect of parathyroid hormone (1–34) on fractures and bone mineral density in postmenopausal women with osteoporosis. N.Engl.J.Med. 344: 1434–1441.

Nguyen, T.V., Blangero, J., Eisman, J.A. (2000) Genetic epidemiological approaches to the search for osteoporosis genes. J.Bone Miner.Res. 15: 392–401.

Pacifici, R. (1996) Estrogen, cytokines, and pathogenesis of postmenopausal osteoporosis. Journal of Bone and Mineral Research: the Official Journal of the American Society for Bone and Mineral Research 11: 1043–1051.

Parfitt, A.M. (1991) Bone Forming Cells in Clinical Conditions. In: B.K.Hall (ed) In Bone, The Osteoblast and Osteocyte. The Telford Press, London, pp 351–426.

Ralston, S.H. (2002) Genetic control of susceptibility to osteoporosis. J.Clin.Endocrinol.Metab. 87: 2460–2466.

Reginster, J.Y., et al. (2005) Strontium ranelate reduces the risk of nonvertebral fractures in postmenopausal women with osteoporosis: TROPOS study. J.Clin.Endocrinol.Metab.

Riggs, B.L., Melton, L.J., III (1986) Involutional osteoporosis. N.Engl.J Med. 314: 1676–1686.

Riggs, B.L., Khosla, S., Melton, L.J., 3rd. (1998) A unitary model for involutional osteoporosis: estrogen deficiency causes both type I and type II osteoporosis in postmenopausal women and contributes to boneloss in aging men. J Bone Miner Res. 13: 763–73.

Rossouw, J.E., et al. (2002) Risks and benefits of estrogen plus progestin in healthy postmenopausal women: principal results From the Women's Health Initiative randomized controlled trial. JAMA 288: 321–33.

Rudman, D., Feller, A.G., Nagraj, H.S., Gergans, G.A., Lalitha, P.Y., Goldberg, A.F., Schlenker, R.A., Cohn, L., Rudman, I.W., Mattson, D.E. (1990) Effects of human growth hormone in men over 60 years old. N.Engl.J.Med. 323: 1–6.

Slovik, D.M., et al. (1981) Deficient production of 1,25-dihydroxyvitamin D in elderly osteoporotic patients. The New England Journal of Medicine 305: 372–374.

Stenderup, K., Justesen, J., Clausen, C., Kassem, M. (2003). Aging is associated with decreased maximal life span and accelerated senescence of bone marrow stromal cells Bone 33: 919–927.

CHAPTER 7

ARTHRITIS AND ITS TREATMENT

ASHIT SYNGLE

Healing Touch City Clinic, Fortis Heart Insitute and Multispeciality Hospital; #547, Sector 16D, Chandigarh-160015, India (Email: ashitsyngle@yahoo.com)

Abstract: Arthritis has afflicted man from prehistoric times and has accompanied him throughout his evolutionary history. However, rheumatology – the discipline of medicine dealing with disorders of joint and connective tissues – is perhaps the youngest of medical specialties. Our understanding of various types of arthritis has improved considerably leading to evolution of specific and effective therapies for most types of arthritis especially in the last few decades. Despite these advancements many myths and ignorance is still prevalent with respect to arthritis. The present write up is aimed to improve our understanding

Keywords: Arthritis, Rheumatology, arthritis treatment

Hippocrates (460–377 BC), the father of medicine, gave an early reference to arthritis as 'a disease with fever, severe joint pain; fixing itself in one joint now, then in another, of short duration, acute, not leading to death, more apt to attack the young than the old'.

Arthritis, today describes more than 100 chronic diseases of the joints, bones and muscles, '*Arthron*' in Greek means joint and '*itis*' means inflammation. Arthritis thus refers to the pain and inflammation of the joints.

Rheumatology refers to the study of medical disorders of joint and connective tissues and doctors who treat these disorders are known as rheumatologists. The connective tissue provides structural support for the cells in the body. Bone, skin, ligaments and tendons are all connective tissue.

Until a few decades ago, a diagnosis of arthritis was deeply discouraging for the patient and doctor alike. Most types of arthritis were considered untreatable and there was little to offer in the medicine chest – a misconception which is still prevalent. However, today there is greater understanding of the disease process and specific and effective therapies are available for most types of arthritis. What is even more is that it is now being recognized that the inflammatory fire kindled in the body by autoimmune

S. Rattan and M. Kassem (eds.), Prevention and Treatment of Age-related Diseases, 105–132.

diseases like rheumatoid arthritis may be the engine that drives many of the most feared illnesses of middle and old age like heart attack, stroke, Alzheimer's disease etc.

Perhaps in no other discipline of medicine as in rheumatology is the arthritis patient vulnerable to influences and counter-influences of modern medicine, homeopathy, ayurveda, and unani system of medicine, not to speak of acupuncture, copper bangles, magnetic therapy etc. However, the stark reality is that there is little understanding of rheumatic diseases and few rheumatologists are available.

This write up is devoted to improve this understanding and discuss some important types of arthritis.

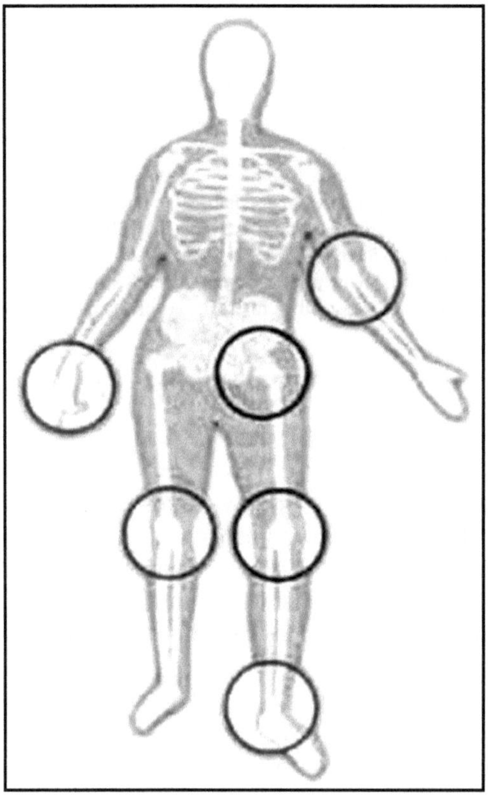

COMMON SYMPTOMS OF ARTHRITIS

1. Pain or tenderness in one or more joints.
2. Swelling in one or more joints.
3. Stiffness around joints that lasts for at least one hour in the early morning.
4. Difficulty in using or moving a joint normally.
5. Warmth or redness in a joint.

Extra-articular manifestations are equally important and often reveal the underlying diagnosis. Fever occurs in rheumatic fever (RF), other connective tissue disorders especially systemic lupus erythematosus (SLE), the vasculitides, Still's disease and infective endocarditis (IE). Alopecia suggests a possibility of SLE. Nail changes like clubbing occurs in IE, interstitial lung disease, hypertrophic osteoarthropathy and nail pitting is reminiscent of psoriatic arthropathy. Nodules occur in disorders like RF, rheumatoid arthritis, gout, erythema nodosum, sarcoidosis and amyloidosis. A photosensitive rash occurs in SLE where as heliotrope and a knuckle rash is seen in dermatomyositis. Eye involvement often provides important clues. The sclera and lacrimals can be involved in RA causing the sicca syndrome, episcleritis and scleritis. In contrast, the uvea and conjunctiva are involved in spondyloarthropathies (SPAs) producing iritis, iridocyclitis, posterior uveitis and conjunctivitis. Mucocutaneous involvement is seen in conditions like SLE, Reiter's Syndrome, Behcet's syndrome, psoriasis and scleroderma. Nose involvement occurs in Wegener's Granulomatosis, Churg-Struass syndrome, Hansen's disease and relapsing polychondritis. Heart, kidneys, lungs, nerves and gastrointestinal system are also involved in a variety of connective tissue diseases.

1. SYSTEMIC LUPUS ERYTHEMATOSUS

SLE is a systemic autoimmune disorder characterized by wide spread inflammation affecting many organ systems of the body. Disease manifestations are protean, ranging in severity from fatigue, malaise, weight loss, arthritis or arthralgias, fever, photosensitivity, rashes, and serositis to potentially life-threatening thrombocytopenia, hemolytic anemia, nephritis, cerebritis, vasculitis, pneumonitis, myositis, and myocarditis. There is no cure for SLE.

1.1 General Measures

These include good nutrition, adequate rest and avoidance of excessive fatigue and stress. Patients are advised to use long sleeved clothes and hat or umbrella and avoid prolonged exposure to sunlight. Sunscreens with high protection factor (SPF 15 or more) should be applied liberally. Isolated skin lesions may respond to topical steroids. Infections should be treated aggressively as they could trigger a disease flair. The blood pressure and lipids should be well controlled especially in the presence of renal disease. Osteoporosis should be prevented in patients likely to require long term steroid therapy and/or with other predisposing factors.

1.2 Specific Treatment

Specific drug therapy is tailored depending on the severity of the disease and the organs system involved.

Mild to moderate disease characterized by constitutional symptoms, mucocutaneous lesions and arthritis is initially treated with NSAIDs and chloroquinine. NSAIDs usually control SLE-associated arthritis, arthralgias and serositis but not fatigue, malaise or major organ system involvement. NSAIDs should be avoided in patients with active nephritis. Antimalarial (Hydroxychloroquinine, chloroquinine and quinacrine) may be effective in the treatment of rash, photosensitivity, arthralgias, arthritis, alopecia and malaise associated with SLE and in the treatment of discoid and subacute cutaneous lupus erythematosus. These drugs are not effective for fever or renal/CNS and hematologic problems. Methotrexate may have a role in the treatment of arthritis and dermatitis but probably not in life-threatening disease.

In any life-threatening or organ-threatening manifestation of SLE, the mainstay of treatment is systemic corticosteroids (0.5–2 mg/kg/d orally or 1000 mg of methyl prednisolone sodium succinate IV daily for 3 days, followed by 0.5–1 mg/kg/d prednisolone or equivalent) for the initial 4–6 weeks followed by a maintenance dose of 5–10 mg/d of prednisolone.

Cytotoxic drugs are another important option for serious SLE. Cyclophosphamide is the drug of choice for life-threatening lupus nephritis alongwith concomitant corticosteroid therapy. Duration of cyclophosphamide therapy is controversial. Azathioprine and mycophenolate mofitil are used often as steroid-sparing agents but may not be as effective as cyclophosphamide in treating lupus nephritis. Apart from pulse cyclophosphamide, renal replacement therapy with renal transplantation has improved the outlook of patients with lupus nephritis. Recurrence of nephritis in the allograft rarely occurs.

It is important to realize that in most patients it is not possible to achieve complete sustained remission. A balance between mild active disease, acceptable drugs side effects is possible, practical and acceptable.

1.3 Emerging Therapies

Potential benefit of UVA1 phototherapy for cutaneous lupus appears promising. There are encouraging reports of the benefit of B-cell depletion with anti-CD 20 antibody, rituximab, for the treatment of SLE. A new immunosuppressant gusperimus may be useful in SLE patients refractory to cyclophosphamide (Lorenz et al., 2004). Tacrolimus may be as effective as monthly pulse cyclophosphamide in patients with lupus diffuse glomerulonephritis (Mok et al., 2004). Immune ablation with high-dose immunosuppressives followed by rescue with autologous haematopoietic stem cell transplantation has been tried for severe and refractory SLE with not very promising results (Traynor et al., 2002; Lisukov et al., 2004).

2. RHEUMATOID ARTHRITIS

RA is a chronic, progressive autoimmune inflammatory disease of unknown etiology that attacks the synovial tissue leading to irreversible joint damage (Figure 1), chronic pain, stiffness and functional impairment. It affects about 1% of adult population. It is prevalent across all ethnic groups and can occur at any age, although most cases are seen in adults between ages 30 and 60 years. Women comprise 75% of all cases.

There has been a radical change in the treatment of RA since early 1990s with initiation of disease-modifying anti-rheumatic drugs (DMARDs) early in the disease rather than late. This has resulted from our understanding that RA is not

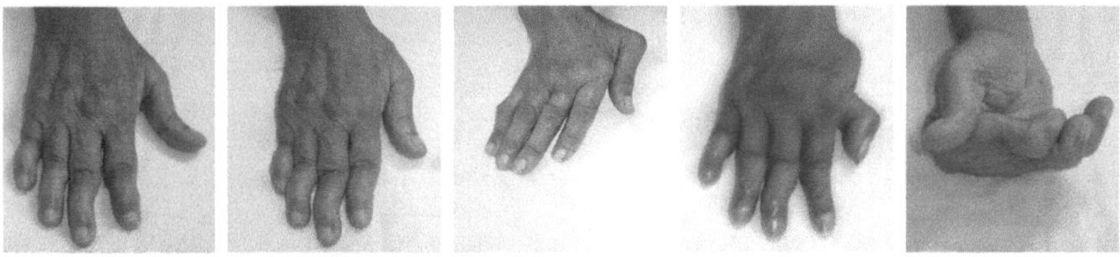

Figure 1. Clinical Spectrum of Rheumatoid Arthritis

a benign but progressive disease and many DMARDs are not prohibitively toxic. In addition to its considerable associated morbidity and economic costs (direct and indirect), RA leads to premature mortality (Reilly et al., 1990). A patient with RA is 2 times more likely to have a myocardial infarction, 70% more likely to have a stroke, 70% more likely to develop an infection, has a 25-fold increased incidence of lymphoma, mortality rates 41% higher for women and life expectancy decreased by 18 years. Since DMARDs alter the disease course in recent onset RA, early and aggressive treatment should be initiated to achieve remission.

There are several options for treating RA. These include traditional DMARDs such as hydroxychloroquine (or chloroquine), sulfasalazine (or salazopyrine), gold salts and methotrexate and newer agents such as lefluonmide and tumor necrosis factor inhibitor etanercept and infliximab. Evidence suggests that treatment with a combination of DMARDs is more effective than monotherapy (O' Dell et al., 1996). It is a challenge to the management skills of the rheumatologist to determine the most efficacious regiment for a particular patient using a combination and even more importantly appropriate timing of pharmacologic therapy, which may include NSAID, DMARD(s), low dose prednisolone, local injection of glucocorticoid, biologic agents, emotional and rehabilitation support and non-narcotic analgesics. An early consultation with a rheumatologist is of paramount importance to create a window of opportunity for early initiation of appropriate treatment. Factors influencing the choice of treatment are efficacy, safety, convenience of treatment regimen, the patient's disease activity, functional status, co-morbidities, life-style (e.g. child-bearing potential), work status and treatment reimbursement issues. Algorithm for achieving therapeutic success in RA is depicted in Figure 2.

2.1 Nonpharmacologic treatment of RA

Management of RA involves more than drug therapy alone. Early in the course of the disease, the patients needs to know and accept to learn to live with RA and will also be required to become involved in decision making for the

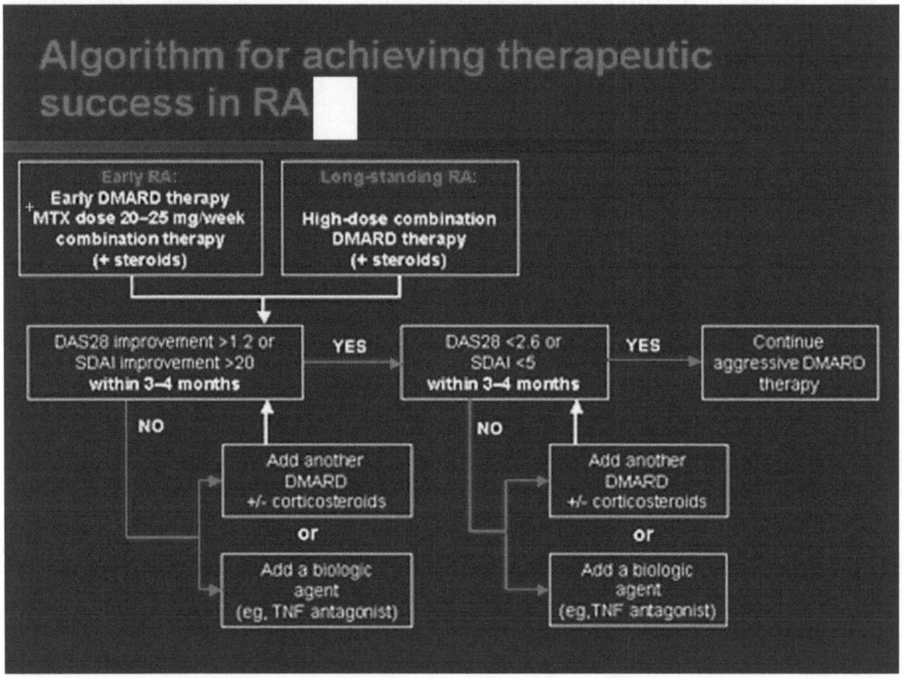

Figure 2. (Is Remission the Mission in RA?)(2005) +Methotrexate is chosen as the initial DMARD quite often, though others may also be used. DAS: Disease Activity Score. TNF: Tumor Necrosis Factor

treatment. Education regarding joint protection, energy conservation and arthritis home exercise programme (one such free exercise programme is available at www.healingtouchwebhelp.net/html/exerci.htm) with range of motion and strengthening exercises are important for maintaining joint function.

Physical therapy and occupational therapy may help the patient who is compromised in activities of daily living. Regular dynamic and aerobic exercises improve joint mobility, muscle strength, aerobic fitness and psychological well being without increasing fatigue or joint symptoms.

2.2 Drug Therapy of RA

Drug therapy of RA often consists of NSAIDs, DMARDs, anti-TNF and/or glucocorticoids. The initial therapy of RA usually involves use of salicyclates or NSAIDs to reduce joint pain and swelling and to improve joint function. NSAIDs have analgesic and anti-inflammatory properties but do not alter the disease course or alter joint destruction.

2.3 DMARDs

DMARDs have the potential to reduce or prevent joint damage, preserve joint integrity and function and reduce total cost of health care and maintain economic productivity of RA patient (ACR Subcommittee on Rheumatoid Arthritis Guidelines, 2002). The initiation of DMARDs therapy should not delayed beyond 3 months for any patient with established diagnosis who, despite adequate treatment with NSAIDs, has ongoing joint pain, significant morning stiffness or fatigue, active synovitis, persistent elevation of ESR or CRP level or radiographic joint damage (ACR Subcommittee on Rheumatoid Arthritis Guidelines, 2002). In fact there is an emerging view that early onset RA should be considered as a medical emergency and early treatment (within 2 weeks of diagnosis) with DMARDs should be initiated as it provides better outcomes at 2 years (Lard et al., 2001). For any untreated patient with persistent synovitis and joint damage, DMARDs should be started promptly to prevent or slow further damage (ACR Subcommittee on Rheumatoid Arthritis Guidelines, 2002) and ultimately to achieve remission.

The DMARDs commonly used in RA include hydroxychloroquine (HCQ), sulfasalazine (SSZ), methotrexate (MTX), and lefluonmide. Those used less frequently include azathioprine (AZA), D-penicillamine (D-Pen), gold salts, minocycline and cyclosporine.

Initial choice of DMARD for a particular patient is influenced by several considerations. Initial DMARD(s) is chosen based on its relative efficacy, convenience of administration, monitoring requirement, cost of drug and monitoring, time until expected benefit and adverse affects. Patient factors like compliance, co-morbidities, life-style (e.g. child-bearing potential), severity and prognosis of the patient's disease also influence the choice. For women of child-bearing age, effective contraception is required with DMARDs.

Based on consideration of safety, convenience and cost many rheumatologists select HCQ or SSZ first, but for patient with very active disease or with indicators of a poorer prognosis MTX or combination therapy would be preferred. For patients in whom MTX is contraindicated or has failed to achieve satisfactory disease control either because of lack of efficacy (in doses upto 25 mg/week) or intolerance, treatment with biologic agents or with other DMARDs, either alone or in combination is indicated (ACR Subcommittee on Rheumatoid Arthritis Guidelines, 2002).

Controversy remains about whether to initiate DMARD in a sequential 'step-up' approach in patients with persistently active disease in whom single agent has failed or whether to initiate combination DMARDs therapy early in the disease course and then apply a 'step-down' approach once adequate disease control is attained (Williams et al., 1992). However, there is emerging data (FIN-RACo trial (Mottosen et al., 1999), COBRA (Boers Met et al., 1997)) that combination therapy may be more effective in early RA and should be considered as induction therapy early especially in those with poor prognostic signs (extra-articular manifestations e.g. cutaneous ulcers, vasculitic rash, neuropathy, scleritis, subcutaneous nodules;

female gender; poor functional score (HAQ >1 at 1 year of disease); multiple joint involvement especially >12 joints; early radiographic evidence of erosive changes; advanced age at onset; high rheumatoid factor titres, anti-cyclic citrullated peptide antibodies; sustained elevation of acute-phase reactants e.g. ESR, CRP, low socioeconomic status/educational level).

2.4 Glucocorticoids

Low dose oral steroids (less than 10 mg/day prednisone) and local steroid injections are highly effective for relieving symptoms in patients with active RA. However, many RA patients become steroid dependent despite DMARD therapy. Evidence suggests that low-dose steroids slow the rate of joint damage and therefore appeared to have disease-modifying potential (ACR Subcommittee on Rheumatoid Arthritis Guidelines, 2002). However, the benefits of low-dose steroids should always be weighed against their adverse effects.

2.5 Biological Agents

These agents bind and neutralize TNF which is an important inflammatory mediator. The three anti-TNF agents used in RA include etanercept, infliximab and adalimumab. The timing of use of biologic agents in treating RA is controversial. In the past, these agents were used only when therapy with DMARDs had failed. However, recent studies (TEMPO, ASPIRE, BeST, PREMIER) have evaluated the role of biologic therapy with anti-TNF agents early in the course of disease, making their early use more favorable, even perhaps as first-line agents. Overall, these studies suggest that early therapy with a combination of methotrexate and an anti-TNF agent is likely lead to improved outcomes in RA over therapies without anti-TNF agents. However, the high-cost of the anti-TNF agents may limit their early use, although we may find that if disease activity is controlled early, long-term costs of RA in terms of disability may be averted.

Also, it may be found that combination therapy with methotrexate and an anti-TNF agent can be used up front to control disease, and once disease activity is minimal, the anti-TNF agent can be discontinued and methotrexate used as monotherapy.

It is now being recognized that radiographic progression can occur even during remission in RA (Esmeralda et al., 2004). If a patient is in constant remission and has no other risk factor (that risk factor is defined as a positive rheumatoid factor or baseline damage already present), then no radiographic progression can be expected (Paco et al., 2004). But in those patients in clinical remission who are either rheumatoid factor positive or have a high baseline radiographic score, radiographic progression is still possible (Paco et al., 2004).

2.6 Is remission in RA a cure?

Using the armamentarium of DMARDs and anti-TNF agents early and aggressively, remission is possible in RA. Does remission mean a cure and can the treatment be discontinued after achieving remission? When patients in remission shifted from active treatment to placebo, they had much higher incidence of flares compared with those who received continued treatment (Saskia et al., 1997). If these patients with flare are again treated with the previously successful regimen they may not achieve the same level of control but actually run a certain risk of getting disease that is worse than they had in the beginning.

2.7 Emerging Concepts in RA Management

It has been found in 2 very large studies that patients destined to develop RA will develop evidence of autoimmunity several years before the onset of their first clinical symptom (Darcy et al., 2003). In fact, up to 40% of RA patients will have either rheumatoid factor and/or anticyclic citrullinated peptide (anti-CCP) antibodies in the preclinical phase of their disease. It's also known that for the 2 years prior to the onset of their first clinical symptom, they will have an increase in their C-reactive protein (CRP), albeit in the normal range, but it will slowly start to increase, indicating subtle inflammation is occurring, much like what is seen with CRP and atherosclerosis patients.

When they do develop their first symptoms of synovitis? If a biopsy is done in asymptomatic joints, one will find evidence of asymptomatic synovitis in those joints that look clinically normal. This actually means that RA is already a chronic disease at the onset of first symptom. The reason this is important to realize is that if we're going to cure RA, we may actually need to develop strategies to identify high-risk patients and intervene before they develop a clinical phase. This in fact is being strategized by several investigators presently.

3. OSTEOARTHRITIS (OA)

OA has afflicted man and other vertebrates from prehistoric times and it has accompanied man throughout his evolutionary history. It is the most common joint disorder in the world and is the leading cause of disability and pain in the elderly.

It is characterized pathologically by localized areas of articular cartilage damage associated with overgrowth of bone at the joint margins (Figure 3), changes in subchondral bone, fibrosis of joint capsule and mild synovitis.

Management of OA needs to be individualized, holistic, patient centered and situational. It is aimed at reducing pain, maintaining mobility and minimizing disability. Various treatment modalities are summarized in Table 1. and treatment algorithm is given in Table 2.

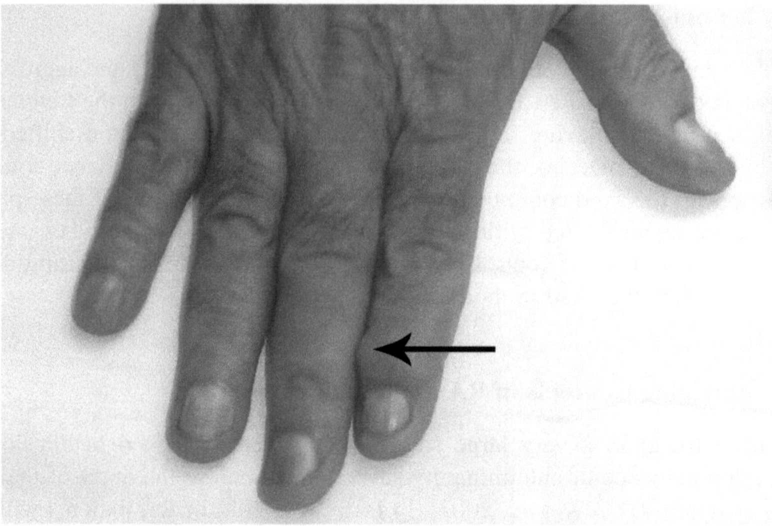

Figure 3. Heberden's Nodule in Osteoarthritis

3.1 Disease-Modifying Osteoarthritis Drugs

Better understanding of the pathogenetic mechanisms underlying the breakdown of articular cartilage in OA and particulary of the mediators involved in tissue breakdown has led the pharmaceutical industry on a search for the "holy grail": a disease modifying osteoarthritis drug (DMOAD). There are several candidates: chemically modified tetracyclines, diacerein, glucosamine and chondroitin sulfate.

Glucosamine and chondroitin sulfate have achieved striking popularity for treatment of OA recently. They are widely sold as nutraceuticals although they are not FDA approved. Several studies have shown glucosamine to be superior to placebo and comparable to NSAIDs with respect to efficacy in patients who have knee OA and it has a better safety profile than NSAIDs. However, the efficacy of glucosamine and chondroitin have not been examined in large, well designed, placebo controlled trials. Much of the beneficial effects are believed to be overestimated in trials most of which were industry sponsored.

Results of two virtually identical randomized controlled trials (Povelka et al., 2002; Reginster et al., 2001) have led to the suggestion that glucosamine not only improves joint pain in patients who have knee OA but are also chondroprotective. Glucosamine/chondroitin Arthritis Intervention Trial (GAIT) (Daniel O. Clegg et al., 2006), a multicenter, double-blind, placebo- and celecoxib-controlled study sponsored by the National Institutes of Health, evaluated the efficacy and safety

Table 1. Treatment modalities available for management of OA

Non pharmacologic treatment
Corner stone of OA management and is recommended in every OA patient to start with.
Educational, lifestyle, and behavioral
Education of patient, spouse, family, and significant others
Empowerment to aid patients in self-management and taking control
Behavioral and environmental changes to reduce the impact of OA
Social support including telephone contact
Alteration of levels of general exercise and activities
Weight loss and other dietary changes
Use of different shoes, orthoses, canes, and other walking aids assistive devices
Other non pharmacologic measures
Exercise to improve muscle strength, joint mobility, fitness, and function and to reduce pain
Weight reduction in obese
Thermal modalities
Physical aids to help joint protection and improve function podiatry
Acupuncture
Transcutaneous nerve stimulation and acupuncture
Dietary additions including glucosamine, chondroitin, vitamins C and D, ginger extracts, avocado, soybean derivatives and combination of esterified fatty acid complex, eicosapentanoic acid and docosahexanoic acid
Pharmacologic measures
Simple analgesics, Acetaminophen (ACET) 1 gm. 3–4 times/d (max 4 gm/d)
Tramadol and opioid analgesics
Antidepressants for analgesia (and for depression)
Systemic nonsteroidal anti-inflammatory drugs (NSAIDs) including coxibs for those not relieved with ACET.
Topical agents including capsaicin and NSAID creams and gels.
Intra-articular (IA) injections including steroids and hyaluronan (HA)
Diacerein
Surgical measures
Tidal irrigation (washout) of the joint (in knee OA)
Arthroscopic debridement
Cartilage transplantation and tissue engineering techniques
Osteotomy
Partial or complete joint replacement
Complementary and alternative therapies
Almost every known type of complementary and alternative medicine has been used in attempts to help people who have OA
Source: Modified from RCNA, Nov 2003, Vol 29, No 4, Pg 692.

of glucosamine, chondroitin sulfate, and the two in combination as a treatment for knee pain from osteoarthritis. In this study, glucosamine and chondroitin sulfate alone or in combination did not reduce pain effectively in the overall group of study patients with osteoarthritis of the knee. The relatively mild degree of pain

Table 2. Algorithm of OA Management

Nonpharmacological treatment Education, support, exercise, weight loss, joint protection *plus* Acetaminophen up to 4 g/d for control of pain and other symptoms, as well as before activity
↓
Add topical capsaicin cream 4 times daily, if needed, to painful joints
↓
If joint effusion is present, consider aspiration and intra-articular injection of a corticosteriod as initial therapy
↓
If more pain and symptom control is needed, add an NSAID in low dosage or a nonacetylated salicylate such as choline magnesium trisalicylate or salsalate. Tramadol may be useful as adjunctive therapy
↓
If more pain or symptom control is needed, use full dosage of an NSAID, plus misoprostol or a proton pump inhibitor if the patient is at risk for upper gastrointestinal tract bleeding or ulcer disease, or subsitute a COX-2-specific inhibitor for the NSAID. Some patients may benefit from intra-articular injections of a hyalronic acid product.
↓
If response is inadequate, consider referring the patient for joint lavage, arthroscopic debridement, osteotomy, or joint replacement.
Source: Mayo Clin Proc, May 2001, Vol 76, Pg 534.

from osteoarthritis among the participants of this study may have limited the ability of GAIT investigators to detect benefits of the treatments. However, the subgroup of study patients with moderate-to-severe pain demonstrated that combination of glucosamine and chondroitin sulfate significantly decreased knee pain related to osteoarthritis.

4. CRYSTAL INDUCED ARTHOPATHY

Arthritis resulting from deposition of different microcrystals is called crystal induced arthritis. "Traditionally" the name "gout" is given to the crystal induced arthritis related to uric acid deposition in joints and tissue. The rheumatic syndromes produced by "other microcrystals" are called "pseudogout."

5. GOUT

GOUT is one of the oldest disease known to humanity. Gout was often known as "the disease of kings and the king of diseases" because it was associated with wealthy men who overindulged in rich food and drink and formerly it was a leading cause of disabling arthritis. Its victims included King Henry VII, Alfred Lord Tennyson, Benjamin Franklin, Immanuel Kant, Samuel Johnson and Thomas Jefferson. Gout even caught Dicken's imagination as numerous of his characters were afflicted with this painful condition. Today, it's known that gout is a complex disorder found exclusively in human species that can affect anyone.

It is really amazing to know that long before *the crystal deposition* phenomenon was ever discovered, the name "gout" was coined by the "ancients", based on an intuition that it is caused by the deposition of some sort of toxin or poison "*noxa*" into the joints "*guta by guta*" (Latin for drop by drop). In fact, the only improvement by modern medicine has been to substitute "*crystals*" in the place of "*drops*" in the ancient description of aetiopathogenesis!

As the ancients suspected, and later science has proved, the central chemical culprit of classical gout is uric acid and its salts. Uric acid is the end product of purine degradation in humans. Purines are substances found naturally in your body as well as in certain foods, including organ meats—such as liver, brains, kidney and sweetbreads—and anchovies, herring, and mackerel. Smaller amounts of purines are found in all meats, fish and poultry.

We humans unfortunately cannot break down the almost insoluble uric acid further to the highly soluble allantoin as in lower animals because, the enzyme uricase is absent in us-one of the prices we are paying for "our superior status in the evolutionary ladder."

Normally, uric acid dissolves in the blood and passes through kidneys into urine. A net total of 90% of filtered uric acid is however reabsorbed in the kidney and only 10% is excreted – another reason responsible for hyperuricemia in humans. But sometimes the body produces too much or excretes too little of this acid. In that case, uric acid can build up, forming sharp, needlelike crystals in a joint or surrounding tissue that cause pain, inflammation and swelling. When the uric acid crystals are released in the joint space, they are primarily phagocytozed by synovial lining cells. These cells in turn release a variety of chemotactic factors that draw in the neutrophils and then draw in the whole variety of the proinflammatory molecules that trigger the inflammatory response.

What precipitates an acute attack of gouty arthritis? The precipitating factors include local trauma; binges of alcohol, overeating or fasting; concurrent medical or surgical illness; acute arise or fall in serum uric acid and seasonal factors.

5.1 Signs and Symptoms

In its natural history, gout passes through four stages:

1. Asymptomatic Hyperuricemia – when the uric acid level is high but there is no symptom of gout.
2. Acute Gouty Arthritis – when there is acute pain in the joints.
3. Intercritical Gout – periods between gouty attacks.
4. Chronic Tophaceous Gout with no pain-free intercritical periods.

The symptoms of gout are almost always acute, occurring suddenly – often at night—and without warning. They include: **Intense joint pain.** Gout usually affects the large joint of big toe but can occur in feet, ankles, knees, hands and wrists. The pain typically lasts 5 to 10 days and then stops. The discomfort subsides gradually over 1 to 2 weeks, leaving the joint apparently normal and pain-free. **Inflammation and redness.** The affected joint or joints will become swollen, tender and red. Figure 4 depicts the common sites of acute flares in gout.

Some people develop uric acid stones. The yearly risk for development of urate stone in people with established gouty arthritis is approximately 1%.

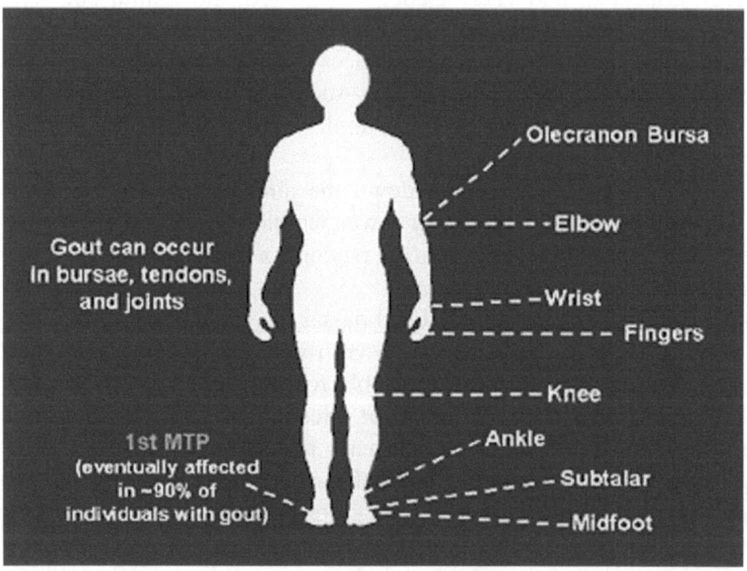

Figure 4. Common sites of acute flares (Marc DC 2005)

5.2 Goals of treatment

1. Terminate the acute attack rapidly.
2. Protect against further attacks.
3. Treat hyperuricemia and prevent disease progression.

5.3 Management of Acute Gouty Arthritis

The first goal is to terminate an acute attack of gout. The drugs used to treat an acute episode control the pain and inflammation but do not have a long lasting benefit in gout. They help to resolve the acute symptoms but the uric acid crystals remain in the joint and the destructive process continues.

5.4 Colchicine

This time tested drug inhibits neutrophil activation by inhibiting crystal-induced protein tyrosine phosphorylation. Oral doses of 0.6 mg are given every hour until improvement occurs, gastroinstestinal side effects developed or 10 doses have been taken without relief (in which case the diagnosis may be questioned).

5.5 NSAIDs

Indomethacin is the usual agent of choice. However, other agents like ibuprofen, naproxen and other short acting NSAIDs or COX-2 inhibitor etoricoxib are all effective. NSAID treatment is usually continued for 3–4 days after all signs of inflammation have resolved.

5.6 Glucocorticoids

This is most useful (1) when patient cannot take oral medication (2) when colchicine and NSAIDs are contraindicated or (3) in refractory cases. An intraarticular injection of glucocorticoid produces rapid dramatic relief. Alternatively, oral prednisolone 40–60 mg/d can be given until relief is obtained and then tapered rapidly.

5.7 Management of intercritical gout

Once the acute attack has been controlled the next aim is to control another such episode. In this asymptomatic intercritical period uric acid-lowering drugs need to be initiated.

Before initiating treatment with urate-lowering agents, the patients should be free of all signs of inflammation and have begun low dose colchicine for prophylaxis. The sudden drop in serum urate with the initiation of allopurinol or uricosuric therapy may prolong or precipitate an acute attack. Colchicine in a dose of 0.6 mg orally one to three times a day is usually effective in preventing gout attacks.

5.8 Managing hyperuricemia and preventing disease progression

Once antihyperperuricemia therapy is initiated, the dose used should maintain the serum urate at or below 5 mg/dl. Successful therapy will prevent future attacks of gout and cause resolution of tophi. Therapy once initiated should be life long. Intermittent therapy or withdrawal of therapy leads to recurrence of acute attack within 6 months and development of tophi within 3 years.

Urate lowering drugs for gout include (1) uricosuric agents (2) xanthine oxidase inhibitors.

5.9 Uricosuric agents

Virtually all uric acid presented to the glomerulus gets filtered. 90% of the filtered urate is reabsorbed and only 10% of the original filtered load of uric acid is excreted.

The uricosuric agents act by blocking reabsorption of uric acid. The most commonly used uricosuric agents are probenecid and sulfinpyrazone. Probenecid therapy is begun at 250 mg twice a day and is increased as necessary upto 3.0 g/d. Sulfinpyrazone therapy is initiated at a dose of 50 mg twice a day. The usual maintenance doses 300–400 mg/d in 3–4 divided doses.

By promoting uric acid excretion, uricosuric agents may precipitate nephrolithiasis. This can occur early in the course of treatment and may be prevented by initiating therapy at low doses, forcing hydration and alkalinizing the urine.

Other drugs that are not typically used as uricosurics but that do have some uricosuric properties include losartan and fenofibrate.

The advantage of uricosurics is that they reverse the most common physiologic abnormality in gout as 80–90% of patients are undersecretors of uric acid. However, they lose effectiveness as the creatinine clearance falls and are ineffective when glomerular filtration falls below 30 ml/min.

5.10 Allopurinol

The other way of lowering uric acid is to block its production. This is achieved by inhibiting xanthine oxidase, the enzyme that converts the relatively soluble hypoxanthine to the less soluble xanthine to the very insoluble uric acid. The commonly used xanthine oxidase inhibitor is allupurinol. It can be given as a single morning dose 300 mg initially and increasing upto 600 mg if needed. The advantage of allupurinol is that it is effective in both over production and undersecretion of urate, has convenience of single daily dose and can be efficacious in renal insufficiency but requires dose reduction in this situation. The most serious side effects includes skin rash with progression to life-threatening toxic epidermal necrolysis, systemic vasculitis, bone marrow suppression, granulomatous hepatitis and renal failure.

Diet modification now plays a minor role in the management of hyperuricemia as modern therapeutic agents are so effective but diet counseling is important and should address obesity, hypertension, hyperlipidemia, diabetes mellitus, alcohol use/abuse and strict avoidance of 'purine rich foods' (sweetbreads, anchovies, sardines, liver and kidney).

5.11 Uric acid nephropathy

Vigorous intravenous hydration and diuresis with frusemide dilute the uric acid in the tubules and promote urine flow to 100 ml/h or more. The administration of acetazolamide, 240 to 500 mg every six to eight hours, and sodium bicarbonate, 89 mmol/L intravenously enhances urine alkalinity and thereby solubilizes more uric acid. It is important to ensure that the urine pH remains above 7.0 and to watch for circulatory overload. In addition, antihyperuricemic therapy in the form of allopurinol in a single dose of 8 mg/kg is administered to reduce the amount of urate that reaches the kidney. If renal insufficiency persists, subsequent daily doses should be reduced to 100 to 200 mg because oxypurinol, the active metabolite of allopurinol, accumulates in renal failure. Despite these measures, hemodialysis may be required.

5.12 Emerging treatments of hyperuricemia

Target uric acid levels are not always achieved with the available drugs. There are also problems of use of these drugs in patients with renal failure, drug interactions and allopurinol intolerance.

The enzyme uricase, which is absent in humans, acts by catabolizing uric acid to a more soluble and readily excreatable form-allantoin. Rasburicase is a recombinant uricase made from Aspergillus flavus and is used for the treatment of tumor lysis syndrome but is not presently indicated in gout. It has blackbox warning for anaphylaxis, hemolysis and methemoglobinemia and has potential immunogenicity. To overcome the latter polyethylene glycol has been added to uricase.

Another drug being developed is febuxostat which is a nonpurine xanthine oxidase inhibitor. It can be given to people with renal insufficiency, mild to moderate hepatic dysfunction and those intolerant of allopurinol.

5.13 Beyond Gout

A host of co-morbidities are associated with hyperuricemia – obesity, metabolic syndrome, diabetes, hypertension, hyperlipidemia, CAD and heart failure. However it needs to be seen whether hyperuricemia is an independent risk factor or is it just a marker for these co-morbidities. Another important issue is whether the treatment of hyperuricemia will assist in the management of these co-morbidities. Although these questions remain, the diagnosis of gout can be used as a signal to search for unrecognized co-morbidities.

6. PSEUDOGOUT

Pseudogout results when calcium pyrophosphate dihydrate (CPPD) crystal deposited in bone and cartilage are released into the synovial fluid and induce acute inflammation. Risk factors include old age, advanced OA, neuropathic joint, gout, hyperparathyroidism, hemochromatosis, diabetes mellitus, hypothyroidism and hypomagnesemia. It may be asymptomatic or present as acute monoarthritis or oligoarthritis mimicking gout or as a chronic polyarthritis resembling RA or OA. Dehydration, acute illness and surgery (especially parathyroidectomy) are common precipitants of an acute attack.

Therapy of choice for most patients is a brief course of NSAID. Oral steroids can be used and colchicine also may relieve symptoms promptly. Daily prophylactic treatment with low doses of colchicine may be helpful in diminishing the number of recurrent attacks. Aspiration of the inflammatory joint fluid often results in prompt relief and intraarticular injection of steroids may hasten the response. Allopurinol or uricosuric agents have no role in the treatment of pseudogout.

Uncontrolled studies suggest that radioactive synovectomy (with yttrium 90) or the administration of antimalarial agents may be helpful in controlling persistent synovitis. Patients with progressive destructive large joint arthropathy usually require joint replacement.

7. APATITE DISEASE

Hydroxyapatite (HA) is the primary mineral in bone and teeth. Abnormal accumulation can occur in areas of tissue damage, in hypercalcemic or hyperparathyroid states. It may present as asymptomatic radiographic abnormality, acute synovitis or tendonitis or chronic destructive arthropathy. HA may be released from exposed bone and cause the acute synovitis occasionally seen in chronic stable osteoarthritis. HA deposition is also an important factor in an extremely destructive arthropathy of the elderly that occurs most often in knees and shoulders (Milwaukee shoulder).

Acute attacks are typically self-limiting, resolving within days to weeks. Therapeutic options are similar to pseudogout and include aspiration of the effusion, NSAIDs or oral colchicine for 2 weeks and intrarticular steroid injection.

8. CALCIUM OXALATE DEPOSITION DISEASE

Primary oxalosis is a rare hereditary metabolic disease. Nephrocalcinosis, renal failure, and death usually occur before age 20. Acute and/or chronic CaOx arthritis and periarthritis may complicate primary oxalosis during later years of illness. Secondary oxalosis is most commonly seen in end stage renal disease.

Clinical features of acute CaOx arthritis may not be distinguishable from those due to sodium urate, CPPD, or HA. Treatment of CaOx arthropathy with NSAIDs, colchicine, intraarticular steroids and/or an increased frequency of dialysis have produced only slight improvement. In primary oxalosis, liver transplantation has produced a significant reduction in crystal deposits.

9. SPONDYLOARTHROPATHIES

The spondyloarthropathies are an inter-related group of disorders characterized by one or more of the following features (1) spondylitis (2) sacroiliitis (3) enthesopathy (inflammation at sites of tendon insertion) and (4) asymmetric oligoarthritis in a genetically predisposed individual. Extra articular features of this group of disorders include: inflammatory eye disease, urethritis and mucocutaneous lesions. The spondyloarthropathies aggregate in females, where they are associated with HLA-B 27.

9.1 Ankylosing Spondylitis (AS)

AS presents with an inflammatory back pain which is characterized by persistent pain of more than 3 months, associated with early morning and rest stiffness and relieved by exercises. There is radiologic evidence of sacroilitis (Figure 5). AS usually affects young individuals below 40 years of age.

9.2 Management of AS

9.2.1 Physiotherapy

Plays an important role in the management of AS. Maintenance of erect posture during sitting, standing and walking is a must. A firm mattress with a single or no pillow is to be used while sleeping. Spinal extension exercises, breathing exercises and swimming are helpful to prevent ankylosis. Cigarette smoking should be discouraged strongly.

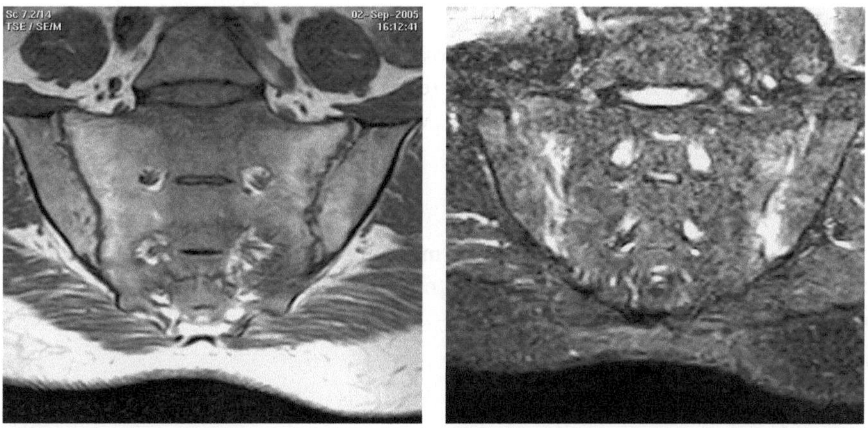

Figure 5. Sacroilitis of ankylosing spondylitis. MRI (axial T1W and STIR) images of a HLA B27+ve patient with inflammatory back pain showing sacroiliac joint irregularity and marrow edema parallel to the joint

9.2.2 *NSAIDs*

The most common and effective drug is indomethacin used upto 150 mg/d in divided doses.

9.2.3 *DMARDs*

Sulfasalazine (1–3 gm/d) is effective in AS. Methotrexate is useful in patients with peripheral arthritis who have not responded to sulfasalazine and NSAIDs. Recalcitrant enthesitis and persistent synovitis may respond to local steroid injection. Improvement has also been reported with mesalazine (Thomson, 2000). It has the advantage over sulfasalazine in not causing oligospermia.

9.2.4 *Anti-TNF Therapy*

Etanercept and infliximab, two drugs which block the inflammatory effect of tumor necrosis factor (TNF), are now licensed for the treatment of patients with severe ankylosing spondylitis whose symptoms have not responded adequately to conventional therapy. Anti-TNF therapy has revolutionized the treatment of AS and other spondyloarthritides. The response to treatment with these agents is rapid, profound and sustained. At present it is not known whether this therapy will halt the disease progression, but it seems likely to do so. Whether this therapy can reverse ankylosis or other damage is less clear but not improbable. However, these agents are quite expensive. They are also associated with side effects – serious infections, including disseminated tuberculosis, haematological side effects, worsening of heart failure, SLE-related antibodies and clinical features and hypersensitivity reactions.

9.2.5 *Surgery*

Total hip arthroplasty overcomes the disability from severe hip disease. Vertebral wedge osteotomy may be needed for correction of severe kyphosis.

Uveitis is managed with local steroid administration alongwith mydriatic agents, although systemic steroids or even immunosuppressive drugs may be occasionally required. Coexistent cardiac disease may require pacemaker implantation and/or aortic valve replacement. Osteoporosis in AS is managed in a similar manner as primary osteoporosis.

9.2.6 *Emerging Therapies*

Pamidronate 60 mg monthly intravenous infusion, thalidomide 200 mg/d, alpha-emitting isotope ^{224}Ra 1 MBq weekly intravenous infusion – all have potential benefit in AS.

9.3 Reiter's Syndrome and Reactive Arthritis

Reiter's syndrome (ReS) is characterized by asymmetric oligoarthritis, urethritis, conjunctivitis and characteristic skin and mucous membrane lesions. Chlamydia infection has been implicated in some patients and may occur with increased frequency in patients infected with HIV. Clinical features in HIV positive patients are similar to those in HIV negative patients.

Reactive arthritis (RA) is an acute arthritis that develops following infection elsewhere in the body, but the organism cannot be isolated or cultured from the joint. Reactive arthritis may follow dysentery caused by Shigella flexneri, Salmonella species, Yersinia enterocolitica or Clostrium difficille infections.

9.4 Treatment

9.4.1 NSAIDs

Arthritis is treated with NSAIDs like indomethacin 25 mg three times a day.

9.4.2 DMARDs

In those cases where NSAIDs do not control arthritis, addition of sulfasalazine in gradual increments upto 2–3 gm/d into divided doses will help. Methotrexate 7.5 to 15 mg/week may be given as an alternative. Azathioprine 1–2 mg/kg/d has also been found to be effective. Sulfasalazine can be safely used and is also beneficial in reiter's patients with HIV infection.

Persistent skin lesions are treated with retinoids or methotrexate.

9.4.3 Corticosteroids

Local steroids injections can be given for plantar fascitis or tendonitis. Topical steroids and keratolytic agents are used for keratoderma blenorrhagicum. Weak topical steroids like hydrocortisone valerate is useful for circinate balanitis.

9.4.4 Antimicrobial therapy

If chlamydia infection is proved doxycycline 100 mg bd for three months is given. Azithromycin is also effective in killing chlamydia. Combination chemotherapy may be needed in occasional cases with persistent disease activity. Antibacterial like trimethoprim-sulphamethoxazole or quinolones can be given in enteric reactive arthritis.

9.5 Psoriatic Arthritis (PsA)

An inflammatory arthritis occurs in 7–42% of patients with psoriasis. Psoriasis may precede, coincide or follow arthritis. There is no direct link between skin disease and arthritis. Five major patterns of joint disease occur (1) asymmetric oligoarticular arthritis (2) distal interphalangeal joint involvement associated with nail changes (Figure 6) like pitting, ridging and onycholysis (3) symmetric rheumatoid-like polyarthritis (4) spondylitis and sacroilitis and (5) destructive arthritis (mutilans) developing deformities like opera glass and telescoping of digits.

Arthritis is treated with NSAIDs and skin lesion with topical application. Intra-articular corticosteroids may be useful in the oligoarticular form of the disease,

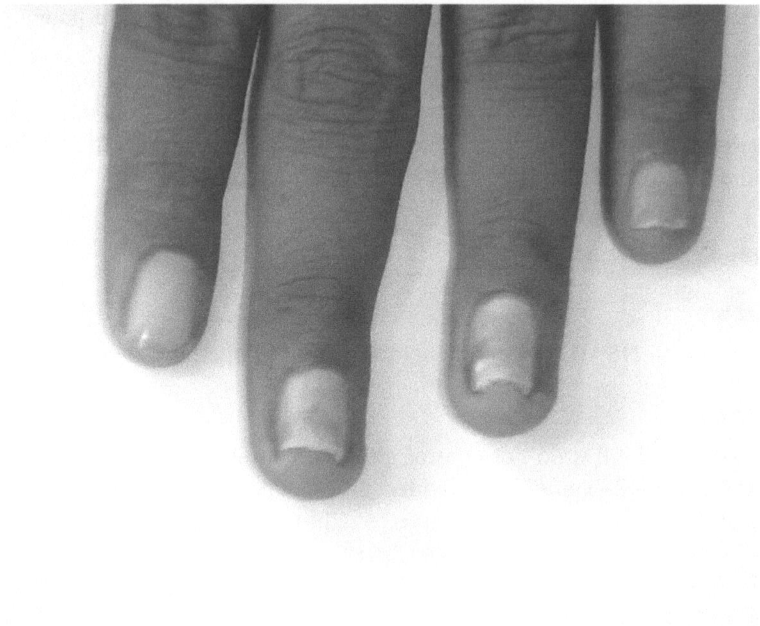

Figure 6. Nail changes in psoriatic arthritis

but injection through a psoriatic plaque should be avoided. Severe skin and joint diseases generally respond well to methotrexate. Sulfasalazine, lefluonmide, and hydroxychloroquine may also have disease-modifying effects in polyarthritis.

9.6 Biologic Therapy

The central role of inflammatory cytokines such as tumor necrosis factor (TNF) and activated T cells have provided new targets for therapy. Placebo-controlled trials of anti-TNF agents, etanercept, infliximab and adalimumab have shown sustained effectiveness of these therapies in their ability to control arthritis and psoriasis, improve quality of life and inhibit disease progression (Mease, 2004). However, because of their high cost, biologic agents are reserved for patients with progressive moderate to severe disease not adequately controlled with DMARDs. Some patients with mild disease, who would be treated with NSAIDs and/or DMARDs, may benefit from biologicals if they cannot tolerate DMARD and have inadequate response to NSAIDs.

What about patients who either have not responded initially to an anti-TNF medication, experience loss of efficacy over time, or for some other reason have had to discontinue such medication? These patients may respond by switching them from one to another biologic agent.

9.7 Emerging Biologic Approaches

These include Alefacept, Efalizumab, Anakira, rituximab, anti IL-1 and anti IL-12 and mitogen-activated protein kinase inhibitors. A single case report of successful treatment of severe PsA with autologous stem cell transplantation has been reported.

9.8 Arthritis of Inflammatory Bowel Disease (IBD)

This occurs in 10–20% of patients with Crohn's disease or ulcerative colitis and is similar to that of AS. Erythema nodosum, pyoderma gangrenosum, aphthous ulcer, clubbing and uveitis are the extraarticular manifestations in IBD.

NSAIDs are the treatment of choice. Sulphasalazine is effective in healing the bowel disease and pyoderma gangrenosum and reducing the arthritis. Local steroid injections and physical therapy are useful adjunctive measures.

9.9 Undifferentiated Spondyloarthropathy

This includes a subset of patients with clinical and radiographic features of Spondyloarthropathy but who fail to meet the criteria for established diseases like AS, PsA, ReA, RS and arthritis associated with IBD. These patients are HLAB 27 +ve and have unilateral sacroilitis. Some of these cases may represent early stages of other specific spondyloarthropathy and need to be followed up to see whether they evolved into a specific spondyloarthropathy.

10. SCLERODERMA

Scleroderma is a chronic autoimmune multisystem disorder of unknown etiology characterized by fibrosis and microvascular injury in the affected organs. Thickening and tightness of the skin and subcutaneous tissue caused by excessive synthesis and deposition of extracellular matrix is the hallmark of the disease. The disease may be confined to the skin (localized) or it may be generalized (systemic sclerosis) when virtually all organ systems can be involved most importantly skin, blood vessels, lungs, kidneys, gastrointestinal tract and heart. Figures 7 & 8 depict some clinical features of scleroderma.

Scleroderma is a fascinating disease but it can be quite frustating disease to treat. Colchicine, D-penicillamine, DMSO, ketotifen, interferon, interavenous pulses of steroids, cyclosporine, azathioprine, methotrexate, chlorambucil, 5-FU, cyclophosphamide, minocycline, thalidomide, etanercept and recombinant human relaxin have all been tried as disease modifying drugs but the results have not been very promising.

D-penicillamine has been the most widely used drugs in diffuse systemic sclerosis. It has been shown to improve the skin thickness. Even 5 years survival was shown to improve in a retrospective study (Steen et al., 1984). However in another placebo-controlled study of high dose D-penicillamine (750–1000 mg/d)

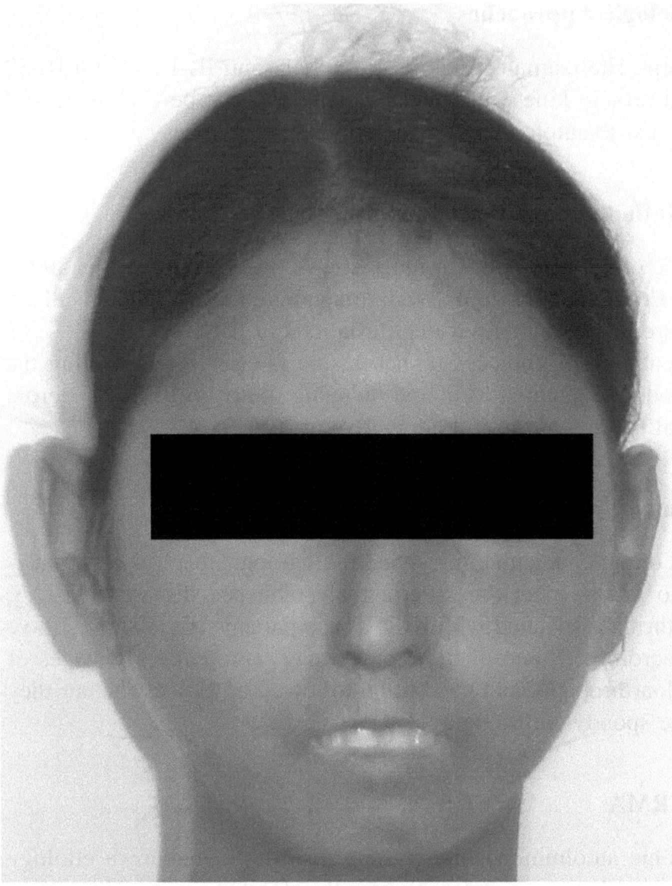

Figure 7. Tightness of skin, loss of wrinkles and facial expression

versus low dose D-penicillamine (125 mg alternate day) failed to show a difference in skin scores or mortality rates (Clements et al., 1997).

Although routinely used in many rheumatologic diseases, corticosteroids have been found to be counter productive in patients with diffuse disease, often precipitating an acute decline in renal function early in the course of the disease. However, corticosteroids may be helpful in certain symptoms such as articular symptoms non-responsive to NSAIDs or muscle inflammation. Short term therapy is recommended in these situations.

Raynaud's phenomenon is a reversible vasospasm of the digital arteries that can result in ischaemia of the digits. For mild to moderate cases, common sense measures such as avoidance of cold exposure, stress, caffeine and sympathomimetic decongestant medications (e.g. pseudoephedrin), abstinence from smoking and protective warm clothing may suffice. Non-selective beta-blockers and vasoconstrictive agents

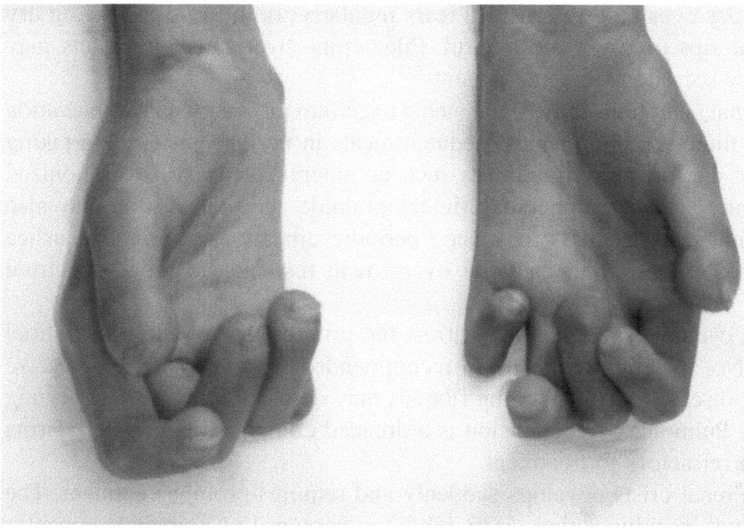

Figure 8. Flexion deformities, tightness of skin, bony resorption and shortening of distal bhalanx

such as ergot alkaloids, nicotine and amphetamine should be strictly avoided. Central body warmth induces peripheral vasodilatation. For severe Raynaud's phenomenon with digital infarcts/ulcers, vasodilator drugs such as nifedine (30–120 mg/d) are recommended. Intravenous infusion of a carboprostacycline (Iloprost) are often successful in refractory ischaemic ulcers (Ceru et al., 1997). Even oral iloprost has been shown to be beneficial (Black et al., 1998). The non-selective endothelin antagonist, bosentan, which has been approved for the treatment of pulmonary hypertension in scleroderma, may have a beneficial effect in digital ischaemia. Anti-platelet therapy, such as low-dose aspirin, can be helpful in preventing the sluggishly flowing blood from thickening and obstructing the partially occluded arterioles and capillaries. Local application of nitroglycerin ointment to the affected digit may improve local blood flow. Other therapies which have been tried include sildenafil, losartan, fluoxetine, pentoxifylline, stellate ganglion block, revascularization, cervical or digital sympathectomy. In addition, local treatment of ischaemic ulcers promotes healing. Occasionally, the debridement and parenteral antibiotics may be needed. Gangrene of distal digit may require surgical amputation.

Skin care is very important in this disease. Scleromatous skin is prone to dryness and pruritus. Excessive detergent use should be avoided. Hydrophilic ointments and oils are useful for dryness. Regular exercises maintain the flexibility and pliability of skin. There is no satisfactory treatment for calcinotic nodules, low-dose warfarin, probenecid and cardizen have all been tried.

Patient with dry eyes require artificial tears regularly. In those experiencing dry mouth, frequent sips of water are helpful. Pilocarpine hydrochloride pellets may increase salivary secretions in some patients.

Gastrointestinal symptoms may be amenable to certain measures such as elevation of head end of the bed, eating small, frequent meals in upright posture and taking an early dinner. Proton pump inhibitors such as omeprazole have revolutionized the management of reflux esophagitis. Metoclopramide and domperidon may also be useful. Esophageal strictures may need periodic dilatations. Chronic diarrhea due to small bowel stasis and bacterial overgrowth responds to broad spectrum antibiotic.

Steroids and cyclophosphamide may arrest the progression of active interstitial lung disease. No specific treatment is recommended for mild non-progressive interstitial lung disease. Advanced lung fibrosis may demand nothing short of lung transplantation. Pulmonary hypertension is a dreaded complication of scleroderma and tends to be refractory to treatment.

Scleroderma renal crisis develops suddenly and requires prompt treatment. The drug of choice is rapidly acting ACE inhibitor captopril. Angiotensin-receptor blockade does not appear to be as effective.

10.1 Experimental Therapy

High-dose immunosuppressive therapy followed by autologous stem-cell transplantation is being tried in scleroderma but is presently experimental.

11. POLYMYOSITIS AND DERMATOMYOSITIS

Polymyositis (PM) is an inflammatory myopathy that presents as weakness and occasionally tenderness of the proximal musculature. Diagnosis is corroborated by an abnormal electromyogram, elevated muscle enzymes (creatinine kinase, aldolase, AST) and muscle biopsy. Dermatomyositis (DM) is PM with a concomitant typical heliotrope rash.

The goal of treatment is to improve muscle strength and ameliorate the extramuscular manifestations (rash, dysphagia, fever). Prednisolone is the initial treatment of choice in a daily dose of 1–2 mg/kg body weight; higher dose is required in case of acute and severe disease. Improvement is usually apparent by 6–8 weeks and the high dose prednisolone should be continued for 12 weeks. In case of significant recovery of muscle power (and not muscle enzymes alone), prednisolone should be reduced at 5 mg/d at weekly intervals till a dose of 0.5 mg/kg/d and thereafter 5 mg/d every fortnight till the daily dose is reduced to 0.25 mg/kg. A maintenance dose of 0.15 mg/kg/d is continued for 6–9 months before reducing by 1 mg every month till it is discontinued.

If there is no improvement at the end of 12 weeks, the diagnosis needs to be reviewed, preferably with the pathologist to look for other metabolic or neurological diseases or the possibility of inclusion body myositis. If the diagnosis is confirmed, then addition of either azathioprine 2–3 mg/kg/d or methotrexate 7.5–15 mg weekly is helpful. These drugs are also helpful as alternatives in patients who do not tolerate the side effects of steroids. In severe cases there may be rapid deterioration at the initiation of therapy with acute respiratory failure or myocarditis. In some cases, IV methylprednisolone 20 mg/kg for 3–5 days can be life saving. In case of respiratory muscle involvement, intubation and ventilatory therapy may be required. IV immunoglobulins and cyclosporine have been useful in juvenile PM/DM. Patients with interstitial lung disease may benefit from aggressive treatment with cyclophosphamide. Plasmapheresis and leukapheresis are not effective in PM/DM.

REFERENCES

ACR Subcommittee on Rheumatoid Arthritis Guidelines. (2002) Guidelines for the management of Rheumatoid Arthritis. Arthritis Rheum, 46: No 2; 328–346.

Black, C.M., Halkier, S.L., et al. (1998) Oral iloprost in Raynaud's phenomenon secondary to systemic sclerosis- A multicentre, placebo-controlled dose comparison study. Br J Rheum, 37: 952–60.

Boers Met, et al. (1997) COBRA trial. Lancet, 350: 309–318.

Ceru, S., Pancreas, P., et al. (1997) Effects of five-day versus one-day infusion of iloprost on the peripheral microcirculation in patients with systemic sclerosis. Clin Exp Rheum, 15: 381–5.

Clements, P.J., Wong, W.K., Seibold, J.R., et al. (1997) High dose versus low-dose penicillamine in early diffuse systemic sclerosis: analysis of trial. Arthritis Rheum, 40: S354.

Daniel, O., Clegg, M.D., Domenic, J., Reda, Ph.D., Crystal, L., Harris, et al. (2006) Glucosamine, Chondroitin Sulfate, and the Two in Combination for Painful Knee Osteoarthritis (GAIT). NEJM 354, 795–808.

Darcy, S., Majka and Michael Holers V. (2003) Can We Accurately Predict the Development of Rheumatoid Arthritis in the Preclinical Phase? Arthritis Rheum, 48: 2701.

Esmeralda, T.H., Molenaar, Alexandre, E., Voskuyl, Huib, J., Dinant, P., Dick Bezemer, Maarten Boers and Ben, A.C. Dijkmans. (2004) Progression of Radiologic Damage in Patients With Rheumatoid Arthritis in Clinical Remission. Arthritis Rheum, 50: 36–42.

Is Remission the Mission in RA? (2005) New Information from the 2005 EULAR Conference. www.medscape.com/viewarticle/508075 10. Last accessed on 8th March, 2006.

Lard, L.R., Visser, H., Speyer, I., et al. (2001) Early versus delayed treatment in patients with recent onset rheumatoid arthritis: comparison of two cohorts who received different treatment strategies. Am J Med, 111: 446–451.

Lisukov, I.A., Sizikova, S.A., Kulagin, A.D., et al. (2004) High-dose immunosuppression with autologous stem cell transplantation in severe refractory systemic lupus erythematosus. Lupus 13:89–94. This series of six SLE patients treated with stem cell transplantation suggests that long-term, steroid-free remissions are possible, but early mortality is not uncommon.

Lorenz, H., Grunke, M., Wendler, J., et al. (2004) Effective treatment of active SLE-associated glomerulonephritis (GN) with 15-deoxyspergualin (15-DSG; gusperimus). Program and abstracts of the American College of Rheumatology/Association of Rheumatology Health Professionals 68th Annual Scientific Meeting; October 16–21; San Antonio, Texas. Abstract 1035.

Marc, D.C. (2005) The Clinical Manifestations of Chronic Hyperuricemia: Focus on Gout. www.medscape.com/viewarticle/496670_12. Last accessed on 8th March, 2006.

Mease, P.J. (2004) Psoriatic Arthritis Therapy. Curr Opin Rheumatol, 16.

Mok, C.C., Tong, K.H., To, C.H., et al. (2004) Tacrolimus for the initial treatment of diffuse proliferative lupus glomerulonephritis: a comparative study with intravenous pulse cyclophosphamide. Program and abstracts of the American College of Rheumatology/Association of Rheumatology Health Professionals 68th Annual Scientific Meeting; October 16–21; San Antonio, Texas. Abstract 1128.

Mottosen, T., et al. (1999) Finnish Rheumatoid Arthritis Combination Therapy trial. Lancet, 353: 1568–1573.

O' Dell, J.R., Haire, C.E., Erickson, N., et al. (1996) Treatment of rheumatoid arthritis with methotrexate alone, sulfasalazine and hydroxycholoroquine or combination of all three medications. N Engl J Med, 334: 1287–1291.

Paco, M.J., Welsing, Robert B.M., Landewé, Piet L.C.M., van Riel, Maarten Boers, Anke, M., van Gestel, Sjef van der Linden, Hilde, L., Swinkels and Dé sirée, M.F.M., van der Heijde. (2004) The Relationship Between Disease Activity and Radiologic Progression in Patients With Rheumatoid Arthritis: A Longitudinal Analysis. Arthritis Rheum, 50: 2082–2093.

Povelka, K., Gatterova, J., Olejarova, M., Machacek, S., Giacovelli, G. and Rovati, L.C. (2002) Glucosamine sulfate use and delay of progression of Knee osteoarthritis: a 3 year, randomised, placebo-controlled, double blind study. Arch Intern Med, 162: 2113–23.

Reginster, J., Deroisy, R., Rovati, Lee, R., Lejeune, E., Bruyere, O., et al. (2001). Long term effects of glucosamine sulfate on osteoarthritis progression. Lancet, 357: 252–6.

Reilly, P.A., Cosh, J.A., Maddison, P.J., et al. (1990) Mortality and survival in rheumatoid arthritis: a 25 year prospective study of 100 patients. Ann Rheum Dis, 49: 363–369.

Saskia ten Wolde, Jo Hermans, Ferdinand, C. and Breedveld and Ben, A.C. Dijkmans. (1997) Effect of resumption of second line drugs in patients with rheumatoid arthritis that flared up after treatment discontinuation. Ann. Rheum. Dis. Apr, 56: 235–239.

Steen, V.D., Medsger, T.A. and Rodnan, G.P. (1984) D-Penicillamine therapy in progressive systemic sclerosis. Ann Intern Med, 97: 652–9.

Thomson, G.T. (2000) Clinical efficacy of mesalamine in the treatment of spondyloarthroathy. J Rheumatol, 27: 714–8.

Traynor, A.E., Barr, W.G., Rosa, R.M., et al. (2002): Hematopoietic stem cell transplantation for severe and refractory lupus. Analysis after five years and fifteen patients. Arthritis Rheum 2002, 46: 2917–2923.

Williams, H.J., Ward, J.R., Reading, J.C., et al. (1992) Comparison of auranofin, methotrexate and combination of both in the treatment of rheumatoid arthritis. Arthritis Rheum, 35: 259–269.

CHAPTER 8

RECENT DEVELOPMENTS IN THE TREATMENT OF DIABETES TYPE 2

JAN O. NEHLIN

Department of Clinical Immunology, Odense University Hospital & University of Southern Denmark, 5000 Odense, Denmark

Abstract: Diabetes type 2 (T2DM) is a life-long metabolic disease that develops commonly in adulthood as a consequence of an unhealthy life style and genetic predisposition. T2DM is the most common form of diabetes, resulting from both insulin resistances in target organs and insufficient insulin production from pancreas beta cells. T2DM is characterized by increased plasma glucose and insulin levels as well as dyslipidemia. If left untreated chronic diseases will develop that result in a higher mortality risk.

The prevalence of type 2 diabetes worldwide has increased dramatically in recent times in part due to changes in diet and physical activity levels. Also, several genes underlying monogenic forms of diabetes as well as polymorphic variants have been identified that can contribute to the etiology of the disease.

A number of treatment strategies exist for T2DM that tackle several of the symptoms. Anti-obesity drugs and PPAR agonists are likely to become efficient pharmacological remedies to prevent further health problems in individuals with T2DM

Keywords: diabetes type 2; obesity; hyperinsulinemia; dyslipidemia; hyperglycemia; PPAR agonists; thiazolidinediones

1. INTRODUCTION

More than 150 million people worldwide suffer from T2DM, also known previously as non-insulin dependent diabetes mellitus (NIDDM). The common problem facing T2DM-afflicted individuals daily is that they are unable to produce sufficient amounts of insulin to stop the rise in blood glucose levels after food intake. T2DM can be defined as a state with hyperglycemia due to insulin resistance (see below) and relative insulin deficiency, showing a heterogeneous group of conditions (English and Williams, 2001; Kahn et al., 2005).

S. Rattan and M. Kassem (eds.), Prevention and Treatment of Age-related Diseases, 133–157.

It is expected that the number of T2DM cases will double within the next 25 years according to the World Health Organization (WHO), representing an enormous social and economic burden. It is generally believed that the dramatic surge of new T2DM cases in some Asian countries correlates with a sudden change in habits, by adopting a Western-like lifestyle characterized by physical inactivity and consumption of energy-rich foods, etc. (Yach et al., 2006; WHO, 2006). The prevalence of T2DM is highest (up to 50 percent) in American Indians and in South Pacific islanders, populations that evolved to survive caloric deprivation but who are now affluent and obese (Press, 2002).

Obesity is considered as a major T2DM risk factor evidenced by a strong correlation between the Body Mass Index (BMI) and T2DM incidence. There has been a marked increase in the percentage of overweight and obese individuals in the American population judged by the BMI index (CDC, 2006). A BMI above 25 kg/m^2 is a risk indicator of T2DM incidence. Risk factors that also can predict obesity include the individual's waist circumference (abdominal fat), physical inactivity, high-blood pressure and a high-fat diet (Wild et al., 2004).

Insulin is a hormone normally made by β (beta) cells in the pancreas whose major role is to promote the conversion of excess blood glucose, into glycogen, a stored form of energy (Kulkarni, 2004). Glycogen is important for providing rapid movement to muscles and maintaining blood glucose levels during fasting (liver glycogen). Excess glycogen can be converted into stored fat in the form of tryglycerides within fat cells (adipocytes) and released as free fatty acids. Excessive accumulation of adipose tissue leads to obesity. Among the other functions of insulin are to stimulate glucose transport into cells by enhancing glucose transporters activity (i.e. GLUT4), glycolysis, glucose oxidation, lipogenesis, and many other processes (Speight and Holford, 1997) (see Figure 1).

T2DM can be divided into stages or phases according to the level of function of pancreatic β-cells. In the first stage of T2DM, a defect(s) primarily in the β cell lead(s) to a drop in insulin levels and the inability to metabolize the excess levels of blood glucose (hyperglycemia). The inability to stimulate sufficiently the cellular uptake of glucose is known as "insulin resistance" that is usually hard to diagnose, leading to a compensatory increase in the production of insulin (hyperinsulinemia). This stage 2 of T2DM can result in heart disease and many other illnesses. The impact of hyperinsulinism has been dubbed syndrome X (Reaven, 2005). Metabolic syndrome (MS) or syndrome X often refers to multiple related clinical disorders including insulin resistance, abdominal obesity, hypertension, a variety of blood sugar abnormalities, high blood levels of triglycerides (hyperlipidemia) and low HDL cholesterol that are risk factors for cardiovascular disease. The metabolic syndrome is an increasingly prevalent disease in industrialized societies (Wang et al., 2003; Kahn et al., 2005).

Eventually, the insulin-producing β cells fail to overcome the defect(s), resulting in a drop in insulin levels leading to impaired glucose tolerance. This pre-diabetic stage 3 of T2DM is often diagnosed through an oral glucose tolerance test (OGGT) and by symptom questionnaires, although measurement of fasting plasma glucose

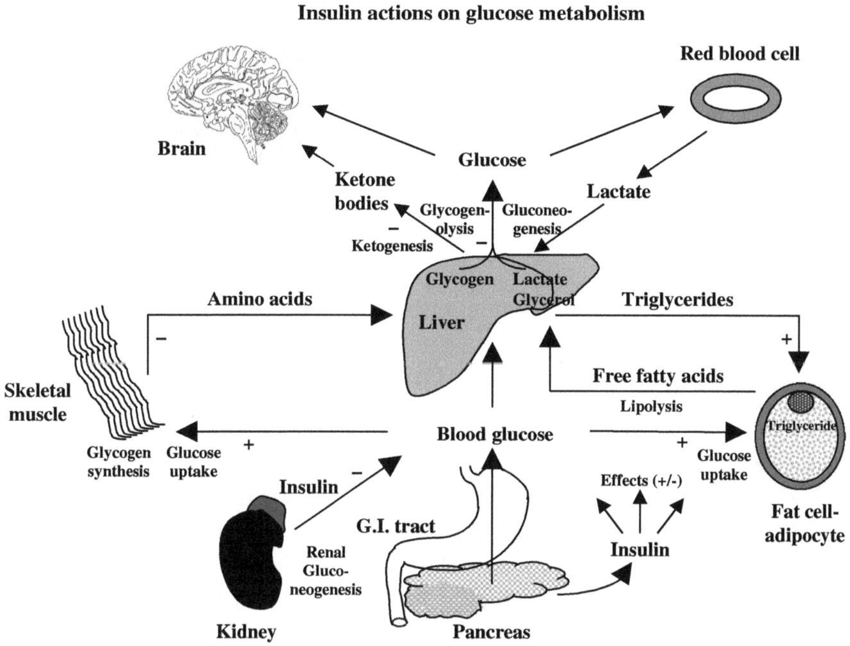

Figure 1. Roles of insulin in energy metabolism. Insulin has multiple actions including activation of glucose uptake into muscle and adipose tissue, activation of glycogen and tryglyceride synthesis, inhibition of lypolysis in adipocytes, inhibition of ketogenesis, glycogenolysis and gluconeogenesis in liver, and inhibition of proteolysis in muscle. Activation by insulin is denoted with (+) and inhibition with (−). Adapted from Taylor, 1999

is the first choice. The insulin disorder affects the blood sugar's response to orally administered glucose, showing e.g. blood glucose rises higher than the "normal" level generally considered 160 mg/dl. Patients commonly exhibit symptoms such as carbohydrate cravings, voracious hunger, excessive tiredness, fluctuation in mood and energy levels that are relieved by food or caffeine (Atkins, 2001; English and Williams, 2001).

In stage 4 T2DM, the chronic hyperglycemia persists generally throughout the day with the underlying insulin resistance and hyperinsulinemia still present. Hyperglycemia is associated with long-term damage to various organs, particularly the retina, nerves and kidney. The elevated plasma triglycerides lead to atherosclerosis and cardiovascular disease, which is the major cause of death. Overall life expectation is reduced between 5–10 years. Only when the levels of insulin fall below subnormal levels, insulin supplements and analogues can be administered (stage 5 T2DM) (Atkins, 2001; English and Williams, 2001).

The clinical symptoms of T2DM are various such as frequent urinating (polyuria), thirst, infections in the urinary tract, polydipsia (swelling of the eye lens – blurring of vision), itching in the legs, difficult healing of small wounds, a random venous plasma glucose concentration of ≥ 11.1 mmol/l or a fasting plasma glucose concentration (FPG) of ≥ 7.0 mmol/l (whole blood ≥ 6.1 mmol/l) or 2 h plasma glucose concentration of ≥ 11.1 mmol/l two hours after an oral load of 75 g anhydrous glucose in an OGTT (English and Williams, 2001; Sorkin et al., 2005; WHO, 2006).

It appears that β-cell dysfunction is the primary cause of T2DM due to an insufficient insulin secretion that prevents the concentration of glucose from being within the normal physiological levels. Below, some of the possible causes of such dysfunction are explained briefly.

2. GENETIC AND ENVIRONMENTAL CAUSES OF T2DM

In order to understand what strategies would be best suited to target the different clinical stages of T2DM, whether it is in its early phases or in its more advanced stage, it is necessary to know more precisely what metabolic functions are defective to be able to administer the most convenient treatment. Even though several of the clinical symptoms overlap between T2DM patients, the precise defects may be different and remain largely unknown without appropriate diagnostic or genotyping tools (see below).

Several studies show strong associations between genetic defects and an increased risk of T2DM. The genetic factors may be divided in two groups: monogenic and polygenic. Monogenic forms of T2DM are a consequence of rare mutations in a single gene, and are characterized by a high phenotypic penetrance and an early age of diagnosis. Many cases show a serious clinical picture including non-pancreas related health problems (Malecki, 2005). There are many examples of genes mutated in monogenic T2DM that affect metabolic functions (Barroso, 2005; Kahn et al., 2005; Stumvoll et al., 2005; Hansen and Pedersen, 2005).

Most T2DM cases are a consequence of polygenic factors whose defective function or functional inability becomes apparent in the presence of a particular lifestyle, often characterized by a high-fat, high-sugar diet and by the lack of physical exercise. Many examples have been reported whereby polymorphic variations are associated with an increase incidence or predisposition to T2DM (Kahn et al., 2005)

Based on findings of a given disease-risk allele it is not yet possible to predict an individual's risk to T2DM. Factors such as age ≥ 45 years, BMI ≥ 30, family history of T2DM, hypertension, dyslipidemia (hyperlipidemia), presence of cardiovascular disease, fasting glucose levels, etc. may be indicative (Diabetes Prevention Program research group, 2005).

Recent progress has been made in understanding the role of various genes in the pathogenesis of T2DM. Additional genes that influence the susceptibility to T2DM will undoubtedly be uncovered in the years to come. Once high-throughput and inexpensive genotyping becomes available, many T2DM susceptibility gene defects/variants will be rapidly identified and characterized leading eventually to the development of future individualized anti-diabetic drugs with higher efficacy and fewer side effects (Hansen and Pedersen, 2005; Kahn et al., 2005).

3. CLINICAL MANAGEMENT OF T2DM AND ITS COMPLICATIONS

Clinical management of T2DM includes lifestyle intervention by means of advisories about exercise and diet, as well as the use of oral or injected hypoglycemic agents. The responsiveness of T2DM patients to pharmacologic therapy varies between individuals due to variability in the clinical course of the disease. Depending on the symptoms and the severity of T2DM in a given patient, insulin and adjunctive therapies may be conferred to treat various consequences of T2DM progression (Table 1).

A review of the pharmacological interventions to date to delay or prevent the onset of T2DM was presented by Padwal et al., 2005. Ten studies of oral hypoglycemic agents and 15 studies of non-oral hypoglycemic agents were analyzed. Studies involving metformin, acarbose, troglitazone and orlistat (see below) showed a decrease in the incidence of T2DM compared with placebo. Genuine diabetes

Table 1. Present therapies in the treatment of T2DM

- **Oral hypoglycemic agents**
  - Biguanides: Metformin
  - Sulphonylureas
  - Nateglinide (D-Phenylalanine)
  - Meglitinide family
  - α-glucosidase inhibitors
  - GLP-1, imidazolines, morpholinoguanidines, etc.
  - Thiazolidinediones
- **Insulin**
- **Adjunctive therapies**
- ***Anti-hypertensive agents***
  - ACE inhibitors
  - Calcium channel blockers
  - α- and β-blockers
  - Angiotensin II receptor antagonists
  - Thiazide diuretics
- ***Lipid lowering agents***
  - Hydroxymethyl glutaryl CoA-reductase inhibitors (statins)
  - Fibrates
- ***Anti-obesity drugs***
  - Lipase inhibitors
  - Serotonin and norepinephrine re-uptake inhibitors
  - Phertermine, other noradrenergic drugs, etc.

Adapted from English and Williams, 2001; Kahn et al., 2005.

prevention studies are being sought and no single agent can be recommended at present for diabetes prevention (Anderson, 2005; Padwal et al., 2005).

A summary of T2DM prevention studies were reviewed by Laakso (2005) and Stumvoll et al., 2005. Therapeutic approaches to T2DM have recently been addressed (Stumvoll et al., 2005; Kahn et al., 2005). Drastic changes in lifestyle seem at the moment the most efficient strategy to tackle the T2DM epidemic. Lifestyle intervention in patients with impaired glucose tolerance results in an impressive reduction in the conversion to overt diabetes, which is greater than the effect of early intervention with drugs such as metformin or acarbose (Hauner, 2004).There are clear benefits between exercise, consumption of a diet rich in fruits and vegetables and the risk of getting T2DM. Daily exercise and a low-glycemic-index nutritional plan have been suggested as palliatives (Anderson, 2006). The progression of T2DM disease stages is preventable through a complete change in dietary habits, replacing refined carbohydrates and sweets, consuming less alcohol, avoiding a sedentary behavior and smoking, and observing proper weight control (Schulze and Hu, 2005). Nutrition therapy guidelines can be found in Kahn et al., 2005 and Shils et al., 2006. A number of general T2DM treatment guidelines can also be found in Lebovitz, 2005 and in Kahn et al., 2005.

Below, potential treatments for individual symptoms are presented such as obesity, hyperglicemia, hyperlipidemia, hypertension and lack of insulin. As with all medications, there might be side effects, and interactions with other drugs taken simultaneously could be harmful. Therefore, the use of T2DM drugs must be prescribed by a physician. Further information about these and other drugs can be found at www.nlm.nih.gov/medlineplus/druginformation.html, www.rxlist.com/, www.drugs.com/, www.drugdigest.org, at the manufacturer's homepages, Kahn et al., 2005, etc.

The thiazolidinediones in the treatment of T2DM will be reviewed separately in section 4.

3.1 Anti-Obesity Agents

Obesity is a state of increased body weight, specifically adipose tissue that can lead to health problems. Obesity develops only if energy intake (feeding) chronically exceeds total body expenditure (Spiegelman and Flier, 2001). Obesity is currently defined using BMI; its etiology and management is described in Hill et al., 2005 and in other chapters in Kahn et al., 2005 and in Shils et al., 2006.

Obesity is the most important modifiable risk factor for T2DM and most patients with diabetes are overweight or obese. It is well known that excess bodyweight induces or aggravates insulin resistance, which is a characteristic feature of T2DM. Thus, bodyweight plays a central role in the prevention and treatment of diabetes.

Agents to treat obesity consist of a) central nervous system agents that affect neurotransmitters or neural ion channels and b) leptin/insulin/central nervous system pathway agents. Efficient anti-obesity drugs must not only reduce fat mass (adiposity) but must also correct fat dysfunction (adiposopathy) (Bays, 2004). FDA approved medications currently available for the treatment of obesity in the elderly is listed in Mathys, 2005. Sibutramine (e.g. Meridia from Abbot) and Orlistat (e.g. Xenical from Roche) are drugs currently approved for the long-term management of obesity.

Orlistat prevents gastrointestinal lipases from breaking down dietary fats into smaller molecules that can be absorbed by the body. Absorption of fat is decreased by about 30 percent. Since undigested triglycerides are not absorbed, the reduced caloric intake may have a positive effect on weight control (FDA, 2006). Orlistat reduces the incidence of T2DM in patients with impaired glucose tolerance and lowers the required dose of metformin, sulfonylureas and insulin in patients with T2DM (Kiortsis et al., 2005).

Sibutramine is used as a short-term supplement to diet and exercise for the treatment of obesity (>30 BMI). Sibutramine works to suppress the appetite primarily by inhibiting the reuptake of the neurotransmitters norepinephrine and serotonin, and promotes thermogenesis (FDA, 2006; Filippatos et al., 2005a; Kaplan, 2005). Rimonabant (i.e. Acomplia from Sanofi-Aventis) is a novel anti-obesity drug in late clinical trials that targets the endocannabinoid receptor CB1 in the brain and in other tissues such as fat, and appears to improve dramatically insulin sensitivity, and blood cholesterol and lipid levels (Despres et al., 2005).

The use of phentermine (e.g. Ionamin from Medeva Pharmaceuticals, Adipex-P from Gate Pharmaceuticals) has also been documented to have some anti-obesity effects. It acts on the central nervous system, affecting either adrenergic or serotoninergic neurotransmission or both. Other related drugs are diethylpropion, phendimetrazine, benzphetamine, and the amphetamines (Kahn et al., 2005).

The increase of energy expenditure by means of enhanced mitochondrial activity could improve metabolism especially in obese T2DM patients. Several therapeutic targets to modulate energy expenditure to treat T2DM have been proposed (Auwerx, 2006). Several other drugs to treat obesity are currently in the development phase (Wadman, 2006). An overview of treatment methods including diet/nutrition, pharmacotherapy, behavioural therapy, weight management, exercise and surgery are revisited in Kahn et al., 2005 and Shils et al., 2006.

Treatment of childhood obesity depends on the severity of the cases but it starts with lifestyle intervention with includes a change of diet, an increase in physical activity and a decrease in sedentary behaviors. Low-energy diets, pharmacological agents and even surgery may be used in severe cases (Steinbeck, 2005; Kahn et al., 2005). Guidelines for the treatment of adolescent obesity have been put forward (Durant and Cox, 2005; Kahn et al., 2005).

The most appropriate recommendation for obese patients with T2DM is a nutritionally balanced, moderately hypocaloric diet with a reduced intake of saturated fat and an increase in physical activity. The control of appetite and suppression of food cravings would also lead to better weight control. Finally, no conclusive evidence has been presented to claim that nutritional supplements or botanicals might have a significant role in weight reduction (Dwyer et al., 2005) but there are many examples of nutritional healing based on the use of natural hypoglycemic substances (Friedman and McLellan, 2006).

3.2 Hypoglycemic Agents

Here are summarized some of the most popular pharmacological interventions used to ameliorate the symptoms of T2DM (English and Williams, 2001). A list of compounds to treat several symptoms of T2DM can be found in Table 1. The action of a group of drugs known as thiazolidinediones will be described in detail in section 4 (see below).

The anti-hyperglycemic drug metformin hydrochloride (e.g. Glucophage from Merck, Fortamet from FHRX, Riomet from Ranbaxy) may be the most usual drug given to overweight or obese subjects with T2DM whose diabetes cannot be controlled by diet alone. Presently, metformin is perhaps the first therapeutic option in the treatment of T2DM, as it may prevent some vascular complications, and mortality. Metformin confers a good control of hyperglycemia, while it only results in moderate effects on weight, lipids, insulinemia and diastolic blood pressure (Lebovitz, 2005). Agents such as the sulphonylureas, the α-glucosidase inhibitors, the thiazolidinediones, the meglitinides, insulin, and diet fail to show more benefit for glycemia control, body weight, or lipids, than metformin (Saenz et al., 2005). Metformin decreases hepatic gluconeogenesis and hepatic glucose output, and increases peripheral glucose uptake, reducing plasma glucose by 3–4 mmol/l. Metformin can inhibit complex 1 of the mitochondrial respiratory chain. However, metformin is not an insulin secretagogue and it does not result in meaningful hypoglycaemia (Owen et al., 2000). Metformin is considered an insulin sensitizer since administration to T2DM-patients results in a decrease in the hepatic insulin resistance (Lebovitz, 2005).

Combination therapies that include metformin consist of metformin and glyburide, a sulphonylurea, such as Glucovance from Merck, metformin and rosiglitazone such as Avandamet from GSK, metformin and glipizide such as Metaglip from Merck, etc. (FDA, 2006; web drug sources). The anti-hyperglycemic effects of agents with different modes of action are additive (Kahn et al., 2005).

The sulphonylureas are insulin segregatogues (stimulators of insulin secretion) that bind to the sulphonylurea receptor (SUR-1) on the K^+-ATP channel on the membranes of β-cells, closing it, and then trigger opening of Ca^{++} channels, increasing intracellular calcium concentration that stimulates insulin release (Speight and Holford, 1997; Lebovitz, 2005). They generally reduce plasma glucose levels by 3–4 mmol/l and are more effective in newly T2DM diagnosed patients (English and Williams, 2001). Several sulphonylureas have been described such as Glibenclamide/Glyburide (e.g. Diabeta from Aventis, Glynase and Micronase from Pfizer), Glipizide (e.g. Glucotrol from Pfizer), chlorpropamide (e.g. Diabenese from Pfizer), gliclazide (e.g. Diamicron from Servier), glimepiride (e.g. Amaryl from Aventis), tolazamide (e.g. Tolinase from Pfizer), tolbutamide (e.g. Orinase from Pfizer), etc.

An insulin segregatogue that does not contain a sulphonylurea moiety is nateglinide. Nateglinide (Starlix from Novartis-Ajinomoto) is a D-phenylalanine derivative that is used as a novel anti-diabetic agent. It also can inhibit the pancreatic β-cell K^+-ATP channel and reduces glucose levels that are inadequately controlled by diet and exercise in T2DM patients (Lebovitz, 2005).

The meglitinides are hypoglycemic agents (non-sulphonylureas) to treat T2DM that are used in mono-therapy or in combination with metformin. These drugs stimulate the pancreas to release insulin, concentrating their effects around meal time glucose load, leveling off spikes in blood sugar levels. An example is Repaglinide (Prandin-Novonorm from Novo Nordisk) (Lebovitz, 2005; web drug sources).

The inhibitors of α-glucosidases, insulin sensitizers commonly known as "starch blockers" are based on the inhibition of intestinal enzymes that participate in the degradation of disaccharides, oligosaccharides and polysaccharides, leading to a dose-dependent delay of carbohydrate digestion, and the glucose released from these molecules enters bloodstream more slowly reducing the glycemic fluctuations during the day. Among the inhibitors of α-glucosidases it is possible to name acarbose, a pseudo-tetrasaccharide (e.g. Precose and Glucobay from Bayer), miglitol (e.g. Glyset from Bayer) and voglibose (e.g. Volix from Ranbaxy) (Lebovitz, 2005; FDA, 2006; web drug sources).

Glucagon-like-peptide-1 (GLP-1) is a hormone secreted by intestinal cells in response to fat and carbohydrate ingestion. GLP-1 and its derivatives (e.g. liraglutide from Novo Nordisk, exenatide-Byetta from Amylin) are sought as T2DM therapeutic agents, adjunct to metformin and/or sulphonylureas. They can stimulate glucose-dependent insulin production and secretion from pancreatic β-cells, as well as the growth and differentiation of β-cells. They are considered as adyuvant therapies in severe forms of T2DM (Sturis et al., 2003; List and Habener, 2004; Gallwitz, 2005).

It appears that combination therapies rather than mono-therapies work more efficiently to reduce plasma glucose levels (Lebovitz, 2005). For several of the available hypoglycemic agents there are generic drugs already available.

3.3 Insulin, Hypolipidemic and Anti-Hypertensive Agents

Failure to respond to oral hypoglycemic agents initially (primary failure) is often due to a severe underlying insulin deficiency. In this situation, insulin therapy is required. However, many patients may not respond to oral agents because of severe initial hyperglycemia. Thus, a much higher dosage of oral anti-diabetic agents may be needed initially, and then reduced subsequently, sparing these particular patients the unnecessary insulin treatment. Among insulin preparations given to T2DM patients with insulin deficiency are the short-acting, the intermediate-acting and the long-acting insulins as well as the insulin analogues. Examples are: Iletin, Humulin and Humalog from Lilly, Lantus from Aventis, Novolin, Novolog and Levemir from Novo Nordisk, etc. (web drug sources).

Angiotensin-converting enzyme (ACE) inhibitors and some angiotensin II receptor blockers (ARBs) may improve insulin sensitivity and decrease the risk for T2DM. One ARB in particular, telmisartan, has been found to effectively activate PPARgamma, a well-known target for insulin-sensitizing, anti-diabetic drugs (see below) (Kurtz and Pravenec, 2004). Anti-hypertensive compounds used in the treatment of T2DM are reviewed by Asfaha and Padwal, 2005.

Diabetic patients have a higher incidence of vascular disease due to elevated plasma triglycerides occurring together with reduced high-density lipoprotein (HDL) cholesterol concentrations. Lipid-lowering treatments should be implemented aggressively in patients with existing clinical vascular disease (Kahn et al., 2005; Shils et al., 2006).

Hypocholesterolemic drugs include fibric acid derivatives (fibrates), bile acid sequestrants (Cholestyramine, Colestipol, Colesevelam), nicotinic acid preparations (Niacin), an intestinal absorption inhibitor (Ezetemibe), and the statins. In T2DM patients with particularly high triglyceride levels and lower levels of LDL-cholesterol, the fibrates should be considered as the initial therapy (Kahn et al., 2005).

The fibrates lower triglycerides and may increase HDL cholesterol levels. The fibrates are effective in lowering blood triglyceride levels to prevent heart disease. By reducing the production of serum triglycerides and increasing HDL cholesterol, they can also reduce the levels of LDL-cholesterol. Two classes of fibrates are commonly being used: fenofibrate (e.g. Tricor from Abbot, Antara from Reliant) and gemfibrozil. Gemfibrozil (e.g. Lopid from Pfizer) (see 4.1.1) is often prescribed early in patients with grossly elevated triglycerides (hypertriglyceridemia), which is associated with an increased risk of acute pancreatitis (Kahn et al., 2005, web drug sources).

Compounds of benefit in the primary prevention of cardiovascular disease (CVD) in T2DM patients (low-density lipoprotein (LDL)-cholesterol-lowering therapy) are the 3-hydroxy-3-methylglutaryl-coenzyme A (HMG-CoA) reductase inhibitors (statins) and aspirin (acetylsalicylic acid) (Hovens et al., 2005). Hypercholesterolaemia is treated by inhibiting the HMG-CoA reductase enzyme which is the rate-limiting step in cellular cholesterol biosynthesis. Cholesterol-reducing medications include statins such as lovastatin (e.g. Mevacor from Merck), pravastatin (e.g. Pravachol from Bristol-Myers-Squibb), simvastatin (e.g. Zocor from Merck), fluvastatin (e.g. Lescol from Novartis), atorvastin (e.g. Lipitor from Pfizer), rosuvastatin (e.g. Crestor from Astra Zeneca), etc.

4. THIAZOLIDINEDIONES TO TREAT DIABETES TYPE 2

The thiazolidinediones are insulin sensitizing agents (improve insulin action), which are able to correct to a certain extent the insulin resistance, hallmark of T2DM. Some early drugs launched into the market were troglitazone, rosiglitazone and pioglitazone (see 4.1.2 PPARγ) that interact primarily with the PPARgamma receptors (see below) to regulate metabolism. As a result of their action, a decrease in hepatic gluconeogenesis and hepatic glucose output takes place as well as an increase in glucose uptake in muscles and a reduction of fatty acid release from adipocytes which decreases insulin resistance. Among their clinical effects it is possible to observe a reduction of glycemia by 2–3 mmol/l, a reduction of glycated hemoglobin (HbA1c) and triglyceride levels, and an increase in HDL-cholesterol levels. Treatment with insulin-sensitizing drugs might be helpful to reduce the progression to both β-cell failure and macrovascular late complications.

Some documented adverse defects include hepatic toxicity, weight gain (subcutaneous fat), rise in LDL-cholesterol levels, fluid retention, heart toxicity and tumorigenicity. In present day therapies, they are often given in combination with metformin to obese subjects with T2DM (see 3.2) (FDA, 2006; English and Williams, 2001; Fajas et al., 2001).

To understand the role of thiazolidinediones in the treatment of T2DM it is important to revisit their targets and their mechanism of action within the cell.

4.1 PPAR family

The peroxisome proliferator-activated receptors (PPARs) alpha (α), gamma (γ) and delta (δ) are ligand-activated transcription factors that belong to the nuclear receptor super family. PPARs are widely recognized as molecular targets for drugs to treat T2DM, with important roles in the regulation of adipogenesis, lipid metabolism, cell proliferation, cell differentiation and inflammatory signaling (Etgen and Mantlo, 2003; Evans et al., 2004; Zhang et al., 2004b; Berger et al., 2005; Kota et al., 2005). The molecular mechanisms underlying the effects of insulin sensitizers in models of insulin resistance are presented in Jiang and Zhang, 2005.

The PPAR receptors form heterodimers with retinoid X receptor (RXR) upon ligand binding and recruit co-factor(s) to modulate expression of target genes by binding to specific peroxisome proliferators response elements (PPRE's). PPARs are activated by fatty acids and their derivatives, and other unknown endogenous ligands whereas RXR is activated by 9-cis retinoic acid. PPARs and RXRs can function independently, in the absence of a hetero-partner, to modulate gene expression, and PPARs can also bind non-PPREs containing genes (Tan et al., 2005). The transcriptional activity and gene specificity of nuclear receptors results from their interactions with co-activators or co-repressors providing the basis for a transcriptional switch to control complex programs of gene expression such as adipocyte differentiation (Puigserver, 2005).

In spite of intensive search for natural ligands, no truly endogenous PPAR ligand has been identified yet. Candidates include free fatty acids, lipid mediators in the arachidonate cascade and polyphenolic compounds such as resveratrol (see below).

The genetic contribution of each PPAR-member can be studied either by the use of pharmacological agents that mimic the effects of *in vivo* PPAR-ligands, and studying their effect on promoter-based reporter gene systems, or by analyzing the effects of overexpression or deletion of each PPAR-member in cells or in animal models.

The defined metabolic properties of each PPAR isotype suggest that the three PPAR isotypes complement each other in the pathophysiology of obesity and the metabolic syndrome. Thus, treatments aimed at targeting all three PPAR isotypes simultaneously are being addressed.

Below is presented a short summary of the role of each PPAR-family member and a summary of some of the recent advances in the treatment of T2DM by using PPAR-binding pharmacological agents.

4.1.1 PPAR-alpha

PPARα regulates hepatic fatty acid metabolism and mediates the effects of the lipid-lowering drugs known as fibrates (reviewed in 3.3). PPARα regulates the expression of genes involved in fatty acid beta-oxidation and is a major regulator of energy homeostasis (van Raalte et al., 2004). PPARα is expressed mainly in the liver, kidney and heart (Ferre, 2004). Two potential natural ligands of PPARα have been described, endocannabinoid oleylethanolamide (OEA) and palmitoylethanolamide (PEA) (Fu et al., 2003; Lo Verme et al., 2005). Oleoylethanolamide (OEA), the naturally occurring amide of ethanolamine and oleic acid, is an endogenous lipid that modulates feeding, body weight and lipid metabolism by binding with high affinity to PPARα (Lo Verme et al., 2005). A compound known as WY-14,643 is widely used as a standard agonist of PPARα.

PPARα knock-out mice were protected from the development of diabetes-induced cardiac hypertrophy while overexpression of cardiac-specific PPARα resulted in a more severe cardiomyopathic phenotype with myocardial long-chain triglyceride accumulation, insulin resistance and reduced cardiac function. If the mice were fed a diet enriched in triglyceride containing long-chain fatty acid the cardiomyopathic phenotype would worsen. This suggests that interventions aimed at lowering serum-lipid levels would be beneficial in the treatment of diabetic cardiomyopathy. Thus, PPARα is a critical regulator of myocardial fatty acid uptake and utilization (Park et al., 2005).

A major study (FIELD) investigating the effects of fenofibrate on cardiovascular disease in T2DM patients showed declines in total and LDL cholesterol (10%) and triglycerides (26%) and an increase in HDL cholesterol (6.5%) during a 6-week trial period (Scott et al., 2005).

In rodents, PPARα agonists increase peroxisome number and volume in conjunction with an increase in peroxisomal β-oxidation enzymatic activities, in addition to ω-oxidation activities by CYP4A enzymes in the smooth endoplasmic reticulum. PPARα agonist treatment of obese rats at high doses results in cellular proliferation and possibly tumour formation (Hoivik et al., 2004). PPARα activators can cause hepatocarcinogenic effects in animals and these effects increase with age. Hepatic tumors are found at a 5–7 fold higher rate in old rats than in young rats treated with such chemicals. This is possibly due to an increase in the oxidative damage exerted by PPARα agonists whilst the expression of antioxidant enzymes in the liver decreases with age (Youssef & Badr, 2005). However, cynomolgus primates are refractory to the pro-mitogenic effects of PPARα agonists (Hoivik et al., 2004).

The PPARα-inducing fibrates have been shown to have immunosuppressive effects and might have potential uses in inflammatory diseases (Cunard, 2005).

4.1.2 PPAR-gamma

PPARγ is a central molecule in obesity and diabetes. It is targeted by the anti-diabetic class of thiazolidinediones (TZD's) known as glitazones such as troglitazone (Rezulin, withdrawn due to liver toxicity), rosiglitazone (BRL49653) (Avandia from GSK) and pioglitazone (Actos from Takeda-Lilly). Treatments of T2DM patients with these potent and selective PPARγ compounds lower blood glucose and insulin levels, and improve insulin sensitivity by decreasing hepatic gluconeogenesis and hepatic glucose output, increasing glucose uptake in muscle, and reducing the release of fatty acids from adipocytes. Thus, TZD's maintain a functional and differentiated adipose tissue and promote lipid storage (Zhang, 2004b; Lazar, 2005; Hammarstedt et al., 2005; Kahn et al., 2005).

PPARγ is highly expressed in adipocytes (adipose tissue) but expressed at lower levels in skeletal muscle and liver, considered the major insulin-target tissues (Fajas et al., 2001; Ferre, 2004). This suggests that TZD's primary insulin sensitizing effect resides within the adipocyte (Zhang et al., 2004b). PPARγ functions primarily in the regulation of glucose homeostasis and adipocyte proliferation and differentiation, inducing genes involved in fatty acid and/or lipid metabolism, and glucose homeostasis (Rosen and Spiegelman, 2001; Rangwala and Lazar, 2004). In addition, PPARγ also has been implicated in anti-inflammatory, antiatherogenic, and antihypertensive effects (Fajas et al., 2001). TZD's are thought to offer protective effects on the cardiovascular system in patients with T2DM (Abdelrahman et al., 2005; Staels, 2005). Obesity and T2DM are associated with a mild, chronic inflammation. The levels of various cytokines, such as TNF-alpha and IL-6, are elevated in the adipose tissue during these conditions. Treatment with TZD's can inhibit cytokine expression and action (Hammarstedt et al., 2005).

PPARγ can be bound by natural ligands such as prostaglandin 15-deoxy-Δ 12,14-prostaglandin J_2 (PGJ_2), synthetic molecules such as the TZD's and certain non-steroidal anti-inflammatory drugs (references in Fajas et al., 2001; Scher and Pillinger, 2005).

Three isoforms of PPARγ exist in humans, γ1, γ2 and γ3, while in mouse there are only 2 (Fajas et al., 2001). Mice lacking both isoforms die in uterus. Heterozygous PPAR gamma-deficient mice were protected from the development of insulin resistance due to adipocyte hypertrophy under a high-fat diet. These phenotypes were abrogated by PPARγ agonist treatment (Kubota et al., 1999). Mice lacking PPARγ2 show impaired adipocyte differentiation and insulin sensitivity, with dramatic decreases in the expression of IRS1 and GLUT4 glucose transporter in the skeletal muscle (Zhang et al., 2004a).

Overexpression of PPARγ results in a large increase in glucose uptake in wild-type C2C12 skeletal muscle cells or in cells resistant to insulin (Nakamichi et al., 2003; Verma et al., 2004) but excess PPARγ in liver promotes adipocyte-specific gene expression and lipid accumulation (hepatic steatosis) (Yu et al., 2003). Thus, pharmacological overexpression of the muscle PPARγ gene in skeletal muscle might be a useful strategy for the treatment of insulin resistance (Verma et al., 2004).

Use of TZDs to treat hypertension-associated syndromes has been reviewed (Sacerdote et al., 2005). PPARγ is a potential new target for the treatment sepsis and inflammation (Zingarelli et al., 2005). The PPARγ agonist rosiglitazone can protect cells against apoptosis and increase mitochondrial potential and cell survival (Wang et al., 2002).

A specific PPARγ coactivator-1 alpha (PGC-1α) is involved in the coordination of gene expression that stimulates a thermogenic program in brown fat, which could become a target for T2DM drugs (Puigserver, 2005).

4.1.3 PPAR-delta

PPARδ controls fatty acid metabolism in several tissues such as skeletal muscle and adipose tissue, by regulating genes involved in fatty acid transport, β-oxidation, and mitochondrial respiration (Tanaka et al., 2003; Wang et al., 2003; Fredenrich and Grimaldi, 2005). PPARδ is ubiquituosly expressed, showing higher expression in the gut, kidney and heart (Ferre, 2004).

The overexpression of PPARδ or treatment of mice with selective PPARδ agonists showed that activation of PPARδ *in vivo* increases lipid catabolism in skeletal muscle, heart and adipose tissue and improves the serum lipid profile and insulin sensitivity in several animal models. PPARδ activation also prevented the development of obesity and improved cholesterol homeostasis in obesity-prone mouse models. Some concerns have been raised as regards to possible tumorigenic effects in gut tissue, but further investigations into PPARδ activation are worthwhile due to its promising cellular effects (Bedu et al., 2005). Also, PPARδ overexpression promoted an increase of muscle oxidative capacity, redistribution of fatty acid flux from adipose tissue to skeletal muscle, and a decrease in adipocyte size leading to a reduction of adipose mass. These results seem to validate the concept that PPARδ is a key component to help cells metabolically adapt to an excess consumption of saturated fat (Fredenrich and Grimaldi, 2005).

Muscle-specific PPARδ overexpression led to the increase of both enzymatic activities and genes implicated in oxidative metabolism. These changes in muscle were accompanied by a reduction of body fat mass, mainly due to a large reduction of adipose cell size. PPARδ plays an important role in muscle development and in the adaptive response to environmental changes, such as exercise training. These observations strongly support the idea that activation of PPARδ could be beneficial in prevention of metabolic disorders, such as obesity or T2DM (Luquet et al., 2003).

Selective PPARδ agonists are not yet available as pharmaceuticals. However, several PPARδ-selective agonist drugs are in the pipeline. One of them, GW501516 (from GSK), was shown to have a significant decrease in the levels of plasma triglycerides and LDL-cholesterol, and an increase in the levels of plasma HDL-cholesterol in obese Rhesus monkeys (Oliver et al., 2001). PPARδ agonists may play a beneficial role in the treatment of lipid disorders, in particular obesity (Zhang et al., 2004b; Luquet et al., 2005). Activation of PPARδ could be beneficial in prevention of metabolic disorders, such as obesity or T2DM.

4.2 Novel PPAR-activators

Novel compounds have been developed that selectively activate the human PPAR receptors, with improved potency and efficacy properties as compared to previously marketed insulin sensitizers such as fenofibrate and rosiglitazone. Single (α or γ or δ), dual (α/γ, α/δ ανδ γ/δ) and triple (α/γ/δ) agonists of PPAR receptors have been generated (Sauerberg et al., 2002, 2003, 2005; reviewed by Nehlin et al., 2006). Full dimeric ligands result in PPARγ agonists with retained or increased potency and have an altered PPAR subtype profile compared to monomeric counterparts. Dimeric design can be used to fine tune the selectivity of PPAR agonists (Sauerberg et al., 2003, 2005).

Targeting simultaneously all three PPAR isoforms with varying degrees of potency and efficacy could represent a viable therapy for the treatment of T2DM. Studies in obese animal models show a significant improvement of the insulin sensitivity (Nehlin et al., 2006).

Synthetic PPAR ligands often constitute better drug candidates than natural ligands due to improved pharmacokinetic properties such as enhanced metabolic stability and better oral bioavailability, better structure-function data, high potency and efficacy, possibility of triple action in one molecule (agonist to all three PPAR isoforms) (Mogensen et al., 2003), and a chemical synthesis process well understood.

Dual PPARα/γ agonists (in development) combine the properties of thiazolidinediones and fibrates, and they hold considerable promise for improving the management of T2DM and providing an effective therapeutic option for treating cardiovascular disease and the metabolic syndrome. Many clinical trials involving PPAR agonists show their therapeutic potential (Staels and Fruchart, 2005).

Muraglitazar (co-developed by Bristol-Myers Squibb and Merck) is a non-thiazolidinedione, oxybenzylglycine dual PPARα/γ agonist for the potential treatment of T2DM and other metabolic disorders. Treatment of hyperglycemic db/db mice resulted in dose-dependent reductions of glucose, insulin, triglycerides, free fatty acids, and cholesterol levels. In addition to its anti-diabetic effects, it preserved pancreatic insulin content, and improved several metabolic abnormalities (Harrity et al., 2006). Muraglitazar is in advanced clinical development for the treatment of T2DM and its associated dyslipidemia. The clinical data on the efficacy and safety of muraglitazar in patients with T2DM is summarized in Cox, 2005.

The combined used of fibrates and insulin sensitizers results in a decrease in the insulin resistance, with reduced blood glucose and triglyceride levels.

4.3 Other uses of PPAR agonists

The use of TZDs is largely limited to the treatment of patients with diabetes, but mounting evidence suggests that TZD's with varying potency and selectivity for the PPAR family may be beneficial to treat other clinical disorders such as hypertension, sepsis, inflammation, immunosuppression, etc. (see 4.1).

PPAR activity and the function of the coactivator PGC-1α can be linked with aging and longevity (Corton and Brown-Borg, 2005). PPAR-independent effects of TZD's on mitochondrial metabolism also have been described (Feinstein et al., 2005).

Finally, future thiazolidinediones could be designed to restore normoglycemia in T2DM individuals with a defective PPAR regulatory system (Stumvoll et al., 2005).

5. OTHER PRESENT AND FUTURE THERAPIES

A number of different strategies are being examined that aim at treating/curing T2DM either by pharmacological means or by gene/cellular therapy. New generation pharmacological agents with improved efficacy and potency will be generated in the years to come, especially suited to treat specific pathogenic defects in T2DM individuals.

5.1 Pharmacological therapies

5.1.1 Combination therapies

A recent study concluded that the combination of orlistat and micronised fenofibrate appears to be safe and may further improve metabolic parameters in overweight and obese patients with metabolic syndrome compared with each monotherapy (Filippatos et al., 2005b). The effects of combination therapies using oral antihyperglycemic agents are presented in Lebovitz, 2005. A triple combination therapy with insulin aspart (a rapid-acting insulin analog) at meals, metformin (which improves hepatic insulin sensitivity), and rosiglitazone (which improves peripheral insulin sensitivity) significantly improved glucose metabolism in T2DM patients (Poulsen et al., 2003). Additional clinical trials combining different therapies will be of great value to improve existing therapies to treat T2DM patients.

5.1.2 Regulators of gluconeogenesis and glycolysis

Fructose 1,6 bisphosphatase inhibitors to control gluconeogenesis could represent a new class of drugs to treat T2DM (Erion et al., 2005). Leptin could be a potent antidiabetic drug in cases of T2DM that are not leptin resistant. Leptin enhances hepatic insulin responsiveness through decreasing gluconeogenesis (Toyoshima et al., 2005). The FOXA2 transcription factor regulates genes involved in fatty acid oxidation, ketogenesis and glycolysis, improving insulin resistance in mouse models (Puigserver and Rodgers, 2006).

5.1.3 GLP-1 pathway

There are now several compounds at different stages of pre-clinical or clinical development for the treatment of T2DM that utilize the GLP-1 signalling pathway; these include GLP-1 receptor agonists with extended half-lives, and inhibitors of DPP-IV that increase circulating levels of endogenous, intact and bioactive GLP-1 (Rotella et al., 2005; Gallwitz, 2005).

5.1.4 PTP1B antagonists

Protein tyrosine phosphatase 1B (PTP1B) is a negative regulator of insulin signalling, and inhibition of its activity has been shown to enhance insulin action in pre-clinical models (Stumvoll et al., 2005).

5.1.5 CNS control of glucose production

Hypothalamic K_{ATP} channels seem to be involved in the regulation of glucose production from the liver and in mechanisms leading to obesity-induced T2DM (Seeley and Tschop, 2006).

5.1.6 Adipocytokines

Adipocytes secrete a variety of bioactive molecules called adipokines (adipocytokines), including TNFα, IL-6, leptin, adiponectin, resistin, etc. Adiponectin receptor agonists and adiponectin sensitizers could serve as versatile treatment strategies for obesity-linked diseases such as T2DM and the metabolic syndrome (Kadowaki and Yamauchi, 2005; Stumvoll et al., 2005). Controlling excessive secretion of TNF or interleukin 6 or blocking their action mediated by serine/threonine kinases would be expected to enhance insulin sensitivity in patients with visceral adiposity (Stumvoll et al., 2005). Resistin, a cysteine-rich protein has been implicated in the pathogenesis of obesity-mediated insulin resistance and T2DM (Kusminski et al., 2005).

5.1.7 Ghrelin antagonists

Ghrelin, an endogenous ligand for growth hormone secretagogue receptor (GHS-R), is an appetite stimulatory signal from the stomach. Antagonists to the ghrelin receptor could be useful as T2DM agents since they contribute to reduce food intake and further weight gains in mice (Asakawa et al., 2003).

5.1.8 Salicylates

Antiplatelet treatment (generally aspirin) decreases the risk of atherosclerotic manifestations. The salicylates can ameliorate insulin resistance by interfering with the inflammatory cascade in insulin signalling (Stumvoll et al., 2005).

5.1.9 Anti-ER stress therapies

Mutations in XBP-1, a mediator of endoplasmic reticulum (ER) stress, cause insulin resistance in mice. Obesity has been shown to increase the level of ER stress. It has therefore been suggested that the ER stress response might be worth targeting by pharmacological interventions to prevent T2DM (Ozcan et al., 2004; de Luca and Olefsky, 2006). Also, mitochondrial defects appear to have an important role in insulin resistance and pancreatic β-cell dysfunction (Lowell and Shulman, 2005). Increased β-cell apoptosis contributes to the onset of T2DM. Insulin receptor substrate 2 (IRS-2) promotes β-cell growth and survival and when inhibited contributes to insulin resistance (Rhodes, 2005).

5.1.10 Islet cell mass inducers

The design of compounds to boost islet mass and function in diabetic patients are sought, since β-cells are lost during T2DM pathogenesis. An inducer of progenitor cell differentiation to generate endocrine cells has been described (Kojima and Umezawa, 2006).

5.1.11 Klotho

Overexpression of Klotho, a hormone that can induce insulin resistance, prolongs life in mouse models, possibly by reducing lipid overload and lipotoxicity (Unger, 2006).

5.2 Cellular/genetic therapies

5.2.1 Islet transplantation

Transplantation protocols have been developed over the years that can successfully restore pancreas functionality. However, two major problems are the lack of donor material and the need for chronic immunosuppression. To circumvent these difficulties, genetically modified animals, such as transgenic pigs expressing human genes have been developed (references in Hakim et al., 2002).

5.2.2 Stem cell therapies

Since purified islets are scarce the possibility of using renewable stem cells for organ or tissue transplantation appears to be a realistic alternative. Insulin-secreting cells have been obtained from undifferentiated embryonic stem cells and transplanted into mouse models to correct hyperglycemia. This opens the way for future treatment of T2DM (Soria et al., 2005).

5.2.3 Gene therapy

Gene and cell therapy has been used to induce tolerance to auto- and alloantigens and to generate the tolerant state in autoimmune rodent animal models of Type 1 diabetes mellitus (T1DM) or in rodent recipients of allogeneic/xenogeneic islet transplants. Examples include viral vector-mediated gene transfer of immunosuppressive cytokines, proteins that block co-stimulation and molecules that prevent apoptotic cell death.

The achievements of gene and cell therapy in T2DM are less evident, but seminal studies promise that this modality can be relevant to treat and perhaps prevent the underlying causes of the disease including obesity and insulin resistance. In T2DM, there are defects both in insulin action and in β-cell function. To deal with the problem of end-organ unresponsiveness, the exact nature of the defect must be understood in order to find specific sites which could be targeted for gene transfer studies. In the case of monogenic forms of T2DM, it would be possible to design an *ex-vivo* gene therapy approach, but in the case of polygenic conditions, that are the most common, with different genotypes underlying T2DM, it is much more cumbersome to apply gene therapy. Muscle and liver cells are the major target of insulin action. Thus, effective transgene delivery systems that remain stable over time need to be further improved. Gene therapy strategies that could have potential in the treatment of T2DM include inhibition of apoptosis, promotion of β-cell regeneration, genetic manipulations prior to β-cell replacement, engineering of β-cells and engineering of non-β-cells (Karanam et al., 2002).

5.2.4 MicroRNA therapy

Short non-coding microRNAs (miRNAs) have been implicated in the control of pancreatic insulin exocytosis and regulation of glucose homeostasis (reviewed by Gauthier and Wollheim, 2006). It is possible that once the biological mechanisms are fully understood, the use of miRNA-based therapies could be a reality for T2DM treatment.

5.3 Phytochemical therapies

A number of natural products exhibit properties that could be used as remedies to improve glucose metabolism (Friedman & McLellan, 2005). Cinnamon extract can significantly reduce blood glucose levels and lipids, improving insulin sensitivity (Kim et al., 2006). Isoflavones can activate PPARs (Ricketts et al., 2005) as well as resveratrol analogues that show lipid and glucose lowering properties mediated by PPARα (Rimando et al., 2005; Corton and Brown-Borg, 2005).

6. CONCLUSIONS

A whole range of pharmacological agents are available to ameliorate the T2DM symptoms by different mechanisms. A reduction in insulin resistance at any stage of T2DM will improve glucose metabolism by allowing the endogenous insulin to be more effective. The use of different insulin sensitizers and segregatogues, either in single therapy or in combination, would help to improve glycemic control, either by increasing peripheral glucose uptake, improving insulin secretion, decreasing hepatic glucose output or reducing the influx of glucose to the body.

It has become evident that T2DM is a complex metabolic disease that requires active management from both individuals and health monitoring agencies. Since there is a high individual variability between T2DM patients it is necessary to establish more personalized therapies to satisfy the precise metabolic needs that are dysfunctional or lacking in T2DM. With the advent of more efficient and less costly ways to diagnose T2DM susceptibility markers as well as measurement of plasma glucose, lipid and insulin levels at various time points throughout the day, it would be possible to apply the most appropriate pharmacological treatments. Further research is required on the causes of obesity in children and adults, and randomized, controlled trials are necessary to establish preventative initiatives.

REFERENCES

Abdelrahman, M., Sivarajah, A. and Thiemermann, C. (2005) Beneficial effects of PPAR-gamma ligands in ischemia-reperfusion injury, inflammation and shock. Cardiovasc. Res., 65(4): 772–781.

Anderson, D.C., Jr. (2005) Pharmacologic prevention or delay of type 2 diabetes mellitus. Ann. Pharmacother., 39(1): 102–109.

Anderson, J.W. (2006) Diabetes mellitus: medical nutrition therapy. In: Shils, M.E., Shike, M., Ross, C.A., Caballero, B. and Cousins, R.J. eds. (2006) Modern nutrition in health and disease. 10th ed. Lippincott, Williams & Wilkins. 2069 pp.

Asakawa, A., Inui, A., Kaga, T., Katsuura, G., Fujimiya, M., Fujino, M.A. and Kasuga, M. (2003) Antagonism of ghrelin receptor reduces food intake and body weight gain in mice. Gut. 52(7): 947–952.

Asfaha, S. and Padwal, R. (2005) Antihypertensive drugs and incidence of type 2 diabetes: evidence and implications for clinical practice. Curr. Hypertens. Rep., 7(5): 314–322.

Atkins, R.C. (2001) Age-defying diet. St. Martin's paperbacks.

Auwerx, J. (2006) Improving metabolism by increasing energy expenditure. Nat. Med., 12: 44–45.

Barroso, I. (2005) Complex disease: pleiotropic gene effects in obesity and type 2 diabetes. Eur J Hum Genet., 13(12): 1243–1244.

Bays, H.E. (2004) Current and investigational antiobesity agents and obesity therapeutic treatment targets. Obes. Res., 12(8): 1197–1211.

Bedu, E., Wahli, W. and Desvergne, B. (2005) Peroxisome proliferator-activated receptor beta/delta as a therapeutic target for metabolic diseases. Expert Opin. Ther. Targets, 9(4): 861–873.

Berger, J.P., Akiyama, T.E. and Meinke, P.T. (2005) PPARs: therapeutic targets for metabolic disease. Trends Pharmacol. Sci., 26(5): 244–251.

CDC, Centers for disease control and prevention, USA, 2006. http://www.cdc.gov/nccdphp/dnpa/obesity/defining.htm

Corton, J.C. and Brown-Borg, H.M. (2005) Peroxisome Proliferator-Activated Receptor {gamma} Coactivator 1 in Caloric Restriction and Other Models of Longevity. J Gerontol A Biol Sci Med Sci., 60(12): 1494–1509.

Cox, S.L. (2005) Muraglitazar: an agent for the treatment of type 2 diabetes and associated dyslipidemia. Drugs Today (Barc)., 41(9): 579–587.

Cunard, R. (2005) The potential use of PPARalpha agonists as immunosuppressive agents. Curr. Opin. Investig. Drugs, 6: 467–472.

de Luca, C. and Olefsky, J.M. (2006) Stressed out about obesity and insulin resistance. Nat. Med., 12(1): 41–42.

Despres, J.P., Golay, A., Sjostrom, L., Rimonabant in Obesity-Lipids Study Group (2005) Effects of rimonabant on metabolic risk factors in overweight patients with dyslipidemia. N. Engl. J. Med., 353(20): 2121–2134.

Diabetes Prevention Program Research Group (2005) Strategies to identify adults at high risk for type 2 diabetes: the Diabetes Prevention Program. Diabetes Care, 28(1): 138–144.

Durant, N. and Cox, J. (2005) Current treatment approaches to overweight in adolescents. Curr. Opin. Pediatr., 17(4): 454–459.

Dwyer, J.T., Allison, D.B. and Coates, P.M. (2005) Dietary supplements in weight reduction. J. Am. Diet Assoc., 105(5 Suppl 1): S80–86.

English, P. and Williams, G. (2001) Type 2 diabetes. Martin Dunitz Ltd. 103 pp.

Erion, M.D., van Poelje, P.D., Dang, Q., Kasibhatla, S.R., Potter, S.C., Reddy, M.R., Reddy, K.R., Jiang, T. and Lipscomb, W.N. (2005) MB06322 (CS-917): A potent and selective inhibitor of fructose 1,6-bisphosphatase for controlling gluconeogenesis in type 2 diabetes. Proc. Natl. Acad. Sci. USA, 102(22): 7970–7975.

Etgen, G.N. and Mantlo, N. (2003) PPAR ligands for metabolic disorders. Curr. Top. Med. Chem., 3: 1649–1661.

Evans, R.M., Barish, G.D. and Wang, Y.X. (2004) PPARs and the complex journey to obesity. Nat. Med., 10: 355–361.

Fajas, L., Debril, M.-B. and Auwerx, J. (2001) PPARγ: An essential role in metabolic control. Nutr. Metab. Cardiovasc. Dis., 11: 64–69.

FDA, Federal drug administration, 2006. http://www.fda.gov

Feinstein, D.L., Spagnolo, A., Akar, C., Weinberg, G., Murphy, P., Gavrilyuk, V. and Russo, C.D. (2005) Receptor-independent actions of PPAR thiazolidinedione agonists: is mitochondrial function the key? Biochem Pharmacol., 70(2): 177–188.

Ferre, P. (2004) The biology of peroxisome proliferator-activated receptors: relationship with lipid metabolism and insulin sensitivity. Diabetes, 53 Suppl 1: S43–S50.

Filippatos, T.D., Kiortsis, D.N., Liberopoulos, E.N., Mikhailidis, D.P. and Elisaf, M.S. (2005a) A review of the metabolic effects of sibutramine. Curr. Med. Res. Opin., 21(3): 457–468.

Filippatos, T.D., Kiortsis, D.N., Liberopoulos, E.N., Georgoula, M., Mikhailidis, D.P., Elisaf, M.S. (2005b) Effect of orlistat, micronised fenofibrate and their combination on metabolic parameters in overweight and obese patients with the metabolic syndrome: the FenOrli study. Curr. Med. Res. Opin., 21(12): 1997–2006.

Fredenrich, A. and Grimaldi, P.A. (2005) PPARdelta: an uncompletely known PPAR nuclear receptor. Diabetes Metab., 31: 23–27.

Friedman, M. and McLellan, A. (2006) Healing diabetes: complementary naturopathic and drug treatments. Ccnm press. 272 pp.

Fu, J., Gaetani, S., Oveisi, F., Lo Verme, J., Serrano, A., Rodriguez De Fonseca, F., Rosengarth, A., Luecke, H., Di Giacomo, B., Tarzia, G. and Piomelli, D. (2003) Oleylethanolamide regulates feeding and body weight through activation of the nuclear receptor PPAR-alpha. Nature, 425: 90–93.

Gallwitz, B. (2005) Glucagon-like peptide-1-based therapies for the treatment of type 2 diabetes mellitus. Treat Endocrinol., 4(6): 361–370.

Gauthier, B.R. and Wollheim, C.B. (2006) MicroRNAs: 'ribo-regulators' of glucose homeostasis. Nat. Med., 12: 36–38.

Gloyn, A.L. (2003) Glucokinase (GCK) mutations in hyper- and hypoglycemia: maturity-onset diabetes of the young, permanent neonatal diabetes, and hyperinsulinemia of infancy. Hum. Mutat., 22(5): 353–362.

Hakim, N., Stratta, R. and Gray, D. (eds.) (2002) Pancreas and islet transplantation. Oxford Univ. Press. 378 pp.

Hammarstedt, A., Andersson, C.X., Rotter Sopasakis, V. and Smith, U. (2005) The effect of PPARgamma ligands on the adipose tissue in insulin resistance. Prostaglandins Leukot. Essent. Fatty Acids., 73(1): 65–75.

Hansen, L. and Pedersen, O. (2005) Genetics of type 2 diabetes mellitus: status and perspectives. Diabetes Obes. Metab., 7(2): 122–135.

Harrity, T., Farrelly, D., Tieman, A., Chu, C., Kunselman, L., Gu, L., Ponticiello, R., Cap, M., Qu, F., Shao, C., Wang, W., Zhang, H., Fenderson, W., Chen, S., Devástale, P., Jeon, Y., Seethala, R., Yang, W.P., Ren, J., Zhou, M., Ryono, D., Biller, S., Mookhtiar, K.A., Wetterau, J., Gregg, R., Cheng, P.T. and Hariharan, N. (2006) Muraglitazar, a novel dual ({alpha}/{gamma}) Peroxisome Proliferator-Activated Receptor activator, improves diabetes and other metabolic abnormalities and preserves {beta}-Cell Function in db/db mice. Diabetes, 55(1): 240–248.

Hauner, H. (2004) Managing type 2 diabetes mellitus in patients with obesity. Treat Endocrinol., 3(4): 223–232.

Hill, J.O., Catenacci, V.A. and Wyatt, H.R. (2006). Obesity: etiology. Chapter 63. pp 1013–1028. In Shils, M.E., Shike, M., Ross, C.A., Caballero, B. and Cousins, R.J. (eds.) (2006) Modern nutrition in health and disease. 10th ed. Lippincott, Williams and Wilkins.2069 pp.

Hoivik, D.J., Qualls, C.W. Jr, Mirabile, R.C., Cariello, N.F., Kimbrough, C.L., Colton, H.M., Anderson, S.P., Santostefano, M.J., Morgan, R.J., Dahl, R.R., Brown, A.R., Zhao, Z., Mudd, P.N. Jr, Oliver, W.B. Jr, Brown, H.R. and Miller, R.T. (2004) Fibrates induce hepatic peroxisome and mitochondrial proliferation without overt evidence of cellular proliferation and oxidative stress in cynomolgus monkeys. Carcinogenesis, 25(9): 1757–1769.

Hovens, M.M., Tamsma, J.T., Beishuizen, E.D. and Huisman, M.V. (2005) Pharmacological strategies to reduce cardiovascular risk in type 2 diabetes mellitus: an update. Drugs, 65(4): 433–445.

Jiang, G. and Zhang, B.B. (2005) Modulation of insulin signalling by insulin sensitizers. Biochem. Soc. Trans., 33(Pt 2): 358–361.

Kadowaki, T. and Yamauchi, T. (2005) Adiponectin and adiponectin receptors. Endocr. Rev., 26(3): 439–451.

Kahn, C.R., Weir, G.C., King, G.L., Jacobson, A.M., Moses, A.C. and Smith, R.J. (eds.) (2005) Joslin's diabetes mellitus, 14th ed. Lippincott Williams and Wilkins. 1209 pp.

Kaplan, L.H. (2005) Pharmacological therapies for obesity. Gastroenterol. Clin. North Am., 34(1): 91–104.

Karanam, M., Song, Z. and Jindal, R.M. (2002) Gene therapy for diabetes. Chapter 21. 291–304. In Hakim, N., Stratta, R. and Gray, D. (eds.) Pancreas and islet transplantation. Oxford Univ. Press. 378 pp.

Kim, S.H., Hyun, S.H. and Choung, S.Y. (2006) Anti-diabetic effect of cinnamon extract on blood glucose in db/db mice. J. Ethnopharmacol. In press.

Kiortsis, D.N., Filippatos, T.D. and Elisaf, M.S. (2005) The effects of orlistat on metabolic parameters and other cardiovascular risk factors. Diabetes Metab., 31(1): 15–22.

Kojima, I. and Umezawa, K. (2006) Conophylline: A novel differentiation inducer for pancreatic beta cells. Int. J. Biochem. Cell. Biol., In press.

Kota, B.P., Huang, T.H. and Roufogalis, B.D. (2005) An overview on biological mechanisms of PPARs. Pharmacol. Res., 51(2): 85–94.

Kubota, N., Terauchi, Y., Miki, H., Tamemoto, H., Yamauchi, T., Komeda, K., Satoh, S., Nakano, R., Ishii, C., Sugiyama, T., Eto, K., Tsubamoto, Y., Okuno, A., Murakami, K., Sekihara, H., Hasegawa, G., Naito, M., Toyoshima, Y., Tanaka, S., Shiota, K., Kitamura, T., Fujita, T., Ezaki, O., Aizawa, S., Kadowaki, T. et al. (1999) PPAR gamma mediates high-fat diet-induced adipocyte hypertrophy and insulin resistance. Mol. Cell, 4(4): 597–609.

Kulkarni, R.N. (2004) The islet β-cell. Int. J. Biochem. Cell Biol. 36: 365–371.

Kurtz, T.W. and Pravenec, M. (2004) Antidiabetic mechanisms of angiotensin-converting enzyme inhibitors and angiotensin II receptor antagonists: beyond the renin-angiotensin system. J. Hypertens., 22(12): 2253–2261.

Kusminski, C.M., McTernan, P.G. and Kumar, S. (2005) Role of resistin in obesity, insulin resistance and Type II diabetes. Clin Sci (Lond)., 109(3): 243–256.

Laakso, M. (2005) Prevention of type 2 diabetes. Curr Mol Med., 5(3): 365–374.

Lazar, M.A. (2005) PPAR gamma, 10 years later. Biochimie, 87(1): 9–13.

Lebovitz, H.E. (2005) Management of hyperglycemia with oral antihyperglycemic agents in type 2 diabetes. Chapter 41.687–710. In: Kahn, C.R., Weir, G.C., King, G.L., Jacobson, A.M., Moses, A.C. and Smith, R.J. (eds.) (2005) Joslin's diabetes mellitus, 14th ed. Lippincott Williams & Wilkins. 1209 pp.

List, J.F. and Habener, J.F. (2004) Glucagon-like peptide 1 agonists and the development and growth of pancreatic beta-cells. Am. J. Physiol. Endocrinol. Metab., 286(6), E875–E881.

Lo Verme, J., Gaetani, S., Fu, J., Oveisi, F., Burton, K. and Piomelli, D. (2005) Regulation of food intake by oleoylethanolamide. Cell. Mol. Life Sci., 62(6): 708–716.

Lowell, B.B. and Shulman, G.I. (2005) Mitochondrial dysfunction and type 2 diabetes. Science, 307(5708): 384–387.

Luquet, S., Lopez-Soriano, J., Holst, D., Fredenrich, A., Melki, J., Rassoulzadegan, M. and Grimaldi, P.A. (2003) Peroxisome proliferator-activated receptor delta controls muscle development and oxidative capability. FASEB J., 17(15): 2299–2301.

Luquet, S., Gaudel, C., Holst, D., Lopez-Soriano, J., Jehl-Pietri, C., Fredenrich, A. and Grimaldi, P.A. (2005) Roles of PPAR delta in lipid absorption and metabolism: a new target for the treatment of type 2 diabetes. Biochim. Biophys. Acta, 1740(2): 313–317.

Malecki, M.T. (2005) Genetics of type 2 diabetes mellitus. Diabetes Res. Clin. Pract., 68 Suppl1: S10–S21.

Mathys, M. (2005) Pharmacologic agents for the treatment of obesity. Clin. Geriatr. Med., 21(4): 735–746.

Mogensen, J.P., Jeppesen, L., Bury, P.S., Pettersson, I., Fleckner, J., Nehlin, J., Frederiksen, K.S., Albrektsen, T., Din, N., Mortensen, S.B., Svensson, L.A., Wassermann, K., Wulff, E.M., Ynddal, L. and Sauerberg, P. (2003) Design and synthesis of novel PPARalpha/gamma/delta triple activators using a known PPARalpha/gamma dual activator as structural template. Bioorg. Med. Chem. Lett., 13: 257–260.

Nakamichi, Y., Kikuta, T., Ito, E., Ohara-Imaizumi, M., Nishiwaki, C., Ishida, H. and Nagamatsu, S. (2003) PPAR-gamma overexpression suppresses glucose-induced proinsulin biosynthesis and insulin release synergistically with pioglitazone in MIN6 cells. Biochem. Biophys. Res. Commun., 306(4): 832–836.

Nehlin, J.O., Mogensen, J.P., Petterson, I., Jeppesen, L., Fleckner, J., Wulff, E.M. and Sauerberg, P. (2006) Selective PPAR agonists for the treatment of diabetes type 2. Annals N.Y. Acad. Sci. In press.

Oliver, W.R. Jr, Shenk, J.L., Snaith, M.R., Russell, C.S., Plunket, K.D., Bodkin, N.L., Lewis, M.C., Winegar, D.A., Sznaidman, M.L., Lambert, M.H., Xu, H.E., Sternbach, D.D., Kliewer, S.A., Hansen, B.C. and Willson, T.M. (2001) A selective peroxisome proliferator-activated receptor delta agonist promotes reverse cholesterol transport. Proc. Natl. Acad. Sci.USA, 98: 5306–5311.

Owen, M.R., Doran, E. and Halestrap, A.P. (2000) Evidence that metformin exerts its anti-diabetic effects through inhibition of complex 1 of the mitochondrial respiratory chain. Biochem J., 348 Pt. 3: 607–614.

Ozcan, U., Cao, Q., Yilmaz, E., Lee, A.H., Iwakoshi, N.N., Ozdelen, E., Tuncman, G., Gorgun, C., Glimcher, L.H. and Hotamisligil, G.S. (2004) Endoplasmic reticulum stress links obesity, insulin action, and type 2 diabetes. Science, 306(5695): 457–461.

Padwal, R., Majumdar, S.R., Johnson, J.A., Varney, J. and McAlister, F.A. (2005) A systematic review of drug therapy to delay or prevent type 2 diabetes. Diabetes Care, 28(3): 736–744.

Park, S.Y., Cho, Y.R., Finck, B.N., Kim, H.J., Higashimori, T., Hong, E.G., Lee, M.K., Danton, C., Deshmukh, S., Cline, G.W., Wu, J.J., Bennett, A.M., Rothermel, B., Kalinowski, A., Russell, K.S., Kim, Y.B., Kelly, D.P. and Kim, J.K. (2005) Cardiac-specific overexpression of peroxisome proliferator-activated receptor-alpha causes insulin resistance in heart and liver. Diabetes, 54(9): 2514–2524.

Poulsen, M.K., Henriksen, J.E., Hother-Nielsen, O. and Beck-Nielsen, H. (2003) The combined effect of triple therapy with rosiglitazone, metformin, and insulin aspart in type 2 diabetic patients. Diabetes Care, 26(12): 3273–3279.

Press, M. (2002) The nature of the problem: why do we need pancreatic transplantation? Chapter 2. In: Hakim, N., Stratta, R. and Gray, D. (eds.) (2002) Pancreas and islet transplantation. Oxford Univ. Press. 378 pp.

Puigserver, P. and Rodgers, J.T. (2006) Foxa2, a novel transcriptional regulator of insulin sensitivity. Nat Med., 12(1): 38–39.

Puigserver, P. (2005) Tissue-specific regulation of metabolic pathways through the transcriptional coactivator PGC1-alpha. Int.J.Obes.(Lond), 29 Suppl 1: S5–S9.

Rangwala, S.M. and Lazar, M.A. (2004) Peroxisome proliferator-activated receptor gamma in diabetes and metabolism. Trends Pharmacol. Sci., 25: 331–336.

Reaven, G.M. (2005) The insulin resistance syndrome: definition and dietary approaches to treatment. Annu. Rev. Nutr., 25: 391–406.

Rhodes, C.J. (2005) Type 2 diabetes-a matter of beta-cell life and death? Science, 307(5708): 380–384.

Ricketts, M.L., Moore, D.D., Banz, W.J., Mezei, O. and Shay, N.F. (2005) Molecular mechanisms of action of the soy isoflavones includes activation of promiscuous nuclear receptors. A review. J. Nutr. Biochem., 16(6): 321–330.

Rimando, A.M., Nagmani, R., Feller, D.R. and Yokohama, W. (2005) Pterostilbene, a new agonist for the peroxisome proliferator-activated receptor alpha-isoform, lowers plasma lipoproteins and cholesterol in hypercholesterolemic hamsters. J. Agric. Food Chem., 53(9): 3403–3407.

Rosen, E.D. and Spiegelman, B.M. (2001) PPARgamma: a nuclear regulator of metabolism, differentiation, and cell Growth. J. Biol. Chem., 276: 37731–37734.

Rotella, C.M., Pala, L. and Mannucci, E. (2005) Glucagon-like peptide 1 (GLP-1) and metabolic diseases. J. Endocrinol. Invest., 28(8): 746–758.

Sacerdote, A., Weiss, K., Tran, T., Rokeya Noor, B. and McFarlane, S.I. (2005) Hypertension in patients with Cushing's disease: pathophysiology, diagnosis, and management. Curr. Hypertens. Rep., 7(3): 212–218.

Saenz, A., Fernandez-Esteban, I., Mataix, A., Ausejo, M., Roque, M. and Moher, D. (2005) Metformin monotherapy for type 2 diabetes mellitus. Cochrane Database Syst. Rev. 3: CD002966.

Sauerberg, P., Pettersson, I., Jeppesen, L., Bury, P.S., Mogensen, J.P., Wassermann, K., Brand, C.L., Sturis, J., Woldike, H.F., Fleckner, J., Andersen, A.S., Mortensen, S.B., Svensson, L.A., Rasmussen, H.B., Lehmann, S.V., Polivka, Z., Sindelar, K., Panajotova, V., Ynddal, L. and Wulff, E.M. (2002) Novel tricyclic-α-alkyloxyphenylpropionic acids: dual PPARalpha/gamma agonists with hypolipidemic and antidiabetic activity. J. Med. Chem., 45: 789–804.

Sauerberg, P., Bury, P.S., Mogensen, J.P., Deussen, H.J., Pettersson, I., Fleckner, J., Nehlin, J., Frederiksen, K.S., Albrektsen, T., Din, N., Svensson, L.A., Ynddal, L., Wulff, E.M., and Jeppesen, L. (2003) Large dimeric ligands with favorable pharmacokinetic properties and peroxisome proliferator-activated receptor agonist activity *in vitro* and *in vivo*. J. Med. Chem., 46: 4883–4894.

Sauerberg, P., Mogensen, J.P., Jeppesen, L., Svensson, L.A., Fleckner, J., Nehlin, J., Wulff, E.M. and Pettersson, I. (2005) Structure-activity relationships of dimeric PPAR agonists. Bioorg. Med. Chem. Lett., 15: 1497–1500.

Scher, J.U. and Pillinger, M.H. (2005) 15d-PGJ2: the anti-inflammatory prostaglandin? Clin. Immunol., 114(2): 100–109.

Schulze, M.B. and Hu, F.B. (2005) Primary prevention of diabetes: what can be done and how much can be prevented? Annu. Rev. Public Health, 26: 445–467.

Scott, R., Best, J., Forder, P., Taskinen, M.R., Simes, J., Barter, P., Keech, A. and FIELD Study Investigators (2005) Fenofibrate Intervention and Event Lowering in Diabetes (FIELD) study: baseline characteristics and short-term effects of fenofibrate. Cardiovasc Diabetol., 4: 13.

Seeley, R.J. and Tschop, M. (2006) How diabetes went to our heads. Nat Med., 12(1): 47–49.

Shils, M.E., Shike, M., Ross, C.A., Caballero, B. and Cousins, R.J. eds. (2006) Modern nutrition in health and disease. 10th ed. Lippincott, Williams and Wilkins. 2069 pp.

Soria, B., Roche, E., Reig, J.A. and Martin, F. (2005) Generation of insulin-producing cells from stem cells. Novartis Found Symp. 265: 158–167; discussion 167–73, 204–11.

Sorkin, J.D., Muller, D.C., Fleg, J.L. and Andres, R. (2005) The relation of fasting and 2-h postchallenge plasma glucose concentrations to mortality: data from the Baltimore Longitudinal Study of Aging with a critical review of the literature.Diabetes Care, 28(11): 2626–2632.

Speight, T.M. and Holford, N.H.G. (eds.) (1997) Avery's drug treatment. 4th ed. Adis International Ltd.

Spiegelman, B.M. and Flier, J.S. (2001) Obesity and the regulation of energy balance. Cell, 104: 531–543.

Staels, B. and Fruchart, J.C. (2005) Therapeutic roles of peroxisome proliferator-activated receptor agonists. Diabetes, 54(8): 2460–2470.

Staels, B. (2005) PPARgamma and atherosclerosis. Curr. Med. Res. Opin., 21 Suppl 1: S13–S20 (2005).

Steinbeck, K. (2005) Childhood obesity. Treatment options. Best Pract. Res. Clin. Endocrinol. Metab., 19(3): 455–469.

Stumvoll, M., Goldstein, B.J.. and van Haeften, T.W. (2005) Type 2 diabetes: principles of pathogenesis and therapy. Lancet, 365: 1333–1346.

Sturis, J., Gotfredsen, C.F., Romer, J., Rolin, B., Ribel, U., Brand, C.L., Wilken, M., Wassermann, K., Deacon, C.F., Carr, R.D. and Knudsen, L.B. (2003) GLP-1 derivative liraglutide in rats with beta-cell deficiencies: influence of metabolic state on beta-cell mass dynamics. Br. J. Pharmacol., 140(1): 123–132.

Tan, N.S., Michalik, L., Desvergne, B. and Wahli, W. (2005) Multiple expression control mechanisms of peroxisome proliferator-activated receptors and their target genes. J. Steroid Biochem. Mol. Biol., 93(2–5): 99–105.

Tanaka, T., Yamamoto, J., Iwasaki, S., Asaba, H., Hamura, H., Ikeda, Y., Watanabe, M., Magoori, K., Ioka, R.X., Tachibana, K., Watanabe, Y., Uchiyama, Y., Sumi, K., Iguchi, H., Ito, S., Doi, T., Hamakubo, T., Naito, M., Auwerx, J., Yanagisawa, M., Kodama, T. and Sakai, J. (2003) Activation of peroxisome proliferator-activated receptor delta induces fatty acid beta-oxidation in skeletal muscle and attenuates metabolic syndrome. Proc. Natl. Acad. Sci. USA, 100(26): 15924–15929.

Taylor, S.I. (1999) Deconstructing type 2 diabetes. Cell, 97: 9–12.

Toyoshima, Y., Gavrilova, O., Yakar, S., Jou, W., Pack, S., Asghar, Z., Wheeler, M.B. and LeRoith, D. (2005) Leptin improves insulin resistance and hyperglycemia in a mouse model of type 2 diabetes. Endocrinology, 146(9): 4024–4035.

Unger, R.H. (2006) Klotho-induced insulin resistance: a blessing in disguise? Nat. Med., 12(1): 56–57.

van Raalte, D.H., Li, M., Pritchard, P.H. and Wasan, K.M. (2004) Peroxisome proliferator-activated receptor (PPAR)-alpha: a pharmacological target with a promising future. Pharm. Res., 21: 1531–1538.

Verma, N.K., Singh, J. and Dey, C.S. (2004) PPAR-gamma expression modulates insulin sensitivity in C2C12 skeletal muscle cells. Br. J. Pharmacol., 143(8): 1006–1013.

Wadman, M. (2006) Rimonabant adds appetizing choice to slim obesity market. Nat. Med., 12, 27 (2006).

Wang, Y.L., Frauwirth, K.A., Rangwala, S.M., Lazar, M.A. and Thompson, C.B. (2002) Thiazolidinedione Activation of Peroxisome Proliferator-activated Receptor γ Can Enhance Mitochondrial Potential and Promote Cell Survival. J. Biol. Chem., 277: 31781–317888.

Wang, Y.X., Lee, C.H., Tiep, S., Yu, R.T., Ham, J., Kang, H. and Evans, R.M. (2003) Peroxisome-Proliferator-activated receptor d activates fat metabolism to prevent obesity. Cell, 113: 159–170.

WHO, World Health Organisation. 2006. http://www.who.int/diabetes/facts/en/

Wild, S., Roglic, G., Green, A., Sicree, R. and King, H. (2004) Global prevalence of diabetes: estimates for the year 2000 and projections for 2030. Diabetes Care, 27: 1047–1053.

Wild, S., Roglic, G., Green, A., Sicree, R. and King, H. (2004) Global prevalence of diabetes: estimates for the year 2000 and projections for 2030. Diabetes Care, 27: 1047–1053.

Yach, D., Stuckler, D., and Brownell, K.D. (2006) Epidemiologic and economic consequences of the global epidemics of obesity and diabetes. Nat. Med., 12(1): 62–66.

Youssef, J.A. and Badr, M.Z. (2005) Aging and enhanced hepatocarcinogenicity by peroxisome proliferator-activated receptor alpha agonists. Ageing Res. Rev., 4: 103–118.

Yu, S., Matsusue, K., Kashireddy, P., Cao, W.Q., Yeldandi, V., Yeldandi, A.V., Rao, M.S., Gonzalez, F.J. and Reddym, J.K. (2003) Adipocyte-specific gene expression and adipogenic steatosis in the mouse liver due to peroxisome proliferator-activated receptor gamma1 (PPARgamma1) overexpression. J. Biol. Chem., 278(1): 498–505.

Zhang, J., Fu, M., Cui, T., Xiong, C., Xu, K., Zhong, W., Xiao, Y., Floyd, D., Liang, J., Li, E., Song, Q. and Chen, Y.E. (2004a) Selective disruption of PPARg2 impairs the development of adipose tissue and insulin sensitivity. Proc. Natl. Acad. Sci. USA, 101: 10703–10708.

Zhang, F., Lavan, B. and Gregoire, F.M. (2004b) Peroxisome proliferator-activated receptors as attractive antiobesity targets. Drug News Perspect., 17(10): 661–669.

Zingarelli, B. and Cook, J.A. (2005) Peroxisome proliferator-activated receptor-gamma is a new therapeutic target in sepsis and inflammation. Shock, 23(5): 393–399.

CHAPTER 9

AGE-RELATED CATARACT: MANAGEMENT AND PREVENTION

MAYANK A. NANAVATY, ABHAY R. VASAVADA AND P.D. GUPTA*

Iladevi Cataract & IOL Research Centre' Raghudeep Eye Clinic, Gurukul Road, Memnagar, Ahmedabad-380 052. India.

Abstract: Cataract is defined as opacity of the crystalline lens. Age is by far the biggest risk factor for cataract, and it is sometimes assumed that cataract is simply an amplification of this aging process. Age-related cataract appears to accompany the latter stages of lifespan inmost cases. With aging, the molecular changes that take place in the crystalline lens that contribute to a gradual reduction in transparency. In many cases, the aging process of the crystalline lens reaches a point where vision is impaired.. However, no method to halt the formation of a cataractous lens has been shown to be effective so far but researches are in progress. Nevertheless, advances in surgical removal of cataracts, including small-incision surgery, use of viscoelastics, and the development of intraocular lenses, have made treatment very effective and visual recovery is rapid in most cases. Despite these advances, cataract continues to be a leading public-health issue with greater life expectancy

1. UNIQUE LENS SYSTEM

The human eye lens is the optically clear structure located behind the iris and in front of the vitreous body and retina. The lens consists of parabolic, anterior and posterior surfaces. It is enclosed by a capsule and is attached to the ciliary processes by the lens zonules. The circumference of the lens is called the equator.

The chief role of lens is to provide an optical component of high transparency and refractive index, which assures that the object may be focused on the retina. Apart from the transparency, the lens has several other unique features: even at birth, it is completely without blood supply and has no innervations; it grows in size and weight throughout the life since no cells are shed; the mass of the cells,

* Present Address: Director, Research and Development, Atmiya Institute of Science and Technology, Rajkot, Gujarat, India. Email: pdg2000@hotmail.com

S. Rattan and M. Kassem (eds.), Prevention and Treatment of Age-related Diseases, 159–174.

at various stages of development and maturation, is completely surrounded by and elastic acellular capsule that has a smooth outer surface. The additional function of accommodation enables near objects to be brought into focus by relaxation of suspensory ligaments.

The lens has a unique molecular make-up as it is two-thirds water with one-third protein; other constituents represent only about 1% of the total lens net weight. This high protein content is necessary for high refractive index, allowing it to bend light rays into focus onto the retina. Glucose is the chief source of energy of lens and although fatty acids can be metabolized in a similar way, the supply of triglyceride or fatty acid does not provide significant energy from this source. Some amino acids are also metabolized, through decarboxylation and deamination, for energy production.

The uniqueness also lies in the fact that the lens is developed from the surface ectoderm overlying the optic vesicle. The development proceeds from lens placode to lens vesicle stage. This is followed by development of nucleus as elongation of the cells in the posterior portion of the lens fills the vesicle, which eventually looses their nuclei. Meanwhile, the cells in the anterior part of the vesicle continue to divide actively to form the lens epithelial cells. The equatorial zone of the lens epithelium continues to divide throughout life, producing the cells that differentiate into the long lens fibers. The embryonic lens is surrounded by blood vessels, the tunica vasculosa lentis. This vascular system regresses at the end of development and it is absent shortly before birth leaving the lens avascular throughout the life.

The lens has a unique property to transmit light throughout the visible spectrum but absorbs heavily in UV at below 400 nm. (Griswold and Stark, 1992) With increasing age there is absorption in the visible light spectrum (Bron et al., 2000) that is exaggerated in presence of nuclear brunescence. The lens becomes increasingly yellow with age, because of the interaction of crystallins with a UV filter compound, 3-hydroxykynurenine glucoside (3-OHKG). Various protein modifications may play a role in human nuclear cataractogenesis (Hood et al., 1999). Apart from its coloration the normal aging lens scatters light after 50 years of age and results in the some of glare in certain conditions, which is likely to be due to increased lens thickness with aging.

The purpose of this chapter is to provide an overview of the age related changes in the structure, biochemistry and physiology of the lens and to discuss the management as well as the preventive aspects of these changes.

2. UNDERSTANDING CATARACT

Cataract (word derived from Greek language, meaning waterfall (Johns et al., 2002; Floyd, 2000)) is the name given to any opacity in the lens, not necessarily with any effect on vision. This definition may be extended to include opacity of the lens capsule and the deposition of material of non-lenticular origin (viz. True exfoliation in glass blowers, pseudoexfoliation, chalcosis of lens in Wilson's disease, siderosis, argyrosis, gold deposits, mercury salts, etc) (Brown, 1999).

Understanding the normal physiology and biochemistry of the lens and the changes that induce cataract formation continues to be an area of active research today. Though some possible risk factors for cataract development have been suggested, there is no confirmed method to prevent cataract formation so far. Cataracts can be caused by a variety of problems, including developmental abnormalities, trauma, metabolic and drug induced changes (Brown, 1999). The main cause of visually significant cataracts is aging, i.e., age-related (senile) cataracts is the focus of this chapter.

Until recently, there has been little need to accurately classify cataract type or severity. Traditionally, clinicians have used anatomical (cortical, nuclear, and posterior subcapsular (PSC)) or etiological (radiation, steroid, and so forth) terms to describe the type of cataract. Descriptors of cataract severity have been base on coarse, subjective scales and have included terms such as immature, advanced immature, and mature. As basic scientists developed means of identifying and quantitating mechanisms of human cataract formation, it became necessary to more accurately and consistently describe or classify cataracts. Also, as pharmaceutical companies encountered drugs with cataractogenic toxicity, and as epidemiologists began to study the risk factors of human cataract formation, better systems of cataract classification were needed. Several have been developed and they include the Lens Opacities Classification System, Versions I to III (LOCS I to III), the Oxford Cataract Classification System, the Wilmer System, and the Wisconsin System.

A number of epidemiological studies have linked UV exposure with the formation of cortical cataract, for the wavelengths UVB (280–315 nm) and UVA (315–400 nm). The preponderance of cortical cataract in the inferonasal quadrant, where levels of solar radiation are said to be highest, has also been offered as indirect evidence of an association between exposure to sunlight and cortical cataract (Schein et al., 1994; Graziosi et al., 1996).

Few studies have consistently demonstrated exposure to UVB light as a risk factor for cortical and perhaps PSC cataract (Bochow et al., 1989; West et al., 1998; Munoz et al., 1993). Calculations of attributable risk based on such work suggest that ocular UVB exposure may explain approximately 10% of the cortical cataract in some populations (VanNewkirk et al., 2002). These calculations and the relatively mild impact of cortical opacity on visual function, suggest that the effect of strategies involving reduced exposure to sunlight, even if practical, might be limited.

The hypothesis that antioxidants nutrients in the serum, lens and aqueous might be protective against lens opacity has attracted much attention. This is, in part, because of the appeal of supplementation as a practical anti-cataract surgery, an approach that has been highly successful in other disorders, as with fluoridated water (Van der Haar, 1997), iodized salt (Krause et al., 1998) and vitamin A (Christen, 1999). However, epidemiological evidence for the antioxidant hypothesis among human subject has been conflicting (Taylor et al., 1995; Bunce et al., 1990; Congdon and West Jr, 1999; Sperduto et al., 1993). Until recently, the majority of studies of antioxidants and lens opacity have been observational, cross–sectional and uncontrolled in design, making it difficult to establish a clear role for any particular agent, and impossible to account for important confounders such as socio-economic status.

Recent controlled trials have largely obviated such concerns, and have cast significant doubt on the role of antioxidants in protecting against lens opacity in nutritionally replete populations. The Linxian Cataract Trial identified a limited protective role against nuclear cataracts among older persons receiving riboflavin and niacin. However, retinal, zinc, ascorbic acid, molybdenum, selenium, a-tocopherol and B-carotene were not protective, and this rural Chinese population appears to have been nutritionally deficient in many ways (Robman et al., 1999). The Vitamin E, Cataract and Age–related Maculopathy study in Australia (AREDS, 2001) and Age–Related Eye Disease Study (Manson et al., 1995) in the US have recently failed to demonstrate any beneficial effect on the progression of lens opacity of giving well nourished persons vitamin E alone, or in a combination of A, C and E (with or without zinc), respectively. Additional prospective studies, which may be expected to offer insight into this question, include the Women's Antioxidant Cardiovascular study (Leske et al., 1999), the Women's Health Study (McCarty et al., 1999) and the Physicians' Health Study II. (Christen, 1999). However, at present, nutritional supplementation is not indicated as an anti-cataract strategy for well nourished populations in the developed world, although a possible role in undernourished populations in the developing world cannot be ruled out.

A set of potentially interrelated personal factors-diabetes, hypertension and body mass index (BMI) –has been implicated as representing an increasing risk for various forms of lens opacity. Diabetes has consistently been associated with increased risk for cortical cataract (Leske et al., 1999; McCarty et al., 1999; Klein et al., 1995), and variably for PSC (Leske et al., 1999; Klein et al., 1995) and nuclear opacities (McCarty et al., 1999). Body mass Index (BMI) has been identified as an independent risk factor for PSC and nuclear cataract (Caulfield et al., 1999; Glynn et al., 1995), and also cortical opacity (Hiller et al., 1998), when controlling for diabetes, age and smoking. Hypertension has also been associated with cortical cataract (Leske et al., 1999). While all of these factors are potentially remediable, suggesting possible avenues for cataract prevention, the effectiveness of such strategies remains to be proven. Although there is some evidence that better diabetic control (demonstrated by lower hemoglobin AI c levels) may reduce the risk of lens opacity (Klein et al., 1998), no controlled, prospective data yet exist to demonstrate that improved treatment of diabetes or hypertension will in fact prevent or delay lens opacity. An added difficulty of intervening on BMI to prevent cataract is that the directionality of the association (e.g. whether elevated or reduced BMI, or both, contributes to lens opacity) has not been definitively established.

Female gender has generally been associated with an increased age-adjusted risk of both nuclear cataract (AREDS, 2001) and cortical cataract (Mitchell et al., 1997) among all races studied, including persons of African (Congdon et al., 2001; Leske et al., 2000), Asian (Cheng et al., 2000) and European (Cumming and Mitchell,

1997) descent. Although gender as a risk factor is clearly not subject to alteration, some studies suggests that post-menopausal use of estrogen may be associated with reduced risk of nuclear cataract (Cumming and Mitchell, 1997). However, other studies have been unable to confirm this finding (McCarty et al., 1999).

Risk factors of importance in certain subpopulations include ocular conditions, such as uveitis and retinitis pigmentosa, both thought to be associated with PSC opacities, perhaps because of breakdown of the blood-ocular barrier and subsequent entry of cataractogenic factors into the eye. Ocular surgery is also an important risk factor, especially trabeculectomy (Klein et al., 1995; Collaborative Normal-Tension Glaucoma Study Group, 1998) and retinal surgery (Wong et al., 2002). It has been suggested that surgically created alternative pathways for the drainage of aqueous from the eye may deprive the lens of aqueous-borne nutrients necessary to preserve normal clarity. A dose dependent association (measured both in terms of concentration and length application) between age-related cataract and mitomycin C, an anti-metabolite used regularly in glaucoma surgery, has also been established in a trial setting (Ramkrishnan et al., 1993). Ocular trauma can clearly be associated with lens opacity in certain individuals, although studies suggest that the impact on the prevalence in the population of lens opacity is probably minimal (Wong et al., 2002). Finally, periocular irradiation with gamma rays (Chen et al., 2001) and proton beams (Brovkina and Zarubei, 1986) can be associated with various forms of lens opacity. These smaller, well-defined subpopulations with a relatively high risk of rapid-onset cataract could ultimately serve as ideal subjects for trials of anti-cataract medications, although the relevance of the findings of such studies to age-related cataract would be unknown.

Finally, there are number of other risk factors for lens opacity which are either poorly understood, or, although they may be of importance for certain groups, do not represent a significant risk for the population as a whole. Several studies (Wong et al., 2001; Lim et al., 1999) have suggested that refractive errors, typically myopia, are associated with age-related cataract, particularly nuclear cataract and PSC (Lim et al., 1999; Wu et al., 1999; Vasavada et al., 2004). It is well known that increased refractive index of the lens in advanced nuclear cataract may cause a secondary myopia; pre-existing myopia may also serve as an independent risk factor (Lim et al., 1999). The mechanism for such as association, if indeed it exists, is not understood.

3. OXIDATION OF LENS MATERIAL AND CATARACT FORMATION

The light-scattering process is the primary factor responsible for the turbidity and wave front distortion by the cataractous lens. The aggregation of lens protein into randomly distributed high molecular weight clusters are thought to produce sufficient fluctuation in protein density to account for the opacification. In fact, protein aggregation results in the development of very high molecular weight aggregates of sufficient size to directly scatter the light and in the creation of protein-rich and protein-poor phases causing local changes in refractive index and thus increased light scattering (Benedek, 1997). Protein aggregation increases with age. The crystallins, which constitute approximately 90% of the total protein content of the lens, accumulate and show many age-related oxidative changes. These include formation of disulfide and other inter- and intramolecular cross-links and methionine oxidation, all of which result in the aggregation of high molecular weight molecules. Therefore, the protein redox status seems to be fundamental to maintain the lens function and transparency. It may be possible that local or systemic conditions affecting the protein redox status, such as myopia and diabetes, influence this process (Altomare et al., 1997).

Recently, it was hypothesized that a threshold of lipid oxidation might exist above which the opacification takes place and that this could be surpassed earlier in some subjects predisposed to cataract formation (Borchman and Yappert, 1998). The assessment of carbonyl and sulfhydril proteins has been suggested as being a valuable index of the protein redox status in the lens (Altomare et al., 1997). In fact, the level of carbonyl proteins, derived from amino acids during metal-catalyzed oxidation of proteins in vitro and in vivo, represents a direct measure of the oxidative injury to these molecules (Stadtman, 1992). The sulfhydryl proteins, known to have structural and functional role in the crystalline lens, contain an elevated number of thiol groups and, therefore are reduced as a result of oxidation. A linear relationship between subject age and the amount of protein carbonyl groups has been found in the human eye lens cortex. It has been already shown that during senile cataract development a progressive decrease in SH content of the crystallins occurs.

It is estimated that the oxygen tension in the vicinity of the lens is low, yet this is sufficient to support some aerobic lens metabolism and is sufficient to act as a source of reactive oxygen species (ROS). A significant proportion of lenses and aqueous humor taken from cataract patients have elevated H_2O_2 levels. Because H_2O_2, at concentrations found in cataract, can cause lens opacification and produces a pattern of oxidation similar to that found in cataract, it is concluded that H_2O_2 is the major oxidant involved in cataract formation (Ramachandran et al., 1991). This viewpoint is further supported by experiments showing that cataract formation in organ culture caused by photochemically generated superoxide radical, H_2O_2, and hydroxyl radical is completely prevented by the addition of a GSH peroxidase mimic. The damage caused by oxidative stress does not appear to be reversible and there is an inverse relationship between the stress period and the time required for loss of transparency and degeneration of biochemical parameters such as ATP, GPD, nonprotein thiol, and hydration. After exposure to oxidative stress, the redox set point of the single layer of the lens epithelial cells (but not the remainder of the lens) quickly changes, going from a strongly reducing to an oxidizing environment (Ito et al., 1993). Almost concurrent with this change is extensive damage to DNA and membrane pump systems, followed by loss of epithelial cell viability and death by necrotic and apoptotic mechanisms (Kleiman et al., 1990). There are evidences suggesting that the epithelial cell layer is the initial site of attack by oxidative stress and that involvement of the lens fibers follows, leading to cortical cataract (Worgul et al., 1989).

Lately it has been shown that Sex Steroid hormones regulate ocular tissues in addition to their conventional target tissues (Gupta et al., 2005). The female gender has been found to display an increased incidence of cataracts, as compared with age-matched men. This increased risk is seen in woman population after menopause only (Gupta et al., 2005; Leske et al., 2004). Protective effect of Sex Steroid Hormones in the perspective of cataractogenesis in females has been substantiated by epidemiological information. The Beaver Dam Eye Study suggests a modest protective effect of estrogen exposure on the lenses of women in the context of age related opacities (Klein et al., 1994). The results indicated that the current use of post-menopausal estrogens is associated with decreased risk of severe nuclear sclerosis. The study also showed that from menarche to menopause the life span of woman is associated with protective effect and decreased risk of nuclear sclerosis and cortical opacities. Recently it is shown that lenses from female rats are more resistant to transforming growth factor β (TGFβ) induce cataract then those from males. In young age estrogen provides protection against cataract by counteracting the damaging effects of TGFβ (Chen et al., 2004; Hales et al., 1997). Proper ionic milieu and hydration of lens cells are essential to maintain transparency of crystalline lens. Estrogen maintains proper ionic composition by its non-genomic action (Singh and Gupta, 1997a). Further, estrogens are known for increasing water imbibitions and retention of hydration in the target tissues (Singh and Gupta, 1997b).

The lens possesses repair mechanism, both at a cellular and at a molecular level and it has an ability to isolate damaged fibers and the histology of this has been shown. At the molecular level, a number of scavenger molecules are present that protects against oxidative stress. Lens membranes contain Vitamin E, which protect against lipid peroxidation. GSH, a patient free redial scavenger, is synthesized in the lens from amino acid precursors. (L-glutamic acid, L-cysteine, and L-glycine). It is present in high concentration in the cortex and in the epithelium, and at a lower concentration in the nucleus (Pau et al., 1990). It is probably important in maintaining lens proton thiols in the reduced state, such as that of Na+, K+ ATPase or of the lens crystalline thiols. It maintains ascorbate in the reduced state and scavenges peroxides and radiation induced free radicals Vitamin C, always in high concentration in the aqueous is actively transported into the lens, where it is at a higher concentration. Like GSH, it is an effective reducing agent. Other compounds, carotenoids, choline, taurine, and thioredoxin-T have been ascribed similar roles.

3.1 Management of Age-related cataract

Cataract surgery is the only remedy of the age related cataract today. Unless the patient presents with an eye threatening condition e.g. hypermautre cataract, where advising immediate surgery is inevitable, this decision to operate should be on patients discretion. If the individual is comfortable in his day to day activities e.g. reading, moving about at home, etc. he can be advised to wait until such time when his present routine activities are curtailed owing to cataract. Glare is another debilitating symptom of cataract for which an active individual needs to be operated. This happens particularly in context of night driving or even in bright sunlight.

The ultimate goal of a cataract surgery is to restore and maintain the precataract vision and to alleviate the other cataract-related symptoms. In the quest for perfection, the techniques and approaches followed by cataract surgeons have constantly evolved over the years from Intra Capsular Cataract Extraction (ICCE) to Aqualase (water jet technique). The phacoemulsification technique, which allows an exquisite intraoperative control and a consistent closed-chamber removal of cataract, undoubtedly reigns supreme in the developed countries. This technique has brought cataract surgery results as close to anatomical perfection as possible with the current technology and skills. In order to increase safety and to achieve faster visual rehabilitation for their patients, many surgeons are now adopting topical anesthesia with an adjunctive intracameral 1% lidocaine (Shah et al., 2004) instead of the peribulbar variety, which is till popular with most surgeons around the world. Incisions have progressed to sub-3 mm size on the temporal clear corneal region, which affords easier access to the cataract under topical anesthesia. Understanding the distinctive uses of the newer dispersive and cohesive viscoelastics has helped ensure better corneal endothelial protection during phacoemulsification. Of the wide range of phaco techniques developed to suit different cataracts and their related conditions, recommendations are for those that ensure endocapsular (posterior plane) phacoemulsification, which ensurse far superior long-term outcome. (Vasavada AR, Raj SM, Nehalani BR, MR Praveen, P @ P = 3P. Video film presented at the symposium of American Society Of Cataract & Refractive Surgeons, 2005, Washington DC, USA).

In the actual phacoemulsification technique, a sub 3 mm clear corneal tunnel is fashioned followed by injection of viscoelastic to form the anterior chamber. Anterior capsular opening (capsulorhexis) is created with the help of a bent needle (cystotome). Hydrodisection procedure is then performed to free the nucleus from the capsule. After ensuring a freely rotating nucleus, a wide trench or crater is created which is confined within the area of the capsulorhexis. After achieving sufficient thinning of the nuclear plate (atleast 90% of the total central depth), the phaco tip is buried at 6 o'clock, using controlled U/S power, to produce a vacuum seal. This results in an effective hold on the nucleus; the "step by step chop in situ and lateral separation" maneuver (Vasavada and Singh, 1998) is then performed by placing the chopper adjacent to the phacotip (Figure 1). The entire nucleus is chopped thus in a step-by-step fashion by rotating the chopped fragment clockwise and repeating the same chop technique. Finally the chopped wedges are consumed in the central space using the "stop, chop and stuff technique" (Vasavada and Desai, 1996) ensuring a completely endocapsular phacoemulsification (Figure 2). After emptying the capsular bag off the nucleus, the cortical matter is aspirated using bimanual irrigation and aspiration system. This is followed by foldable intraocular lens implantation in the capsular bag.

Figure 1. "Step by step chop in situ and lateral separation" maneuver to divide the nucleus of cataract into small wedges

Cataract surgeons today, in their relentless pursuit of perfection and excellence, are still looking into the probable advantages of other available options like ultrasound assisted by a secondary energy source such as PhocoTimesis, and fluid-assisted cataract removal like Aqualase™, pulsed hot water technology and the LASER-assisted cataract removal. Although some of these alternative futuristic techniques are available today, they have not been extensively adopted.

Posterior capsule opacification (PCO) is the prime deleterious consequence of cataract surgery. This aphoristic concern over the clarity of the posterior capsule shall undoubtedly dominate the future arenas of research and innovation. Presently, improving the IOL design and material appears to be a more practical means of reducing the incidence of PCO. The use of accommodative material also has a bright future if the absence of capsular opacification can be ensured. The current experimentation and innovation to perfect the chemoemulsification technique may turn out to be and easier alternative. The concept of implanting an intraocular drug delivery device at the end of cataract surgery is in its infancy. Its routine use in future may definitely bring significant relief to a surgeon from the worries of patient compliance and ensure an excellent round the clock postoperative medical control.

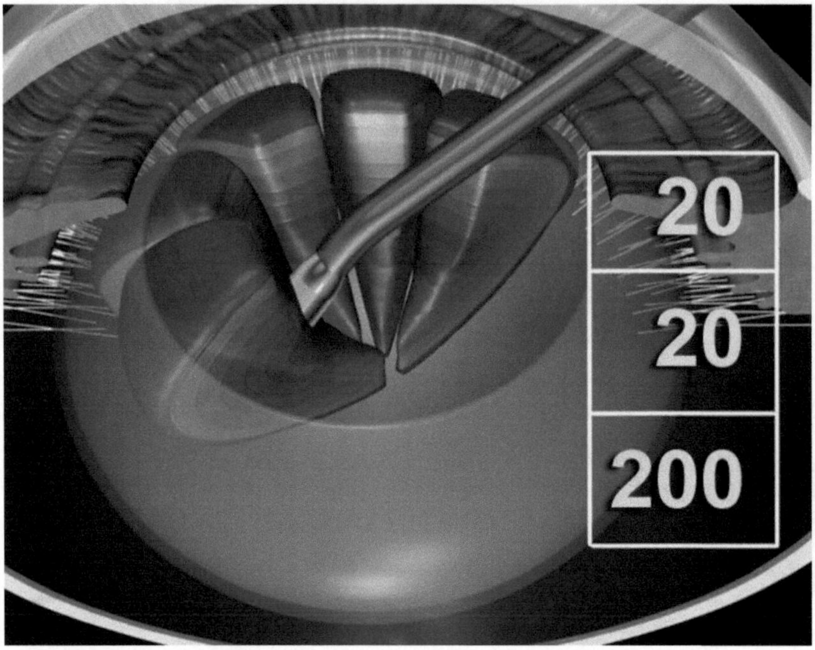

Figure 2. Consumption of wedges in the central space by "stop, chop and stuff technique"

4. PREVENTION STRATEGIES: PRESENT LIMITATIONS AND FUTURE POSSIBILITIES

Though surgery may be an effective means to reverse cataract blindness, visual outcomes will be poor where experienced surgeons and appropriate postoperative care, including refraction, are not available (He et al., 1999; Dandona et al., 1999). Moreover, even where high quality surgery is readily accessible, it may be expensive. It has been estimated that a delay in cataract onset of only 10 years could reduce the need for cataract surgery by as much as half. At present, no proved methods exist to effect such a result. This section will review existing and possible future strategies to prevent or delay age related cataract.

Reduction of sun exposure is an attractive means of preventing cataract related visual disability. Unfortunately, the proportion of risk attributable to sunlight exposure is small, and the type of lens opacity most consistently associated with UV-B is cortical opacity, a form which has generally been shown to be less visually disabling and less likely to require surgery than nuclear or PSC cataract (Klein et al., 1997).

There has recently been much interest in the impact of nutrients with antioxidant potential. In vitro and animal research has suggested that antioxidant substances present in the diet (Rose et al., 1998), in particular vitamins A, C (Delamere, 1996), and E (Fryer, 1993), may have a protective role from activated oxygen species.

Epidemiological evidence for the antioxidant hypothesis among human subjects, however, has been conflicting (Christen, 1999; Taylor et al., 1995; Congdon and West Jr, 1999). This is due to the large number of different antioxidants that have been examined, levels for many of which are likely to be highly colinear across individuals. A prospective follow up after a specific intervention may allows the role of different nutritional factors to be distinguished more readily.

A set of potentially interrelated personal factors like diabetes, hypertension, and body mass index are potentially remediable, suggesting possible avenues for cataract prevention but the effectiveness of such strategies remains to be proved. Although there is some evidence that better diabetic control (demonstrated by lower haemoglobin A_{1c} levels) may reduce the risk of lens opacity (Klein et al., 1998), no controlled, prospective data yet exist to demonstrate that improved treatment of diabetes or hypertension will in fact prevent or delay lens opacity. An added difficulty of intervening on BMI to prevent cataract is that the directionality of the association (for example, whether elevated or reduced BMI, or both, contribute to lens opacity) has not been definitively established.

An alternative strategy to risk factor reduction in the prevention of cataract would be pharmacological intervention. Compounds receiving attention as potential anticataract agents include aldose reductase inhibitors (Bron et al., 1998), pantethine (Congdon et al., 2000), and aspirin-like drugs such as ibuprofen (Harding, 1998). Population studies have also revealed a decreased risk of nuclear sclerosis among current users of oestrogen replacement therapy (Klein et al., 1994; Cumming and Mitchell, 1997; McCarty et al., 1999). However, none of these agents has demonstrated efficacy in the prevention of human lens opacity in a trial setting. A number of new drugs and pharmacological strategies remain under investigation (Ito et al., 2000; Spector et al., 2000; Takikawa et al., 1999). It is clear, however, that challenges to development of a practical anticataract agent for wide human distribution will be substantial: such an agent would need to be sufficiently safe for (presumably) long term use, and sufficiently inexpensive to compete with increasingly cheap cataract surgery. It appears very unlikely that a pill or eye drops requiring regular, long term use would be practical or sufficiently inexpensive.

Very great differences in the prevalence between racial groups are an evidence for a genetic effect on the distribution of age related cataract. Lens opacity was also found to develop on average 12 years earlier among the Indian subjects. The prevalence of previous cataract surgery among Indian people 40 years and above in Hyderabad, India, was 13.7% (Dandona et al., 1999a; 1999b), as opposed to 3.79% for the same age group in Melbourne (McCarty et al., 2000). These observed differences could be due to environmental factors rather than genetic. These include differences in nutrition, exposure to ultraviolet light (Burton et al., 1997; Javitt and Taylor, 1995), and rates of dehydrating episodes of diarrhoea (Javitt and Taylor, 1995). However, migrant studies of Indians living in Great Britain, where environmental differences with the local dwelling population might be expected to be reduced over time, have continued to demonstrate elevated rates of lens opacity and cataract surgery among people of sub continental descent (Bhatnagar et al., 1991;

Gray, 1996; Rauf et al., 1994). The data is in favour of a hereditary tendency for cataract among Indians.

Although evidence of a genetic effect on the development and progression of lens opacity is growing, to date no genes have yet been identified which are clearly associated with any form of isolated, adult onset cataract. Moreover, age related cataract is a complex trait, and it is likely that multiple loci will be involved. Among strategies currently being employed are the "candidate gene" approach, which seeks to identify mutations or sequence variants in well characterized genes thought likely to be associated with age related cataract. Candidate genes of current interest include those affecting crystallins (Stephan et al., 1999), structural proteins (Conley et al., 2000), gap junction proteins (Mackay et al., 1999), and aquaporins (Berry et al., 2000).

In summary, it must be said that those cataract prevention strategies for which adequate evidence exists namely, avoidance of ocular sun exposure, are not likely to result in large reductions in visual disability. Other strategies, which have been considered, involving nutritional, pharmacological, and specific medical interventions (against diabetes, for example), remain of unproved benefit. It seems likely that at least one fruitful avenue of investigation will be the genetics of age related cataract, an area which has as yet been little studied.

5. CONCLUSION

Cataract, opacification of the lens, is one of the commonest causes of loss of useful vision, with an estimated 16 million people worldwide affected. Several risk factors have been identified in addition to increasing age–genetic composition, exposure to ultraviolet light, and diabetes. However, no method to halt the formation of a cataractous lens has been shown to be effective. Nevertheless, advances in surgical removal of cataracts, including small-incision surgeries, use of viscoelastics, and the development of intraocular lenses, have made treatment very effective and visual recovery rapid in most cases. Despite these advances, cataract continues to be a leading public-health issue that will grow in importance as the population increases and life expectancy is extended worldwide.

REFERENCES

The age related eye disease study (AREDS) group. (2001) Randamized, placebo controlled, clinical trial of high dose supplementation with Vitamin C and E and Beta Carotene for age-related cataract and vision loss: AREDS report no. 9 Arch Ophthalmol, 119: 1439–1452.

Age-Related Eye Disease Study (AREDS) group. (2001) Risk factors associated with age-related nuclear and cortical cataract: a case-control study in the Age-related Eye Disease Study, AREDS Report NO. 5. Ophthalmology, 108: 1400–1408.

Altomare, E., Grattagliano, I., Vendemiale, G., et al. (1997) Oxidative protein damage in human diabetic eye: evidence of a retinal participation. Eur J Clin Invest, 27: 141–147.

Benedek, G.B. (1997) Cataract as a protein condensation disease: the Proctor Lecture. Invest Ophthalmol Vis Sci, 38: 1911–1921.

Berry, V., Francis, P., Kaushal, S., et al. (2000) Missense mutations in MIP underlie autosomal dominant 'polymorphic' and lamellar cataracts linked to 12q. Nat Genet, 25: 15–17.

Bhatnagar, R., West, K.P., Vitale, S., et al. (1991) Risk of cataract and history of severe diarrheal disease in southern India. Arch Ophthalmol, 109: 696–699.

Bochow, T.W., West, S.K., Azar, A., et al. (1989) Ultraviolet light exposure and risk of posterior subcapsular cataracts. Arch Ophthalmol, 107: 369–372.

Borchman, D., Yappert, M.C. (1998) Age-related lipid oxidation in human lenses. Invest Ophthalmol Vis Sci, 39: 1053–1058.

Bron, A.J., Brown, N.A.P., Harding, J.J., et al. (1998) The lens and cataract in diabetes. Int Ophthalmol Clin, 38: 37–67.

Bron, A.J., Vrensen, G.F., Koretz, J. (2000) The ageing lens. Ophthalmologica, 214: 86–104.

Brovkina, A.F., Zarubei, G.D. (1986) Ciliochoroidal melanomas treated with anarrow medical proton beam. Arch Ophthalmol, 104: 402–404.

Brown, N.P. (1999) Classification and pathology of cataract. In: Easty, D.M., Sparrow, J.M., editors. Oxford textbook of ophthalmology. Oxford: Oxford university press. vol., 1, p. 474.

Bunce, G.E., Kinoshita, J., Horwitz, J. (1990) Nutritional factors in cataract. Annu Rev Nutr, 10: 233–254.

Burton, M., Fergusson, E., Hart, A., et al. (1997) The prevalence of cataract in two villages of northern Pakistan with different levels of ultraviolet radiation. Eye, 11: 95–101.

Caulfield, L., West, S.K., Baron, Y., Cid-Ruzafa, J. (1999) Anthropometric status and cataract: The Salisbury Eye evaluation project. Am J Clin Nutr, 69: 237–242.

Chen, W.L., Hwang, J.S., Hu, T.H., Chen, M.S., Chang, W.P. (2001) Lenticular opacities in populations exposed to chronic low-dose-rate gamma radiation from radiocontaminated buildings in Taiwan. J Radiat Res (Tokyo), 156: 71–77.

Chen, Z., Johan, M., Subramanian, S., et al. (2004) 17-Beta-estradiol confers a protective effect against transforming growth factor-beta2-induced cataracts in female but not male lenses. Exp Eye Res, 78: 67–74.

Cheng, V.Y., Liu, J.H., Chen, S.J., Lee, F.L. (2000) Population-based study on prevalence and risk factors of age-related cataracts in Peitou, Taiwan. Zhonghua Yi Xue Xa Zhi (Taipei), 63: 641–648.

Christen, W.G. (1999) Antioxidant vitamins and age-related eye diseases. Proc Assoc AM Physicians, 111: 16–21.

Christen, W.G. (1999) Antioxidant vitamins and age-related eye disease. Proc Ass Am Phys, 111: 16–21.

Collaborative Normal-Tension Glaucoma Study Group. (1998) Comparison of glaucomatous progression between untreated patients with normal-tension glaucoma and patients with therapeutically reduced intraocular pressures. Am J Ophthalmol, 126: 487–497.

Congdon, N.G., West, K.P., Jr. (1999) Nutrition and the eye. Curr Opin Ophthalmol, 10: 464–473.

Congdon, N.G., West, S.K., Duncan, D., et al. (2000) The effect of pantethine and ultraviolet-B radiation on the development of lenticular opacity in the Emory mouse. Curr Eye Res, 20: 17–24.

Congdon, N., West, S.K., Buhrmann, R.R., et al. (2001) Prevalence of the different types of age-related cataract in an African population. Invest Ophthalmol Vis Sci, 42: 2478–2482.

Conley, Y.P., Erturk, D., Keverline, A., et al. (2000) A juvenile-onset, progressive cataract locus on chromosome 3q21-22 is associated with a mis-sense mutation in the beaded filament structural protein-2. Am J Hum Genet, 66: 1426–1431.

Cumming, R.G., Mitchell, P. (1997) Hormone replacement therapy, reproductive factors, and cataract. The Blue Mountains Eye Study. Am J Epidemiol, 145: 242–249.

Dandona, L., Dandona, R., Naduvilath, T.J., et al. (1999) Burden of moderate visual impairment in an urban population in southern India. Ophthalmology, 106: 497–504.

Dandona, L., Dandona, R., Naduvilath, T.J., et al. (1999) Population-based assessment of the outcome of cataract surgery in an urban population in southern India. Am J Ophthalmol, 127: 650–658.

Delamere, N. (1996) Ascorbic acid and the eye. Subcell Biochem, 25: 313–329.

Floyd, R.P. (2000) History of cataract surgery. In: Albert, D.M., Jakobiec, F.A., editors. Principles and Practice of Ophthalmology. 2nd ed. Philadelphia: Saunders, p. 1463–76.

Fryer, M.J. (1993) Evidence for the photoprotective effects of vitamin E. Photochemistry and Photobiology, 58: 304–312.
Glynn, R.J., Christen, W.G., Manson, J.E., et al. (1995) Body mass index. An independent predictor of cataract. Arch Ophthalmol, 113: 1131–1137.
Gray, P.J. (1996) The prevalence of eye disease in elderly Bengalis in Tower Hamlets. J R Soc Med, 89: 23–26.
Graziosi, P., Rosmini, F., Bonacini, M., et al. (1996) Location and severity of cortical opacities I different regions of the lens in age-related cataract. Invest Ophthalmol Visc Sci, 37: 1698–1703.
Griswold, M.S., Stark, W.S. (1992) Scotopic spectral sensitivity of phakic and aphakic observers extending into the near ultraviolet. Vis Res, 32: 1739–43.
Gupta, P.D., Johar Kaid, S.R., Nagpal, K., Vasavada, A.R. (2005) Sex hormone receptors in the human eye. Surv Ophthalmol. 50(3): 274–84. Review.
Hales, A.M., Chamberlain, C.G., Murphy, C.R., McAvoy, J.W. (1997) Estrogen protects lenses against cataract induced by transforming growth factor-beta (TGFbeta). J Exp Med, 185: 273–80.
Harding, J.J. (1998) Can cataract be prevented? Eye, 13: 554–556.
He, M., Xu, J., Li, S., et al. (1999) Visual acuity and quality of life in patients with cataract in Doumen County China. Ophthalmology, 106: 1609–1615.
Hiller, R., Podger, M.J., Sperduto, R.D., et al. (1998) A longitudinal study of body mass index and lens opacities. The Framingham Studies. Ophthalmology, 105: 1244–1250.
Hood, B.D., Garner, B., Roger, J.W.T. (1999) Human Lens Coloration and Aging: Evidence for crystalline modification by the major ultraviolet filter, 3-hydroxy-kynurenine O-β-D-Glucoside. J Biol Chem (communication), 274: 46; 32547–32550.
Ito, K., Inoue, S., Yamamoto, K., Kawanishi, S. (1993) Hydmxycleoxy guanosine formation at the 5' site of 5'-GS-3' sequences in double-stranded DNA by UV radiation with riboflavin. Biol. C/tern, 268: 13221–13227.
Ito, Y., Cai, H., Koizumi, Y., et al. (2000) Effect of lipid composition on the transcorneal penetration of liposomes containing disulfiram, a potential anti-cataract agent, in the rabbit. Biol Pharm Bull, 23: 327–333.
Javitt, J.C., Taylor, H.R. (1995) Cataract and latitude. Doc Ophthalmol, 88: 307–325.
Johns, K.J., Feder, R.S., Hammill, B.M., Miller-Meeks, M.J., Rosenfeld, S.I., Perry, P.E., editors (2002). Lens and Cataract: Section 11, Basic and Clinical Science Course. San Francisco: American Academy of Ophthalmology.
Kleiman, N.J., Wang, R.-R., Spector, A. (1990) Hydrogen peroxide-induced DNA damage in bovine lens epithelial cells. Mutation Res, 240: 35–45.
Klein, B.E., Klein, R., Ritter, L.L. (1994) Is there evidence of an estrogen effect on age-related lens opacities? The Beaver Dam Eye Study. Arch Ophthalmol, 112: 85–91.
Klein, B.E., Klein, R., Wang, Q., et al. (1995) Older-onset diabetaes and lens opacities: the Beaver Dam Eye Study. Ophthalmic Epidemilo, 2: 49–55.
Klein, B.E., Klein, R., Moss, S.E. (1997) Incident cataract surgery: the Beaver Dam eye study. Ophthalmology, 104: 573–580.
Klein, B.E., Klein, R., Lee, K.E. (1998) Diabetes, cardiovascular disease, selected cardiovascular risk factors and the 5-year incidence of age-related cataract and progression of lens opacities: the Beaver Dam Eye Study. Am J Ophthalmol, 126: 782–790.
Krause, V.M., Delisel, H., Solonons, N.W. (1998) Fortified foods contribute one half of the recommended Vitamin A Intact in poor urban Guatemalan toddlers. J Nutr, 128: 860–864.
Leske, M.C., Wu, S.Y., Hennis, A., et al. (1999) Diabetes, hypertension, and central obesity as cataract risk factors in a black population. The Barbados Eye Study, Ophthalmology, 106: 35–41.
Leske, M.C., Wu, S.Y., Nemesure, B., Li, X., Hennis, A., Connell, A.M. (2000) Incidence and progression of lens opacities in Barbados Eye Studies. Ophthalmology, 107: 1267–1273.
Leske, M.C., Wu, S.Y., Nemesure, B., et al. (2004) Nine-year incidence of lens opacities in the Barbados eye studies. Ophthalmology, 111: 483–90.
Lim, R., Mitchell, P., Cumming, R.G. (1999) Refractive associations with cataract: the Blue Mountains Eye Study. Invest Ophthalmol Vis Sci., 40: 3021–3026.

Livingston, P.M., Guest, C.S., Stanislavsky, Y., et al. (1994) A population-based estimate of cataract prevalence: the Melbourne Visual Impairment Project experience. Dev Ophthalmol, 26: 1–6.

Livingstone, B.I., Bourke, R.D. (1999) Retrospective study of macular holes treated with pars plana vitrectomy. Aust NZ J Ophthalmol, 27: 331–341.

Mackay, D., Ionides, A., Berry, V., et al. (1999) Connexin-46 mutations in autosomal dominant congenital cataract. Am J Hum Genet, 64: 1357–1364.

Manson, J.E., Gaziano, J.M., Spelsberg, A., et al. (1995) Secondary prevention trial of antioxidant vitamins and cardiovascular disease in woman. Rationale, design and methods. Ann Epidemiol, 5: 261–269.

McCarty, C.A., Mukesh, B.N., Fu, C.L., et al. (1999) The epidemiology of cataract in Australia. Am J Ophthalmol, 128: 446–465.

McCarty, C.A., Nanjan, M.B., Taylor, H.R. (2000) Operated and unoperated cataract in Australia. Clin Exp Ophthalmol, 28: 77–82.

Mitchell, P., Cumming, R.G., Attebo, K., Panchapakesan, J. (1997) Prevalence of cataract in Australia: the Beaver Dam eye study. Ophthalmology, 104: 581–588.

Munoz, B., Tajchman, U., Bochow, T., et al. (1993) Alcohol use and risk of posterior subcapsular opacities. Arch Ophthalmol, 111: 110–112.

Pau, H., Graf, P., Sies, H. (1990) Glutathione levels in human lens: regional distribution in different forms of cataract. Exp Eye Res, 50: 17–20.

Ramachandran, S., Morris, S.M., Devamanoharan, P.S., et al. (1991) Radio-isotopic determination of hydrogen peroxide in aqueous humor and urine. Exp. Eye Res, 53: 503–506.

Ramkrishnan, R., Michon, J., Robin, A.L., Krishnadas, R. (1993) Safety and efficacy of mitomycin C trabeculectomy in southern India. A short-term pilot study. Ophthalmology, 100: 1619–1623.

Rauf, A., Ong, P.S., Pearson, R.V., et al. (1994) A pilot study into the prevalence of ophthalmic disease in the Indian population of Southall. J R Soc Med, 87: 78–79.

Robman, L.D., Tikellis, G., Garrett, S.K., et al. (1999) Baseline ophthalmic findings in the vitamin E, cataract and the age-related maculopathy (VECAT) study. Aust NZ J Ophthalmol, 27: 410–416.

Rose, R.C., Richer, S.P., Bode, A.M. (1998) Ocular oxidants and anti-oxidant protection. Proc Soc Exp Biol Med, 217: 397–407.

Schein, O.D., West, S.K., Monoz, B., et al. (1994) Cortical lenticular opacification: distribution and location in longitudinal study. Invest Ophthalmol Visc Sci, 35: 363–366.

Shah, A.R., Diwan, R.P., Vasavada, A.R., Keng, M.Q. (2004) Corneal endothelial safety of intracameral preservative-free 1% xylocaine. Indian J Ophthalmol, 52(2): 133–8.

Singh, S., Gupta, P.D. (1997) Mechanism of action of estradiol; Non-genomic events, in Sengupta J, Ghosh D: Cellular and molecular signalling in reproduction. New Delhi, New-age International (P) Ltd., pp 69–83.

Singh, S., Gupta, P.D. (1997) Induction of phosphoinositide-mediated signal transduction pathway by 17 beta-oestradiol in rat vaginal epithelial cells. J Mol Endocrinol, 19: 249–57.

Spector, A., Zhou, W., Ma, W., et al. (2000) Investigation of the mechanism of action of microperoxidase-11 (MP11), a potential anti-cataract agent, with hydrogen peroxide and ascorbate. Exp Eye Res, 71: 183–194.

Sperduto, R.D., Hu, T.S., Milton, R.C., et al. (1993) The Linxian Cataract Study. Two nutrition intervention trials. Arch Ophthalmol, 111(9): 1246–1253.

Stadtman, E.R. (1992) Protein oxidation and aging. Science, 257: 1220–1224.

Stephan, D.A., Gillanders, E., Vanderveen, D., et al. (1999) Progressive juvenile-onset punctate cataracts caused by mutation of the gamma-D crystallin gene. Proc Nat Acad Sci USA, 96: 1008–1012.

Takikawa, O., Littlejohn, T., Jamie, J.F., et al. (1999) Regulation of indoleamine 2,3-doxygenase, the first enzyme in UV filter biosynthesis in the human lens. Relevance for senile nuclear cataract. Adv Exp Med Biol., 467: 241–245.

Taylor, A., Jacques, P.F., Epstein, E.L. (1995) Relations among aging-antioxidant status and cataract. Am J Clin Nutr, 62(suppl): 1439–1447.

Taylor, A., Jacques, P.F., Epstein, E.M. (1995) Relations among aging, anti-oxidant status, and cataract. Am J Clin Nutr, 62(suppl): 1439S–1447S.

The AGIS investigators. (2000) The advanced glaucoma intervention study, 6: effect of cataract on visual field and visual acuity. Arch Ophthalmol, 118: 1639–1652.

Van der Haar, F. (1997) The challenge of the global elimination of iodine deficiency disorders. Eur J Clin Nutr, 51: 53–58.

VanNewkirk, M., Alfonso, C.E., Chuang, E.L., Collins, M.L., Isenberg, S.J., Klein, R., Lietman, R.M., editors. (2002) International Ophthalmology: Section 13, Basic and Clinical Science Course. San Francisco: American Academy of Ophthalmology, p. 160–1.

Vasavada, A.R., Desai, J.P. (1996) Stop, chop, chop, and stuff. J Cataract Refract Surg, 22: 526–9.

Vasavada, A.R., Singh, R. (1998) Step-by-step, chop in situ and separation of very dense cataracts. J Cataract Refract Surg, 24: 156–9.

Vasavada, A.R., Mamidipudi, P.R., Sharma, P.S. (2004) Morphology of and visual performance with posterior subcapsular cataract. J Cataract Refract Surg, 30: 2097–2105.

West, S.K., Duncan, D.D., Munoz, B., et al. (1998) Sunlight exposure and risk of lens opacities in a population-based study: The Salisbury Eye evaluation progect. JAMA, 280: 714–718.

Wong, T.Y., Klein, B.E., Klein, R., Tomany, S.C., Lee, F.L. (2001) Refractive errors and incident cataracts:; the Beaver Dam Eye Study. Invest Ophthalmol Vis Sci, 42: 1449–1454.

Wong, T.Y., Klein, D.E.K., Klein, R., Tomany, S.C. (2002) The relation of ocular trauma to cortical, nuclear, and posterior subcapsular cataracts: The Beaver Dam Eye Study. Br J Ophthalmol, 86: 152–155.

Worgul, B.V., Merriam, C.R., Medveclovsky, C. (1989) Cortical cataract development-an expression of primary damage to the lens epithehum. Lens Eye Toxicity Res, 6: 559–571.

Wu, S.Y., Nemesures, B., Leske, M.C. (1999) Refractive errors in a black adult population: the Barbados Eye Study. Invest Ophthalmol Vis Sci, 40: 2179–2184.

CHAPTER 10

SKIN AGING: PATHOGENESIS, PREVENTION AND TREATMENT

MARY S. JUNG*, KRISTEN M. KELLY* AND JERRY L. McCULLOUGH*
* *Department of Dermatology, University of California, Irvine, California*

Abstract: Skin aging is a consequence of genetically programmed processes of intrinsic aging and extrinsic aging caused by ultraviolet light and other environmental insults. There are many different approaches to reduce or postpone the untoward effects of intrinsic programmed aging and extrinsic environmental injury. The prevention of extrinsic aging utilizes various methods of photoprotection and antioxidants. Treatments of aged skin are not limited to a single procedure but may consist of a combination of many adjuvant treatments each of which offers different degrees of effectiveness, risk, duration and cost. It is important to have an understanding of the likely benefits and limitations of available treatments. Scientific evaluation and education about the modalities available for treatment of aged skin can help to achieve these goals

Keywords: aging; anti-aging; photoaging; skin rejuvenation; photoprotection; antioxidants; retinoids; hydroxyl acids; microdermabrasion; chemical peels; botulinum toxin; soft tissue fillers; laser resurfacing; ablative resurfacing; intense pulsed light; light-emitting diode photomodulation; radiofrequency devices; fractional photothermolysis; cosmetic surgery

1. INTRODUCTION

The aging process begins at birth and cutaneous manifestations of aging generally begin to be visible in the second decade of life (Oikarinen, 1994). Aging is gradual, but persistent and irreversible and occurs at different rates in individuals. Cutaneous changes associated with aging include decrease of skin elasticity, sagging secondary to gravity, and fat atrophy which result in facial wrinkles and jowls.

The aging process of the skin can be divided into chronological or intrinsic aging and extrinsic aging. Intrinsically aged skin is generally smooth, pale, more evenly pigmented, and finely wrinkled (Chung, 2003). The histologic findings of intrinsic aging include a decrease in the extracellular matrix characterized by

S. Rattan and M. Kassem (eds.), Prevention and Treatment of Age-related Diseases, 175–192.

reduced elastin and elastic fiber disintegration. In contrast, photoaged skin is sallow, coarsely wrinkled, and associated with irregular pigmentation and telangiectasias (Chung, 2003). Dermatoheliosis is the term used to describe these photoaging-associated clinical changes. Histologically, photoaged skin has an atrophic epidermis, thinned spinous layer, loss of rete ridges, and decreased numbers of Langerhans' cells. There is condensed collagen beneath the basement-membrane zone, basophilic degeneration of deeper dermal collagen, and telangiectasia in the upper dermis. Collagen deficiency in chronically photodamaged skin may result from increased, repetitive degradation of collagen by ultraviolet (UV)-induced matrix metalloproteinases (Chung et al., 2001).

Chronic sun exposure is widely accepted as the principal environmental cause of extrinsic skin aging. Ultraviolet B (UVB) radiation is mainly responsible for sunburn, suntanning, and photocarcinogenesis following sun exposure (Afaq et al., 2005). Ultraviolet A (UVA) is suspected of playing a proportionately larger role in photoaging because of its greater abundance in the sunlight reaching the earth's surface, greater year-round and day-long exposure, and greater depth of penetration into the dermis compared with UVB (Lavker et al., 1995).

Photoaging depends primarily on the degree of sun exposure and skin pigment. Individuals who have outdoor lifestyles, live in sunny climates, and are lightly pigmented will experience the greatest degree of photoaging (Fisher et al., 2002).

Synergistic with sun exposure, cigarette smoking may further contribute to extrinsic aging, particularly in women, with a direct correlation between the number of pack-years smoked and the severity of wrinkling and grayish discoloration (Smith and Fenske, 1996).

Premature aging of the skin is observed in several hereditary disorders and has been associated with mutations of genes that code for proteins involved in repair of DNA damage (Pesce and Rothe, 1996). For example, patients with Cockayne syndrome and Werner syndrome display mutations in DNA helicases. This suggests that decreased DNA repair capacity is associated with accelerated aging and that cellular injury, particularly cumulative DNA damage, plays a major role in the aging process (Furuichi, 2001).

Modern society's increasing emphasis on a youthful image and physical beauty has resulted in soaring demand for and resultant development of a wide range of skin care products and procedural interventions for use by the aging population. Available products to prevent aging include sunscreens and antioxidants. The most commonly utilized interventions to "treat' aged skin include topical pharmaceuticals and a wide range of surgical procedures.

2. PREVENTION OF SKIN AGING

2.1 Sun Protection

In the absence of adequate protection from the sun, other treatments will be less effective and may even be detrimental. Good protection strategies include wearing broad-brimmed hats, protective clothing, and sun avoidance, particularly during midday hours. In addition, tanning must be discouraged (Stern, 2004).

One of the main pharmaceutical approaches to prevention of photoaging is sunscreen. In a randomized trial in humans, the use of a sunscreen with a sun protection factor (SPF) of 29 for two years stabilized histologic changes in the skin, as compared to the placebo group where photoaging-associated changes increased (Boyd et al., 1995). Furthermore, in large multicenter studies investigating topical tretinoin as a treatment for photoaging, patients in the control groups who used only daily sunscreen and moisturizer for 6 months were found to have statistically significant improvement in fine wrinkling, roughness, dyspigmentation, and overall appearance as compared with their own baseline status (Gilchrest, 1996). Avoidance of sun exposure and use of sunscreen also leads to regression of skin pre cancers, actinic keratoses (Thompson et al., 1993), which indicates the skin has an intrinsic repair capacity. These studies underscore the importance and mandate the inclusion of photoprotection in any treatment regimen.

2.2 Antioxidants

Antioxidants are another pharmaceutical approach to prevention and also treatment or reversal of skin photoaging. They neutralize reactive oxygen species generated by ultraviolet (UV) light exposure (Kullavanijaya and Lim, 2005). Substances marketed as antioxidants include vitamins C and E, coenzyme Q10, idebenone, ferulic acid, and cytokinins. Objective evidence to support the role of these substances is available but limited.

Fitzpatrick and Rostan (Fitzpatrick and Rostan, 2002) documented statistically significant increased skin hydration, increased collagen production and wrinkle reduction in 4 of 10 subjects who applied 10% vitamin C for 12 weeks. Average improvement on the treatment side was 25% compared to 7.7% on the control side. Biopsies showed increased Grenz zone collagen, and increased type I collagen mRNA.

Topical vitamin E provides photoprotection by both antioxidant and UV absorptive properties (Krol et al., 2000).

Coenzyme Q10 (CoQ10) occurs naturally in human cells and is believed to prevent oxidative stress-induced apoptosis by inhibiting lipid peroxidation in plasma membranes (Baumann, 2004). CoQ10 levels decrease naturally with age as well as with stress and illness.

Idebenone, an analog of CoQ10, and ferulic acid, a plant extract, are recently available antioxidant ingredients in topical formulations.

Cytokinins are plant-growth substances that promote cell division and play a role in cell differentiation (Barciszewski et al., 1999). Most commonly cytokinins are N6-substituted adenine derivatives. Kinetin (N6-furfuryladenine), a cytokinin which is naturally occurring in DNA and cell extracts, retards senescence of plants (Van Staden et al., 1988) and delays age-related changes in human skin fibroblasts in culture (Rattan and Clark, 1994). Studies of the molecular pathways through which kinetin brings about its biological effects have shown that kinetin prevents oxidative damage to DNA (Olsen A et al., 1999) and glycoxidation-mediated damage to proteins (Verbeke P et al., 2000). In a 52-week study in 96 subjects with photodamaged facial skin, twice daily application of kinetin improved skin roughness (63%), mottled hyperpigmentation (32%) and fine wrinkles (17%) (McCullough, 1999). Treatments also improved skin-barrier function as measured by a decrease in transepidermal water loss. Extended treatment with kinetin was well tolerated and did not cause clinical signs or subjective symptoms of irritation (McCullough and Weinstein, 2002).

Recent studies have demonstrated that trans-zeatin (6-[4-hydroxy-3-methyl-but-2-enylamino]adenine, a cytokinin isolated from plants (Letham, 1963) and present in the tRNA of a wide variety of organisms (Mok and Mok, 1994) also has gerontomodulatory, youth preserving and anti-aging effects on human fibroblasts undergoing aging in culture (Rattan, 2005). Zeatin and other cytokinins or their derivatives may provide useful compounds with applications in aging prevention, intervention and therapy for the future.

3. TREATMENT OF SKIN AGING

As noted above, a wide range of treatments are available for aged skin including topical pharmaceuticals, microdermabrasion, chemical peels, botulinum toxin (BTX), soft tissue fillers, dermabrasion, ablative resurfacing, non-ablative light-based rejuvenation, radiofrequency, fractional photothermolysis and traditional cosmetic surgery (Table 1).

3.1 Topical Interventions

Topical pharmaceuticals available for treatment of photoaged skin include antioxidants (see above), retinoids and alpha- and beta-hydroxy acids. Of these approaches, only topical retinoids, particularly tretinoin (all-*trans* retinoic acid), have a well-documented ability to repair photoaged skin at the clinical, histological and molecular level.

3.1.1 Topical retinoids

A large number of controlled clinical trials have been published demonstrating that the topical application of 0.025% to 0.1% tretinoin (Retin-A®, Renova® (OrthoNeutrogena, Skillman, NJ, USA); Avita® (Mylan Laboratories, Inc., Canonsburg, PA, USA)) improves the appearance of photoaged skin by significantly reducing fine wrinkling, skin roughness, and mild to moderate hyperpigmentation (Kang and Voorhees, 1998). The histologic changes correlating to these clinical improvements include epidermal thickening, increased granular layer thickness, stratum corneum compaction, and decreased melanin content (Gilchrest, 1999). At the molecular level, topical tretinoin has been shown to induce type I and type III procollagen gene expression in photoaged human skin. Because procollagen is the

Table 1. Treatments for Skin Aging

Treatment	Recovery/ *(Discomfort)	Onset of Improvement	Effect Duration	**Expected Improvement	Risks/ Disadvantages
Topical Treatment					
Retinoids	None/(0–1)	2–3 months	All topical	0–1	Irritation
Hydroxy acids	None/(0)	2–3 months	treatments require	0–1	Irritation
Antioxidants	None/(0)	2–3 months	continuous use	0–1	None
Microdermabrasion	0–1 day/ (0–1)	0–1 day	Requires repeat treatments	0–1	Hyperpigmentation, Prolonged erythema
Chemical Peels					
Superficial	0–4 days/ (1–2)	2–3 peels	2–4 months	1–2	Hyperpigmentation
Medium	7–12 days/ (2–3)	2–4 weeks	1–2 years	1–3	Infection, scarring
Deep	2–4 weeks/ (3)	4–8 weeks	2–5 years	2–3	Hypopigmentation, infection, scarring
Botulinum Toxin	0–3 days/ (1–2)	1–3 days	3–6 months	1–3	Headache, bruising, ptosis
Dermal Fillers	0–1 week/ (1–2)	Immediate	Variable	1–3	Bruising, allergic reaction
Ablative Resurfacing	1–4 weeks/ (3)	1–4 weeks	3–7 years	2–3	Scarring, infection
Dermabrasion					
Dermasanding					
Ablative Laser Skin Resurfacing					
Non-ablative Light-based Rejuvenation	1–4 hours/ (0–1)	2–9 months	Variable	1–2	Pigment change
Radiofrequency Devices	1–24 hours/ (1–3)	Immediately	Unknown	1–2	Scarring, pain
Fractional Photothermolysis	1–2 weeks/ (1–3)	1–3 months	Unknown	1–2	Erythema, mild edema
Cosmetic surgery	1–6 weeks/ (2–3)	1–6 weeks	5–7 years	2–3	Invasive, prolonged recovery

* Discomfort (0 = none; 1 = mild; 2 = moderate; 3 = severe).
** Expected Improvement (0 = none; 1 = subtle; 2 = moderate; 3 = major).

precursor to collagen, it is likely that increased production of procollagen results in increased deposition of collagen (Griffiths et al., 1993).

In addition to tretinoin, another topical retinoid, tazarotene (Tazorac®)(Allergan, Inc., Irvine, CA, USA), has been approved by the Food & Drug Administration (FDA) for the improvement of fine wrinkles and irregular pigmentation associated with photoaging. In a multicenter, randomized trial evaluating the efficacy of 0.1% tazarotene cream for photodamage, clinically and statistically significant improvements were noted in a variety of skin characteristics (ie, fine wrinkling, mottled hyperpigmentation, lentigines, elastosis, pore size, irregular depigmentation, tactile roughness, and coarse wrinkling) (Phillips et al., 2002).

3.1.2 Alpha- and beta-hydroxy acids

Alpha-hydroxy acids (AHAs) and beta-hydroxy acids (BHAs) are naturally found in foods, including dairy products (lactic acid), fruit (citric acid), and sugar cane (glycolic acid). Hydroxy acids in low concentrations (typically 4 to 12 percent) are components of nonprescription creams and lotions that are promoted as ameliorating the signs of aging. In higher concentrations, these preparations are used as "peels."

The topical treatment of photodamaged skin with AHAs results in subtle clinical improvements in wrinkling, roughness, and dyspigmentation within months of daily application (Stiller et al., 1996). Histological improvement has been reported after 6 months of daily applications of products containing 25% glycolic, lactic, or citric acid (Ditre et al., 1996). Bernstein et al. (2001) demonstrated that epidermal and dermal hyaluronic acid and collagen gene expression were increased in skin treated with 20% glycolic acid (twice daily for 3 months) as compared to vehicle-treated controls.

3.2 Surgical Interventions

An array of surgical approaches are available for treatment of photoaging.

3.2.1 Microdermabrasion

Microdermabrasion is used to treat individuals with early photodamage and other skin imperfections. The procedure involves using tiny particles of either aluminum oxide, sodium chloride, or sodium bicarbonate crystals directed at the skin through a vacuum tube causing mechanical removal of the superficial epidermis and stimulation of new cell growth. Studies demonstrate small but quantifiable improvements post-microdermabrasion. Shim et al. (2001) evaluated clinical and histopathologic effects of microdermabrasion. In 14 subjects with photoaging, acne, and acne scarring who underwent 6–7 treatments over 12–14 weeks, there was significant decrease in roughness, mottled pigmentation, and enhancement of overall skin appearance but only minimal improvement in rhytides as judged by patient assessment. Microdermabrasion can also be used as an adjuvant therapy to facilitate the efficacy of other rejuvenation procedures including photodynamic therapy (Sadick and Finn, 2005).

3.2.2 Chemical peels

Chemical peels have a long history of safety and efficacy and are relatively easy to perform. Chemical peels are classified as superficial, medium-depth, and deep. Superficial peels cause epidermal injury and occasionally extend into the papillary dermis; medium-depth peels injure through the papillary dermis to the upper reticular dermis; and deep peels injure to the mid reticular dermis. Degree of clinical improvement, length of recovery period, and risk of complications are all proportionate to the depth of tissue injury.

Superficial chemical peels, often referred to as "lunch time" peels, are used in the management of mild photoaging. Superficial peeling agents include salicylic acid, glycolic acid, low-dose trichloroacetic acid (10–20% TCA), and Jessner's solution (resorcinol, salicylic acid, lactic acid, and ethanol). In a double-blind, vehicle-controlled study with 41 subjects, either glycolic acid (50%) or vehicle was applied topically for 5 minutes to one side of the face, forearms, and hands, once weekly for four weeks (Newman N et al., 1996). There was a statistically significant decrease in rough texture, fine wrinkling, number of solar keratoses, and slight lightening of solar lentigines on areas treated with glycolic acid. This corresponded histologically to thinning of the stratum corneum, granular layer enhancement, and epidermal thickening. Some specimens showed increased collagen thickness in the dermis (Newman J et al., 1996). Superficial peeling agents require multiple procedures to obtain results. All of them share the advantages of only mild stinging and burning during application as well as minimal time needed for recovery. However, noted improvements are usually subtle because there is little to no effect on the dermis. Thus, the results of repetitive superficial chemical peels never approach the effect obtained with a single medium-depth or deep peel.

Most medium-depth chemical peels are performed utilizing 35% TCA in combination with either 70% glycolic acid or Jessner's solution (Tse et al., 1996). These latter agents both weaken the epidermal barrier and allow deeper, more uniform, and controlled penetration of the 35% TCA. Medium depth chemical peels can be repeated at 6 months intervals (Monheit, 2001) but frequently one procedure achieves the desired effect. Potential complications include skin discoloration or scarring.

Deep peeling can be achieved with TCA in concentrations above 50% (Matarasso and Glogau, 1991) or a phenol-containing preparation, such as the Baker-Gordon phenol formula (3 mL Phenol, USP, 88%, 2 mL tap or distilled water, 8 drops septisol liquid soap, 3 drops croton oil). The use of phenol results in new collagen formation, leading to wrinkle reduction, but its cardio-toxic profile also increases the procedure's associated risks. Patients with liver and renal impairment can quickly accumulate toxic levels and develop cardiac arrhythmias. Therefore, careful monitoring is required throughout the procedure. Other disadvantages of this procedure include having a longer recovery period and greater risk of adverse effects, mainly permanent hypopigmentation and scarring.

3.2.3 Botulinum toxin

Different strains of the bacterium *Clostridium botulinum* produce distinct types of botulinum toxins (A,B,C1,D,E,F,and G), all of which block the release of acetylcholine and relax muscles. In 1992, Drs. Jean and J. Alastair Carruthers noted smoothing of the glabellar brow furrow in a patient who had been treated with botulinum toxin injection for blepharospasm (Carruthers and Carruthers, 1992). Open-label studies and two double-blind, placebo-controlled studies documented the safety and efficacy of botulinum toxin injections (Keen et al., 1994; Lowe et al., 1996) for cosmetic purposes, and in 2002 the FDA granted approval of Botox® Cosmetic (botulinum toxin type A, Allergan, Inc., Irvine, CA, USA) for "temporary improvement in the appearance of moderate to severe glabellar lines in adult patients 65 or younger". One large randomized, multicenter, double-blind, placebo-controlled trial of 264 patients found at least moderate improvement in 50 to 75 percent of patients treated for glabellar lines (Carruthers et al., 2002). Improvement was rapid (nearly peak effect by day 7 with a small degree of continued enhancement up to one month post-injection) and effects lasted 3–4 months. Botox® Cosmetic is the most studied brand of botulinum toxins, although other forms are commercially available. Ipsen Ltd (UK) markets BTX-A in Europe under the brand name Dysport® and Solstice Neuroscience, Inc. (San Diego, CA, USA) produces MYOBLOC®, a formulation of botulinum toxin type B.

The most common use is treatment of dynamic expression lines of the upper third of the face (glabellar brow furrow, horizontal forehead frown lines and periocular "crow's feet" rhytides); however, in recent years, BTX has been increasingly used in the mid and lower face and neck for "bunny lines" (downward radiating lines on the sides of nose), perioral rhytides, dimpled chin, and platysmal bands (Matarasso et al., 1999; Semchyshyn and Sengelmann, 2003). Consensus treatment guidelines were developed in 2004 (Carruthers et al., 2004). BTX can be used alone or in combination with other cosmetic procedures such as soft tissue augmentation and laser resurfacing, to enhance and prolong effects (Patel et al., 2004; West and Alster, 1999). BTX injections are minimally invasive, well tolerated, and do not require a lengthy recovery period.

Side effects are uncommon and generally mild but can include bruising, eyelid & brow ptosis, and headaches (Klein, 2004). Peripheral motor neuron disease is a relative contraindication to treatment because this condition can be potentiated by the toxin.

3.2.4 *Soft tissue fillers*

Soft tissue fillers (Table 2) are used to smooth and correct wrinkles, non-dynamic furrows, and hollows in the face. Other indications include lip augmentation and replacement of lost subcutaneous fat. Products have previously been categorized as either temporary or permanent (Werschler and Weinkle, 2005). Recently the number of available products has increased greatly and "semi-permanent" fillers have emerged that provide augmentation on the face for 2–5 years (Stegman et al., 1988).

3.2.4.1 Temporary Products The below described products last 3–6 months and as such, require frequent re-administration to maintain desired results. While the transient nature of these products can be frustrating to patients, there is the advantage that any adverse effects are also generally temporary.

Bovine collagen products were the first FDA approved fillers and achieve correction for approximately 3 months (Stegman et al., 1988). There are three

Table 2. Soft Tissue Fillers

Products	Company	Description	Skin Test	Results
Temporary Products				
Zyderm® I/II and Zyplast®	Inamed (Santa Barbara, CA, USA)	Bovine collagen	Yes	Immediate, lasts 3–6 months
Cosmoderm® I/II and Cosmoplast®	Inamed (Santa Barbara, CA, USA)	Human collagen derived from fibroplast cell cultures	No	Immediate, lasts 3–6 months
Isolagen®	Isologen (Houston, TX, USA)	Autologous collagen grown in culture	No	Variable
Dermalogen®	Collagenesis (Beverly, MA, USA)	Allograft material of human tissue collagen matrix from cadavers	No	Variable
Cymetra®	LifeCell (Branchburg, NJ, USA)	Acellular dermal graft material from cadavers	No	Variable, lasts 3–6 months
Restylane Fine Line®, Restylane®, Perlane®	Q-Med (Uppsala, Sweden)	Non-animal derived hyaluronic acid	No	Immediate, lats between 6–9 months
Captique™	Genzyme (UK)	Non-animal derived	No	Immediate, lasts up to 6 months
Juvéderm® 18, Juvéderm® 24, Juvéderm® 30	Inamed (Santa Barbara, CA, USA)	Non-animal derived, cross-linked hyaluronic acid	No	Immediate, lasts up to 1 year
Hylaform®, Hylaform® Plus	Genzyme (UK)	Hyaluronic acid extracted from rooster combs	Yes	Immediate, lasts up to 6 months
Fascian®	Fascia Biosystems (Beverly Hills, CA, USA)	Derived from donor fascia, stimulates collagen formation	No	Lasts up to 6 months
Autologous fat		Harvest fat & reinject it beneath the facial skin	No	Variable
Semi-Permanent Products				
Sculptra™	Dermik (Berwyn, PA, USA)	Synthetic polylactic acid	No	Prgressive results over time, lasts 2–4 years
Radiesse®	Bioform Medical (Franksville, WI, USA)	Calcium hydroxylap-atite	No	Immediate, lasts 2–7 years
Permanent Products				
Artecoll®/Artefill™	Artes Medical (San Diego, CA, USA)	75% percent bovine collagen & 25% polymethyl-methacrylate	Yes	Immediate, permanent
Silicone		Liquid silicone	No	Immediate, permanent

available forms (Zyderm I®, Zyderm II®and Zyplast®, Inamed, Santa Barbara, CA, USA) all of which are derived from the skin of an American cattle herd that is carefully monitored to prevent contamination with prion-mediated disease. Prior to initiating therapy, double skin testing is required to evaluate potential for an allergic response to the products. Localized hypersensitivity has been found in approximately 3% of patients and indicates a pre-existing allergy to bovine collagen (Stegman et al., 1988). The issue of whether injection of collagen is associated with an increased risk of developing connective tissue disease is controversial (Drake et al., 1996).

In the last few years, human-derived collagen fillers (Cosmoderm I®, Cosmoderm II®, Cosmoplast®, Inamed, Santa Barbara, CA, USA; Isologen®, Houston, TX, USA; Dermalogen® Collagenesis, Beverly, MA, USA; Cymetra® LifeCell, Branchburg, NJ, USA) have become available in order to address the issue of hypersensitivity associated with bovine collagen. Skin testing is not required prior to use of these products.

More recently hyaluronic acid (HA) derived fillers have gained favor. HA is a ubiquitous natural polysaccharide produced by many cell types which resides in the ground substance, functioning as a space-filling, stabilizing molecule. HA is reduced in aged skin (Piacquadio et al., 1997). HA's enormous ability to bond water, assists in hydration and provides skin turgor and unlike collagen, it is identical across all species, which minimizes the risk of foreign body reactions (Duranti et al., 1998). Products from non-animal (Restylane®, Q-Med, Uppsala, Sweden; Captique®, Genzyme, UK; Juvéderm®, Inamed, Santa Barbara, CA, USA) and animal derived sources (Hylaform®, Genzyme, UK) are available. HA fillers are well tolerated but have been associated with self-limited mild-moderate swelling, erythema, and tenderness at the implant site, with an average duration of 2 weeks (Duranti et al., 1998). Acne has also been noted. Sensitivity is uncommon but can occur. In 709 patients who were treated with either Hylaform® or Restylane®, 3 patients (0.42%) developed delayed skin reactions (Lowe et al., 2001).

A study conducted in 177 patients who received Hylaform® injections found that a two-thirds level of initial correction was maintained by 78% of patients at 3 months, 44% of patients at 6 months, and 8% of patients at 12 months (Duranti et al., 1998).

Preserved particulate fascia lata from cadavers (Fascian™, Fascia Biosystems, Beverly Hills, CA, USA), was introduced in 1999 for use as a soft tissue filler (Schecter and Sadick, 2005). By inducing the production of endogenous collagen, preserved fascia grafts have the potential to produce longer-lasting tissue augmentation (Burres, 1999). Burres followed 81 subjects after implantation of fascia grafts (mostly lip augmentation) and observed effects for at least 3–4 months in all patients. No extrusion, allergic reactions, or infection was observed (Burres, 1999).

Autologous fat transplantation is advantageous because it has no risk of immunologic reaction. Disadvantages include the necessity for two procedures (harvesting and insertion of the fat) and the inability to predict the percentage of graft survival (Ersek, 1991). Potential donor areas with easy access, limited postoperative morbidity, and relative insensitivity to dietary fluctuation include the abdomen, medial knee, and upper outer buttock areas (Drake et al., 1996). Areas of aging skin amenable to autologous fat transfer include the dorsal hands, depressed temples, hollow cheeks, deep nasolabial grooves, and defects caused by lipodystrophy (Drake et al., 1996).

3.2.4.2 Semi-Permanent Products Substances categorized as semi-permanent fillers include poly-L-lactic acid (Sculptra™, Dermik, Berwyn, PA, USA) and calcium hydroxylapatite (Radiesse®, Bioform Medical, Franksville, WI, USA). In August 2004, the FDA approved Sculptra™ for treatment of human immunodeficiency virus (HIV) facial lipoatrophy. It is an immunologically inert polymer derived from lactic acid, which achieves gradual volume enhancement. The precise mechanism is unknown but it may stimulate new collagen production through a normal foreign-body reaction (Werschler and Weinkle, 2005). Its durability is thought to range from 2 to 4 years (Werschler and Weinkle, 2005).

Calcium hydroxylapatite is the principal mineral component of bone. Radiesse® (formerly known as Radiance®) is presently approved in Europe for subdermal augmentation. The product has received FDA approval for vocal cord injection, as a radiographic tissue marker, and for oral maxillofacial defects, but is not presently approved for wider cosmetic applications. Radiesse® was evaluated in a trial of 64 patients undergoing a total of 101 treatments for cosmetic improvement of a wide variety of facial defects (Sklar and White, 2004). Aesthetic correction was immediate and well-tolerated. The most common complication was palpable, non-visible nodules reported in 20% of patients who underwent lip augmentation. The longevity of the product is also to be determined.

3.2.4.3 Permanent Products Permanent products do not degrade with time and seemingly have the advantage of long-term correction. Longevity may however be detrimental as long-term complications can occur. Artecoll® (Europe)/ Artefill™ (US) (Artes Medical, San Diego, CA, USA) is a solution that contains polymethylmethacrylate (PMMA) microspheres suspended in bovine collagen and lidocaine. Once the collagen is degraded, the remaining inert, non-biodegradable PMMA beads remain intact and provide permanent augmentation. A randomized, controlled, multicenter trial of 251 patients treated with either Artecoll® or a collagen filler demonstrated significantly greater maintained augmentation with Artecoll® as compared to collagen at 6 months (Cohen and Holmes, 2004). Twelve month follow-up was obtained for 87% who sustained improvement with Artecoll® at 1 year (Cohen and Holmes, 2004).

Silicone was used for years as a tissue filler before the FDA prohibited marketing of injectable liquid silicone for cosmetic purposes because of safety issues, including development of potentially severe foreign-body-type silicomas up to 11 years after implantation (Ellenbogen et al., 1975). Monitored clinical trials are permitted.

3.2.5 Ablative Resurfacing Procedures

Ablative resurfacing procedures including dermabrasion, dermasanding and laser skin resurfacing (LSR) injure or remove superficial cutaneous layers resulting in formation of a new epidermis and promoting synthesis of dermal collagen. Dermabrasion uses wire brushes, diamond fraises and serrated wheels attached to a dermabrader to remove the upper layers of the skin while dermasanding uses sand paper. In LSR, collimated light is absorbed by tissue water and converted to heat to precisely remove tissue. Two lasers are commonly utilized: 1) the carbon dioxide (CO_2) laser with a wavelength of 10,600 nm; and 2) the erbium:yttrium aluminum garnet (Er:YAG) laser with a wavelength of 2940 nm. Combination devices are also utilized.

The wrinkle reduction and skin tightening potential of ablative procedures is second only to plastic surgery and ablative procedures have an advantage of also improving skin surface texture. Areas most amenable to wrinkle reduction during ablative procedures are perioral and periorbital regions, which are traditionally unresponsive to face-lifting procedures. However, epidermal removal creates an open wound which requires extensive care and puts the patient at risk for the development of infections, dyspigmentation, and scarring. Re-epithelialization occurs over 5–7 days but residual erythema commonly lasts 4 weeks (Gold, 2003) or more. Local anesthesia and sedation, regional nerve blocks, or general anesthesia is generally used secondary to significant intra-operative discomfort.

Holmkvist et al. (2000) treated half of the perioral area of 15 patients with a pulsed CO_2 laser and the other half with dermabrasion using a hand engine-drive diamond fraise or a medium-grade drywall sanding screen. Dermabrasion resulted in more bleeding during the immediate post-operative period. Significantly more crusting and initial erythema (up to 1 month post-treatment) was noted on the CO_2 laser-treated side. Both treatment methods resulted in statistically significant improvement in rhytid score. The mean decrease in rhytid score was 1.09 for laser-treated skin and 0.94 for dermabrasion-treated skin but the difference was not statistically significant. Fine wrinkles were more responsive than deep wrinkles with both treatments.

3.2.6 Non-Ablative Light-Based Rejuvenation

Ablative procedures offer significant rejuvenation; however, they are associated with prolonged healing times, potential complications such as scarring, infection, and pigmentary alteration, and moderate discomfort (Fitzpatrick, 1997). As such, non-ablative light rejuvenation systems were developed and are associated with minimal down time and less patient discomfort.

3.2.6.1 Non-ablative Rejuvenation Lasers Non-ablative laser rejuvenation is designed to confine selectively thermal injury, avoiding epidermal injury while achieving fibroblast activation and synthesis of new collagen and extracellular matrix material (Nelson et al., 2002). The skin surface is not removed or modified which minimizes or eliminates downtime but also eliminates any improvement in surface textural and chromatic irregularities. Wrinkle reduction varies with device and technique, but in general, improvement is significantly less as compared to LSR.

Kelly et al. (1999) evaluated periorbital rhytid improvement in 35 adults who were given 3 treatments with the 1320 nm CoolTouch® Neodymium Yttrium Aluminum Garnet (Nd:YAG) laser used in combination with cryogen spray cooling. Small but statistically significant improvements were noted in the mild, moderate, and severe rhytid groups 12 weeks after the final laser treatment. A final assessment performed 24 weeks after the last treatment showed statistically significant improvement only in the severe rhytid group. The procedure was found to be safe, although four sites (5.6%) developed transient hyperpigmentation and two sites (2.8%) developed barely perceptible pinpoint-pitted scars. Subsequent device improvements minimized the risk of adverse effects.

3.2.6.2 Intense Pulsed Light Intense pulsed light (IPL) is a noncoherent filtered flashlamp that emits broadband light in the 500 to 1200 nm range (Raulin et al., 2003). A multi-center study evaluated IPL for non-ablative rejuvenation of 93 patients with photoaged skin (Sadick et al., 2004). Up to five treatments were performed at 4-week intervals with follow-up visits at 4 and 6 months after the last treatment. Patients received full-face treatments with the Quantum SR/HR (Lumenis Inc., Santa Clara, CA, USA) and results were based on physicians' assessments as well as patient satisfaction. Wrinkling score improved significantly by 1.39 and 1.32 units at 4 and 6 months, respectively, correlating to improvements noted for 82% and 75% of the patients at each of these time points. Significant improvement was also seen using the investigators' assessment of overall improvement in facial appearance, which reflected pigmentary, vascular, and rhytid reduction. The first IPL treatment improved overall appearance in 61% of the study population. Number of patients with improvement were 98% and 90% respectively, four and six months after the last treatment.

The use of short-incubation topical 5-aminolevulinic acid (5-ALA) (Levulan® Kerastick™, DUSA Pharmaceuticals, Wilmington, MA, USA) enhances the effectiveness of IPL for facial rejuvenation, reducing the number of treatments required and enhancing the clinical effects (Alster et al., 2005). A retrospective review demonstrated that one ALA-IPL treatment was equal in efficacy to 3 IPL treatments alone (Carcamo et al., 2005). A variety of lasers, including blue light (405–420 nm), red light (635 nm), and pulsed dye lasers (585 nm), used with 5-ALA photodynamic therapy, have also shown safety and efficacy in photorejuvenation (Gold and Goldman, 2004). Recently, the application of 5-ALA photodynamic therapy with a combined IPL and radiofrequency device has been reported (Hall et al., 2004).

3.2.6.3 Light-emitting diode photomodulation Light-emitting diode (LED) photomodulation is a process which modifies cell activity using low energy light delivered in a specific pattern without thermal effect (Weiss et al., 2005). Weiss et al. evaluated 90 patients after a series of 8 treatments with 590 nm LED photomodulation delivered over 4 weeks. Ninety percent of subjects demonstrated some improvement in skin texture or reduction of periorbital rhytids, erythema or pigmentation. Histologic analysis of biopsies demonstrated a 28% average increase in collagen density and a 4% average reduction of matrixmetalloproteinase (MMP)-1. No side effects or pain were noted. The GentleWaves LED Photomodulation® System (Light BioScience, Virginia Beach, VA, USA) received FDA clearance for the treatment of periorbital rhytids in 2005.

3.2.7 Radiofrequency devices

Radiofrequency (RF) is a newer skin rejuvenation method which has generated significant interest over the last 5 years. The first radiofrequency device designed for skin rejuvenation was the monopolar ThermaCool TC™ System (Thermage, Inc., Hayward, CA, USA) which utilizes two electrodes on the skin to produce an electric field (Kelly et al., 1999). Ions and charged molecules within the electric field move and/or rotate and inherent resistance to this movement causes heat. The epidermis is protected by a proprietary tip which utilizes cooling spray. The ThermaCool™ System received 510K clearance for non-invasive treatment of periorbital wrinkles and rhytids but has also been used for cheek, (Alster and Tanzi, 2003) neck, (Tanzi and Alster, 2003) and brow (Fitzpatrick et al., 2003) lifting.

A multicenter study evaluated 86 subjects up to 6 months after a single treatment to the periorbital area with the ThermaCool TC™ System (Fitzpatrick et al., 2003). Objective photographic analysis showed that 61.5% of eyebrows were lifted by at least 0.5 mm. Three independent reviewers noted improvement of at least 1 Fitzpatrick wrinkle score (a 9-point scale) in 83.2% of subjects and fifty percent of subjects were satisfied or very satisfied with periorbital wrinkle improvement. Three patients had small areas of residual scarring at the 6-month follow-up (Fitzpatrick et al., 2003). Subsequent device and technique improvements have significantly reduced the incidence of scarring. Another study demonstrated that 14/15 patients had facial skin tightening induced by the ThermaCool TC™ System. No scarring was noted in this study (Ruiz-Esparza and Gomez, 2003).

The Aurora™ and the Polaris™ WR (Syneron, Inc., Richmond Hill, Ontario, Canada) combine bipolar RF and IPL (Aurora™) or bipolar RF and a 900 nm diode laser (Polaris™ WR) to tighten collagen fibers and reduce wrinkles and pigmentation. Monopolar and biopolar RF have different mechanisms of action and likely different clinical effects.

3.2.8 Fractional photothermolysis

In 2004, fractional photothermolysis (Fraxel™, Reliant Technologies, Palo Alto, CA, USA) was introduced. This novel 1550 nm laser creates localized microscopic treatment zones (MTZs) of thermal injury in the skin which are surrounded by zones of normal tissue, limiting damage and allowing more rapid recovery (Manstein et al., 2004). MTZs typically have a diameter of 100 μm and penetrate 300 μm into the skin. In one study, periorbital treatment of 30 subjects using moderate MTZ density (pattern density with spacing of 250 μm or more) improved wrinkle score by 18% at 3 months, and histology revealed enhanced undulating rete ridges and increased mucin in the papillary dermis. The procedure was also well tolerated, with minimal erythema and edema. The study concluded that both efficacy and side effects are dependent on the shape and location of individual MTZs and on the pattern in which the MTZs are arranged (Manstein et al., 2004).

Rokhsar et al. (2005) evaluated 12 patients who received 4–5 Fraxel™ treatments to rhytids of the face, neck, or chest at 1–4 week intervals. Improvement was seen in texture, dyschromia, and wrinkles, and biopsies demonstrated new collagen formation. Side effects were minimal and were limited to post-treatment erythema lasting a few days, mild edema, and small linear abrasions which healed uneventfully.

3.2.9 Cosmetic surgery

The greatest improvement in wrinkles and skin laxity can be achieved with cosmetic surgery. Natural-looking appearance enhancement is the goal which can be achieved through a variety of procedures including face-lifts, brow lifts and eyelid surgery. Enhanced effect is accompanied by increased risk and prolonged recovery. A more thorough discussion of plastic surgery options is beyond the scope of this manuscript, but it is useful to note that endoscopically-assisted cosmetic surgery reduces invasiveness and minimizes recovery time.

4. CONCLUSION

An array of topical and procedural treatments are available to benefit the aging face. Perhaps the optimal method lies in a combination of treatments which are complementary and can together achieve an enhanced result. Patients need to be aware that there are no "quick fixes" or "miracle cures" and that use of cosmeceuticals, medications, and certainly performance of any procedure, is associated with some risk. Anyone seeking treatment for the aging face should be informed about their options in order to determine the best approach which will meet their needs and goals.

REFERENCES

Afaq, F., Adhami, V.M., Mukhtar, H. (2005) Photochemoprevention of ultraviolet B signaling and photocarcinogenesis. Mutat Res, 571: 153–73.

Alster, T.S., Tanzi, E.L. (2003) Treatment of prominent nasolabial folds and cheek laxity with a nonablative radiofrequency device. Lasers Surg Med, 15S: 34.

Alster, T.S., Tanzi, E.L., Welsh, E.C. (2005) Photorejuvenation of facial skin with topical 20% 5-aminolevulinic acid and intense pulsed light treatment: a split-face comparison study. J Drugs Dermatol, 4: 35–38.

Barciszewski, J., Rattan, SIS., Siboska, G., Clark, B.F.C. (1999) Kinetin – 45 years on. Plant Science, 148: 37–45.

Baumann, L.S. (2004) A refresher on antioxidants. Skin and Allergy News, 35(5): 31.

Bernstein, E.F., Lee, J., Brown, D.B., Yu, R., Van Scott, E. (2001) Glycolic acid treatment increases type I collagen mRNA and hyaluronic acid content of human skin. Dermatol Surg, 27: 429–33.

Botox® Cosmetic (botulinum toxin type A) purified neurotoxin complex (Allergan). Prescribing information. www.botoxcosmetic.com.

Boyd, A.S., Naylor, M., Cameron, GS., Pearse, A.D., Gaskell, S.A., Neldner, K.H. (1995) The effects of chronic sunscreen use on the histologic changes of dermatoheliosis. J Am Acad Dermatol, 33: 941–6.

Burres, S. (1999) Preserved particulate fascia lata for injection: a new alternative. Derm Surg, 25: 790–94.

Carcamo, A.S., Ehrlich, M., Goldman, M.P. (2005) Enhanced photorejuvenation with combination ALA + intense puled light. Poster presentation. Annual meeting of the American Society for Laser Medicine and Surgery.

Carruthers, J.D., Carruthers, J.A. (1992) Treatment of glabellar frown lines with C. botulinum-A exotoxin. J Dermatol Surg Oncol, 18: 17–21.

Carruthers, J.A., Lowe, N.J., Menter, M.A., et al. (2002) A multicenter, double-blind, randomized, placebo-controlled study of the efficacy and safety of botulinum toxin type A in the treatment of glabellar lines. J Am Acad Dermatol, 46: 840–9.

Carruthers, J., Fagien, S., Matarasso, S.L. (2004) Botox Consensus Group Consensus recommendations on the use of botulinum toxin type a in facial aesthetics. Plast Reconstr Surg, 114(6 Suppl): 1S–22S.

Chung, J.H., Seo, J.Y., Choi, H.R., et al. (2001) Modulation of skin collagen metabolism in aged and photoaged human skin in vivo. J Invest Dermatol, 117: 1218–1224.

Chung, J.H. (2003) Photoaging in Asians. Photodermatol Photoimmunol Photomed, 19: 109–121.

Cohen, S.R., Holmes, R.E. (2004) Artecoll: a long-lasting injectable wrinkle filler material: report of a controlled, randomized, multicenter clinical trial of 251 subjects. Plast Reconstr Surg, 114: 964–75.

Ditre, C.M., Griffin, T.D., Murphy, G.F., et al. (1996) Effects of alpha-hydroxy acids on photoaged skin: a pilot clinical, histologic, and ultrastructural study. J Am Acad Dermatol, 34: 187–95.

Drake, L., Dinehart, S., Farmer, E., et al. (1996) Guidelines of care for soft tissue augmentation: fat transplantation. J Am Acad Dermatol, 34: 690–4.

Drake, L., Dinehart, S.M., Farmer, E.R., et al. (1996) Guidelines of care for soft tissue augmentation: collagen implants. American Academy of Dermatology J Am Acad Dermatol, 34: 698–702.

Duranti, F., Salti, G., Bovani, B. (1998) Injectable hyaluronic acid gel for soft tissue augmentation: a clinical and histologic study. Derm Surg, 28: 1317–22.

Ellenbogen, R., et al. (1975) Injectable fluid silicone therapy: human morbidity and mortality. JAMA, 234: 308.

Ersek, R.A. (1991) Transplantation of purified autologous fat: a 3-year follow-up is disappointing. Plast Reconstr Surg, 87: 219.

Fisher, G.J., Kang, S., Varani, J., Bata-Csorgo, Z., Wan, Y., Datta, S., Voorhees, J.J. (2002) Mechanisms of photoaging and chronological skin aging. Arch Dermatol, 138: 1462–70.

Fitzpatrick, R.E., Rostan, E.F. (2002) Double-blind, half-face study comparing topical vitamin C and vehicle for rejuvenation of photodamage. Derm Surg, 28: 231–236.

Fitzpatrick, R.E., Geronemus, R., Goldberg, D., Kaminar, M., Kilmer, S., Ruiz-Esparza, J. (2003) First multi-center study of a new non-ablative radiofrequency device to tighten facial tissue. Lasers Surg Med, 15S: 35.

Fitzpatrick, R., Geronemus, R., Goldberg, D., Kaminer, M., Kilmer, S., Ruiz-Esparza, J. (2003) Multi-center study of noninvasive radiofrequency for periorbital tissue tightening. Lasers Surg. Med, 33: 232–242.

Fitzpatrick, R.E. (1997) Laser resurfacing of rhytides. Dermatol Clin, 15: 431–447.

Furuichi, Y. (2001) Premature aging and predisposition to cancers caused by mutations in RecQ family helicases. Ann N Y Acad Sci, 928: 121–131.

Gilchrest, B.A. (1996) A review of skin ageing and its medical therapy. Br J Dermatol, 135: 867–75.

Gilchrest, B.A. (1999) Treatment of photodamage with topical tretinoin: an overview. J Am Acad Dermatol, 36: S27–S36.

Gold, M.H. (2003). Dermabrasion in dermatology. Am J Clin Dermatol, 4: 467–471.

Gold, M.H., Goldman, M.P. (2004) 5-aminolevulinic acid and photodynamic therapy: where we have been and where we are going. Dermatol Surg, 30: 1077–1083.

Griffiths, C.E., Russman, A.N., Majmudar, G., Singer, R.S., Hamilton, T.A., Voorhees, J.J. (1993) Restoration of collagen formation in photodamaged human skin by tretinoin (retinoic acid). N Engl J Med, 329: 530–535.

Hall, J.A., Keller, P.J., Keller, G.S. (2004) Dose response of combination photorejuvenation using intense pulsed light-activated photodynamic therapy and radiofrequency energy. Arch Facial Plast Surg, 6: 374–8.

Holmkvist, K.A., Rogers, G.S. (2000) Treatment of perioral rhytides a comparison of dermabrasion and superpulsed carbon dioxide laser. Arch Dermatol, 136: 725–731.

Kang, S., Voorhees, J.J. (1998) Photoaging therapy with topical tretinoin: an evidence-based analysis. J Am Acad Dermatol, 39: S55–S61.

Keen, M., Blitzer, A., Aviv, J., Binder, W., Prystowsky, J., Smith, H., Brin, M. (1994) Botulinum toxin A for hyperkinetic facial lines: results of a double-blind, vehicle-controlled study. Plast Reconstr Surg, 94: 94–9.

Kelly, K.M., Nelson, J.S., Lask, G.P., Geronemus, R.G., Bernstein, L.J. (1999) Cryogen spray cooling in combination with nonablative laser treatment of facial rhytides. Arch Dermatol, 135: 691–694.

Klein, A.W. (2004) Complications with the use of botulinum toxin. Dermatol Clin, 22: 197–205.

Krol, E.S., Kramer-Stickland, K.A., Liebler, D.C. (2000) Photoprotective actions of topically applied vitamin E, Drug, Metab Rev, 32: 413–20.

Kullavanijaya, P., Lim, H.W. (2005) Photoprotection. J Am Acad Dermatol, 52: 937–958.

Lavker, R.M., Gerberick, G.F., Veres, D., Irwin, C.J., Kaidbey, K.H. (1995) Cumulative effects from repeated exposures to suberythemal doses of UVB and UVA in human skin. J Am Acad Dermatol, 32: 53–62.

Letham, D.S. (1963) Zeatin, a factor inducing cell division isolated from Zea mays (1963) Life Sciences, 41: 569–573.

Lowe, N.J., Maxwell, A., Harper, H. (1996) Botulinum A exotoxin for glabellar folds: a double-blind, placebo-controlled study with an electromyographic injection technique. J Am Acad Dermatol, 35: 569–72.

Lowe, N.J., Maxwell, C.A., Lowe, P., Duick, M.G., Shah, K. (2001) Hyaluronic acid skin fillers: adverse reactions and skin testing. J Am Acad Dermatol, 45: 930–3.

Manstein, D., Herron, S.H., Sink, R.K., Tanner, H., Anderson, R.R. (2004) Fractional photothermolysis: a new concept for cutaneous remodeling using microscopic patterns of thermal injury. Lasers Surg Med, 34: 426–438.

Matarasso, S.L., Glogau, R.G. (1991) Chemical face peels. Dermatol Clin, 9: 131–50.

Matarasso, A., Matarasso, S.L., Brandt, F.S., Bellman, B. (1999) Botulinum A exotoxin for the management of platysma bands. Plast Reconstr Surg, 103: 645–52.

McCullough, J.L. (1999) Furfuryladenine-A New Antiaging Topical: Research and Clinical Experience. Skin and Allergy News: Developments in Topical Skin Treatments: An Update (Skin Disease Education Foundation Symposium), 3–5.

McCullough, J.L., Weinstein, G.D. (2002) Clinical study of safety and efficacy of using topical kinetin 0.10% (Kinerase®) to treat photodamaged skin. Cosmetic Dermatology, 15: 29–32.

Mok, D.W.S., Mok, M.C. (1994) Cytokinins – Chemistry, Activity, and Function, CRC Press, Boca Raton, FL.

Monheit, G.D. (2001) Medium-depth chemical peels. Dermatol Clin, 19: 413–25.

Nelson, J.S., Majaron, B., Kelly, K.M. (2002) What is non-ablative photorejuvenation of human skin? Sem Cutan Med Surg, 21: 238–250.

Newman, J., Newman, A., Moy, L.S., Babapour, R., Harris, A.G., Moy, R.L. (1996) Clinical improvement of photoaged skin with 50% glycolic acid: a double-blind vehicle-controlled study. Derm Surg, 22: 455–460.

Oikarinen, A. (1994) Aging of the skin connective tissue: how to measure the biochemical and mechanical properties of aging dermis. Photodermatol Photoimmunol Photomed, 10: 47–52.

Olsen, A., Siboska, G.E., Clark, BFC., et al. (1999) N6-furfuryladenine, kinetin, protects against Fenton reaction-mediated oxidative damate to DNA. Biochem. Biophys Res Commun, 265: 499–502.

Patel, M.P., Talmor, M., Nolan, W.B. (2004) Botox and collagen for glabellar furrows: advantages of combination therapy. Ann Plast Surg, 52: 442–7.

Pesce, K., Rothe, M.J. (1996) The premature aging syndromes. Clin Dermatol, 14: 161–170.

Phillips, T.J., Gottlieb, A.B., et al. (2002) Efficacy of 0.1% tazarotene cream for the treatment of photodamage: a 12-month multicenter, randomized trial. Arch Dermatol, 138: 1486–1493.

Piacquadio, S., Jarcho, M., Goltz, R. (1997) Evaluation of hylan b gel as a soft tissue augmentation implant material. J Am Acad Dermatol, 36: 544–549.

Rattan, S.L. (2005) Gerontomodulatory and youth-preserving effects of zeatin on human skin fibroblasts undergoing aging in vitro. Rejuvenation Res, 8: 46–57.

Rattan, S.I., Clark, B.F. (1994) Kinetin delays the onset of aging characteristics in human fibroblasts. Biochem Biophys Res Commun, 201: 665–672.

Raulin, C., Greve, B., Greme, H. (2003) IPL technology: a review. Lasers Surg Med, 32: 78–87.

Rokhsar, C.K., Tse, Y., Lee, S., Fitzpatrick, R. (2005) The treatment of photodamage and facial rhythides with Fraxel (fractional photothermolysis). Poster presentation. Annual meeting of the American Society for Laser Medicine and Surgery.

Ruiz-Esparza, J., Gomez, J.B. (2003) The medical face lift: a noninvasive, nonsurgical approach to tissue tightening in facial skin using nonablative radiofrequency. Dermatol Surg, 29: 325–32.

Sadick, N.S., Finn, N. (2005) A review of microdermabrasion. Cosm Dermatol, 18: 351–354.

Sadick, N.S., Weiss, R., Kilmer, S., Bitter, P. (2004) Photorejuvenation with intense pulsed light: results of a multi-center study. J Drugs Dermatol, 3: 41–9.

Schecter, A.K., Sadick, N.S. (2005) Preserved particulate fascia lata for soft tissue augmentation: a review and early results of comparative studies using bovine collagen. Cosm Dermatol, 18: 337–340.

Semchyshyn, N., Sengelmann, R.D. (2003) Botulinum toxin A treatment of perioral rhytides. Dermatol Surg, 29: 490–5.

Shim, E.K., Barnette, D., Hughes, K., et al. (2001) Microdermabrasion: a clinical and histopathologic study. Dermatol Surg, 27: 524–530.

Sklar, J.A., White, S.M. (2004) Radiance FN: a new soft tissue filler. Dermatol Surg, 30: 734–768.

Smith, J.B., Fenske, N.A. (1996). Cutaneous manifestations and consequences of smoking. J Am Acad Dermatol, 34: 717–32.

Stegman, S.J., Chu, S., Armstrong, R. (1988) Adverse reactions to bovine collagen implant: clinical and histologic features. Derm Surg, 14: 439–48.

Stern, R.S. (2004) Treatment of photoaging. N Engl J Med, 350: 1526–1534.

Stiller, M.J., Bartolone, J., Stern, R., Smith, S., et al. (1996) Topical 8% glycolic acid and 8% lactic acid creams for the treatment of photodamaged skin: a double-blind vehicle-controlled clinical trial. Arch Dermatol, 132: 631–636.

Tanzi, E.L., Alster, T.S. (2003) Improvement of neck laxity with a nonablative radiofrequency device: a lifting experience. Lasers Surg Med, 15S: 34.

Thompson, S.C., Jolley, D., Marks, R. (1993) Reduction of solar keratoses by regular sunscreen use. N Engl J Med, 329: 1147–51.

Tse, Y., Ostad, A., Lee, H.S., Levine, V.J., Koenig, K., Kamino, H., Ashinoff, R. (1996) A clinical and histologic evaluation of two medium-depth peels. Glycolic acid versus Jessner's trichloroacetic acid. Dermatol Surg, 22: 781–6.

Van Staden, J., Cook, E.L., Nooden, L.D. (1988) Cytokinins and senescence. In: Nooden, L.D., Leopold, A.C. (eds). Senescence and Aging in Plants. Pages 281–328, Academic Press, New York, NY.

Verbeke, P., Siboska, G.E., Clark, B.F.C., et al. (2000) Kinetin inhibis protein oxidation and glyoxidation in vitro Biochem Biophys Res Commun, 276: 1265–1267.

Weiss, R.A., McDaniel, D.H., Geronemus, R.G., Weiss, M.A. (2005) Clinical trial of a novel non-thermal LED array for reversal of photoaging: clinical, histologic, and surface profilometric results. Lasers Surg Med, 36: 85–91.

Werschler, W.P., Weinkle, S. (2005) Longevity of effects of injectable products for soft-tissue augmentation. J Drugs in Dermatol, 4: 20–27.

West, T.B., Alster, T.S. (1999) Effect of botulinum toxin type A on movement-associated rhytides following CO2 laser resurfacing. Dermatol Surg, 25: 259–61.

CHAPTER 11

AGING AND PERIODONTAL DISEASE

R. SURESH
Sri Ramachandra Dental College and Hospitals; Sri Ramachandra Medical College and Research Institute (DU); Chennai, India. (Email: chennai_dentist@yahoo.co.in)

Abstract: Periodontal disease is the most prevalent disease of the oral cavity. The role of aging in periodontal disease is debatable, but the means of preventing periodontal disease are available. This article gives an overview of the role of aging on the periodontium, prevention and therapy of age-related periodontal diseases

Keywords: Aging, periodontal disease, human

> "There are no diseases peculiar to old age and very few from which it is exempt" – Alfred Worcester (1855–1951).

Apt to the above quotation, age seems to take the blame for many diseases. Periodontal disease is one such disease where the role of age is still debatable. Though there are many age-related changes in the oral cavity, by its sheer prevalence rate and association with adults, chronic periodontitis (inflammation of the supporting structures of the tooth) is very highly equated with age. The tooth supporting structures consist of cementum – a hard tissue covering the root, bone – forming a socket within which the tooth is placed, the periodontal ligament fibers connecting the cementum to the bone, and gingiva. Gingiva (gum) is that part of the oral mucosa that covers the jaws and surrounds the necks of the teeth providing protection to the above mentioned structures.

One of the earliest proposals was that, the periodontal disease is of a degenerative nature. Egyptian, Hebrew and Chinese writings from ancient times mentioned "long teeth" as an indicator of old age. Some therefore argued, that periodontitis was a natural consequence of aging. Many other local factors were introduced later as possible causes for periodontal pathology. In the 1950s and 1960s, plaque (an organized microbial matrix) and age were suggested as the primary etiological

S. Rattan and M. Kassem (eds.), Prevention and Treatment of Age-related Diseases, 193–200.

factors. The 1998 classification (Armitage, 1999) of periodontal disease made an extensive list of conditions (which are independent of age) out of which, aggressive periodontitis and chronic periodontitis formed different ends of the spectrum. Aggressive periodontitis is one where genetically based defective host factors play a major role, whereas chronic periodontitis, with genetic factors as a baseline, requires both microbes and host factors.

Though age was initially proposed as the primary cause of chronic periodontitis, it has now been proved beyond doubt that plaque is the primary cause. Is age an associated factor? No, since plaque present at any age, can cause chronic periodontitis. Is age a modifying factor? In favor of this is the shift in the microbial flora of plaque from predominantly *actinobacillus* to *porphyromonas gingivalis* with advancing age (Rodenburg et al., 1990). Finally the question remains "can age be a risk factor for chronic periodontitis?"

1. AGE: A RISK FACTOR FOR PERIODONTITIS?

Risk is the probability that an individual gets the specific disease in a given period. Risk factors may be environmental, behavioral or biological in nature. The extents to which physiological and pathological changes that accompany aging are due to the aging process itself or caused by concomitant pathosis, medication usage or social and environmental changes is debatable (Locker et al., 1998). Nevertheless, since numerous age-associated changes can be observed in the biochemical, immunological and physiological processes of periodontal tissues, there are reasonable grounds to suspect that aging could potentially be a risk factor for periodontal disease (Papapanou et al., 1989; Ismail et al., 1990).

Periodontal status worsens with age in the general population (Schurch et al., 1988; Beck, 1996; Papapanou et al., 1988). Degenerative changes related to aging are due to prolonged exposure to the primary factor (plaque) and other risk factors over a period of time, which have a cumulative effect. Therefore, periodontitis is not an inevitable result of only aging: on the contrary it may be a contributing factor.

Other factors determining susceptibility and severity of periodontitis are: (i) microbial infection, (ii) host parasite interactions, (iii) external socio-economic influences, (iv) smoking, (v) systemic diseases, and (vi) stress.

2. MICROBIAL ECOLOGY

Studies have shown a correlation between severity of periodontal disease and composition of the sub gingival (below the gum) microbiota. Age induced environmental changes may influence the attachment, growth and metabolisms of microorganisms. The adhesion of microbes to a surface depends on physical and chemical reactions. As age advances, surfaces changes take place due to chemical and physical factors. As the gingiva recedes root dentin, furcations and developmental anomalies get exposed increasing areas vulnerable to plaque attachment.

In addition, as age advances, an increase in restorations and prosthesis favor more plaque accumulation. Periodontal disease hence becomes a cumulative one due to different factors mentioned above making it possible for a shift in the microbial composition.

3. THE IMMUNE SYSTEM IN THE ELDERLY

In the adaptive immune system, with increasing age cytokines like IL2 are decreased and cytokines like IL1, TNF alpha and IL6 are increased. This shift in cytokine levels along with a reduced cellular immunity may be the reason for the increased periodontitis in older adults. Meydani et al proposed that alterations in T cell functions identified in elders were in part due to increased levels of free radicals and membrane lipid per oxidation in cells (Meydani et al., 1995).

4. SYSTEMIC DISEASE AND MEDICATION

Majority of older adults have one or more systemic disease increasing their chances of being on medication. There is a psychological impact of these conditions on interest and attention to oral health. The effect of systemic diseases on periodontitis could be a direct influence on the pathogenesis of the disease, e.g. Diabetes Mellitus type II or indirectly compromising oral hygiene maintenance by handicapping the patient's motor skills, e.g. Stroke/ Parkinsonism. Also the intake of medicines especially anti-hypertensives, hypoglycemic agents and anti-depressants may induce mouth dryness (Xerostomia) that may also be associated with increased risk for periodontitis. According to Beck and Hunt (1985) of the 160 million prescription drugs, 47% could have direct effects on the oral cavity and 34% may have an indirect effect.

5. PERIODONTAL MEDICINE

Recent studies have resulted in the development of periodontal medicine, which indicates that having periodontitis may be a risk indicator for developing systemic diseases (Mattila, 1993; Taylor et al., 1996). So, it becomes imperative that the dentist identifies individuals with periodontitis and recommends systemic /medical evaluation.

6. PREVENTIVE PERIODONTICS IN OLDER ADULTS

Though the objective of periodontal therapy is to get a perfect functional and aesthetic dentition, sometimes considering the age, it may not be possible to undertake all the procedures especially extensive surgical procedures to attain this objective. So under such circumstances, it would be ideal to reduce the microbial load to make the patient asymptomatic and prevent further damage to the supporting tissues.

So the baseline prevention in treatment of chronic periodontitis depends on elimination of microbes. The gingival inflammation initiated in the second/third decade of life if left untreated may lead to periodontitis. Though in old age chronic periodontitis may already be present, preventive procedures can still be carried out to maintain the inactive stage of the disease, prevent the exacerbation and ultimately prevent any further loss of periodontal structures.

The following aspects are considered in planning for preventive periodontal procedures for elderly:

A. Systemic
1. **Systemic diseases**
2. **Medications**
3. **Attitude**
4. **Knowledge (awareness)**
5. **Motor skills of the patient**

B. Local
1. **Salivary secretion**
2. **Prosthesis and restorations**
3. **Exposure of root surfaces**
4. **Interdental spaces**

For the prevention of periodontal disease the following steps are carried out:
1. **Motivation**
2. **Education**
3. **Tooth Brushing**
4. **Oral hygiene aids**
5. **Chemotherapy**
6. **Dietary functions**
7. **Professional Help**

6.1 Motivation and education

This is to make the patient understand the concepts of the disease, to change the habits of the lifetime to adjust the hierarchy of ones belief and practice. In periodontics, home care by the patients should be emphasized. It is mandatory that the patients get familiarized with the technique they are supposed to perform routinely. To educate the patients on the importance of oral hygiene is essential in the prevention of periodontal disease and indirectly systemic diseases.

6.2 Tooth brushing and oral hygiene aids

Most of the patients would require a soft variety of toothbrush (battery operated ones could be suggested for stroke patients with lack of dexterity), toothpaste with fluoride, and desensitizing toothpastes. The use of interdental brushes can help patients with complaints of food impaction.

6.3 Chemotherapy

Though brushing is an effective means of removal of plaque, salivary pellicle formation succeeds 2 hours after brushing. Hence, in between brushing, mouth rinses would be useful. Several anti plaque agents such as chlorhexidine, Listerine and their generic counter parts are available for use in different formulations. Chlorhexidine, a cationic bisguanide with potent bactericidal and bacteriostatic efficiency has been suggested to aid plaque control in older adults. Chlorhexidine could be particularly useful for older individuals who take phenytoin, nifedipine & cyclosporine and are at a risk for gingival hyperplasia. Listerine has also been shown to be an efficient anti plaque agent but its active ingredients, which comprise of three essential oils (eucalyptol, thymol & menthol) that are alcohols contraindicates its use in older adults who suffer from xerostomia.

Fluorides with their anti cariogenic potential are available in several formulations. Topical fluorides are recommended in the treatment and prevention of dental caries. Older adults are advised to use fluoridated toothpaste and are sometimes prescribed topical fluoride gels.

Saliva substitutes, such as artificial saliva are of great use to older adults who suffer from dry mouth (xerostomia). They are easily used as sprays or swab sticks and could be used even in patients with compromised psychomotor skills. Patients with dry mouth may also benefit from sugarless candies and sugarless gums, which stimulate the flow of saliva.

6.4 Dietary counseling and professional help

Older adults may need diet counseling to aid them take high fiber diet with detergent action and to discourage them from taking refined soft sticky foods. In addition professional help may be sought on a regular basis.

7. TREATMENT OF PERIODONTAL DISEASE IN OLDER ADULTS

The National Health and Nutrition Examination Survey (NHANES) III study has suggested that the prevalence and severity of periodontitis increase with advancing age (Albander et al., 1999). Moderate levels of attachment loss are seen in a higher proportion of older adults: however, severe loss is detected in only a small proportion of older adults (Locker et al., 1998). Studies have shown an increased annual rate of destruction of periodontal bone support in individuals of age over 70, which shows aging and its related problem on their own may marginally increase the destruction process (Papapanou et al., 1989). Whether it is an age related loss of tissue or an active disease or a change in the severity, the degree and type of treatment may differ but all the same, treatment is essential. Healing is not compromised due to age unless complicated by systemic factors.

Treating periodontal disease in older adults needs a careful approach since in addition to biological factors, other systemic and socio economic factors are also

Table 1. Factors to be considered and treatment options

Factors to be considered	Treatment options
1. Systemic diseases	1. Surgery
2. Medications	2. Non-surgical
3. Active/inactive state of the disease	3. Antibiotics
4. Socioeconomic	4. Local drug delivery
5. Motor skills	

involved? Before starting treatment in older adults, knowledge of their individual past medical and dental histories is important and a careful examination of the intra and extra oral structures is also essential. For patients with systemic conditions, medication for the same would not only influence the treatment plan but also give an idea about the priority for oral hygiene procedures and the motor skills to perform the same. Perception, knowledge, socio economic status and attitude may contribute to it.

Though the objective of the treatment is to improve the esthetics and function, the ultimate goal of the treatment is to reduce the microbial load since they are the primary cause for initiation and recurrence of disease. The Table 1 shows the factors, which have to be considered for treating older adults and the different treatment options. It can be postulated that with more the factors, lesser are the treatment options, with treatments of least intensity predominating. In other words, the factors are inversely proportional to the treatment options.

In the majority of incidences, the four treatment options are desirable though medically compromised conditions do not contraindicate surgical and regenerative procedures per se, however when one takes into consideration all the factors by confounding effect, the treatment options become restricted.

In assessing the risk for undergoing surgical procedures, one can follow the guidelines laid down by the American Society of Anesthesiologists (ASA) Classification of Physical Status (American Society of Anesthesiologists: New classification of physical status, 1963). Though it is meant for general anesthesia, it can also be followed for out patient periodontal surgery under local anesthesia (Table 2).

Table 2. American Society of Anesthesiologist (ASA) classification of physical status

1.	A normal healthy patient
2.	A patient with mild systemic disease
3.	A patient with severe systemic disease that limits activity but is not incapacitating
4.	A patient with incapacitating systemic disease that is constant threat to life
5.	A moribund patient not expected to survive 24 hours with/without operation
E.	Precede an emergency operation with an E

8. SUPPORTIVE PERIODONTAL TREATMENT IN OLDER ADULTS

Supportive periodontal treatment forms an important part of the treatment plan. Most failures of periodontal therapy such as recurrence of disease are due to the non-execution of the maintenance program. It is mandatory that the patients be informed about the significance of supportive periodontal treatment. In one study, it has been found that in treated cases tooth loss was found to be three times more in patients who did not return for recall visits. The maintenance phase starts immediately after the active phase of treatment. According to Kerry (1995) there are three therapeutic objectives of supportive periodontal treatment:

1. **to prevent the progression and recurrence of periodontal disease among patients who have been previously treated;**
2. **to reduce the incidence of tooth loss; and**
3. **to increase the probability of recognizing and treating other diseases and conditions found in the oral cavity.**

These recall visits give an opportunity for dentist to assess the patient's ability to follow oral hygiene instructions. Also the dentist is able to carry out non-surgical procedures to arrest the recurrence and progression of disease and minimize further loss of tissues.

From the different treatment options given above, in the majority of incidences, the first four treatment options are desirable though medical compromise does not contra indicate surgical and regenerative procedures. But when one takes into consideration all the factors by confounding effect, the treatment options become restricted. So the periodontal disease starting as plaque- induced gingivitis at regular intervals are aggravated by different factors till old age at which time the factors become accumulated ones. In spite of this healing following treatment between younger and older people do not show any difference.

ACKNOWLEDGEMENT

I thank the staff and post graduate students of Sri Ramachandra Dental College and Hospital, Chennai, India.

REFERENCES

Albander, J.M., Brunelle, J.A. and Kingman, A. (1999) Destructive periodontal disease in adults 30 years of age and older in the United States.1988–1994. J Periodontol., 70: 13–29.

American Society of Anesthesiologists: New classification of physical status. (1963). Anesthesiology, 24: 111.

Armitage, G.C. (1999) Development of a classification system for periodontal diseases and conditions. Ann Periodontal., 4: 1–6.

Beck, J.D. and Hunt, R.J. (1985) Oral health status in the United States: Problems of special patients. J Dent Educ., 49: 407–425.

Beck, J.D. (1996) Periodontal implications: Proceedings of the 1996 World Workshop in Periodontics Ann Periodontol., 322–357.

Ismail, A.I., Morrison, E.C., Burt, B.A., Caffesse, R.G. and Kavanagh, M.T. (1990) Natural history of periodontal disease in adults: finding from the Tecumseh Periodontal Disease Study 1959–1987. J Dent Res., 69: 430–435.

Kerry, G.J. (1995) Supportive periodontal treatment. Periodontol 2000, 9: 176–185.

Locker, D., Slade, G.D. and Murray, H. (1998) Epidemiology of periodontal disease among older adults: a review. Periodontol 2000, 16: 16–33.

Mattila, K.J. (1993) Dental infections as a risk factor for acute myocardial infarction. Eur Heart J., 14(Suppl K): 51–53.

Meydani, S.N., Wu, D., Santos, M.S. and Hayek, M.G. (1995) Antioxidants and immune response in aged persons: overview of present evidence. Am J Clin Nutr., 62: 1462s–1476s.

Papapanou, P.N., Wennstrom, J.L. and Grondahl, K. (1988) Periodontal status in relation to age and tooth type: A cross-sectional radiographic study. J Clin Periodontol., 15: 469–478.

Papapanou, P.N., Wennstrom, J.L. and Grondahl, K.A. (1989) A 10-year retrospective study of periodontal disease progression J Clin Periodontol., 16: 403–411.

Rodenburg, J.P., van Winkelhoff, A.J., Winkel, E.G., Goene, R.J., Abbas, F. and de Graff, J. (1990) Occurrence of *Bacteroides gingivalis,Bacteriodes intermedius* and *Actinobacillus actinomycetum-comitans* in severe periodontitis in relation to age and treatment history. J Clin Periodontol., 17: 392–399.

Schurch, E.J., Minder, C.E., Lang, N.P. and Geering, A.H. (1988) Periodontal conditions in a randomly selected population in Switzerland. Community Dent Oral Epidemiol., 16: 181–186.

Taylor, G.W., Burt, B.A., Becker, M.P., Genco, R,J., Shlossman, M., Knowler, W.C. and Pettitt, D.J. (1996) Severe periodontitis and risk for poor glycemic control in patients with non-insulin dependent diabetes mellitus. J Periodontol., 67: 1085–1093.

CHAPTER 12

MOLECULAR DIAGNOSIS OF BREAST CANCER

LISE LOTTE HANSEN

Institute of Human Genetics, The Bartholin Building, University of Aarhus, DK-8000 Aarhus C, Denmark

Abstract: Breast cancer is the most prevalent disease and cause of death among women in Northern Europe and the USA. The incidence rate is still increasing, and despite early diagnosis and improved treatment, the mortality is still high. Breast cancer is a very heterogeneous disease and less than 10% of the diagnosed cases are believed to be caused by an inherited factor. The information on tumor specific genomic alterations has dramatically increased during the past decade, and seen in relation to the effect on survival and treatment efficiency, these genomic changes may prove to act as prognostic and predictive factors. The introduction of methods to screen the entire genome for alterations has led to important knowledge of tumor biology, progression and targets of therapy. This chapter describes the different kinds of genomic alterations found in the tumor, the methods to assess them and examples of correlations between the changes and prognostic or predictive parameters

Keywords: Breast Cancer, Genomic Alterations, Prognostic Marker, Predictive Marker, Genome-wide Screening

1. INTRODUCTION

Breast cancer is the most common malignancy and second leading cause of death among women in Europe and the USA. The annual incidence has increased over the past two decades to an estimated 1 million new cases worldwide and has not yet stagnated. Especially after the menopause, the breast cancer incidence, is five to ten folds higher in Northern Europe and Northern America than in Africa, South America and the Far East (Parkin et al., 1999). The mortality is presently declining, due to screening programs leading to early diagnosis and improved, efficient treatment (Jatoi and Miller, 2003).

S. Rattan and M. Kassem (eds.), Prevention and Treatment of Age-related Diseases, 201–233.

So far, there is evidence that less than 10% of breast cancer incidents are of inherited origin in which the disease segregates in Mendelian way, leaving the vast majority of incidents to be caused by other factors. Two high penetrant genes of BRCA1 and BRCA2 have been characterized, and together with the contribution of breast cancer incidences from other inherited cancer syndromes like: Li-Fraumeni (p53), Ataxia-telangiectasia (ATM), the Cowden disease (PTEN), Peutz-Jeghers syndrome (LKB1/STK11) and mutations in CHK2 they all account for 20–30% of the familiar aggregation of breast cancer (Heikkinen et al., 2005). Still, there are families with an accumulation of breast cancer incidences, in which no disease-causing mutation has been identified. Therefore, other genes must be involved in the inherited form of breast cancer, and these genes are likely to be of low penetrance, recessive inheritance, and the loss of function could be dependent on other/secondary genomic variations (Weber and Nathanson, 2000; Pharoah et al., 2002). Due to low penetrance these genes may prove useful as diagnostic, prognostic and predictive markers, also in the group of patients suffering from primary somatic breast cancer.

The incidence of somatic breast cancer is still increasing and several risk factors have been identified through epidemiological studies. Living in Northern Europe or Northern America, age, height, socioeconomic status, history of benign breast disease and high mammographic breast density, reproductive events (early age of menarche, late first pregnancy, no breast feeding, late menopause), exogenous hormones (menopausal hormone replacement, oral contraceptives) and life-style (lack of exercise, alcohol intake, obesity) are well-documented risk factors (Waard and Thijssen, 2005). A comprehensive study of 99,500 premenopausal women showed no significant effect of exercise on the risk of breast cancer, indicating that the positive effect may increase by age (Margolis et al., 2005).

It has become increasingly clear that the individual genetic profile is a strong risk factor, and low-penetrance cancer susceptibility genes influenced by endogenous and life-style risk factors may account for the majority of the somatic breast cancer incidences. The rapid growing amount of information about genomic variations, within and between ethnic populations, correlated to known risk factors and information on tumor specific genomic variations will prove a powerful tool in the diagnosis and treatment of cancer patients.

Screening programs and the high level of information on cancer in general have contributed to the diagnosis of an increased number of early-stage breast tumors. There is, though, an urgent need for new strong prognostic markers based on the genetic profile of the individual breast tumor not only to predict the outcome of the disease, but also to link the genetic profile to the tissue affected by the distant metastases. Patients with tumors classified by classical risk assessment including tumor size, axillary node involvement, estrogen receptor (ER) status, grade, and HER-2 status may present a completely different outcome than patients undergoing genetic profiling as risk assessment.

The generation of large-scale gene expression profiles of breast tumors has made it possible to divide tumors into subgroups, each with a distinct phenotype and prognostic outcome. Prospective studies have shown that node negative patients could be divided into two distinct groups based on the gene expression profile of their tumor. The group with a "low-risk" profile had a 96% probability of survival and a 87% likelihood of disease-free survival for 10 years without receiving adjuvant therapy. In contrast, the group with a "high-risk" profile had a 50% probability of overall survival and a 48% likelihood of disease-free survival for 10 years without treatment (van de Vijver et al., 2002). The genetic profile of a tumor will eventually become strong prognostic and predictive markers in the selection of patients who will benefit from therapy, especially in the light of current international guidelines recommending systemic adjuvant therapy for up to 85–90% of the node negative patients (Eifel et al., 2001; Goldhirsch et al., 2003).

As breast cancer is a very heterogeneous disease, it is of major importance to have strong prognostic and predictive markers to characterize each tumor and to assess its ability to metastasize, despite the lack of local metastases at the time of diagnosis, and to select the optimal treatment for each individual patient. There is a special need to identify strong prognostic markers to evaluate the outcome of patients with node negative tumors and to divide this group of patients into long-term survivors or early disease-related deaths on the basis of tumor specific genomic aberrations. This is of importance to provide the most efficient treatment for the group with poor prognosis as well as to limit unnecessary treatment to a minimum.

It is recommended that prognostic markers based on the individual genetic profile of a tumor are combined with current clinico-patological and histological markers. The laboratory techniques should be straightforward and low cost to apply in hospital regis.

The aim of this chapter is to discuss future perspectives towards the establishment of an individual tumor specific genetic profile. Prior to this, a short introduction will be made to the currently available prognostic and predictive markers with the main focus on the molecular genetic markers and the methods applied to determine them.

2. PROGNOSTIC AND PREDICTIVE MARKERS IN BREAST CANCER

A prognostic marker is defined as a characteristic of the patient or the tumor correlated with the natural history of the disease. The prognostic marker must be measurable at the time of diagnosis and before the systemic adjuvant therapy is applied. By correlation with disease-free or overall survival the marker can be used to predict the risk of recurrence in the absence of therapy.

The nature of a prognostic marker is highly variable, from the age of the patient, the size of the tumor to the presence of hormone receptors at the cell surface, spread of the cancer to the lymph nodes or distant organs and the histological characteristics of the tumor. These traditional markers are well established and have been used clinically over decades. The risk assessments derive from large randomized prospective trials with a sufficient follow-up of 10 years or more.

A predictive marker is defined as any measurement that can be correlated to the outcome of a given therapy. The classical example in breast cancer is ER and HER2/neu status, both being prognostic of survival and predictive of hormone receptor targeting therapies, since the expression level provides information on both aggressiveness and tumor specific treatment.

In clinical practice there are examples of economic limitations that prevent markers from being used even if the prognostic or predictive marker is fulfilling all criteria for implementation. Where implemented, it is of major importance that the technology is adequate enough to provide uniform results under different conditions in laboratories worldwide. The ideal marker is analyzed by use of standard hospital equipment, unambiguous to evaluate and economically feasible for a hospital budget. The source could be a blood sample or tumor tissue from a biopsy taken prior to surgery, which makes the result available for evaluation together with the clinical, pathological markers. DNA is easily extracted from both blood and tumor tissue, it is stable and remains undegraded for days in a crude blood sample. A very low amount of either DNA or RNA is needed for the majority of analyzes and could be from only a few cells. Contamination of the tumor sample by non-malignant cells is an important issue that can be avoided by micro dissection and subsequent isolation of malignant cells. The procedure demands specialized equipment and the outcome is limited but of a high quality.

3. PROGNOSTIC AND PREDICTIVE MOLECULAR GENETIC MARKERS

3.1 Selected molecular markers in clinical use

The genes described below are characterized by being both prognostic and predictive markers for aggressive tumor growth and poor prognosis in association with overexpression of the gene. The HER-2 marker is described in detail, whereas the characteristics of Cyclin E, COX-2, TOP2A, uPA and PAI-1are briefly mentioned.

3.1.1 HER-2/neu or c-erbB-2

The proto-oncogene HER-2 encodes a transmembrane tyrosine kinase cell surface growth receptor with substantial homology to the epidermal growth factor receptor and is expressed on normal epithelial cells. The gene is located at chromosome 17q12–q21.1 (www.ncbi.nlm.nih.gov and http://genome.ucsc.edu). HER-2 is overexpressed in 10–34% of primary breast carcinomas, due to a 2 to >20 fold amplification of the gene. (summarized in a review of 47 published studies comprising 15,248 breast cancer patients) (Ross and Fletcher, 1998). The amplification and overexpression of HER-2 can, in addition, be linked to the disease outcome of other neoplasms like ovarian, gastrointestinal, pulmonary, genitourinary and adenocarcinomas of the salivary gland.

Overexpression of HER-2 is a strong prognostic and predictive marker for both node-negative and node-positive breast cancer patients. HER-2 overexpression correlates with markers that define an aggressive tumor phenotype like: positive axillary lymph nodes, high tumor grading, lack of estrogen receptors, DNA aneuploidy, a high S-phase fraction, vascular invasion, p53 positive and a short overall survival (Pinto et al., 2005; Fusun et al., 2005). Importantly, the HER-2 expression level is an independent marker for recurrence in node-negative patients. Any level of overexpression in node-negative patients increased the risk of recurrence by a factor of 3.0 and 9.5 ($p = 0.0001$) for the group with a high level of overexpression, when compared with node-negative patients without HER-2 overexpression (Press et al., 1993).

HER-2+/ER+ patients are less likely to respond to hormone treatment and have a shorter survival time than HER-2-/ER+ patients ($p = 0.0001$) (Leitzel et al., 1995). Furthermore, overexpression of HER-2 correlates with chemosensitivity and resistance towards the antiestrogen tamoxifen in advanced disease (Pegram et al., 1997; De Placido et al., 2003).

Over the years, many different techniques have been used to evaluate the level of HER-2 expression in tumor tissue from Southern blot, PCR amplification, immunohistochemistry (IHC), chromogenic in situ hybridization (CISH) and fluorescence in situ hybridization (FISH) (Dressler et al., 2005; Dandachi et al., 2004; Press et al., 2002). The influence of the choice of method is reflected by the highly variable results obtained from different studies on the correlation between HER-2 overexpression and prognosis (Ross and Fletcher, 1998). Selection of the method should be based upon evaluation of the available tumor material. If paraffin embedded tissue is used, the age of the sample, temperature and time of tissue fixation are important for accurate measurement of quality and amount of the HER-2 protein. A wide variation in antibody sensitivity is seen among different commercially available antibodies as well as in relation to how the paraffin embedded tissue is fixed (Busmanis et al., 1994; Press et al., 1994). Optimal results are obtained with IHC methods using fresh or frozen tumor tissue. Consensus between different studies and large prospective studies are particularly reached using FISH, which is highly reproducible and reliable, because the DNA is more resistant than proteins to the different preservation technologies of the tumor material (Press et al., 2002).

For breast cancer FISH analysis indicates that HER-2 amplification status is consistent in the primary tumor, in locoregional or distant metastasis. Furthermore, Gong et al. found that chemotherapy did not change the HER-2 status in the metastases which is important in relation to analysis of the malignant tissue available for diagnosis (Gong et al., 2005).

Herceptin (trastuzumab), a humanized monoclonal antibody to target the HER-2 receptor launched in 1998, is implied in a large number of clinical studies. Trastuzumab is applied to patients with overexpression of HER-2 either as a single-agent therapy or in combination with chemotherapy in which the effect is significantly higher (Rueckert et al., 2005; Emens, 2005). The effect on metastatic breast cancer results in an improved response rate, time until progression and duration of response and overall survival as well as improved quality of life. Unfortunately, a subset of patients suffered from myocardial toxicity (reviewed by Gasparini et al., 2005), otherwise trastuzumab is generally well-tolerated.

3.1.2 COX-2

Cyclooxygenases (COX) are key enzymes in the conversion of arachidonic acid to prostaglandins, and the expression of cyclooxigenase-2 (COX-2) is related to angiogenesis and associated with tumor aggressiveness like: tumor size, axillary node metastasis, hormone receptor negative tumors and HER-2 amplification. Increased expression of COX-2 is detected in preinvasive and invasive tumors with a poor prognosis (Arun and Goss, 2004). COX-2 inhibitors are being used in clinical trials with promising results.

3.1.3 Cyclin E

Cyclin E forms a complex with cdk2 regulating the G1 to S-phase transition. Cyclin E is found overexpressed in breast tumors (up to 64-fold), and the protein was found to be the strongest, independent prognostic marker for survival in stages I–III tumors (Keyomarsi, 2002). Cyclin E is a predictive marker for the response to chemotherapy and to hormone treatment, since overexpression of Cyclin E increased the sensitivity of the tumor to cisplatin in combination with paclitaxel (Smith and Seo, 2000). Resistance has been detected towards antiestrogens in tumors overexpression cyclin E, but these tumors may benefit from therapy with cdk2 inhibitors (Akli and Keyomarsi, 2004; Hunt and Keyomarsi, 2005).

3.1.4 TOPO II/TOP2A

Topoisomerase II alpha (TOPO II/TOP2A) is situated close to HER-2 at chromosome 17q21-q22 and catalyzes the relaxation of supercoiled DNA molecules, catenation, decatenation, knotting and unknotting of circular DNA. TOP2A is commonly coamplified with HER-2, and the amplification level is a predictive marker for patients with advanced breast cancer (Hicks and Tubbs, 2005). Treatment with doxorubicin rather than docetaxel provides a higher probability of response in tumors with more than 10% cells expressing TOP2A (Durbecq, 2004).

3.1.5 uPA and PAI-1

Urokinase-Type Plasminogen Activator (uPA) and its inhibitor Plasminogen Activator Inhibitor Type 1 (PAI-1) play an essential role in solid tumor growth, invasion and metastasis. A low level of uPA and PAI-1 correlates with a very favorable prognosis, whereas a high level denotes reduced recurrence-free survival and overall survival. The expression of uPA and PAI-1 is especially suitable to distinguish between the groups of node-negative patients who could be spared from the adjuvant therapy and those who have a high risk of recurrence that would clearly benefit from early therapy (Harbeck et al., 2004; Manders et al., 2004).

4. SINGLE NUCLEOTIDE ABERRATIONS

4.1 Tumor suppressor genes (TSG)

The existence of genes able to suppress tumor growth was suggested by Harris et al. (1969) after fusion of malignant and non-malignant cells which resulted in suppression of the malignant phenotype in the hybrid cells.

The recessive trait of a tumor suppressor gene (TSG) was based on Knudson's hypothesis that both alleles of a gene should be silenced to induce tumorigenesis.

Studies of the rare disease of Retinoblastoma that causes eye tumors in young children, led to the first TSG two hit model (revised by Knudson, 2001) (Knudson, 2001). An inherited mutation silencing one allele of *RB1* was the initial hit. During embryogenesis or early in life a deletion or mutation affecting the second allele in one somatic cell may functionally silence *RB1* and lead to retinoblastoma. The two hit events are confirmed for TSGs involved in inherited cancer diseases as for example: *BRCA1, BRCA2* in breast cancer, *MSH2, MLH1* in hereditary non-polyposis colon cancer, *APC* in familial adenomatosis polyposis colon cancer and *p53* in Li Fraumeni syndrome. These diseases are characterized by early onset and a high life time risk of cancer due to a dominant inheritance. Retinoblastoma patients with a family history of the disease had a high risk of developing bilateral retinoblastoma as well as secondary malignancies. In contrast, patients with non-inherited retinoblastoma usually presented an unilateral disease and no secondary malignancies. This supports the model, instead of a germ-line mutation RB1 was hit by two different events in just one somatic cell, silencing both alleles and initiating tumor growth.

A model for a multi-step development of colon cancer was proposed by Fearon and Vogelstein in which allelic loss or a mutation in one tumor suppressor gene may initiate a chain of genetic events eventually leading to uncontrolled cell growth (Fearon and Vogelstein, 1990; Fodde and Smits, 2001).

Over the years it became evident that not all tumor suppressor genes could fit into this two-hit model.

For an increasing number of genes sufficient expression (to obtain a normal function in the cell) cannot be obtained from only one intact allele (a hemizygous state). A functional transcript should be provided from both alleles to produce a normal phenotype (haploinsuiffiency) (Quon and Berns, 2001).

TSGs are described as guardians of the genome as well as gatekeepers and caretakers of cell cycle check points. This refers to the important function in different cell maintenance processes like DNA repair, cell adhesion and apoptosis.

TSGs involved in the maintenance of the genome by DNA repair and preservation of both chromosome number and integrity are named "caretakers". Disruption of the function as caretaker gene increases genomic instability, allowing additional mutations in other TSGs and thereby pushing the cell further towards uncontrolled growth. Functional failure of a caretaker gene cannot be reverted by reconstitution of the gene in contrast to another group of TSGs, the gatekeepers. Gatekeeper genes are important for keeping a constant number of cells in a specific tissue, and they play an important part in controlling the cell cycle. Mutations in a gatekeeper gene, abolishing one pathway to apoptosis, will lead to a displaced balance between cell renewal and cell death. A balanced cell number is maintained, despite mutations in other genes, if the function of a gatekeeper gene is normal (Kinzler and Vogelstein, 1996).

TSGs are difficult to target with anticancer therapy due to their loss of function in tumor cells in contrast to oncogenes characterized by gain of function. Instead, TSGs are suitable for genetic profiling and prognostic markers as their loss of function can be determined directly on the genomic level by numerous well-established methods.

4.2 Mutations

Mutation analysis for diagnostic purpose exists for BRCA1 and BRCA2, to establish the inherited predisposition to breast cancer. A substantial amount of tumor suppressor genes have been analyzed for mutations in somatic breast tumors. The effect of a disease-causing mutation can be evaluated by correlation studies with prognostic parameters, association studies with control populations and by functional studies of the effect on protein level.

A large spectrum of methods is available for mutation detection. The methods can roughly be divided into two categories depending on whether the mutation is known or not. A known mutation could be a single, rare nucleotide substitution or polymorphism (SNP), a minor deletion or insertion. The semiautomatic primer extension method is widely used for genotyping of a single nucleotide substitution, since multiple analyses can be performed in one reaction, the results can be assessed by automated capillary electrophoresis and easily evaluated via specific software. The method is based upon the principles behind the Sanger sequencing, one primer is constructed to anneal the 3' terminal nucleotide to the nucleotide preceding the mutation. The primer is extended by a polymerase reaction containing dideoxynucleotides labeled with base-specific fluorescent dye. The primers to assess each mutation differ by length, and after separation by electrophoresis the genotypes are determined by the fluorescent color of each extended primer.

Matrix-assisted laser desorption/ionization time-of-flight mass spectrometry (MALDI-TOF) is an efficient method that allows a high throughput, accurate SNP discovery and sequence validation (Smylie et al., 2004; Nelson et al., 2004). DNA fragments up to 450 bp can be analyzed. The principle is based on primer extension: once the primer is annealed, the strand is extended by incorporating dNTPs. The 5' phosphodiester bonds of each newly incorporated pyrimidine nucleotide are replaced by acid-labile phosphoamidite (P-N) bonds. The template strand is attached to magnetic beads through biotin-streptavidin binding. The P-N bonds are cleaved by hydrolysis and the small fragments are subjected to MALDI-TOF, separating the fragments according to size. The technique requires specific equipment, but is hereafter cost-effective.

A large variety of methods are available for detection of unknown mutations in a DNA fragment. Most methods to detect an unknown mutation are based on the formation of heteroduplexes and the denaturing conditions necessary to separate either of the two DNA strands from the hetero- and homoduplexes. Denaturing Gradient Gel Electrophoresis (DGGE) and Denaturing High Performance Liquid Chromatography (DHPLC) are based upon the principle that separation of the two DNA strands in a heteroduplex is faster than in a homoduplex. The DHPLC is automated and software is available to calculate the denaturing conditions suitable for each DNA fragment. The method has several pitfalls, it is important to analyze the same DNA fragment under different denaturing conditions, especially the temperature is important, as two different mutations in the same analyzed fragment may require different conditions to denature. DHPLC is very efficient, once the analysis conditions for specific DNA fragments and mutations are established, a large amount of samples can be analyzed within a limited time.

Chemical Cleavage of Mismatches (CCM) is based on the principles of Maxam and Gilbert sequencing. Initially, OsO_4 was used to modify a mispaired thymine and hydroxylamine a mispaired cytosine in a heteroduplex. The sugar phosphate backbone was cleaved at the modified bases, and the length of each fragment measured by gel electrophoresis. The length of each fragment combined with the chemical that successfully modified a base led to a very precise prediction of the nature and location of the mutation (Cotton et al., 1988). The method is now modified; $KMnO_4$ substitutes OsO_4, the DNA fragments are labeled by fluorescent dyes to detect the cleaved fragments by automated electrophoresis (Hansen et al., 2003). CCM has proved very reliable, close to 100% of all mutations are found, even if the mutated DNA accounts for only 5% of the total DNA content in the sample, and more than one mutation can be detected at the same time in the same DNA fragment (Hansen et al., 1996).

DNA sequencing is often referred to as the ultimate mutation detection method, but for most tumor samples the presence of non-malignant tissue may reduce the tumor-specific mutation to being indistinguishable from the background (noise). Therefore, it is very important to use two different methods to assess a new mutation, a sensitive mutation detection method followed by verification of the mutation by DNA sequencing.

In research it is important to use the most sensitive mutation detection method to be sure to find all tumor-specific and germ-line mutations independent of the nature of the variation to evaluate the impact of the mutation on the development and progression of the tumor. Once the mutation is characterized, the DHPLC is an efficient method to screen a large number of samples for the known mutation. The optimal analysis conditions are stored in the software and are easy to access also for diagnosis of only a few samples.

4.3 Mutations in a TSG with a prognostic value

The tumor suppressor gene p53 is located at chromosome 17p13.1, a region commonly deleted in breast tumors. p53 is probably the most well-described TSG and has undergone intensive research through more than 25 years. p53 is called the guardian of the genome, a multifunctional protein involved in cell cycle arrest, DNA repair, apoptosis and differentiation. The gene is activated in cells under stress like irradiation, hypoxia, DNA damage and virus infection thereby leading to the protection of the cell by inducing a long-range of genes involved in different pathways.

p53 is mutated in 50% of cancers and germline mutations in p53 results in the multi-cancer Li-Fraumeni Syndrom disease. Mutations in p53 are found in 25% of all breast tumors and are associated with an aggressive tumor phenotype involving: a high histological grade, aneuploidy, a high mitotic index, ER and PgR negative cells.

p53 status has been evaluated by mutation detection, direct sequencing and immunohistochemistry methods. The results are in disagreement as to the prognostic value of p53 mutations, especially when the immunohistochemical methods have been used, reviewed in (Ross et al., 2004). Discrepancies are also found when p53 is evaluated as a predictive marker. Studies on metastatic breast cancer have led to an association between p53 mutations and resistance to hormone and adjuvant, neoadjuvant and combination chemotherapy whereas other studies find no association (reviewed in Ross et al., 2004).

These results reflect the difficulty of using mutation status as a prognostic or predictive marker. To evaluate the effect of a mutation on the protein level, an extensive number of different mutations, spread along the entire gene must each be associated with the impact on the protein activity. Even if the protein is detected by immunohistochemistry methods, nothing is known about the activity/efficiency of this protein, or if it is capable of withholding the normal functions in the cell. Studies comparing the nature of the mutation to prognosis have found subgroups of sequence variations, which correlates significantly to disease-free survival (Bergh et al., 1995; Alsner et al., 2000; Borresen et al., 1995)

Overexpression of p53 induces apoptosis, which points at p53 as an interesting target for therapy.

A long-range of proto-oncogenes and TSGs have been analyzed for mutations. Despite extensive studies a limited number of mutations have a prognostic value. de Jong et al. reviewed 34 polymorphisms in 18 different genes and found an association to breast cancer risk for 13 polymorphisms in 10 genes (de Jong et al., 2005). Among these specific polymorphisms in p53, ER and PgR are associated with a decreased risk of breast cancer, whereas mutations in HRAS, GSTM1 and CYP19 increased the risk of breast cancer. Polymorphisms in genes involved in DNA repair like XRCC1, XRCC3, ERCC4/XPF, BRCA2 and RAI either alone or in combination are associated with an increased risk (Dumitrescu and Cotarla, 2005; Nexo et al., 2003).

4.4 Single nucleotide polymorphism (SNP)

The most frequent variation within the genome is the SNPs. The estimated frequency for SNPs with an allelic frequency exceeding 1% is one in 300 bp (Kruglyak and Nickerson, 2001; Judson and Stephens, 2001; Reich et al., 2003. Of the predicted

30 million genomic SNPs it is estimated that 100–300,000 are nonsynonymous and that each person carries between 24 and 40,000 nonsynonymous SNPs (Cargill et al., 1999). More than two million SNPs have been identified and reported to databases, and more than one million SNPs have been fully characterized via genotypes in 269 DNA samples from four different populations as part of the HapMap project (Altshuler et al., 2005), (http://www.ncbi.nlm.nih.gov/SNP/).

When a population based study is set up with the purpose of finding new susceptibility genes for the disease or new genotypes significantly associated with disease risk, progression or survival, the difficulty is how to choose among this growing number of SNPs. The majority of SNPs in public databases are not validated in large populations, the level of polymorphism therefore being unknown. A genotype variation ranging between 0–24% was found in a case-control study of the HER-2 SNP I655V in different populations (Ameyaw et al., 2002). It may be necessary to screen the selected SNPs in a small number of individuals to distinguish between true polymorphic SNPs and rare nucleotide substitutes. For large-scale studies the selection of highly polymorphic SNPs is important to assure the most optimal result. Nelson et al. constructed arrays to analyze SNPs selected for a minor allele frequency >2% and for being located within 10 kb of 66% of all known or predicted genes in the human genome (Nelson et al., 2004). Of the initial 204,200 SNPs extracted from public databases, fulfilling the criteria and providing a result, only 125,799 were polymorphic in the analyzed population (61.6%).

To establish selective criterias for choosing informative SNPs, 166 molecular epidemiological studies of 46 SNPs in 39 different cancer-related genes were evaluated, including 355 nonsynonymous SNPs from 90 DNA repair genes in which 103 SNPs were found to alter an amino acid in a position which is highly conserved among species. The authors found a significant association between the odds ratio for cancer risk and the conservation level among different species of the SNP (Zhu et al., 2004). This study is important, and implemented in the choice of targets for array-based SNP analyses the obtained results may be highly specific and informative.

The methods available for mutation detection can be used to assess the SNP genotypes. The methods can be divided into two groups depending on whether the SNP is well characterized or the search is for new SNPs. Methods used to search for unknown mutations will eventually identify new SNPs, and especially in large-population based studies it is evident if the frequency of the mutation exceeds 1%. The primer extension and mass spectrometry-based method of MALDI-TOF are widely used for detection of known SNPs and described in the mutation section. The MALDI-TOF is suitable for large-scale analysis.

The increasing numbers of SNPs require large-scale genotyping. The oligonucleotide-based array methodology meets this demand, a large number of SNPs can be genotyped at a time, using highly specific arrays directed towards prognosis and therapy.

A single nucleotide deletion or insertion is also considered a SNP if the polymorphism rate exceeds 1%. These variations can be measured by the difference in length of the affected region. One example of a fragment length polymorphism is the presence of one or two guanines in the promotor region of MMP-1. This SNP is adjacent to an acceptor protein (AP-1) site and the presence of two guanines enhances the transcription of MMP-1, whereas the same AP-1 site mediates a decrease in the transcription level when only one guanine is present. The high level of MMP-1 transcription is likely to contribute to the invasive potential of the analyzed breast cell lines. (Tower et al., 2003). A case-control study could provide interesting results on the prognostic effect of this SNP.

Tri-nucleotide repeat expansion is well known from diseases as Fragile X, Myotonic Dystrophy and Chorea Huntington. A number of tri-nucleotide repeats are positioned in the coding region of the gene, and expansion or shrinkage of such a repeat has a very dramatic effect on the protein. Especially huntingtin, the Chorea Huntington disease gene, is known to expand with several hundred additional repeats. Some tri-nucleotide repeats are positioned outside the coding region in the 3' or 5' end of the gene. Expansion of these repeats may affect regulatory elements and thereby the transcription of the mRNA level. The androgen receptor contains two coding tri-nucleotide repeats, a CAG and a GGC repeat. Expansion to more than 28 repeats is associated with early onset in BRCA1 or 2 carriers, and breast cancer patients in this study all carried one allele with more than 29 CAG repeats. This assocation could not be confirmed in Jewish BRCA1-2 carriers (Rebbeck et al., 1999; Dagan et al., 2002). In sporadic breast and prostate cancer a weak association is found between short alleles of the CAG repeat, positive lymph nodes and reduced survival (Yu et al., 2000). A short repeat length of the GGC sequence can be associated with a reduced risk of breast cancer in young women (Suter et al., 2003).

Despite the growing number of epidemiologic based studies on the association between genotype variations within coding and regulatory regions of the genome and prognostic parameters, only a few SNPs have been found to be statistically strong predictors of breast cancer risk and survival. In a large-scale case-control study 25,000 SNPs were selected from the 125,799 SNP array previously described in (Nelson et al., 2004). These SNPs were located within 10 kb of 13,735 genes and 95% had a minor allele frequency larger than 0.1. The genotypes were analyzed in 254 German breast cancer patients and 268 age-matched women without malignant disease. One marker at 14q24.3-q31.1 was weakly associated with breast cancer status. High density mapping of the region defined a SNP in intron 1 of the zinc-finger gene *DPF3/CERD4* for which the genotype correlated significantly to breast cancer status (OR = 1.6, $P = 0.003$), increased lymph node metastases ($p = 0.006$), age of onset ($P = 0.01$) and tumor size ($P = 0.01$). (Hoyal et al., 2005). A similar study of 25,000 SNPs in approximately 16,000 genes identified a strong association between the SNP variations in a 20 Kb region at chromosome 19p13.2 and risk of breast and prostate cancer. The association was strongest in individuals with a family history of breast cancer (OR = 3.4, $P = 0.001$). A detailed mapping of the region identified one SNP within *ICAM5* that associated strongly with disease progression and prognosis (Kammerer et al., 2004).

The creation of a haplotype within a single gene or spanning a narrow chromosomal region may prove to be a strong prognostic marker. The genotype frequencies of F31I in the Aurora-A gene were predicted to have a functional impact, but no variation was found between a breast cancer and a control population. When combined with multiple SNPs in the Aurora-A gene a specific haplotype associated strongly with breast cancer risk. Within this haplotype, the putative at risk genotype Ile31 was more frequent in the subgroup of women carrying a higher risk of breast cancer than in the low risk group (Lo et al., 2005). The genotype frequencies in a small breast cancer and control cohort from Taiwan were determined for three silent SNPs in the ER-alpha gene. The genotype frequencies were significantly different in the two groups and associated to the presence of lymph node metastasis (Hsiao et al., 2004).

The importance of creating a haplotype instead of focusing on a single SNP is illustrated by the extensive studies in different breast cancer populations of the HER-2 SNP I655V, in which no conclusive results have been obtained. In order to reach a statistically significant conclusion six of 29 polymorphic SNPs were chosen in the HER-2 gene, including the missense mutations of I655V and A1170P, due to a high degree of polymorphism established for each SNP in a control cohort. The six SNPs could be assigned to one haplotype block due to strong linkage disequilibrium. Alone, each SNP genotype did not correlate with any prognostic parameter, but the tumor specific protein expression of HER-2 was increased 1.5 fold ($p = 0.009$) and the disease outcome was worse ($p = 0.032$) in the patients carrying the specific haplotype (Han et al., 2005). Five common SNPs in HER-2 were analyzed in large British breast cancer and control cohorts leading to the conclusion that these polymorphisms are not contributing to the predisposition of breast cancer in this population. Only two missense SNPs were alike in the two studies (Benusiglio et al., 2005).

SNPs are highly valuable as risk predictors, prognostic markers and as a tool to discover new tumor suppressor genes via haplotype determination of linkage disequilibrium. They are easy to assess, either as single SNPs or in large-scale studies, and they contain a high level of information. Haplotypes created from SNPs present in the same pathway may become strong prognostic markers and lead to further identification of new genes possessing a strong prognostic potential for breast cancer.

The attempt to use SNPs in HER-2 as prognostic markers is a good example of how difficult it can be to identify new statistically convincing prognostic markers. One SNP genotype may be a strong predictor in one population but have no statistical effect in another. The allele frequency of one SNP may vary tremendously in different populations from a conserved homozygous to minor allele frequencies of 0.25. Instead of analyzing one candidate SNP, the result may have a strong significance if a haplotype across the susceptibility gene is analyzed. Large-scale studies have proven successful to identify new SNPs with a strong prognostic value; the extensive number of results derived from thousands of SNPs analyzed in large cohorts makes the statistical calculations very strong.

5. CHROMOSOMAL DELETIONS OR AMPLIFICATIONS

5.1 Loss of heterozygosity (LOH)

Somatic LOH or allelic imbalance (AI) has been used to search for new tumor suppressor gene loci, based on Knudson's theory (1971) that both alleles of a TS gene are transcriptionally silenced by two different events to exert the tumorigenic effect (Knudson, 2001). One possibility could be an inherited mutation to knockout the functional product from one allele in all cells of the body followed by a tumor-specific deletion of the second allele. The picture, though, is probably (likely to be) much more complex; the primary event does not have to be an inherited mutation as initially suggested, but a hit to the genome in one or few cells could be caused by an environmental factor, as chemicals damaging the DNA (chemotherapy, etc) or radiation as UV radiation from the sun. This hit may not affect the (over all) function of the cell but slightly increase the instability of its genome leading to secondary lesions to the genome.

LOH analysis is excellent to either screen a whole chromosome for regions that may contain new tumor suppressor genes or to map small well-defined regions in detail to further locate the susceptibility gene. There are, though, several pitfalls to consider using the different methods to measure the allelic imbalance in a locus.

The principle of assessment of LOH is based on a measurement of the quantitative difference between two alleles in the tumor when compared with the same alleles in non-malignant tissue from the same patient (Hansen and Justesen, 2003). Highly polymorphic microsatellite markers (simple tandem repeats, STRs), preferentially di-, tri- or tetra nucleotide repeats or SNPs are useful markers for this analysis. Mononucleotide repeats should be avoided since the profile after electrophoresis makes it difficult to interpret each allele. STRs are scattered over the genome with a very high frequency.

Different approaches can be used to select new STRs. The Human Genome Browser (http://genome.cse.ucsc.edu) provides a map of polymorphic STRs at their genomic position, all information concerning primer sequences, allele number and frequency.

For screening whole chromosome arms, the markers can be selected with a mutual distance of 5–10 cM. Once a region has been identified and the search is for the susceptibility gene, the sequence of each gene and the close flanking region can be screened for STRs. LOH analysis of intragenic STRs provides direct information on the genomic lesions of the genes in the region, and in association with prognostic parameters of the patient cohort single genes can be picked and further analyzed for the implication in carcinogenesis.

The STR and flanking region is PCR amplified, using DNA from both malignant and non-malignant tissue from each patient. One primer is labeled with a fluorescent dye and the product can be analyzed via capillary electrophoresis. These PCR reactions can be multiplexed with 4–5 reactions in one tube and further pooled for electrophoresis with 10–15 other STRs. Software is available for calculation of the ratio between the alleles from the tumor and wild type (Hansen and Justesen, 2003). The final conclusion depends on where the cut-off level between LOH and retention of the alleles is defined. The optimal sample is the micro-dissected tumor tissue without traces of non-malignant cells, but the majority of studies are made on tumor tissue containing a certain fraction of non-malignant cells. The cut-off level should be evaluated for each tumor type and for each analyzed panel since the amount of non-malignant cells may vary between different panels. The cut-off level described in the literature varies from a 50% to 16% decrease in allele intensity (Gaki et al., 2000; Skotheim et al., 2001). The choice of cut-off value influences the conclusion tremendously when correlated with prognostic parameters of the patient cohort. Use of a high cut-off level may reduce the amount of information considerably and a too low cut point may dilute a possible significance of the study. The pit-falls of using LOH analysis are well described by (Tomlinson et al., 2002; Miller et al., 2003).

The level of information is also dependent on the polymorphic level of the STR. Especially in studies on the association to prognosis it is important to obtain information on each tumor from each loci. At least half of the information is lost, due to uninformative tumors, in a LOH study. This can in part be overcome by searching the entire genomic sequence for all STRs in the region choosing STRs in a very close proximity, for instance inside the same gene or within the same region of a large gene. During the LOH calculations all information from closely situated STRs can be pooled, thereby enriching the level of information considerably (Figure 1).

The LOH analysis provide only information on whether there is an imbalance between the allele quantities in the tumor, but not if one allele is deleted or the second allele is amplified in the genome. Comparison with Comparative Genome Hybridization (CGH) results may provide this information, especially if a small chromosomal region is analyzed, since the resolution under normal conditions is low. If a more detailed picture is needed for a narrow region with few genes, results from expression arrays may provide an answer. In case of uncertainty the phrase "allelic imbalance" (AI) should be used.

During tumor progression the DNA repair system may be impaired to different degrees, and for a small number of breast tumors a third, and occasionally more, alleles are seen when STRs are PCR-amplified. STRs are by nature sensitive to mutations affecting the length of the nucleotide repeat, and the presence of additional alleles in the tumor genome provides information of a decreased function of the DNA repair system.

Searching a chromosomal region for LOH using a panel of STRs with a precise location will provide additional information on possible chromosomal breakpoints.

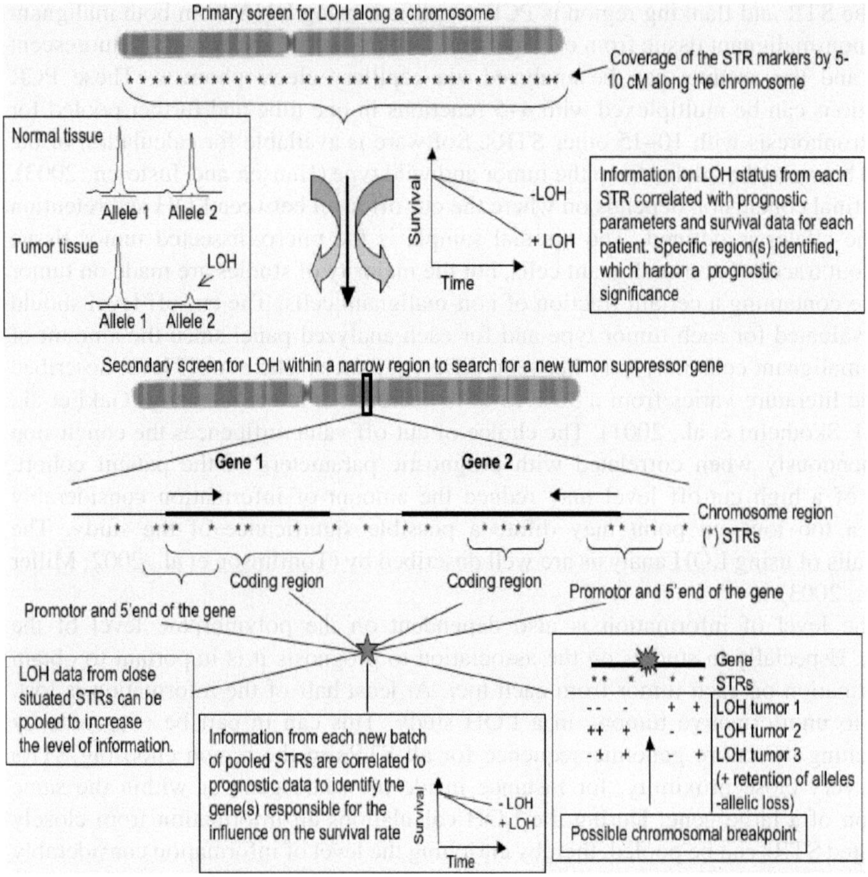

Figure 1. Screening for Loss of heterozygosity (LOH) along a chromosome, in the search for new tumor suppressor genes. Tha initial screen is performed with highly polymorphic microsatellite markers along the chromosome. Chromosomal regions with a high rate of LOH, and in which the allelic loss correlates to prognostic factors, are selected for a detailed scan. New markers are selected with preference to intragenic positions to map the genomic alterations affecting susceptibility genes

A breakpoint can be defined if one chromosomal site is flanked by LOH on the one side and by retention of both alleles on the other. The exact position of a breakpoint is important especially when it affects the transcription unit of a gene.

When retention of alleles is flanked by LOH over a short distance, it may reflect a small homozygous deletion in which PCR amplification of the wild type DNA appears as allelic retention.

The extensive search over the past decade for new tumor suppressor genes has mapped many susceptibility loci. In breast cancer LOH has been reported to be a frequent event on chromosome arms 1p, 1q, 3p, 6q, 7q, 8p, 11p, 13q, 16q, 17p, 17q, 18q and 22q (Devilee and Cornelisse, 1994; Callahan et al., 1993).

Table 1. Example of correlations between LOH/Allelic imbalance and prognostic parameters in different cohorts of breast cancer patients. (EIC, extensive intraductal component; PALI, peritumoral angiolymphatic invasion)

Prognostic parameter	Chromosomal regions affected by LOH	Significance	Number of patients	Ref.
ER-:	1p22, 3p25.1, 3p14.3, 17q21.1		504	Nagahata et al., 2002
	17p13		51	Seitz et al., 1997
PR:	13q12-13, low PR content		139	Eiriksdottir et al., 1998
	18q22 (D18S51)		228	Huiping et al., 1998
	D10S583	p = 0.01	105	Garcia et al., 1999
Tumor grade:	17p13		51	Seitz et al., 1997
	PTEN at 10q23	p = 0.02	105	Garcia et al., 1999
Tumor size:	17p13		51	Seitz et al., 1997
Age:	PTEN at 10q23	p = 0.02	105	Garcia et al., 1999
Lymph node metastasis:	11q23-24	p = 0.0042	504	Nagahata et al., 2002
	13q12	p = 0.0207	504	Nagahata et al., 2002
	17p13.3	p = 0.0478	504	Nagahata et al., 2002
	22q13	p = 0.0162	504	Nagahata et al., 2002
	D13S1699 (local recurrence)	p = 0.024	39	Regitnig et al., 2002
	D17S855 (BRCA1)	p = 0.019	39	Regitnig et al., 2002
	PTEN at 10q23	p = 0.02	105	Garcia et al., 1999
S-phase fraction:	RB1, high S-phase fraction (no BRCA2 mutation)	p = 0.0001	139	Eiriksdottir et al., 1998
	18q22 (D18S51)		228	Huiping et al., 1998
Early local recurrence:	D17S5 and retention of alleles of TP53 locus versus no LOH at D17S5.	p = 0.007	67	Nagai et al., 1994
	TP53 at 17p13.3	p = 0.018	39	Regitnig et al., 2002
	1q21-23	P = 0.01 (EIC) 0.04 (PALI)	50	Gaki et al., 2000
	PEM at 1q21	p = 0.006	89	Borg et al., 1992
Distant metastasis:	13q12-13, increase of risk by a factor 4 (no BRCA2 mutation)	p = 0.001	139	Eiriksdottir et al., 1998
Prognostic parameter	Chromosomal regions affected by LOH	Significance	Number of patients	Ref.
Mortality:	16q23.2-24.2 (D16S511) freedom from distant metastasis	P = 0.002)	199	Hansen et al., 1998
	8p22 (patients received high dose adjuvant chemotherapy)	p = 0.0354	150	Tsuneizumi et al., 2002
	8p22	p = 0.017	298	Utada et al., 2000

(Continued)

Table 1. (Continued)

Prognostic parameter	Chromosomal regions affected by LOH	Significance	Number of patients	Ref.
	3p25.1 + 17p13.3, mortality risk increased by factor 4.9	p = 0.0006	298	Haga et al., 2001
	3p25.1 + 13q12, mortality risk increased by a factor 2.9	p = 0.0441	298	Haga et al., 2001
	3p24-25	p = 0.0014	504	Matsumoto et al., 2000
	11q24.1-25 (D11S387) age below 37	p = 0.028	102	Gentile et al., 1999
	16q23.2-24.2 (D16S511) disease-free survival and overall survival	P = 0.002	199	Hansen et al., 1998
	11q23+/– LOH of 11p15 (aggressive post metastatic disease)	p = 0.0005	86	Winqvist et al., 1995
	D11S387 at 11q24.1-25 (below age 37)	p = 0.028	102	Gentile et al., 1999
	1p (overall survival)	p = 0.001	238	Ragnarsson et al., 1999
	D1S435 at 1p31.1	p = 0.0022	238	Ragnarsson et al., 1999

Despite the huge effort to characterize the LOH pattern in breast tumors only few new tumor suppressor genes have been identified in this way. Several explanations can be mentioned:

1. Lack of fine mapping, down to single genes by LOH.
2. Haploinsufficiency, LOH affecting one allele of the susceptibility gene is the prime cause of the reduced function. No mutations or promotor hypermethylation are present to affect the protein function.
3. Before the release of the Human Genome sequence the precise location of the STRs were uncertain and depended on how precise the markers/landmarks of the genome were mapped. An incorrect position of just one marker could influence the entire flanking linkage map of the genome and the target gene is overseen.

LOH affecting specific regions of the genome acts as strong predictors of either favorable prognosis (LOH of 16q23.3-24.2 is an independent marker of long overall survival) or poor prognosis (LOH of 13q12-13 is a marker of high risk of recurrence and LOH of 1p for short overall survival). As can be seen from the table a few regions turn up from several studies showing the strongest association with prognosis. These regions should be further analyzed in large cohorts, the regions should be further narrowed to isolate the region or the gene that carries the strongest prognostic potential.

For clinical use, a few cells from a needle biopsy of the tumor and a blood sample can be used. A panel of several STRs can be analyzed at the same time and the answer concerning predictors of favorable or poor prognosis can be provided within 1 to 2 days and be considered as part of the entire picture of prognostic factors.

Large-scale LOH studies can be performed using oligonucleotide SNP arrays. The method is described in the section below.

6. AN OVER-ALL VIEW OF ENTIRE GENOMIC CHANGES AS PROGNOSTIC PREDICTORS

The micro array technology can be applied to a wide spectrum of large-scale analysis of the genome. SNP arrays provide information on the genotypes of each selected polymorphism at the array, and in addition it can be used for LOH analysis at each SNP loci in which the test person is heterozygous. A picture of the global genomic methylation pattern can be generated and the CGH analysis can be performed using arrays instead of immobilized metaphase chromosomes on glass slides. Tissue specific arrays are analyzed via immunohistochemical techniques and provide information on the protein expression level of selected proteins within each tumor. Gene expression arrays determine the level of mRNA in the tumor cell compared with the level in a homologues non-malignant cell.

The overall advantage of arrays is the ability to screen the entire tumor or wild type genome for specific variations. Especial launching of the tiling BAC-arrays that cover the genome several times in overlapping fragments is a powerful tool for an initial screening for methylation or a CGH analysis (Ishkanian et al., 2004). Targets providing a statistical correlation to any prognostic marker in a representative cohort, from the initial whole-genome screening, can be selected for the design of new arrays directed specifically towards a prognostic or predictive diagnosis.

6.1 Microarray based gene expression studies

A substantial amount of data is now generated by microarray-based gene expression analysis combined with the correlation of differently expressed genes to prognostic and predictive parameters. Still, there are strong contradictions among the results obtained from the literature, possibly due to different platforms (cDNA or oligonucleotide derived), variable quality of the samples, different hybridization protocols and to the final evaluation/preparation of the results. The microarray technology is without doubt a very powerful tool to define new prognostic markers (prognostic profiles consisting of multiple up or down regulated genes), refine the tumor classification, generation of a personalized genetic profile useful for the determination of optimal type of treatment, and eventually in developing new targets of therapy.

A large variety of genetic changes can influence the expression level of a single gene. A decrease may be due to allelic loss affecting the whole gene or the promotor region, chromosomal breaks, nonsense mutations, methylation of the promotor region and lesions affecting enhancer elements. Overexpression may be due to amplification of whole chromosomes or minor regions, silencing of silencer

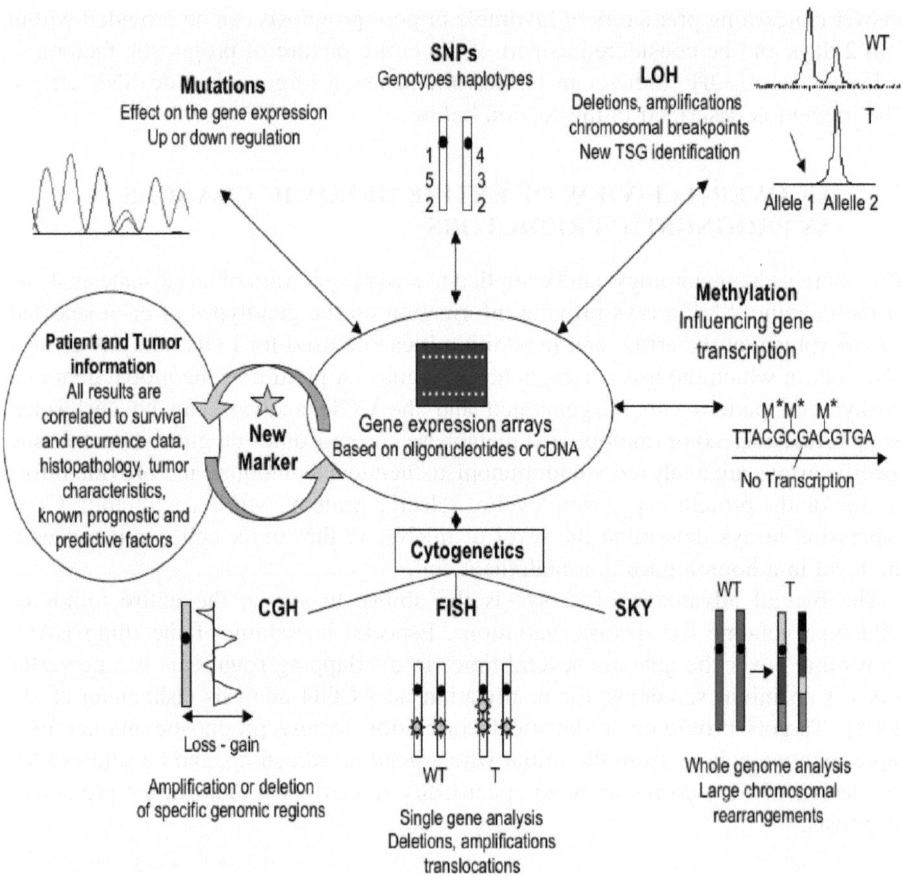

Figure 2. Different pathways to identify new prognostic markers for the outcome of breast cancer

elements, hypomethylation of the promotor region, and gain of function mutations. Therefore, the results from gene expression arrays are keys to numerous of other genomic analysis, see Figure 2.

The principle behind the method is the initial immobilization of DNA, representing single genes, onto a glass slide or a chip. It is possible to place up to 60.000 samples/items on one glass slide. RNA is purified from the tumor, PCR amplified and thereby converted to cDNA, and simultaneously labeled with a fluorescent dye. RNA from non-malignant tissue of the same origin as the tumor tissue is treated likewise and labeled with a different color.

Tumor and control cDNA are hybridized to the complementary DNA fragments on the array. The chip or glass slide is automatically scanned, the resulting fluorescence from each spot providing information on the relative rate between tumor and control cDNA. If the tumor cDNA is labeled with red and control cDNA with green, then overexpression of a gene is visualized as a red spot, and if the gene is down-regulated, the spot will be green due to the control cDNA. An orange spot indicates no change in expression of that particular gene since there is equal hybridization efficiency between tumor and control cDNA. Specific algorithms are developed to calculate the differences in expression level of each analyzed gene (spot). The tumors are thereby divided into hierarchical clusters defined by the expression pattern across the chip.

The clinical information and survival data from each patient can be correlated to the expression profile, relating each cluster to a specific tumor developmental stage. The differently expressed genes of a prognostic or predictive significance can be selected for the construction of a new chip directed specifically towards diagnosis or choice of treatment.

The analysis is based on measurement of the cellular level of RNA and therefore the purification of high quality, intact RNA is one of the most critical steps.

Based on large-scale studies on gene expression profiles the breast tumors can be divided into subgroups with clinically different outcome as normal breast-like, basal-like, ERBB2/HER-2 positive and luminal subtypes A, B and C (Sorlie et al., 2001).This classification is solely based upon the gene expression pattern with no inclusion of any clinical endpoints and is designated "unstructured cluster analysis". The outcome of the luminal A subtype, with tumors primarily ER positive, is distinctively better than for both B and C. The luminal subtype and C presents the worst outcome, with a short time to recurrence. In addition, the ER negative tumors can be divided into the basal-like and the HER-2 overexpressing subtypes, both with a poor prognosis (Sorlie et al., 2001, 2003). Hierarchical clustering of the protein expression profile from a tissue microarray study comprising 1,076 invasive breast tumors, divided the tumors into five subgroups characterized by ER status, HER-2 expression level, p53 positivity, expression of MUC1 and E-cadherin, luminal epithelial cell phenotype characteristics and luminal epithelial cytokeratin expression level (Abd El-Rehim et al., 2005). These five subgroups represent significantly different correlations to the established prognostic parameters and to survival and illustrate how heterogeneous breast tumors are.

The patients without metastasis to the lymph nodes at the time of the initial surgery can roughly be divided into two groups, one in which the patients suffer from relapse within 5 years and one without secondary disease. Tsumagari et al. analyzed the gene expression profile in 12 patients without relapse and 12 who developed metastasis within 5 years after surgery, using a cDNA array with 25,344 human genes. Fifty-eight genes were differentially expressed in the two groups, and the separation of the two groups of patients was 100% accurate (Tsumagari et al., 2005).

From an initial pool of 25,000 genes, 70 genes involved in cell cycle regulation, invasion, metastasis and angiogenesis were identified to predict disease recurrence in node negative women under the age of 55 years, whose tumors were smaller than 5 cm (van't Veer et al., 2002). The patients were divided into two groups, based upon their gene expression pattern, one with a short interval to distant metastasis and one without relapse within the follow-up period of at least five years. Further validation of the gene expression profile (prognosis classifier) led to the prediction of a 28-fold odds ratio (CI 95%7-107, $p = 1.0 \times 10^{-8}$) risk of distant metastasis, for a node negative patient below the age of 55 with the poor prognosis profile, when compared with patients with the good prognosis profile. Further validation of the prognosis classifier was performed on 295 patients both node positive and node negative. The profile turned out to be a strong independent predictive marker for outcome and more efficient than standard markers based on clinical and histological criteria (van de Vijver et al., 2002). A prognostic profile including 76 genes was derived using 115 node negative tumors and validated via 171 new tumor samples (Wang et al., 2005). Despite similar clinical material only few genes were the same in the prognosis classifier profiles from these two studies.

This illustrates the importance of reaching consensus in terms of results and conclusions. Brenton et al. suggest a three step analysis comprising:

1. Data from already existing predictive gene expression studies should be analyzed with different algorithms to find overlapping consensus sets of genes to be further validated by PCR based methods.
2. Large retrospective studies using a substantial number of tumors from each of the subtypes defined by nodal status and status of ER, PR and HER-2, should be analyzed to generate a more definitive breast tumor taxonomy and to validate the prognostic classifier prospectively or in tumors from completed clinical trials.
3. Prospective systemic-therapy clinical trails should be designed with predictive marker validation in mind (Brenton et al., 2005).

Gene expression profiling of breast tumors is a very powerful tool. Consensus is hopefully reached between the large number of studies carried out world-wide, and patients will eventually benefit from a diagnosis and a treatment resulting in increased long-term survival and lack of unnecessary treatment.

6.2 Comparative Genome Hybridization (CGH)

The information obtained from Comparative Genomic Hybridization (CGH) analysis is a map of amplification or deletion of entire chromosomes or chromosomal regions (Kallioniemi et al., 1992). From these data several chromosomal abnormalities can be deciphered like aneuploidy, interstitial deletions, non-reciprocal translocations, amplification of small regions like insertions or double minutes (Albertsen et al., 1994). One advantage over LOH analysis is that information on deletion or amplification is obtained along the entire chromosome independently of the selection of specific STRs. It is thereby possible to detect aberrations as small interstitial and homozygous deletions that otherwise may be left out by LOH studies.

Experimental methods. Tumor DNA extracted from tissue with a high proportion of tumor cells (>60%) and control DNA is labeled with two different fluorescent dyes of Cy5-dCTP and Cy3-dCTP. Both sets of labeled DNA are simultaneously hybridized to metaphase chromosomes from normal cells immobilized on glass slides. The addition of Cot-1 DNA prevents repetitive sequences from hybridization.

The different levels of fluorescence intensity between the two colors detected along the chromosomes represent deletions (excess of control DNA), amplifications (excess of tumor DNA) or a normal level of the tumor genome (equal mixture of the two dyes).

Arrays with large genomic fragments of BAC clones are highly suitable for CGH analysis. The array-based CGH has several advantages as compared with the chromosome spread, aberrations are mapped directly to the genome with a high resolution and the procedure is automated thereby allowing a high throughput.

Initially, breast cancer cell lines were used as test material to screen for chromosomal aberrations via CGH. A considerable number of regions were found to have an altered copy number. In one study, analysis of 38 different breast cell lines revealed aberrations at 19 chromosome arms, as gain in decreasing frequency at: 8q, 1q, 20q, 7p, 3q, 5p, 7q, 17q, 1p and 20p and loss at: 8p, 18q, 1p, Xp, Xq, 4p, 11q, 18p, 10q and 19p (Forozan et al., 2000). To be used as prognostic

Table 2. Genomic alterations from CGH studies on different breast carcinomas. IDC, Invasive ductal carcinomas, ILC, invasive lobular carcinomas

Tumor classification:	Gain:	Loss:	Ref.
G1 (highly differentiated)	1q, 8q	16q	Buerger et al., 1999
G1/ER+	5q13-q23	6q, 16q, 22q	Richard et al., 2000
G2	1q, 3q, 8q	8p, 13q 16q	Buerger et al., 1999
G3/ER- (highly undifferentiated)	2p, 3q21-qter, 6p, 8q21-qter, 10p, 18p11-q11, 20q	2q35-q37, 3p12-p14, 4p15-p16, 5q, 7p15, 8p22-p23, 10q, 11p, 14q21-q31, 15q	Richard et al., 2000
Tumors from node negative patients	1q31-q32, 3q26-q27, 8q22-q23, 11q13, 17q11-q21, Xq13-q21	1p32-pter, 8p22-p23, 11q23-pter, 16q22-q23, 17p12, 22q11-q12	Janssen et al., 2003
Summary:			
High risk of recurrence and short time survival	3q, 8q, 11q13, 17q, 20q	13q, 17p	Janssen et al., 2003; Hermsen et al., 1998; Blegen et al., 2003; Aubele et al., 2002
Higher incidence in ILC than IDC	4, 5q13-q23	6q, 11q14-qter, 12p12-pter, 16q, 17p, 18q12-q21, 19, 22q	Richard et al., 2000
Alterations in ER- but not in ER+ carcinomas	1p31-p34, 2p, 3q, 5p15, 6p, 7q32-qter, 8q, 9p23-p24, 10p, 16q22-qter, 17q, 18p11.2-q11.2, 22q12	2q35-q37, 4p15-p16, 4q12-q13, 5q, 7p, 8p11-p12, 10q23-q25, 12q13-q23, 13q, 14q12-q31, 15q14-qter	Richard et al., 2000

markers or to identify new candidate genes for breast cancer via CGH analysis these regions had to be considerably narrowed, from whole chromosome arms to highly specific regions. This was achieved using high resolution CGH microarrays where the copy number was directly compared to the mRNA expression level of 13,824 genes. By screening of 14 breast cancer cell lines, 24 independent amplicons, each spanning from 0.2–12 Mb on 12 chromosome arms were defined and 270 abnormally amplified genes identified (Hyman et al., 2002).

To produce useful strong prognostic markers based upon genomic aberrations and differently expressed genes panels of different subgroups of breast tumors with full information on histopathology and follow-up are used. Results from CGH analysis of the tumors within each category are then compared to see if tumors exhibiting the same phenotype share some of the same chromosomal abnormalities, which eventually can be correlated to clinically prognostic parameters. This comparison has in addition led to speculations on the connection between different chromosome lesions and the pathway leading from a normal somatic cell to the different stages of malignant growth and proliferation.

The genomic changes between invasive tumors like invasive ductal carcinoma (IDC), invasive lobular carcinoma (ILC), well and poorly differentiated tumors Grade (G) 1–3 and ER+ and ER− tumors were compared via CGH analysis and revealed a striking genomic difference (Buerger et al., 1999; Buerger et al., 2001; Richard et al., 2000). The highly differentiated low-grade tumors (G1) show few alterations as gain of 1q, 8q and a loss of 16q, and there is a clear association between a high number of genomic alterations and a poor prognosis of the disease (see Table 1).

6.3 Epigenetic transcriptional silencing

The dysfunction of a cell is determined not only by genetic lesions but also decisively by epigenetic changes as hyper- or hypomethylation of specific regions of the genome (Epigenetic changes). Abnormal changes in the methylation pattern of a cell may cause severe inherited diseases, and is found implicated in (all) cancers and in aging. Furthermore, each neoplastic lesion seems to have a specific genomic methylation pattern, the epigenotype.

Epigenetic traits are inheritable but do not affect the primary DNA sequence. The methylation of cytosine residues within the symmetric CpG dinucleotide is one of the most frequent epigenetic alternations of the DNA sequence. Roughly, the human genome contains 30,000 CpG islands, which are characterized by a high density of CpG dinucleotides, spanning from 200 bases to several kilobases. The CpG islands are spread in a non-random pattern throughout the genome with a preference to the promoter region and the first exon of housekeeping genes, imprinted genes, some tissue specific genes, and genes inactivated on the female X chromosome. Methylation of CpGs in a promotor region may inhibit the transcription, and changes in the hypo- or hyper-methylation pattern can initiate or block transcription, respectively.

Hyper-methylation of CpG islands within other structural parts of the genome is capable of repressing silencer elements and preventing the insulator protein CTCF to bind to enhancer-blocking elements, thereby causing overexpression of the gene (Bell et al., 2001).

Acetylation and deacetylation of the N-terminal tail of histones is an additional epigenetic aberration. Acetylation is linked to high transcriptional activity, whereas deacetylation creates a tight chromatin structure preventing transcriptional factors, activators, repressors and other regulatory factors to access the DNA strand.

Recent findings suggest a strong connection between the two above-mentioned epigenetic events, proposing that methylation of the DNA sequence is the initial event, leading to deacethylation of the histones within the nucleosome core of the methylated region, thereby creating a permanent transcriptional silencing of the local gene by chromatin remodeling (Cameron et al., 1999).

The methodology used is dependent on the analysis of the methylation pattern affecting either one single gene or the entire genome. The analysis of a single CpG island is based upon the design of PCR amplification primers distinguishing between methylated and unmethylated DNA. The DNA is treated with bisulfite prior to PCR amplification, thereby deaminating unmethylated cytosine to uracil. The primer design is the critical step and varies with the method of detection. For simple gel electrophoresis, two sets of primers are designed to distinguish between the methylated and the unmethylated bisulfite treated template and to

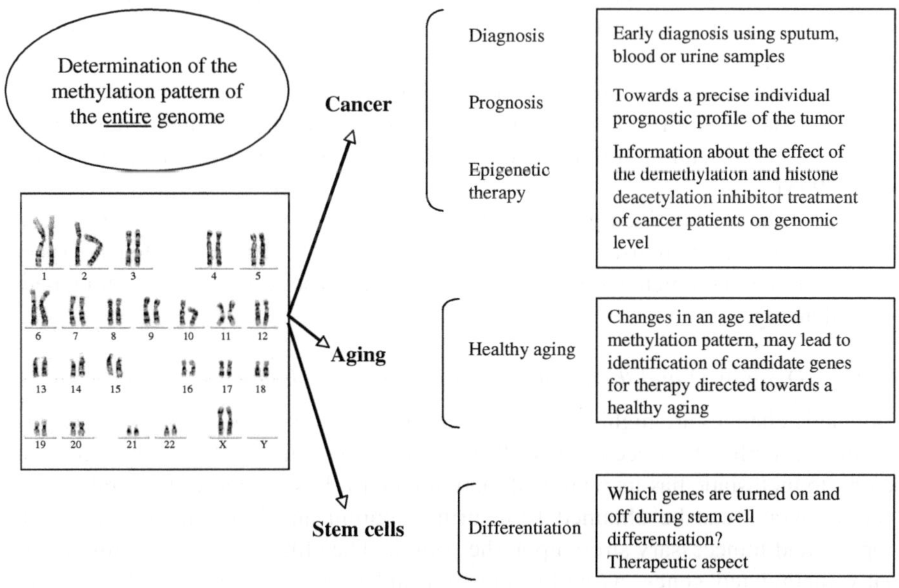

Figure 3. The implication of changes in the global methylation pattern in relation to cancer, aging and differentiation

produce products of varying size. The melting properties differ from a methylated and an unmethylated DNA fragment, amplified with the same primer set, after bisulfite treatment, due to the changes from unmethylated cytosine to uracil.

The products are visualized as separate peaks by Real-time PCR amplification followed by generation of a melting curve from the methylated and unmethylated products (Worm et al., 2001). Despite software to design methylation specific primers thorough experience and the inclusion of positive and negative controls are crucial to avoid false positives. These methods are reviewed in (Dobrovic, 2005).

The genome-wide analysis is based upon the microarray technology. Different approaches have been published, one is based upon oligonucleotides representing CpG islands from promotor regions of genes selected due to a changed expression pattern in tumor cells (Gitan et al., 2002; Shi et al., 2003). The oligonucleotides attached to the arrays represent both the methylated and unmethylated CpG islands. The test DNA is bisulfite treated, PCR amplified and labeled with a fluorescent dye. The methylated and unmethylated amplicons differ at the methylated sites by either cytosine or thymine and will thereby hybridize to different targets on the array. This method is both quantitative and qualitative, but the limitation is that the gene-specific CpG islands are selected and not genome-widely represented. Differential methylation hybridization (DMH) is based upon the isolation of CpG islands, and available as a library enriched for CpG islands within the size range of 0.2–2 kb (Cross et al., 1994). The test DNA, apart from the CpG islands, is cut by Mse1 into small fragments. After linker ligation and PCR amplification the test DNA is cleaved by methylation-sensitive restriction enzymes (BstU1) and hybridized to the array (Huang et al., 1999; Yan et al., 2002).

Combining tiling BAC arrays with full coverage of the genome, methylation sensitive restriction enzymes and CGH, provides a quantitative methylation assay and allows identification of new affected CpG islands (Ishkanian et al., 2004; Ching et al., 2005).

Shi et al. have combined gene expression, DNA methylation and histone acetylation in a triple microarray system (Shi et al., 2003). This integrated approach provides a more complete picture of the complicated processes leading to epigenetic gene silencing.

6.3.1 Diagnostic and therapeutic use of epigenetic changes

Epigenetic changes are highly suitable as predictive and prognostic markers. Unlike mutations, methylation occurs in well-defined regions, and each tumor stage from benign to metastatic has its own methylation pattern. DNA samples for methylation specific assays can be obtained from urine, sputum and blood thereby avoiding biopsies and unnecessary stress upon the patient. The clinical value of using abnormally methylated genes, as early detection and prognostic markers, has already been confirmed (Miyamoto et al., 2005; Palmisano et al., 2000; Hoque et al., 2004; Ichikawa et al., 2004; Topaloglu et al., 2004). However, the present technology needs further improvement and validation before a screening program for early breast cancer detection can be implemented.

In contrast to genetic abnormalities the methylation state is potentially reversible. Methylation directed treatment is already in clinical trial in USA despite the fact that the effect of the demethylation agents at the cellular level is largely unexplored. Thus, a fast and reliable method for examination of drug induced genome-wide methylation changes is crucial both for the design of clinical trial procedures and for monitoring the outcome of the therapy in clinical use.

Methylation of CpG islands blocking transcription in breast tumor cells has been reported for genes involved in cell cycle regulation (p16), DNA repair (BRCA1, hMLH1), hormone sensitivity (ER, PgR), cell adhesion (CDH1) and apoptosis (TMS1) (reviewed in Esteller, 2002).

7. CONCLUSION

Breast cancer is a very heterogeneous disease in which a large variety of genomic aberrations has been identified. Only a few high-penetrant genes have been found implicated in the development of inherited and sporadic breast cancer, but a highly comprehensive number of studies report correlations between genomic lesions like chromosomal deletions, amplifications, rearrangements, and mutations and tumor and patient characteristics.

Over the past decade, a series of new methods to analyze genomic variations has been developed rapidly, ranging from focusing on a single variation to large-scale analysis of the entire genome automated for high through-put. The amount of published results is increasing exponentially, and the important task is now to establish a consensus between all these studies. Especially, screening for aberrations across the entire tumor genome is interesting, and we have proceeded a step further towards making a diagnosis based upon the individual genomic profile of the tumor. Large comparative studies are now required to establish a link between the individual genetic profile, based upon germ-line and tumor specific variations, and the optimal treatment for each patient. At present, single genomic characteristics as the presence of hormone receptors and the expression level of HER-2 are used both as prognostic and predictive markers, but future diagnosis and therapy will be based upon the extensive information on the connection between the genome-wide alterations and progression of the disease.

Further characterization of individual genomic aberration remains important to evaluate the possibilities of developing new therapies directed towards these specific alterations.

ACKNOWLEDGEMENT

M. Nordsmark and J, Justesen are thanked for critical review of the manuscript.

REFERENCES

Abd El-Rehim, D.M., et al. (2005) High-throughput protein expression analysis using tissue microarray technology of a large well-characterised series identifies biologically distinct classes of breast cancer confirming recent cDNA expression analyses. Int J Cancer, 116: 340–50.

Akli, S. and Keyomarsi, K. (2004) Low-molecular-weight cyclin E: the missing link between biology and clinical outcome. Breast Cancer Res, 6: 188–91.

Albertsen, H.M., et al. (1994) A physical map and candidate genes in the BRCA1 region on chromosome 17q12-21. Nat.Genet., 7: 472–479.

Alsner, J., et al. (2000) Heterogeneity in the clinical phenotype of TP53 mutations in breast cancer patients [In Process Citation]. Clin Cancer Res, 6: 3923–31.

Altshuler, D., et al. (2005) A haplotype map of the human genome. Nature, 437: 1299–320.

Ameyaw, M.M., et al. (2002) Ethnic variation in the HER-2 codon 655 genetic polymorphism previously associated with breast cancer. J Hum Genet, 47: 172–5.

Arun, B. and Goss, P. (2004) The role of COX-2 inhibition in breast cancer treatment and prevention. Semin Oncol, 31: 22–9.

Aubele, M., et al. (2002) Chromosomal imbalances are associated with metastasis-free survival in breast cancer patients. Anal Cell Pathol, 24: 77–87.

Bell, A.C., et al. (2001) Insulators and boundaries: versatile regulatory elements in the eukaryotic. Science, 291: 447–50.

Benusiglio, P.R., et al. (2005) Common ERBB2 polymorphisms and risk of breast cancer in a white British population: a case-control study. Breast Cancer Res, 7: R204–R209.

Bergh, J., et al. (1995) Complete sequencing of the p53 gene provides prognostic information in breast cancer patients, particulary in relation to adjuvant systemic therapy and radiotherapy. Nature Medicine, 1: 1029–1034.

Blegen, H., et al. (2003) DNA amplifications and aneuploidy, high proliferative activity and impaired cell cycle control characterize breast carcinomas with poor prognosis. Anal Cell Pathol, 25: 103–14.

Borg, A., et al. (1992) Chromosome 1 alterations in breast cancer: allelic loss on 1p and 1q is related to lymphogenic metastases and poor prognosis. Genes Chromosomes Cancer, 5: 311–20.

Borresen, A.L., et al. (1995) TP53 mutations and breast cancer prognosis: particularly poor survival rates for cases with mutations in the zinc-binding domains. Genes Chromosomes.Cancer, 14: 71–75.

Brenton, J.D., et al. (2005) Molecular classification and molecular forecasting of breast cancer: ready for clinical application? J Clin Oncol, 23: 7350–60.

Buerger, H., et al. (1999) Different genetic pathways in the evolution of invasive breast cancer are associated with distinct morphological subtypes. J Pathol, 189: 521–6.

Buerger, H., et al. (2001) Ductal invasive G2 and G3 carcinomas of the breast are the end stages of at least two different lines of genetic evolution. J Pathol, 194: 165–70.

Busmanis, I., et al. (1994) Analysis of cerbB2 expression using a panel of 6 commercially available antibodies. Pathology, 26: 261–7.

Callahan, R., et al. (1993) Genetic and molecular heterogeneity of breast cancer cells. Clin Chim Acta, 217: 63–73.

Cameron, E.E., et al. (1999) Synergy of demethylation and histone deacetylase inhibition in the re-expression of genes silenced in cancer. Nat Genet, 21: 103–7.

Cargill, M., et al. (1999) Characterization of single-nucleotide polymorphisms in coding regions of human genes. Nat Genet, 22: 231–8.

Ching, T.T., et al. (2005) Epigenome analyses using BAC microarrays identify evolutionary conservation of tissue-specific methylation of SHANK3. Nat Genet, 37: 645–51.

Cotton, R.G.H., et al. (1988) Reactivity of cytosine and thymine in single-base-pair mismatches with hydroxylamine and osmium tetroxide and its application to the study of mutations. Proc Natl. Acad. Sci. USA, 85: 4397–4401.

Cross, S.H., et al. (1994) Purification of CpG islands using a methylated DNA binding column. Nat Genet, 6: 236–44.

Dagan, E., et al. (2002) Androgen receptor CAG repeat length in Jewish Israeli women who are BRCA1/2 mutation carriers: association with breast/ovarian cancer phenotype. Eur J Hum Genet, 10: 724–8.

Dandachi, N., et al. (2004) Evaluation of the clinical significance of HER2 amplification by chromogenic in situ hybridisation in patients with primary breast cancer. Anticancer Res, 24: 2401–6.

de Jong, M.M., et al. (2005) No increased susceptibility to breast cancer from combined CHEK2 1100delC genotype and the HLA class III region risk factors. Eur J Cancer, 41: 1819–23.

De Placido, S., et al. (2003) Twenty-year results of the Naples GUN randomized trial: predictive factors of adjuvant tamoxifen efficacy in early breast cancer. Clin Cancer Res, 9: 1039–46.

Devilee, P. and Cornelisse, C.J. (1994) Somatic genetic changes in human breast cancer. Biochim.Biophys.Acta, 1198: 113–130.

Dobrovic, A. (2005) Methods for analysis of DNA methylation. In: Molecular Diagnostics: For the clinical Laboratorian, Sec ed. (Eds.: Coleman, W.B. and Tsongalis, G.J.) Pages 149–160, Humana Press Inc., Totowa, NJ.

Dressler, L.G., et al. (2005) Comparison of HER2 status by fluorescence in situ hybridization and immunohistochemistry to predict benefit from dose escalation of adjuvant doxorubicin-based therapy in node-positive breast cancer patients. J Clin Oncol, 23: 4287–97.

Dumitrescu, R.G. and Cotarla, I. (2005) Understanding breast cancer risk – where do we stand in 2005? J Cell Mol Med, 9: 208–21.

Durbecq, V., et al. (2004) Topoisomerase-II alpha expression as a predictive marker in a population of advanced breast cancer patients randomly treated either with single-agent doxorubicin or single-agent docetaxel. Mol Cancer Ther, 3: 1207–14.

Eifel, P., et al. (2001) National Institutes of Health Consensus Development Conference Statement: adjuvant therapy for breast cancer, November 1–3, 2000. J Natl Cancer Inst, 93: 979–89.

Eiriksdottir, G., et al. (1998) Mapping loss of heterozygosity at chromosome 13q: loss at 13q12-q13 is associated with breast tumour progression and poor prognosis. Eur J Cancer, 34: 2076–81.

Emens, L.A. (2005) Trastuzumab: targeted therapy for the management of HER-2/neu-overexpressing metastatic breast cancer. Am J Ther, 12: 243–53.

Esteller, M. (2002) CpG island hypermethylation and tumor suppressor genes: a booming present, a brighter future. Oncogene, 21: 5427–40.

Fearon, E.R. and Vogelstein, B. (1990) A genetic model for colorectal tumorigenesis. Cell, 61: 759–67.

Fodde, R. and Smits, R. (2001) Disease model: familial adenomatous polyposis. Trends Mol Med, 7: 369–73.

Forozan, F., et al. (2000) Comparative genomic hybridization analysis of 38 breast cancer cell lines: a basis for interpreting complementary DNA microarray data. Cancer Res, 60: 4519–25.

Fusun, T., et al. (2005) Association of HER-2/neu overexpression with the number of involved axillary lymph nodes in hormone receptor positive breast cancer patients. Exp Oncol, 27: 145–9.

Gaki, V., et al. (2000) Allelic loss in chromosomal region 1q21-23 in breast cancer is associated with peritumoral angiolymphatic invasion and extensive intraductal component. Eur J Surg Oncol, 26: 455–60.

Garcia, J.M., et al. (1999) Allelic loss of the PTEN region (10q23) in breast carcinomas of poor pathophenotype. Breast Cancer Res Treat, 57: 237–43.

Gasparini, G., et al. (2005) Therapy of breast cancer with molecular targeting agents. Ann Oncol, 16 Suppl 4: iv28–iv36.

Gentile, M., et al. (1999) Frequent allelic losses at 11q24.1-q25 in young women with breast cancer: association with poor survival. Br J Cancer, 80: 843–9.

Gitan, R.S., et al. (2002) Methylation-specific oligonucleotide microarray: a new potential for high-throughput methylation analysis. Genome Res, 12: 158–64.

Goldhirsch, A., et al. (2003) Meeting highlights: updated international expert consensus on the primary therapy of early breast cancer. J Clin Oncol, 21: 3357–65.

Gong, Y., et al. (2005) Comparison of HER-2 status determined by fluorescence in situ hybridization in primary and metastatic breast carcinoma. Cancer, 103: 1763–9.

Haga, S., et al. (2001) Association of allelic losses at 3p25.1, 13q12, or 17p13.3 with poor prognosis in breast cancers with lymph node metastasis. Jpn J Cancer Res, 92: 1199–206.

Han, W., et al. (2005) A haplotype analysis of HER-2 gene polymorphisms: association with breast cancer risk, HER-2 protein expression in the tumor, and disease recurrence in Korea. Clin Cancer Res, 11: 4775–8.

Hansen, L.L., et al. (1996) Sensitive and fast mutation detection by solid-phase chemical cleavage. Human Mutation, 7: 256–263.

Hansen, L.L., et al. (1998) Allelic loss of 16q23.2-24.2 is an independent marker of good prognosis in primary breast cancer. Cancer Res, 58: 2166–9.

Hansen, L.L., et al. (2003) Sensitive and fast mutation detection by solid-phase chemical cleavage method. In: PCR Primer. A laboratory manual. (Eds.: Dieffenbach, C.W. and Dveksler, G.S.) Pages 265–278, Cold Spring Harbor Laboratory Press, New York, USA.

Hansen, L.L. and Justesen, J. (2003) Loss of heterozygosity, a multiplex PCR method to define narrow deleted chropmosomal regions of a tumor genome. In: PCR Primer. A laboratory manual. (Eds.: Dieffenbach, C.W. and Dveksler, G.S.) Pages 223–236, Cold Spring Harbor Laboratory Press, New York, USA.

Harbeck, N., et al. (2004) Urokinase-type plasminogen activator and its inhibitor type 1 predict disease outcome and therapy response in primary breast cancer. Clin Breast Cancer, 5: 348–52.

Heikkinen, K., et al. (2005) Mutation analysis of the ATR gene in breast and ovarian cancer families. Breast Cancer Res, 7: R495–R501.

Hermsen, M.A., et al. (1998) Genetic analysis of 53 lymph node-negative breast carcinomas by CGH and relation to clinical, pathological, morphometric, and DNA cytometric prognostic factors. J Pathol, 186: 356–62.

Hicks, D.G. and Tubbs, R.R. (2005) Assessment of the HER2 status in breast cancer by fluorescence in situ hybridization: a technical review with interpretive guidelines. Hum Pathol, 36: 250–61.

Hoque, M.O., et al. (2004) Quantitative detection of promoter hypermethylation of multiple genes in the tumor, urine, and serum DNA of patients with renal cancer. Cancer Res, 64: 5511–7.

Hoyal, C.R., et al. (2005) Genetic polymorphisms in DPF3 associated with risk of breast cancer and lymph node metastases. J Carcinog, 4: 13.

Hsiao, W.C., et al. (2004) Estrogen receptor-alpha polymorphism in a Taiwanese clinical breast cancer population: a case-control study. Breast Cancer Res, 6: R180–6.

Huang, T.H., et al. (1999) Methylation profiling of CpG islands in human breast cancer cells. Hum Mol Genet, 8: 459–70.

Huiping, C., et al. (1998) High frequency of LOH at chromosome 18q in human breast cancer: association with high S-phase fraction and low progesterone receptor content. Anticancer-Res, 18: 1031–6 issn: 0250-7005.

Hunt, K.K. and Keyomarsi, K. (2005) Cyclin E as a prognostic and predictive marker in breast cancer. Semin Cancer Biol, 15: 319–26.

Hyman, E., et al. (2002) Impact of DNA amplification on gene expression patterns in breast cancer. Cancer Res, 62: 6240–5.

Ichikawa, D., et al. (2004) Detection of aberrant methylation as a tumor marker in serum of patients with gastric cancer. Anticancer Res, 24: 2477–81.

Ishkanian, A.S., et al. (2004) A tiling resolution DNA microarray with complete coverage of the human genome. Nat Genet, 36: 299–303.

Jatoi, I. and Miller, A.B. (2003) Why is breast-cancer mortality declining? Lancet Oncol, 4: 251–254.

Janssen, E.A., et al. (2003) In lymph node-negative invasive breast carcinomas, specific chromosomal aberrations are strongly associated with high mitotic activity and predict outcome more accurately than grade, tumour diameter, and oestrogen receptor. J Pathol, 201: 555–61.

Judson, R. and Stephens, J.C. (2001) Notes from the SNP vs. haplotype front. Pharmacogenomics, 2: 7–10.

Kallioniemi, A., et al. (1992) Comparative genomic hybridization for molecular cytogenetic analysis of solid tumors. Science, 258: 818–21.

Kammerer, S., et al. (2004) Large-scale association study identifies ICAM gene region as breast and prostate cancer susceptibility locus. Cancer Res, 64: 8906–10.

Keyomarsi, K., et al. (2002) Cyclin E and survival in patients with breast cancer. N Engl J Med, 347: 1566–75.

Kinzler, K.W. and Vogelstein, B. (1996) Life (and death) in a malignant tumour. Nature, 379: 19–20.

Knudson, A. (2001) Alfred Knudson and his two-hit hypothesis. (Interview by Ezzie Hutchinson). Lancet Oncol, 2: 642–5.

Kruglyak, L. and Nickerson, D.A. (2001) Variation is the spice of life. Nat Genet, 27: 234–6.

Leitzel, K., et al. (1995) Elevated serum c-erbB-2 antigen levels and decreased response to hormone therapy of breast cancer. J Clin Oncol, 13: 1129–35.

Lo, Y.L., et al. (2005) Breast cancer risk associated with genotypic polymorphism of the mitosis-regulating gene Aurora-A/STK15/BTAK. Int J Cancer, 115: 276–83.

Manders, P., et al. (2004) Complex of urokinase-type plasminogen activator with its type 1 inhibitor predicts poor outcome in 576 patients with lymph node-negative breast carcinoma. Cancer, 101: 486–94.

Margolis, K.L., et al. (2005) Physical activity in different periods of life and the risk of breast cancer: the Norwegian-Swedish Women's Lifestyle and Health cohort study. Cancer Epidemiol Biomarkers Prev, 14: 27–32.

Matsumoto, S., et al. (2000) Loss of heterozygosity at 3p24-p25 as a prognostic factor in breast cancer. Cancer Lett, 152: 63–9.

Miller, B.J., et al. (2003) Pooled analysis of loss of heterozygosity in breast cancer: a genome scan provides comparative evidence for multiple tumor suppressors and identifies novel candidate regions. Am J Hum Genet, 73: 748–67.

Miyamoto, K., et al. (2005) Identification of 20 genes aberrantly methylated in human breast cancers. Int J Cancer, 116: 407–14.

Nagahata, T., et al. (2002) Correlation of allelic losses and clinicopathological factors in 504 primary breast cancers. Breast Cancer, 9: 208–15.

Nagai, M.A., et al. (1994) Allelic loss on distal chromosome 17p is associated with poor prognosis in a group of Brazilian breast cancer patients. Br J Cancer, 69: 754–8.

Nelson, M.R., et al. (2004) Large-scale validation of single nucleotide polymorphisms in gene regions. Genome Res, 14: 1664–8.

Nexo, B.A., et al. (2003) A specific haplotype of single nucleotide polymorphisms on chromosome 19q13.2-3 encompassing the gene RAI is indicative of post-menopausal breast cancer before age 55. Carcinogenesis, 24: 899–904.

Palmisano, W.A., et al. (2000) Predicting lung cancer by detecting aberrant promoter methylation in sputum. Cancer Res, 60: 5954–8.

Parkin, D.M., et al. (1999) Global cancer statistics. CA Cancer J Clin, 49: 33–64, 1.

Pegram, M.D., et al. (1997) The effect of HER-2/neu overexpression on chemotherapeutic drug sensitivity in human breast and ovarian cancer cells. Oncogene, 15: 537–47.

Pharoah, P.D., et al. (2002) Polygenic susceptibility to breast cancer and implications for prevention. Nat Genet, 31: 33–6.

Pinto, A.E., et al. (2005) Correlations of cell cycle regulators (p53, p21, pRb and mdm2) and c-erbB-2 with biological markers of proliferation and overall survival in breast cancer. Pathology, 37: 45–50.

Press, M.F., et al. (1993) Her-2/neu expression in node-negative breast cancer: Direct tissue quantitation by computerized image analysis and association of overexpression with increased risk of recurrent disease. Cancer Res., 53: 4960–4970.

Press, M.F., et al. (1994) Sensitivity of HER-2/neu antibodies in archival tissue samples: potential source of error in immunohistochemical studies of oncogene expression. Cancer Res, 54: 2771–7.

Press, M.F., et al. (2002) Evaluation of HER-2/neu gene amplification and overexpression: comparison of frequently used assay methods in a molecularly characterized cohort of breast cancer specimens. J Clin Oncol, 20: 3095–105.

Quon, K.C. and Berns, A. (2001) Haplo-insufficiency? Let me count the ways. Genes Dev, 15: 2917–21.

Ragnarsson, G., et al. (1999) Loss of heterozygosity at chromosome 1p in different solid human tumours: association with survival. Br J Cancer, 79: 1468–74.

Rebbeck, T.R., et al. (1999) Modification of BRCA1-associated breast cancer risk by the polymorphic androgen-receptor CAG repeat. Am J Hum Genet, 64: 1371–7.

Regitnig, P., et al. (2002) Microsatellite analysis of breast carcinoma and corresponding local recurrences. J Pathol, 198: 190–7.

Reich, D.E., et al. (2003) Quality and completeness of SNP databases. Nat Genet, 33: 457–8.

Richard, F., et al. (2000) Patterns of chromosomal imbalances in invasive breast cancer. Int J Cancer, 89: 305–10.

Ross, J.S. and Fletcher, J.A. (1998) The HER-2/neu oncogene in breast cancer: prognostic factor, predictive factor, and target for therapy. Stem Cells, 16: 413–28.

Ross, J.S., et al. (2004) Targeted therapy in breast cancer: the HER-2/neu gene and protein. Mol Cell Proteomics, 3: 379–98.

Rueckert, S., et al. (2005) A monoclonal antibody as an effective therapeutic agent in breast cancer: trastuzumab. Expert Opin Biol Ther, 5: 853–66.

Seitz, S., et al. (1997) Deletion mapping and linkage analysis provide strong indication for the involvement of the human chromosome region 8p12-p22 in breast carcinogenesis. Br J Cancer, 76: 983–91.

Shi, H., et al. (2003) Oligonucleotide-based microarray for DNA methylation analysis: principles and applications. J Cell Biochem, 88: 138–43.

Shi, H., et al. (2003) Triple analysis of the cancer epigenome: an integrated microarray system for assessing gene expression, DNA methylation, and histone acetylation. Cancer Res, 63: 2164–71.

Skotheim, R.I., et al. (2001) Evaluation of loss of heterozygosity/allelic imbalance scoring in tumor DNA. Cancer Genet Cytogenet, 127: 64–70.

Smith, M.L. and Seo, Y.R. (2000) Sensitivity of cyclin E-overexpressing cells to cisplatin/taxol combinations. Anticancer Res, 20: 2537–9.

Smylie, K.J., et al. (2004) Analysis of sequence variations in several human genes using phosphoramidite bond DNA fragmentation and chip-based MALDI-TOF. Genome Res, 14: 134–41.

Sorlie, T., et al. (2001) Gene expression patterns of breast carcinomas distinguish tumor subclasses with clinical implications. Proc Natl Acad Sci U S A, 98: 10869–74.

Sorlie, T., et al. (2003) Repeated observation of breast tumor subtypes in independent gene expression data sets. Proc Natl Acad Sci U S A, 100: 8418–23.

Suter, N.M., et al. (2003) Androgen receptor (CAG)n and (GGC)n polymorphisms and breast cancer risk in a population-based case-control study of young women. Cancer Epidemiol Biomarkers Prev, 12: 127–35.

Tomlinson, I.P., et al. (2002) Loss of heterozygosity analysis: practically and conceptually flawed? Genes Chromosomes Cancer, 34: 349–53.

Topaloglu, O., et al. (2004) Detection of promoter hypermethylation of multiple genes in the tumor and bronchoalveolar lavage of patients with lung cancer. Clin Cancer Res, 10: 2284–8.

Tower, G.B., et al. (2003) The 2G single nucleotide polymorphism (SNP) in the MMP-1 promoter contributes to high levels of MMP-1 transcription in MCF-7/ADR breast cancer cells. Breast Cancer Res Treat, 82: 75–82.

Tsumagari, K., et al. (2005) Postoperative prognosis of node-negative breast cancers predicted by gene-expression profiling on a cDNA microarray of 25,344 genes. Breast Cancer, 12: 166–77.

Tsuneizumi, M., et al. (2002) Association of allelic loss at 8p22 with poor prognosis among breast cancer cases treated with high-dose adjuvant chemotherapy. Cancer Lett, 180: 75–82.

Utada, Y., et al. (2000) Allelic loss at the 8p22 region as a prognostic factor in large and estrogen receptor negative breast carcinomas. Cancer, 88: 1410–6.

van't Veer, L.J., et al. (2002) Gene expression profiling predicts clinical outcome of breast cancer. Nature, 415: 530–6.

van de Vijver, M.J., et al. (2002) A gene-expression signature as a predictor of survival in breast cancer. N Engl J Med, 347: 1999–2009.

Waard, F.D. and Thijssen, J.H. (2005) Hormonal aspects in the causation of human breast cancer: Epidemiological hypotheses reviewed, with special reference to nutritional status and first pregnancy. J Steroid Biochem Mol Biol,

Wang, Y., et al. (2005) Gene-expression profiles to predict distant metastasis of lymph-node-negative primary breast cancer. Lancet, 365: 671–9.

Weber, B.L. and Nathanson, K.L. (2000) Low penetrance genes associated with increased risk for breast cancer. Eur J Cancer, 36: 1193–9.

Winqvist, R., et al. (1995) Loss of heterozygosity for chromosome 11 in primary human breast tumors is associated with poor survival after metastasis. Cancer Res, 55: 2660–4.

Worm, J., et al. (2001) In-tube DNA methylation profiling by fluorescence melting curve analysis. Clin Chem, 47: 1183–9.

Yan, P.S., et al. (2002) Applications of CpG island microarrays for high-throughput analysis of DNA methylation. J Nutr, 132: 2430S–2434S.

Yu, H., et al. (2000) Shorter CAG repeat length in the androgen receptor gene is associated with more aggressive forms of breast cancer. Breast Cancer Res Treat, 59: 153–61.

Zhu, Y., et al. (2004) An evolutionary perspective on single-nucleotide polymorphism screening in molecular cancer epidemiology. Cancer Res, 64: 2251–7.

CHAPTER 13

PROSTATE DISEASE IN THE AGING MALE

Prevention, diagnosis and treatment of prostate cancer

ANNE R. SIMONEAU

Associate Clinical Professor of Urology, University of California, Irvine, 101 The City Drive Rt 81, Orange, CA 92868

Abstract: Prostate cancer is commonly found in older men. Whether its presence is clinically significant, requiring screening or treatment has been intensely debated, fueled by the indolent nature of many cancers as well as the competing cardiovascular mortality of this age group. Risk factors include age, race and family history. Diet has been linked to prostate cancer risk and is being investigated both for understanding the pathogenesis of prostate cancer and for use in supplements in preventing prostate cancer. In the past two decades the advent of serum marker, Prostate Specific Antigen [PSA], and definitions of pre- neoplastic lesions have brought new understandings and questions to the etiology and epidemiology of prostate cancer. This chapter will focus on the function of the prostate, pathological definitions and grading, PSA and its role in the debate, epidemiology and risk factors of prostate cancer. Current treatments and prevention trials will be reviewed

Keywords: Cancer, prostate, aging, old age, neoplasia

Prostate cancer: Is it a disease needing to be cured or a facet of aging- much like wrinkles and gray hair? Opinions are varied, and strong, on the clinical implications of prostate cancer. From the benign view that all men will eventually have prostate cancer if they live long enough- though few will be clinically affected by their cancer- to the opposing view that prostate cancer is second only to lung cancer in cancer mortality and thus an important and critical issue in men's' health; proponents can be found for both views, and despite their seemingly disparate outcomes these two sides of prostate cancer are not mutually exclusive of each other. Prostate specific antigen [PSA] has intensified this debate as the incidence of prostate cancer has increased since PSA's introduction to men's health in 1986 with subsequent screening protocols. But is cancer detected by PSA screening clinically relevant

S. Rattan and M. Kassem (eds.), Prevention and Treatment of Age-related Diseases, 235–270.

cancer? For the clinician, the critical and difficult task is to predict for the individual standing in front of them – is there benefit to screening and if cancer is detected which prostate cancer scenario will take place, one of latency or one of progression? For the researcher the critical and difficult task is to adequately categorize the case or tissue before them to determine if the case will be informative to the genetics or biology of prostate cancer or obscure the findings of the whole. (Platz et al., 2004) An example of this would be the difficulty in determining the genes involved in hereditary prostate cancer. Definitions which would seem immune to screening practices and indicative of significant disease, such as hereditary or familial prostate cancer, are still impacted by screening. A case of sporadic prostate cancer might lead several family members to become screened, discovering a few small incidental tumors, which may otherwise never be diagnosed. When this family's genetic profile is added to other familial and hereditary cancer cases instead of adding strength to the genetic association, their genetic information may obscure what might otherwise be a genetic site of interest.

The face of prostate cancer is changing (Cooperberg et al., 2005). From the past when men presented with back pain and a positive bone scan, to the present when men wonder if their PSA discovered prostate cancers are clinically significant, new knowledge and challenges have occurred these past two decades. In this chapter an overview of the history and epidemiology of prostate cancer, especially as it relates to the prevention and detection of prostate cancer will be undertaken. A brief overview of the prostate, prostate cancer grading, pathological nomenclature, and prostate specific antigen [PSA] will begin the discussion.

1. THE PROSTATE

Simplistically, the prostate is an accessory sex gland influenced by androgens found at the base of the bladder, surrounding the urethra in men. It is responsible for providing fluid rich in polyamines, prostaglandin, citrate, and phosphorylcholine as well as other components, which are produced by individual prostate glands, and then transported through 15 to 30 secretory ducts before being deposited into the urethra. The prostatic component of the ejaculate composes less than half of the total seminal fluid (Mann, 1974; Marker et al., 2003). Anatomically the prostate is divided into zones. McNeal has elegantly written descriptions of five zones (McNeal, 1981), but in day to day clinical practice the prostate is referred to as two zones. The peripheral zone is where the majority of prostate cancers arise, and the posterior aspect of the peripheral zone can be examined by a digital rectal exam [DRE]. Prostate cancer is generally multifocal (Sakr and Grignon, 1998). The peripheral zone is targeted by trans-rectal needle biopsy of the prostate. The transition zone surrounding the urethra is where benign prostatic hyperplasia [BPH] predominates (McNeal, 1981) causing urinary obstructive symptoms, which on occasion are treated by transurethral resection of the prostate [TURP]. Though some cancers are seen in the transition zone.

2. GLEASON GRADING

Gleason grading has established itself as a predictor of prostate cancer aggressiveness and is now the preferred grading system for prostate cancer. Pathologist, Donald Gleason wrote his description of prostate cancer in 1966 (Gleason, 1966). His grading system was unique in that it focused on the glandular architecture, not the cytological features of individual cells. In addition Gleason recognized the importance of heterogeneity of tumors and assigned a grade to the predominant pattern as well as a secondary pattern to arrive at a Gleason score or sum. Thus as the architectural changes are graded from a 1 to 5, with 5 being the most aggressive, the Gleason score or sum can range from 2 to 10. A typical cancer is either referred to as a Gleason score of 7 or can be written as 3 + 4, the first number being the predominant pattern. Occasionally there will be three patterns. If the third pattern is the least predominant but the highest grade it has been suggested that the higher Gleason grade be reflected in the total sum. An example is if a cancer has a predominant pattern of 3, the second pattern a 2, but also has minimal component of a 4 that the score be written 3 + 4. Gleason scores have been proven to be prognostic with patients with tumors demonstrating components of Gleason grade 4 or 5 having poorer outcomes. (Narain et al., 2001; Lin et al., 2005) Gleason grade is used in predictive prostate cancer nomograms such as Partin tables (Partin et al., 1993; Partin et al., 2001) or Kattan probability of indolent tumor (Kattan et al., 2003) which are used to guide therapy or the need for therapy based on, in addition to Gleason score on the biopsy material, the clinical exam, and serum PSA levels. Though the grading is based on architectural changes, there are cytological differences in the prostate cancer cells with changes in nucleoli that can be noted. Important in the pathological identification of prostate cancer is the loss of the basal cell layer of the glandular acini that occurs in prostate cancer. (Gleason, 1966) Staging is based on the Tumor Node Metastasis system. (Taylor et al., 2005) The most common presentation today is T1c, a man with a normal prostate exam but elevated PSA.

3. PRENEOPLASTIC LESIONS

Prostatic Intraepithelial Neoplasia, [PIN] is considered to be a precursor to moderate and high grade Gleason [Gleason score 6 and higher] cancers in the periphery of the prostate. In PIN the architecture of the glands is normal, but the individual cells lining the ducts are cytologically abnormal and almost indistinguishable to Gleason grade 3 cancers. The basal cell layer, though present, can have disruptions as the lesions progress to higher grade PIN. (Bostwick and Brawer, 1987) PIN was accepted as the preferred terminology over CIS, dysplasia or atypia, terms previously used to describe these findings, by the Workshop on Prostatic Dysplasia in the 1989 (Drago et al., 1989). The group further simplified the classification of PIN from three grades to two; low grade- grade 1, which is not currently commented upon in pathology reports, and high grade, which includes grades 2 and 3 (Drago et al., 1989). The evidence that PIN was a precursor to prostate cancer was initially by association. The lesions were seen in the same vicinity where cancers arise, and they were identified more often in prostates that also demonstrated cancer, and not seen as often in glands without cancer (Bostwick and Brawer, 1987; McNeal and Bostwick, 1986). Men with PIN were more likely to have cancer on subsequent biopsy, and African American men had higher prevalence and earlier demonstration of these lesions (Sakr and Partin, 2001; Sakr, 1999). Subsequently many molecular associations between PIN and prostate cancer have been made, and PIN has been established as a precursor lesion (Sakr and Partin, 2001; Sinha et al., 2004; Sakr et al., 2000; Montironi et al., 2004). Many have estimated that PIN predates prostate cancer by 5 to10 years (Bostwick, 1988; Sakr et al., 1993,1996). Examples of genetic changes seen in both prostate cancer and PIN are alterations in racemase (Wu et al., 2004), CDGF (Pan et al., 2004), 8p, GSTP1 CpG island hypermethylation, as well as methylation in other genes linked to prostate cancer (Nakayama, M., et al., 2004). FISH has demonstrated that PIN and prostate cancer have similar cytological aberrations. Gain of 7, particularly 7q31; loss of 8p and gain of 8q; loss of 10q, 16q, 18q have been described (Qian et al., 1999; Qian et al., 1998).

In reviewing the literature on PIN since 1987 when PIN became a standardized term there has been a wide variability in the incidence reported for PIN, and on the clinical significance, i.e. subsequent cancers after the initial diagnosis of PIN. Besides the obvious variable of different pathologist interpretations between institutions and countries, other causes of discrepancies between reports on the incidence of PIN and subsequent cancer detection are due to different patient populations; is it a report based on a hospital based practice, a clinic population or a screening population? Feneley et al. reported the differences in the incidence of PIN between these three populations in England though all slides were read by the same pathologist. The prevalence was respectively 11%, 25%, and 20% based on which population was being reported upon (Feneley et al., 1997). In addition racial distribution of the cohort may influence the reported prevalence (Sakr et al., 1996). The range of incidence of PIN on PNB is reported to be from 0.7% to 25%, with an average of 9% (Feneley et al., 1997). Epstein has reported the incidence of PIN to be 5.5%. This report was published in 1997 based on review of sextant biopsies (Wills et al., 1997). In 1998 a reference pathology laboratory published its results of first time biopsies received from office based urologists. 62,537 biopsies over a two year period were assessed. The rate of isolated PIN was 4.1% (Orozco et al., 1998). The Rotterdam section of the ERSPC reports a low rate of PIN as a stand alone lesion, but did note an statistically significant increase in PIN when their screened populations was rescreened 4 years later [0.8% to 2.5%] (Postma et al., 2004).

One of the first reports in 1991 on the significance of PIN on subsequent biopsy demonstrated a 100% incidence of prostate cancer in the 10 men rebiopsied (Brawer et al., 1991).

It is important to remember that it was during this same time period when PIN incidence and consequence were being studied PSA was introduced generating a shift towards more localized, smaller volume of tumors (Cooperberg et al., 2005; Orozco et al., 1998; Sakr, 2004; Montironi et al., 2005). In addition the sextant biopsy schemata which was standard throughout until the late 1990s has been altered to increase the number of cores, generally to 10 or 12 cores [sometimes more] taken at a biopsy setting by most institutions. In comparing two reports published in 2001, one of a Naval Medical Center where sextant biopsies where performed (Borboroglu et al., 2001), to a Veterans Hospital where 12 core biopsies were performed the incidence of cancer detection of those who agreed to rebiopsy was 20/45 for the sextant group versus 1/43 in those with 12 core biopsies (Lefkowitz et al., 2001). The Veterans group was subsequently followed up at 3 years, men with initial PIN and a second biopsy not showing cancer, were contacted 3 years later for another biopsy. Of 72 men identified by records, 31 men underwent a biopsy which demonstrated 8 cancers (Lefkowitz et al., 2002).

Table 1 demonstrates that in subsequent years as stage migration was occurring and the number of cores initially taken was increasing the clinical significance of PIN for subsequent cancer detection decreased. Mian et al. (Mian et al., 2002), and Fowler et al. (Fowler, Jr., et al., 2000) report the presence of PIN was not predictive of subsequent cancer, and Postma (Postma et al., 2004) goes so far as to say PIN is never predictive. Lefkowitz reported that early repeat biopsy after a 12 core biopsy rarely detected cancer, but cancer can develop at 3 years so follow up should be considered (Lefkowitz et al., 2001, 2002).

Thus it may not be surprising that in 2003, 2004 there have been published differing opinions on the management of PIN. Steiner advised saturation biopsies followed by interval biopsies at 3 to 6 month intervals to manage PIN (Steiner, 2003). Others have felt that PIN after a 12 core biopsy was not a greater indicator for subsequent cancer on immediate rebiopsy, and with the benefit of rebiopsy over 1 year still to be identified. (Postma et al., 2004; Mian et al., 2002; Fowler, Jr., et al., 2000) And San Francisco et al. suggest rebiopsy if DRE or PSA changes occur during the every 6 month follow up (San Francisco et al., 2003).

Table 1. Significance of PIN for subsequent prostate cancer detection by prostate needle biopsy. Migration of significance of PIN may be from cancer volume migration and increase in initial number of cores taken at biopsy session

Author /Year/Reference	Number of men in study	% with cancer on subsequent biopsy
Brawer et al., 1991	10	100%
Weinstein and Epstein, 1993	19	53%
Davidson et al., 1995	100	35%
Raviv et al., 1996	48	48%
Kronz et al., 2001	245	32%
Lefkowitz et al., 2001	43	2.3%
Borboroglu et al., 2001	100	47%
San Francisco et al., 2003	21	24%
Gokden et al., 2005	190	30.5%
Moore et al., 2005	22	4.5%

Atypical Adenomatous Hyperplasia [AAH] is the name given to the circumscribed proliferation of small round glands, with no nucleoli or cytological atypia which is usually found in the transition zone (Bostwick, 1996). The lesion mimics Gleason grade 1 cancer, only the presence of the basal cell layer, which at times is attenuated and difficult to discern, distinguishes these lesions apart. Because of their physical similarities AAH has been suggested as a precursor of Gleason grade 1 or 2 cancer (Bostwick, 1996). Because AAH has architectural change with cytological blandness others feel it is an intermediate between BPH and cancer (Helpap et al., 1995). Histologically there is considerable evidence linking AAH with low grade cancers, but molecularly by the evaluation of proliferation rate, a few markers, and cytological abnormalities on chromosome 8, there was no convincing evidence linking AAH to cancer by one report (Grignon and Sakr, 1996). No additional follow up is considered necessary for this lesion. (Bostwick, 1996) Adenosis is sometimes cross-referenced as AAH.

'Suspect for but not diagnostic for prostate cancer' is one of many terms used in the literature for a suspicious lesion on biopsy. Though not considered a premalignant lesion, these lesions have a high rate of cancer detection on subsequent biopsy and thus should be followed closely. The reasons the for the unequivocal diagnosis of cancer not to be given in these scenarios is usually either the small size of the lesion, small number of cells with enlarged nucleoli, a clustered growth pattern, and/or the presence of PIN within or adjacent to the lesion (Cheville et al., 1997). Other names in the literature for suspicious lesions have included 'focal glandular atypia', 'atypical, suspicious for cancer', 'borderline lesions', 'atypical small acinar proliferation' [ASAP], 'atypical acinar proliferation' [AAP], 'atypia', or 'lesions suspicious for prostate cancer' [LSP] (Postma et al., 2004; Iczkowski et al., 1997).

The incidence of these types of lesions has been reported to range from 0.5% to 18%, and in recent publications the range is narrower from 1.9 to 5.2% (Postma et al., 2004; Moore et al., 2005). This diagnosis is potentially affected by more external variables such as the transportation and processing of the cores. Fragmentation of the cores could lead to disruption of the architecture of the specimen making the diagnosis more difficult. 'Suspect' should not be more than 5% of the diagnosis at a given institution, and can be used as a measure of internal quality control (van der Kwast et al., 2003). Less divergent than the names given to this lesion are the subsequent cancer detection rates. Cancer detection on subsequent biopsy is 40 to 50% (Iczkowski et al. 1998; Chan and Epstein, 1999; Allen et al., 1998; Iczkowski et al., 1997). An Italian group published in 2004 its experience with radical prostatectomy for ASAP lesions and found cancer in all 9 prostates removed for ASAP, this has not been accepted as practice in the United States (Brausi et al., 2004).

Proliferative inflammatory atrophy [PIA] is under investigation as a precursor lesion to prostate cancer. Inflammation is a component of carcinogenesis in other tumor systems, such as stomach and liver, and may be in prostate cancer. Prostatitis is common, and some studies show a relationship with prostatitis and sexually transmitted diseases with prostate cancer (Palapattu et al., 2005). Some focal atrophic lesions of the prostate have been shown to have high proliferation rates with signs by molecular analysis of oxidative stress. These lesions can be located next to PIN and cancer lesions in the periphery and have similar genetic changes. Work continues in the area (Nelson et al., 2004; Palapattu et al., 2005).

4. LATENT VERSUS CLINICAL PROSTATE CANCER

Autopsy series have demonstrated a high percentage of men who have died of causes other than prostate cancer that upon sectioning these men's prostate clinically unsuspected prostate cancer was found. The rates of unsuspected prostate cancer increase with increasing age (Sakr et al., 1994; Sanchez-Chapado et al., 2003) giving rise to the often repeated clip- "Men are more likely to die with, then of, prostate cancer." Intriguing, countries with low mortality rates of prostate cancer have similar autopsy prevalence rates for prostate cancer as the countries with higher mortality rates from prostate cancer (Breslow, N., et al., 1977). Generally though, the extent of cancer is much less between these 'incidental', 'latent', 'microcarcinoma' tumors in the countries of low risk compared to the autopsy tumors of high risk countries; with fewer foci of cancer, smaller volumes of cancer and well differentiated histology (Breslow, N., et al., 1977; Jackson et al., 1981; Yatani, R., et al., 1982; Dhom, 1983).

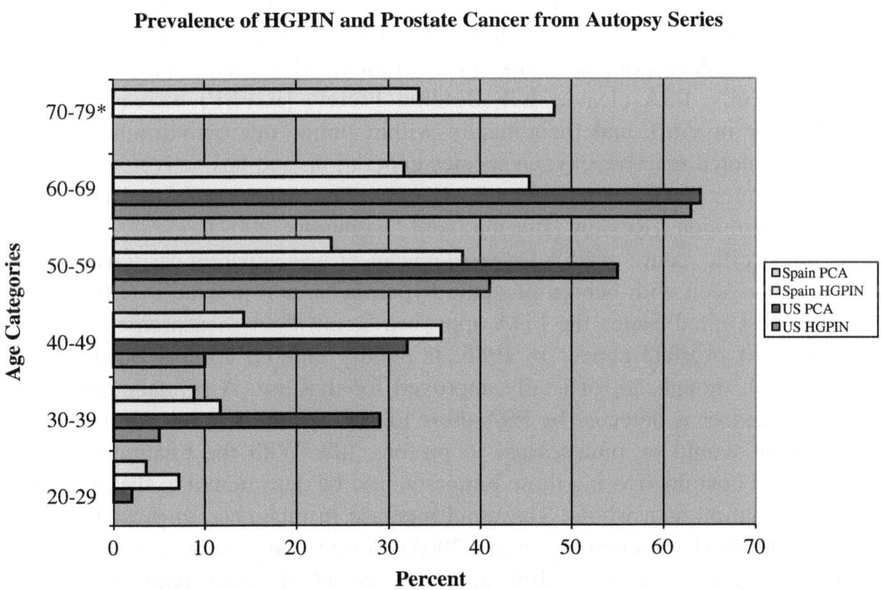

Figure 1. Prevalence of HGPIN and prostate cancer in two populations, a European Mediterranean and a mixed race American, per decade demonstrates the early appearance of lesions per autopsy (Sakr et al., 1994; Sanchez-Chapado et al., 2003). * The prevalence in the 8th decade in the U.S. population was not reported.

Not all series demonstrate similar rates of cancer, a series from Spain found lower rates compared to series published in the U.S. Figure 1 graphically describes the prevalence of PIN and prostate cancer per decade in these two recent autopsy series (Sakr et al., 1994; Sanchez-Chapado et al., 2003) demonstrating the increasing rates of HGPIN and cancer with age. This underlying prevalence of latent cancer skews prostate cancer incidence data between countries and between decades as differences in medical access, procedures such as TURPs for benign disease and screening policies will alter prostate cancer incidence for individual countries and decades. Prostate cancer mortality may give insight into the impact of the disease on a particular community. Even so infrastructure for reporting cancer cases and deaths is lacking in some countries and may make comparisons between countries difficult.

5. PROSTATE SPECIFIC ANTIGEN

Any current discussion on prostate cancer needs an understanding of prostate specific antigen, [PSA]. PSA is a serine protease, a member of the kallikrein family. Initially a protein identified in ejaculate for forensics in the 1970's, subsequent serum isolation was documented and an association with prostate disease was made. The gene responsible for PSA was cloned in the 1990s and localized to chromosome 19q13.4. PSA, as is prostate growth, is under androgen regulation. PSA has clinical relevance as a marker, but it also has functional relevance. In addition to its recognized role to liquefy the coagulum there are other possible functions which are being investigated though not completely understood. Kallikreins, including PSA, are felt to be regulators of Insulin-like Growth Factor [IGF], an important mitogen for prostate cells. PSA cleaves IGF Binding Protein [IGFBP] increasing IGF's bioavailability in vitro, and theoretically within tumor microenvironments. PSA has also stimulated reactive oxygen species generation, and has activated protease-activated receptor [PAR] in experimental systems. This theoretically may contribute to tumor progression. [Review (Borgono and Diamandis, 2004).

The nonspecific nature of PSA for prostate cancer was apparent early with serum elevations also seen with benign prostatic hyperplasia, and prostatitis (Chan et al., 1987). In the United States the FDA approved serum PSA measurements for the surveillance of prostate cancer in 1986. It rapidly entered clinical practice as a screening tool, though not officially approved for that use. A primary concern is that even if cancer is detected by PSA these tumors would be clinically latent and any treatment would be unnecessary to prolong life. With the treatments' given morbidity and cost uncovering these tumors would be detrimental to the individual and the population as a whole. The rapid increase in incidence in prostate cancer from 1986 to 1991 (Cooperberg et al., 2005) caused many to be concerned PSA was diagnosing latent cancers. But since the advent of widespread PSA use in the U.S, advanced prostate cancer at presentation has decreased, prostate cancer deaths have decreased (Cooperberg et al., 2005) and a few authors have published that PSA screening has caused a decrease in cancer and overall mortality. (Labrie et al., 1999) Randomized PSA screening trials are in progress with the primary endpoint of improved survival to answer whether PSA screening improves survival. Currently the United States preventative task force has given PSA an "I" rating for insufficient evidence as a cancer screening tool.

PSA as a surveillance and prognostic tool is well accepted. Post radical prostatectomy elevation in PSA signals prostate cancer return. PSA doubling is used to determine if further therapy is needed after primary therapy (Pound et al., 1997; D'Amico et al., 2004a). Prognostically PSA velocity prior to treatment has been linked to death from prostate cancer (D'Amico et al., 2004b). In 2004 Kuller et al. reported on the ability of PSA determination from stored frozen serum obtained in 1973 to 1975 when men where 35 to 57 years old to predict death from prostate cancer, mean follow up was 17 years. Men who died from prostate cancer had higher levels of PSA than controls, 2.84 vs. 1.10, $p=0.002$ (Kuller et al., 2004).

PSA as a single blood test is rapid and inexpensive- but its sensitivity and specificity are each about 70% (Catalona et al., 1994). Digital rectal exam is recommended by some advocate groups to be done in addition to the PSA (Catalona et al., 1994). Recently a prostate cancer prevention trial reported on the number and type of prostate cancers found in the control [placebo] arm on the end of study biopsy. Of the 9459 men on placebo there were 2950 men with an end of study biopsy, normal DRE and PSA <4. Four hundred and forty nine men had cancer [15.2%]. Sixty seven men had a Gleason 7 or higher which was 14.9% of the cancers, and accounted for 2.27% of all men with a biopsy (Thompson et al., 2004). Suddenly the debate switched from over detection of latent tumors to inability to detect aggressive cancers.

The lack of specificity of PSA elevation for cancer because of benign enlargement or infection can lead to expensive biopsies, and persistent worry that the cancer was missed when biopsies do not detect cancer. Percent free PSA has been used to improve the specificity, decrease the need for unnecessary biopsy and is advocated by some. Age specific PSA ranges, and PSA isoforms have been proposed to improve testing (Partin and Carter, 1996). ProPSA has been reported to be associated with higher Gleason score cancers (Catalona et al., 2004) and BPSA has been associated with the benign enlargement of the prostate (Mikolajczyk and Rittenhouse, 2004).

6. EPIDEMIOLOGY

The use of PSA has exacerbated prostate cancer incidence differences between countries with different medical practices, but even prior to 1986 there were known geographic variations for clinical prostate cancer which still persist (Breslow, N., et al., 1977; Dhom, 1983; Zaridze et al., 1984). African American and black men from the Caribbean have the highest rates for prostate cancer (Dhom, 1983; Jackson et al., 1980; Mallick et al., 2005). Asian countries have extremely low rates of prostate cancer (Donn and Muir, 1985). Figure 2 demonstrates the mortality from prostate cancer of selected countries illustrating the range in mortality from the period of 1986-1988 (Boring et al., 1992) prior to PSA use and estimates for 2002 (Ferlay et al., 2004). Figure 3 plots the incidence and mortality through out the

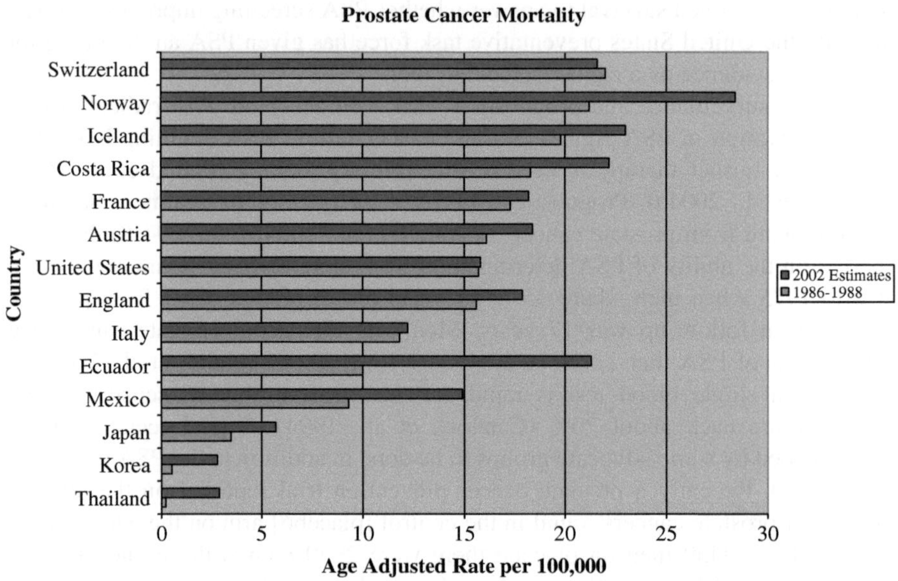

Figure 2. Prostate cancer mortality from 1986–1988 prior to the routine use of PSA, and estimates from 2002 after the introduction of PSA (Ferlay et al., 2004). The use of PSA has been most embraced by the US, followed by Western Europe

world by region as reported by Globocan for 2002 estimates (Ferlay et al., 2004). Developed countries with their access to health care have much higher incidences of reported cancer than developing countries. The differences in mortality are striking between African countries to Asian regions. Historically the rates for prostate cancer in Africa were reported as low, but African Americans and the Caribbean have well established higher mortality (Angwafo et al., 2003).

There are 3 risk factors, age, family history, and being African American [Africans and Africans living in other geographic regions have not been as well studied]. Several other dietary/environmental risk factors have been suggested due to observations from world cancer incidence rates. The strongest risk factor for prostate cancer is age. As highlighted previously, autopsy series demonstrate histological prostate cancer increasing in each decade, starting at a remarkably early time (Sakr et al., 1994). Rates of clinical prostate cancer also increase each decade. A rare event before the age of 40, with an incidence less than 1 in 40,000, prostate cancer's peak incidence increases in the mid 70's, but varies between countries (Jemal et al., 2005; Baade et al., 2005).

The late age at diagnosis, prolonged development and slow progression has implications for treatment and prevention. Treatment, as the clinician is asked to judge competing causes of mortality for an individual- will death be from the patient's moderate grade prostate cancer or cardiovascular disease. Prevention, in that to incorporate prostate healthy diets for the prevention of prostate cancer men may

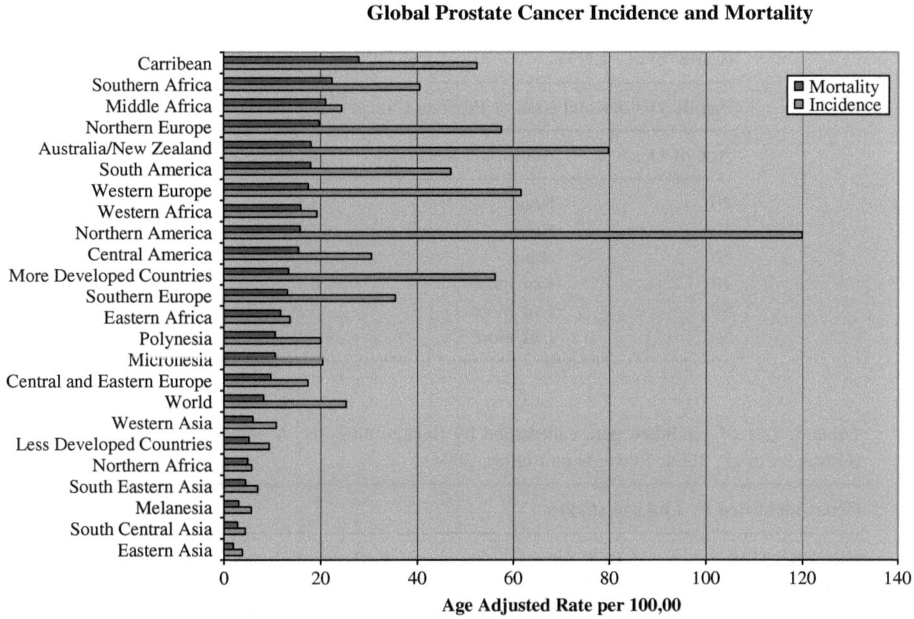

Figure 3. Global prostate cancer mortality by region (Ferlay et al., 2004). The racial and global distribution of prostate cancer has given rise to numerous etiologies; genetics, diet, and sun exposure [vitamin D metabolism]

need to begin in their 20s and 30's. To prevent the progression of the disease from an indolent disease to clinically aggressive disease with diet or chemopreventive agent.

Family history is important, but as prostate cancer is a common disease the number of affected individuals and age of onset of the disease are important variables to predicting an individual's risk. Table 2 demonstrates the increasing risk with increasing the number of relatives and decreasing the age of onset of the disease (Carter et al., 1993). Several recent publications have placed the relative risk for family history at 2 to 3 when there is a first degree relative. A meta-analysis reported the risk to be higher among brothers [RR 3.9] than among father sons [RR 2.5] (Johns and Houlston, 2003).

The search for the gene that causes prostate cancer, familial or sporadic, has been elusive. Several groups have reported their findings for a potential prostate cancer gene determined from hereditary [3 generations affected] or familial families [first degree relatives affected], only to have other groups unable to validate the findings using separate test groups, or to have the assessed contribution of that gene to the risk for familial prostate cancer considered minimal (Ostrander et al., 2004). Table 3 outlines the candidate genes proposed for prostate cancer by linkage analysis.

It is intriguing that some of the genes that have been identified with prostate cancer through linkage involve the inflammatory or infectious process. Macrophage

Table 2. The relative risk of prostate cancer based on number of relatives and age of presentation of the relatives affected (Carter et al., 1993)

Family History and Risk of Prostate Cancer		
Age of Onset	Additional Relatives	Relative Risk
70	None	1.0
60	None	1.5
50	None	2.0
70	1 or more	4.0
60	1 or more	5.0
50	1 or more	7.0

Table 3. List of candidate genes identified by linkage analysis. A composite from sources (Ostrander et al., 2004; Edwards and Eeles, 2004)

Genes Identified by Linkage studies			
HPC1/RNASEL	1q24-25	Early onset	1996/2002
PCaP	1q42-43	Early onset	1998
HPCX	Xq27-28	No male to male	1998
CAPB	1p36	Brain	1999
HPC20	20q13	Late onset	2000
MRS1	8p22	Late onset	2001/2003
HPC2/ELAC2	17p11	Early onset	2001

scavenger receptor 1 [MSR1] is induced in macrophages by oxidative stress and may modify amounts of Reactive Oxygen Species [ROS] (Ostrander et al., 2004; Xu et al., 2002). RNASEL is an endoribonuclease involved in antiviral and proapoptotic activities of interferon regulated 2–5A system- RNA decay pathway (Carpten et al., 2002). Genes involved in other tumor systems such as breast cancer, BRCA genes have been evaluated and there is an increase of BRCA2 mutations in men with prostate cancer compared to controls but again these mutations have been suggested to account for only a small proportion of genetic prostate cancer (Edwards and Eeles, 2004).

That the recent meta analysis gave higher risk assessment to brothers than sons of men with prostate cancer could point to different screening practices in the generations, or environmental factors (Johns and Houlston, 2003). It also suggests multiple low penetrance genes or recessive or X linked inheritance rather than dominant high penetrant pattern of inheritance. Mitochondrial DNA [mtDNA] mutations have been recently assessed. The mitochondria, inherited from the mother, have their own separate genetic code. Mitochondria as the energy producer for the cell and its role in apoptosis are critical for proper cellular function. The energy machinery of the cell requires proteins from both nuclear DNA and mtDNA. Mutations in either cause a spectrum of clinical manifestations and have been shown to cause an increase in reactive oxygen species. Mitochondrial gene and nuclear DNA encoded mitochondrial gene mutations have been linked to cancer. Recently Petros et al. reported on the increase of mutations in the mtDNA Cytochrome Oxidase subunit I [COI] gene in prostate cancer cases with laser capsure microdisection. Twelve percent of the prostate cancer specimens had mutations in the cytochrome oxidase subunit 1, whereas the general population had 7.8% mutations, and the no cancer controls demonstrated less than 2%. The authors created a mouse model with the mitochondrial mutation in which the tumors with the mutation created more ROS and had tumors which were seven times larger than the non mutated mtDNA (Petros et al., 2005).

Several polymorphisms of common genes in the androgen pathway (Latil et al., 2001), or DNA repair genes (Goode et al., 2002), steroid biosynthesis (Gsur et al., 2004) or PSA (Gsur et al., 2004) are being assessed as contributors of prostate cancer risk. As there has not been a single dominant gene yet identified, multiple low penetrance genes with modulation from the environment may dictate prostate cancer progression. Polymorphisms in the CYP3A4 gene, which is responsible for the oxidative deactivation of testosterone, have been studied. In one study, older men with no family history of prostate cancer were more likely to have the CYP3A4-V allele if they presented with advanced disease. Forty six percent of these men with > T3 disease had the CYP3A4-V allele compared to 5% in the T1 group. [OR = 9.45 p= 0.001] (Rebbeck et al., 1998) Again confirmation from other data bases is not fully consistent. One of many examples of the interaction of genetic polymorphisms in 2 pathways with an environmental toxin is outlined in Table 4. Here polymorphisms in the ornithine decarboxylase [ODC] gene coupled with polymorphisms in the androgen receptor [CAG repeats] and smoking history give rise to different relative risks for prostate cancer (Visvanathan et al., 2004).

Table 4. One example of the numerous proposed interactions between multiple genetic polymorphisms with environmental factors which could account for the genetic variability in prostate cancer incidence (Visvanathan et al., 2004)

Example of genetic polymorphism with environmental factor	
Low Risk	
ODC GG with AR ≥22 CAG repeats	Age Adjusted Odds Ratio
Nonsmoker/smoker	1.0/1.01
Intermediate Risk	
ODC AG or AA with AR ≥22 CAG repeats or ODC GG with AR ≤22 CAG repeats	
Nonsmoker/smoker	1.48/1.31
High Risk	
ODC AA or AG with AR ≤22 CAG repeats	
Nonsmoker/smoker	1.43/2.77

7. RACE

African Americans and blacks living in the Caribbean have the highest rates of prostate cancer. A recent review of prostate cancers in Jamaica reported that the incidence to be 304 per 100,000 men- higher than the US African American rate of 249/100,000 during a similar period, the US Caucasian rate was 187/100,000 at that time. The authors also report more clinical symptoms at presentation in Jamaica (Glover, Jr., et al., 1998). In the U.S. the differences in prostate cancer rates between African Americans, Caucasians, and Asian Americans has been studied to elucidate the essential promoters in clinical cancer- no definitive answer is available. Circulating androgen levels, genetic differences in the androgen receptor and zinc transporter (Rishi et al., 2003), vitamin D metabolism, body mass index, diet, socioeconomic class, and access to health care have been accessed to explain incidence and death disparities with no definitive answer as yet (Danley et al., 1995; Bianco, Jr., et al., 2002). Historically Sub-Saharan Africa has been reported to have low rates of prostate cancer, but a recent screening program set up in Dibombari, Cameroon for 111 men led to 24 biopsies due to abnormal DRE or PSA, which diagnosed 6 cancers and 2 HGPIN, with 6 LGPIN. The authors conclude that the low rate reported may reflect cultural and economic barriers to health care versus the previous theory that better diet was the etiology of the low rates of cancer (Angwafo et al., 2003). A comparison between high incidence area Washington DC, US to low incidence area in West Africa published in 1980 after consecutive necropsy cases were performed in Nigeria, Ghana, and Washington DC, suggested that the age adjusted rates for both areas were similar; 36.7 for West Africa versus 40.6 per 100,000 in DC. The African tumors were more advanced stage, 75% versus 49% were stage III and IV on necropsy. (Jackson et al., 1980) As improvement in cancer registries throughout the world occurs clearer comparisons can be made.

8. DIET

Additionally diet is considered by some to be a risk factor. Epidemiological trends between countries, and migration studies define differences in risk of clinical prostate cancer which could be institutional [differences in health care systems or reporting], environmental or dietary (Rose et al., 1986). Incidence of prostate cancer foci on histological section have been found to be similar between Asian countries and the West, but the size and aggressiveness of the tumors are much smaller and well differentiated in the Asian countries, leading to theory that it is promotion, not initiation of carcinogenesis that leads to the differing clinical scenarios between countries (Dhom, 1983). That the differences may be more than genetic have been evaluated with migration studies. After migration to the U.S., Chinese and Japanese men have substantial increases in prostate cancer rates (Muir et al., 1991; Nguyen, 2003). Those men who maintain a more traditional Asian diet have lower rates of prostate cancer, which some authors have attributed to the phytoestrogens in the traditional more vegetarian diet (Vij and Kumar, 2004). Others have noted the increase in prostate cancer risk in foreign born Asian Americans increased independently with length of residence in North America and saturated fat intake (Whittemore et al., 1995). In addition the Westernization of diet in Asian countries has led to increase in prostate cancer incidence in those countries (Sim and Cheng, 2005; Pu et al., 2004; Lee et al., 1998).

The difference between western and eastern diets in prostate cancer clinical cancer incidence has had some of the greatest interest. Certainly countries with diets rich in cereals, soybeans, other nuts and oilseeds, and fish that are also associated with less energy intake, less total fat, and less animal products [milk and meat] have lower prostate cancer rates (Hebert et al., 1998). The rates change with migration patterns or as Asian countries adopt western dietary practices, but is it the loss of a protective factor-fish, vegetables or soy, or the addition of a promoting factor-red meat or fat, that accounts for the incidence change? Cohort and case control studies, give additional, though sometimes conflicting, evidence with respect to which dietary factors have harmful or protective effects. Some of the inconsistencies come from inadequate measures or stratification of dietary elements. The complexity of food products; an example is fat-animal, vegetable, saturated, or essential, are not always well delineated on food questionnaires and can account for some of the inconsistent results seen in the literature.

9. DIETARY FAT

Dietary fat has been one of the earliest elements linked to prostate cancer. Several epidemiological studies have reported on increased odds ratio or relative risk with increased consumption of fat. Comparing cancer mortality with national food consumption reported a positive association with animal fat in 1986 (Rose et al., 1986) and again in 1998 (Hebert et al., 1998). Case control and cohort studies have not been as consistent with the association of fat (Dagnelie et al., 2004), though a case control study in China (Lee et al., 1998) demonstrated a increased risk in this low risk country with higher consumption of animal fat, the adjusted odds ratio for highest to lowest consumption was 3.6. Prentice and Sheppard reviewed the epidemiological data, and performed regression analyses of the international variation and determined RR for fat intakes (Prentice and Sheppard, 1990), they suggest practical reduction in fat intake could lower cancer disease. The regression rates for prostate cancer with disappearance of fat calories was significant [$p = 0.0001$], with a relative risk estimate of "essentially zero" for a 60% fat reduction in the diet.

Recent publications have studied individual components of fat, such as individual fatty acids. A cohort of 47,866 U.S. men followed for 14 years showed an increase risk of advanced prostate cancer with fatty acid alpha linolenic [ALA, 18:3n-3] fatty acid [they looked at variations from all sources- meat, dairy and vegetable oil], but a decreased risk of total and advanced prostate cancer with the fatty acids from fish, eicosapentaenoic [EPA: 20:5n-3] and docosanhexaenoic [DHA: 22:6n-3]. Fish oil supplements had no relationship, which may suggest that fish contain other

Table 5. Compilation of polymorphisms being investigated for a role in prostate carcinogenesis. Compiled from (Goode et al., 2002; Gsur et al., 2004; Visvanathan et al., 2004)

Polymorphism of common genes	
• Androgen Receptor	• cytochrome 450
– CAG	– CYP17 A2 allele
– GGC	– CYP3A4*1B
• 5 alpha reductase	• 3Beta-hydroxysteroid dehydrogenase
– V89L	• Glutathione S transferases
– A49T	• N-acetyl transferases
• PSA	• Ornithine decarboxylase
– Androgen Response Element	• CDKN1B [p27]
	• Base Excision Repair
• HPC2/ELAC2	– OGG1
– Ala541Thr	– XRCC1

protective agents such as vitamin D or retinol other than the fatty acids (Leitzmann et al., 2004). A review of several studies with fatty acids by Attar-Bashi et al. also commented on the possible positive association of prostate cancer with alpha linolenic fatty acid- but suggest further investigation as the cardiovascular health benefits of ALA are documented and cardiovascular deaths are a major concern in this age group (Attar-Bashi et al., 2004).

10. OTHER FOOD SOURCES

Figure 4 is a composite from a meta-analysis by Dagnelie et al. Using only prospective studies- randomized or cohort – they reviewed the dietary evidence for prostate cancer associations. The x axis gives the number of studies reporting either inverse, null or positive associations on the y axis with particular dietary component (Dagnelie et al., 2004).

The authors concluded that despite the prospective nature of the trials limitations in study size, measurements and validation were apparent, but they did suggest some consistent associations with selenium, possibly vitamin E, pulses [soy], and tomatoes as protective. Other dietary factors were inconclusive, though high levels of calcium [>2000 mg/day] appeared to be adverse (Dagnelie et al., 2004). Which particular compound in the foods, and the amount needed to be protective is under investigation.

Dairy has been reported to be either null or demonstrating a risk for prostate cancer (Dagnelie et al., 2004). Advanced cancers had a stronger association (Kristal et al., 2002; Chan et al., 2005). Whether it was fat, or hormonal contamination or other cause was unknown. Recent studies have hypothesized that the calcium in the milk products lower circulating levels of vitamin D, which may be protective (Giovannucci, 2005; Chan et al., 2001).

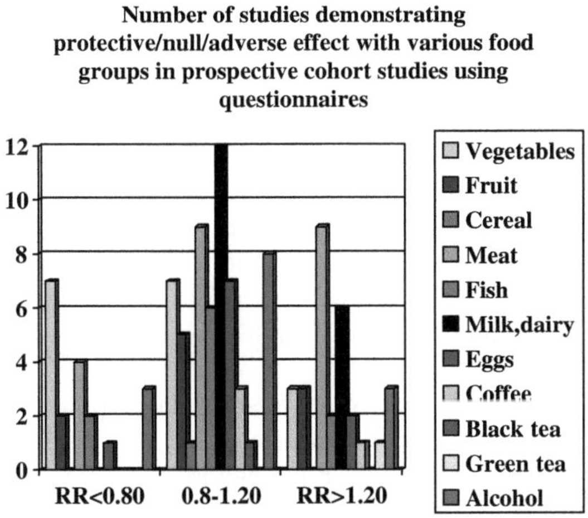

Figure 4. Meta-analysis of prospective cohort and intervention trials with diet and prostate cancer (Dagnelie et al., 2004)

Noting the association between groups identified as having lower circulating levels of vitamin D, [those living in areas with less UV B radiation, African American race, or being overweight], with higher prostate cancer mortality there has been a hypothesis generated that vitamin D protects against prostate cancer. The evidence is not entirely consistent and is further being studied in the lab and with clinical trials (Giovannucci, 2005). Studies on cigarettes have been mixed, a recent study has documented a moderate risk [O.R. 1.4] for current smokers, a dose effect was seen [trend $p = 0.03$] and a stronger association with aggressive disease was seen [O.R. 2] (Plaskon et al., 2003). Alcohol has not shown to be consistently associated with risk(Dagnelie et al., 2004), a recent study demonstrated a modest risk reduction only with red wine; for each glass consumed per week there was a 6% reduction in relative risk (Schoonen et al., 2005). Aspirin and non-steroidal anti inflammatory drug consumption has had mixed results as to whether there is a null or modest protective association (Habel et al., 2002; Platz et al., 2005). Vigorous exercise in one prospective study was associated with less fatal tumors in men over 65, but had no association with all cancer incidences or in younger men (Giovannucci et al., 2005).

11. RANDOMIZED TRIALS WITH PROSTATE CANCER PREVENTION AS A SECONDARY ENDPOINT

The earliest large cancer prevention trials were not carried out on prostate cancer, but with two trials in particular, analysis of secondary endpoint gave rise to candidate agents for prostate cancer prevention. Alpha- tocopherol, beta carotene study, ATBC, conducted a randomized, 4 arm, double blind, lung cancer prevention trial with 20 mg of beta carotene and 50 mg of vitamin E. The trial accrued 29,133

Finnish male smokers followed for 5 to 8 years. The primary endpoint of lung cancer prevention was not realized, in the beta carotene arm there were more lung and prostate cancers with a higher total mortality of 8%. In the alpha-tocopherol [Vitamin E] arm but there was a reduction in prostate cancer, 99 versus 151 cases, a reduction by approximately one third [34%]. The protective effect was observed by 18 months. There was also a higher total mortality of 2%. Hemorrhagic strokes in men with uncontrolled hypertension contributed to the higher mortality in the vitamin E arm, there was a 45% increased risk during the trial (Albanes et al., 1996). In a post trial analysis there was a persistent protective effect of vitamin E on prostate cancer after intervention, but diminished fairly rapidly- by the third year (Virtamo et al., 2003).

A second trial, Nutritional Prevention of Cancer Study, testing the hypothesis that selenium 200ug would decrease the rate of skin cancer also did not validate the primary hypothesis, but there was a 63% reduction in the incidence of prostate cancer in the men receiving selenium. 1312 subjects of which 974 were men treated for a mean of 4.5 years and followed for 6.5 years (Clark et al., 1996). There were 13 prostate cancers in the treated group and 35 in the placebo group [RR 0.37, $p = 0.002$]. For the 843 men who entered the trial with a PSA less than 4 there were 4 cancers in the treated group and 16 in the placebo group. [RR 0.26, $p = 0.009$] (Clark et al., 1998) Giovannucci and colleagues correlated the selenium levels in toe nail clippings, a measure of long term selenium intake and calculated the OR from highest to lowest quartile to be 0.35 $p = 0.03$, once controlling for family history and other dietary factors. For selenium concentration alone the OR was 0.49 with trend $p = 0.11$ (Yoshizawa et al., 1998). Later using a nested case control from a cohort study analysis of serum selenium levels with prostate cancer the authors found an inverse relationship with advanced prostate cancer [OR 0.52 $p = 0.5$], and in those with baseline PSA levels >4 [OR 0.49 $p = 0.002$ (Li et al., 2004). As another example of gene/environment interaction Li et al reported that a polymorphism in the superoxide dismutase [MnSOD], the primary antioxidant enzyme in the mitochondria, did not affect prostate cancer risk, but when coupled with baseline serum levels of antioxidants, selenium, lycopene and alpha tocopherol, polymorphisms in MnSOD modified risk stratifications (Li et al., 2005). The earlier findings lead to the trial design of the Selenium and Vitamin E Cancer Prevention Trial, [SELECT], for prostate cancer prevention.

12. RANDOMIZED TRIALS FOR PROSTATE CANCER PREVENTION

There are numerous trials for prostate cancer prevention in various phase development and with a wide variety of agents. The larger Phase III trials have involved changing the hormonal milieu of the prostate or increasing antioxidant consumption through supplements.

Finasteride [Proscar], is a 5 alpha reductase inhibitor which blocks the conversion of testosterone to dihydrotestosterone and grossly causes a reduction of prostate volume by 30%. Finasteride has been used to treat bladder outlet obstruction from prostate enlargement since 1992. The Prostate Cancer Prevention Trial [PCPT] funded by the National Cancer Institute, accrued 18,882 men starting in 1993 for a randomized double blind placebo controlled trial of finasteride 5 mg for 7 years with an end of study biopsy to determine prostate cancer prevalence. The trial was closed early as the primary endpoint of 25% prostate cancer reduction was achieved in the arm treated with finasteride. There were 803 cancers [18.4%] compared to 1147 [24.4%] in the placebo arm. Sub stratification of the cancer demonstrated that in the finasteride arm 280 men had Gleason 7 or higher [37% of cancer, 6.4% of men] compared to the placebo arm which had 237 men with Gleason >7 cancer [22% of cancers, 5.1% of men]. Despite the overall reduction of cancer, the use of finasteride has not been embraced because of concern over the increase in higher Gleason grade cancers. It has been reported there is potential for grading bias due to changes in architecture, nuclei and nucleoli seen in hormonally treated prostate cancers that could potentially falsely up grade disease (Bostwick et al., 2004). Another explanation is that finasteride prevents low grade lesions, but not high grade lesions, and coupled with the prostate volume reduction [up to 30%] from finasteride there is an improved biopsy efficiency for higher grade lesions (Carver et al., 2005; Rubin et al., 2005; Andriole et al., 2005). The concern is that finasteride may alter biology and induce cells to become higher grade. Further study and longer follow up is needed.

These issues of grading or detection bias will be further addressed and perhaps clarified with the current ongoing trial, Reduction by DUtasteride in prostate Cancer Events -REDUCE. Sponsored by GlaxoSmithKline the trial will enroll 8000 men with an elevated PSA- but less than 10- and a negative biopsy, from around the world to be randomized into a double blind placebo controlled trial with dutasteride 0.5 mg for 4 years. Biopsies will be taken at 2 and 4 years (Andriole et al., 2004). Dutasteride [Avodart] is the dual 5 alpha reductase inhibitor, inhibiting type 2, as does finasteride, but also type 1. Though type 2 is the predominant enzyme in the prostate, there is some evidence that type 1 is upregulated in prostate cancer as demonstrated by more intense immunohistochemistry staining of type 1 in cancer cells but not BPH (Thomas et al., 2003). In addition microarray gene analysis and semiquanitative RT-PCR has demonstrated that type 2 expression is decreased in prostate cancer compared to BPH and normal prostate cells (Luo et al., 2003).

The Physicians' Health Study-II [PHS II] is a large ongoing trial. testing vitamin C, vitamin E, beta-carotene, and a multivitamin for the primary prevention of cardiovascular disease, total cancer, and prostate cancer. Since August 1997, 14,642 men have been randomized. There are 16 possible combinations of vitamin C (500 mg synthetic ascorbic acid), vitamin E (400 IU of synthetic alpha-tocopherol), beta-carotene (50 mg Lurotin), a multivitamin (Centrum Silver), or their placebos. Vitamin C and the multivitamin or their placebos are taken daily, while vitamin E and beta-carotene or their placebos are taken every other day (Christen et al., 2000).

SELECT opened in 2001 and quickly achieved its accrual goal of 32,000 but results are not yet available. Men were randomized to one of four arms, either 200ug Selenium [L-selenomethionine] or 400 mg of [dl-alpha-tocopheryl acetate] or neither or both. Intervention is a minimum of 7 years for the last participants and up to 10–11 years for those who entered early. Clinical cancer detection is the endpoint. Since its inception reports of toxicity with vitamin E, a meta-analysis demonstrating increasing all cause mortality with higher doses of vitamin E, and a report from a cardiovascular trial demonstrating increased heart failure with vitamin E have caused for participant notification and review by the data safety and monitoring committee. Because the ATBC trial had demonstrated an increased stroke risk, all men on SELECT had to have a baseline blood pressure below 140/90 before randomization, and are a healthy population (Lippman et al., 2005).

The possible chemopreventive action of anti-inflammatory agents was to be studied on a large scale with the ViP Study. Sponsored by Merck the study was to involve 15,000 men between the ages of 50 and 75 with no history of prostate cancer, but a PSA between 2.5 and 10. These men were to be randomized to placebo versus 25 mg of VIOXX daily for 6 years. The end point was clinical cancer detection and accrual opened June 2003. In the summer of 2004 VIOXX was removed from the market due to cardiotoxicity reported from another prevention trial with VIOOX for colon cancer (*Merck VIOXX Timeline*, *2004*). This illustrates two points-one that toxicity in healthy people with a drug to prevent a possible cancer in the future is unacceptable. Second, as with the ATBC trial these large trials will document toxicity otherwise under assessed.

PIN as a preneoplastic lesion with an association to future cancer development has seen much enthusiasm as for marker for chemopreventive trials. There are difficulties with such an approach. Firstly as a stand alone lesion it is not common, secondly PIN may be late in the molecular transformation to cancer and may not respond to intervention efforts. Despite the difficulties several clinical trials have targeted PIN for treatment in an effort to reduce prostate cancer incidence. Southwest Oncology Group activated a phase III randomized, double blind trial comparing placebo with selenium for 3 years for men with PIN alone on biopsy (Stratton et al., 2003). Another trial sponsored by the National Cancer Institute of Canada randomizing men with PIN into two groups one group receiving selenium, soy protein isolate, and vitamin E , the other arm placebo.

Toremifene [GTx-006, Acapodene] a selective estrogen receptor modulator has been evaluated in a small phase II trial, an open label trial in 21 men with PIN who received 4 months of toremifene 60 mg/day orally before undergoing a second biopsy with 8 cores. The results were compared to historical controls, and there was felt to be a reduction in PIN on follow up (Steiner, 2003). Subsequently a larger phase II trial sponsored by GTx has been completed with 500 men with PIN on biopsy randomized into a double blind trial with four arms-3 different doses of toremifene versus placebo. Men were followed every 3 months and had biopsies at 6 and 12 months. As presented in an abstract, the placebo arm had a cancer detection rate of 31% at 1 year, with 20 mg of toremifene reducing the cancer rates to 24.4%, $p < 0.05$] (Steiner et al., 2004).

13. ARE THERE BENEFITS OF SCREENING OR TREATMENT

Screening for prostate cancer has become synonymous with PSA screening. It should be clarified that screening implies performing a test on an asymptomatic man in the general population. For men who present with symptoms or have a very strong family history the use of PSA could be considered more case finding than screening. For screening to be beneficial several factors should be present. First the disease to be screened for should be a major health concern, second there should be a health benefit for early intervention, third the screening test should be rapid and inexpensive, and fourth the test should have high sensitivity and specificity. For each of these factors in PSA screening there is debate. Prostate cancer is prevalent, but debate about its clinical impact is ongoing. Autopsy series demonstrating large numbers of men with clinically insignificant prostate disease give some concern that aggressive treatment of prostate cancer is not indicated when so many men die of other causes (Sakr et al., 1994). It is estimated that there will be 230,000 cases of prostate cancer diagnosed in the U.S., but only 29,900 deaths due to prostate cancer (Jemal et al., 2005). Which of those 230,900 men would benefit from treatment-who would not? Because of the slow growing nature of most prostate cancers and the age of the men at diagnosis with other co morbidities, many feel early treatment does not improve overall survival. To study whether treatment impacts survival randomized trials of observation versus surgical treatment have been implemented. The Scandinavian Prostate Cancer Group in 1989 randomized 695 men with T1b, T1c, T2 tumors to surgery or observation. In 2002 they reported on the progress of this group. The mean age was 64.7 years with a median 6.2 years of follow-up. There was a 50% reduction in prostate cancer mortality in the arm treated with surgery- but no overall improvement in survival (Holmberg et al., 2002). In 2005 they presented further follow- up- now with a median 8.2 years follow-up and there was a significant improvement with surgery in overall survival, 83 deaths versus 106 [$p = 0.04$], prostate cancer deaths 30 versus 50 [$p = 0.01$] as well as less distant metastasis and use of hormonal therapy [$p = 0.01$]. Men younger than 65 seemed to benefit more- but the study was not powered to stratify for age (Bill-Axelson et al., 2005). This study does seem to illustrate the rationale that PSA screening should be limited to men with a 10 year life expectancy (Smith et al., 2005). The U.S has a similar ongoing trial- PIVOT- Prostate Intervention versus Observation which began in 1994 with 700 men. The median age is 68 years, and there are more men with T1c stage cancers- PSA found tumors- than the Swedish study. No results are yet available (Wilt and Brawer, 1997).

To study whether PSA screening improves men's health are two large prospective screening trials. The European Randomized Screening Prostate Cancer begun in 1992 consists of seven centers in Europe where 163,126 men age 55 to 69 years were randomized to screening or not. Pathology reports have demonstrated a migration to smaller tumors, results if there is a survival benefit are to be in 2007–2008 (Hoedemaeker et al., 1997; van der Cruijsen-Koeter et al., 2005). The U.S. Prostate, Lung, Colorectal, Ovary began in 1993 and enrolled 37,000 men and 37,000 women into two arms with planned follow-up of 13 years. To date 1.4% screened men have been diagnosed with PCA, but mortality results are pending (Andriole et al., 2005). Several authors have published on their trials with PSA screening. Labrie in Quebec, Canada randomized from an electoral directory men to receive screening or not. They reported the death rate decreased with screening [15 versus 48.7/ 100,000 man years], but the analysis was not on an intent to treat basis and lead to criticisms of the study (Labrie et al., 1999). Further follow up of this trial in 2004 reported a 62% reduction in prostate cancer deaths in the screened group (Labrie et al., 2004). Bartsch in Tyrol Austria reported on the 'natural experiment' where PSA screening was free to Tyrol and not implemented in the rest of Austria. Beginning in 1993 for men age 45–75, two thirds of the men were tested in first 5 years. There was a stage migration with a 33% reduction in PCA deaths. Difference in PCA mortality was significant at $p = 0.006$ (Bartsch et al., 2001; Horninger et al., 2005) Not all studies show overall survival, a Swedish study enrolling every sixth man in the age range of 50–69 to a trial screening for prostate cancer every third year was performed. In 1987 and 1990 only DRE was used and in 1993 and 1996 PSA was also implemented. The screened group had a higher rate of cancer detection [5.7% versus 3.8%], but half the cancers were detected between screenings. The screened group also had more organ confined [56% versus 26%] disease and were more likely to undergo curative treatment. The overall and cause specific survival was not statistically different (Sandblom et al., 2004).

As stated previously PSA screening has been given an 'I' rating for insufficient evidence to promote its use as a screening agent by the U.S. Preventative Task Force. The American Cancer Society has published its recommendations encouraging men to be informed on the risks and benefits of screening. For those choosing screening, screening should begin at age 50 and be reserved for men with a 10 year life expectancy. For men of African decent, especially from the Sub-Saharan, or an affected first degree relative the recommendation is to begin at 45, and if multiple first degree relatives age 40 is recommended. If the PSA is <1.0 the option to return at age 45 is given. If >2.5 a biopsy should be done (Smith et al., 2005). As a practical matter, discussing when you plan to stop obtaining a PSA- such as age 70 or 75 and screen with DRE only early on makes acceptance of no longer checking an annual PSA later easier.[Personal Observation]

The dilemma with PSA and prostate cancer points to areas of necessary research. When PSA was introduced, causing a rise in incidence, the concern was it was uncovering latent, incidental prostate cancer. Fifteen years later when the PCPT end of study biopsy on the placebo arm demonstrated 15% had cancer despite normal DRE and PSA the concern was PSA was not detecting cancer. Current efforts are aimed to detect, or to predict of those tumors detected, the potentially lethal cancers, leaving undiagnosed or untreated those tumors with limited ability to progress.

14. TREATMENT

Because of the age of the individuals involved, many men with other co morbidities, and the indolent nature of many prostate cancers, as well as the potential for significant morbidity of treatment- impotence and incontinence; treatment decisions for prostate cancer are an analysis of competing risks for the individual in question. The age, health and particular morbidity concerns of the individual as well as the PSA level, clinical stage, Gleason score, number and extent of affected cores will factor into the decision as to what treatment regiment should be adopted. The first question- will this man live long enough to benefit from treatment should be asked prior to PSA screening in the majority of men being diagnosed today.

The second question whether the cancer requires treatment for localized or advanced disease can be facilitated by clinical exam as well as tables and nomograms incorporating PSA levels, clinical exam, and Gleason score. PSA levels over 20, or >10 if the Gleason score is 7, or any PSA value if the Gleason score is ≥ 8 trigger a bone scan to rule out metastasis disease to the bone- a common metastasis site. Men with PSA levels below these parameters have minimal risk for bone disease. Whether CT Scan and MRI should be used to rule out lymph node involvement can be determined from utilization of the Partin Tables (Carroll et al., 2001; Wolf, Jr., et al., 1995; Allen et al., 2004). Partin first published his nomogram in 1997 and later updated in 2001 (Partin et al., 2001; Partin et al., 1997). Comparing 3 presurgical parameters with the surgical pathology specimens of 4133 men after radical prostatectomy with pelvic lymph node dissection, tables have been generated outlining the percent of men who will have organ confined disease, extraprostatic disease, seminal vesicle disease or positive lymph nodes given their PSA, clinical stage and Gleason score. A CT scan maybe indicated, given its limitations in detecting positive lymph nodes, if a man's risk of having positive pelvic lymph nodes is high by Partin Table. Others have calculated that a CT scan maybe helpful in detecting positive nodes if the PSA is > 25 (Carroll et al., 2001; Wolf, Jr., et al., 1995). This may change if sensitivity of imaging tests improves. The difficulty arises in that the Tables are helpful in general categories and discussing risks, but are not specific enough to an individual to be definitive.

Once it has been determined if the disease is most likely localized several options are available. A third question is whether the localized disease has potential to progress. Kattan nomogram published in 2003 expands on the same parameters of clinical stage, PSA, ultrasound volume and findings from the pathology report to give a probability of indolent disease, based on correlation of these factors to clinically insignificant cancers on radical prostatectomy specimens. The nomogram, as it defines indolent, excludes any cancer with a Gleason component of 4, and gives the probability of a cancer with a presumed volume of <0.5 cc volume of well or moderately differentiated tumors (Kattan, Eastham et al. 2003). There are several other nomograms to predict individual outcomes after primary treatment with either surgery or radiation therapy, and they and tend to out perform experts and risk stratification. But again one is given a probability of being disease free at particular intervals; improvement for individual survival is needed (Diblasio and Kattan, 2003).

15. LOCALIZED DISEASE

15.1 Expectant Management

Observation or watchful waiting has long been a mainstay of prostate cancer treatment. No curative intent is performed- but rather once progression has been documented generally to bone metastasis or symptomatic bone metastasis systemic hormonal therapy is instituted. Several older studies and one recent have reported that the survival from prostate cancer up to 10 years is 60 to 87%. Unless cancer is poorly differentiated- Chodak reported 34% 10 year cancer specific survival. But at 15 years cancer specific survival is 53–56% (Hugosson et al., 1996). Thus for older men with less than 10 year life expectancy no treatment gives similar outcomes as treatment for moderate and low grade tumors. A population based study of long term nonrandomized 10 year cancer survival using the U.S. based SEER data, (Table 6) again demonstrates that for low grade tumors 10 year survival data is similar between treatment options. The authors extracted data from the 10 participating centers across the U.S. There were 59,876 cancer registry patients from 1983 to 1992, age 50 to 79 years. True comparisons between treatment can not be made as men undergoing radiation were older than those receiving surgery (Lu-Yao and Yao, 1997). Again 10 year survival in men with poorly differentiated tumors is poor in this series also.

Albertsen et al. reported on the competing risk analysis of death of men aged 55 to 74 at time of diagnosis, managed conservatively and generated insightful graphs outlining the death from prostate cancer versus other causes stratified by Gleason score and age of diagnosis. First published in 1998 it demonstrated chance of dying from prostate cancer was linked to Gleason score versus age. (Albertsen et al., 1998) Table 7. Follow up publication in 2005 concluded mortality rates were stable after 15 years, low grade cancers have minimal risk during 20 year follow

Table 6. Stratification of 10 year survival by tumor grade and treatment from SEER data base, a non randomized data base (Lu-Yao and Yao, 1997)

Grade and Treatment	% Survival at 10 years
Grade 1	
Radical Prostatectomy	98%
Radiation Therapy	89%
Watchful Waiting	92%
Grade 2	
Radical Prostatectomy	91%
Radiation Therapy	74%
Watchful Waiting	76%
Grade 3	
Radical Prostatectomy	76%
Radiation Therapy	52%
Watchful Waiting	43%

Table 7. Probability of dying from prostate cancer within 15 years based on presenting Gleason Grade (Albertsen et al., 1998)

Gleason Grade	2–4	5	6	7	8–10
Probability of dying from prostate cancer by 15 years	4–7%	6–11%	18–30%	42–70%	60–87%

up [6 deaths per 1000 person years] and high grade cancers have high probability during the first 10 years. [121 deaths per 1000 person years] (Albertsen et al., 2005). One discernment with this particular study is that many of these tumors were from TURP specimens [60%] or open prostatectomy [11%] and only 26% were diagnosed by transrectal biopsy. It is debatable if this series accurately reflects the tumors diagnosed by transrectal biopsy in the periphery of the prostate.

Expectant management differs from watchful waiting in that intervention, if needed, would be instituted when curative intent is still possible. Thus protocols with monitoring PSA rate, follow up biopsy are being implemented to determine who needs invasive treatment and who can be followed (Klotz, 2005).

16. SURGICAL MANAGEMENT

Treatment with surgery is relegated to presumed localized disease in men with 10 year life expectancy. Several centers of excellence have reported on excellent 5, 10 and 15 year overall survival and cancer specific survival after radical prostatectomy, respectfully 99%, 96% and 90% (Roehl et al., 2004; Han et al., 2001; Walsh et al., 1994; Kundu et al., 2004; Khan et al., 2003). Centers of excellence also report good urinary control and potency in younger men with adequate erections (Khan et al., 2003). Series from community based data collection and quality of life questionnaires do not demonstrate equal continence and potency rates, though 89% of men who chose surgery would do so again (Fowler, Jr., et al., 1995). With the advent of PSA- biochemical recurrence is another measure of cancer control. The significance of PSA recurrence is difficult to estimate at times. Pound et al reported on the outcome of PSA recurrence after radical prostatectomy in men not treated with adjuvant therapy. Fifteen percent of men of the 1997 men had a biochemical recurrence, 34% of these men with PSA recurrence developed metastatic disease in the study period. Median time from PSA recurrence to bone metastasis was 8 years, and from bone metastasis to death median time was 5 years. PSA doubling time, Gleason score, and time to biochemical recurrence were predictive of progression (Pound et al., 1999) Recent studies have confirmed PSA velocity or doubling time after prostatectomy as a prognostic factor. Comparison with radiation is difficult as there has not been a prospective trial in the current PSA era. Just recently the Sweedish prospective comparison of surgery with WW for T1c to T2 tumors has shown a survival advantage with surgery at median follow up of 8.2 years, but not in an earlier analysis at 6.9 years (Bill-Axelson et al., 2005). The PIVOT trial should be helpful with additional information.

Newer surgical techniques with laparoscopy and robotic surgery are making inroads on open surgery with advocates touting less blood loss and better visualization, it will take time to determine if these advantages translate to improved continence and potency (Smith, Jr. and Herrell, 2005).

17. RADIOTHERAPY [RT]

There are several modalities of radiation; conventional RT, three dimensional conformal RT [3DCRT], intensity-modulated RT [IMRT], conformal proton beam RT [CPBRT], brachytherapy with permanent iodine or palladium seeds [PPI] or with high dose temporary implants [HDR]. In addition, varying doses of radiation have been given in conjuncture with varying time courses of hormonal therapy. The scope of the treatment options with radiation therapy are beyond this review. For an excellent summary of evidence based direction with radiation therapy the reader is referred to recent review by Speight and Roach (Speight and Roach, 2005). Though there are many unanswered questions on the optimum type, dose, and timing of RT, through the use of prospective clinical trials some questions have been answered. Points made by the authors include, doses less $< 70\,\text{Gy}$ are suboptimal for curative intent, but it is unclear if doses greater than 78 are beneficial. Androgen deprivation therapy is beneficial in conjunction with radiation therapy and combinations and timing can be optimized to patient populations. Pelvic irradiation to the pelvic lymph nodes is debated, but should be considered for intermediate and high risk patients (Ryan and Eisenberger, 2005). A prospective trial with radiation versus observation for localized disease has not been reported upon, as also there is no current prospective trial of radiation versus surgery. This makes it difficult for patients to compare survival outcomes between the three therapies. Radiation therapists feel the outcomes are similar to prostatectomy, and urologist generally feel surgery is the better treatment option. Both are more likely to recommend their form of therapy (Fowler, Jr., et al., 2000).

Comparisons of morbidity are more readily available from community data bases. From the Medicare data base with treatment prior to 1991 men undergoing surgery are more likely to wear pads than those receiving radiation, 32% versus 7%, and have a higher rate of impotence 56% versus 23%, but less side effects with bowel dysfunction 4% versus 10%. Radiation patients were less likely to say they were cancer free and had more cancer worry than surgical counterparts (Fowler, Jr., et al., 1996). At a 5 year follow up men undergoing radiation had better urinary control, but had declined in sexual function from the second year to 5 years so both groups had similar erectile function (Potosky et al., 2004). Litwin's group compared quality of life function from men receiving external beam, brachytherapy and surgery. Each group reported sexual decline compared to controls, surgery was associated with urinary bother, external beam with bowel dysfunction and brachytherapy with all three domains impaired (Wei et al., 2002).

Cryotherapy is making a resurgence as treatment for local therapy. Improvements in freeze delivery, urethral warming and imaging have made complications such as urethral sloughing, or rectoprostatic fistula which limited earlier enthusiasm for cryotherapy to acceptable levels and renewed enthusiasm, but the long term efficacy is not yet available (Pareek and Nakada, 2005).

18. SYSTEMIC THERAPY

Prior to the advent of PSA men presenting with metastatic disease accounted for 14.4% of men presenting with prostate cancer, this dropped to 3.3% in 1998 (Cooperberg et al., 2005). Initiation of androgen ablation by castration was instituted generally with initial relief of symptomatic bone pain followed by androgen independence and death. A VA cooperative trial with timing of hormonal therapy early versus late when symptoms developed was not reported to impact survival, but had to contend with cardiovascular toxicity of DES (Messing, 2003). With PSA use and stage migration, men presenting with bone disease has decreased, it would seem the use of androgen ablation should be diminishing. But expansion of indications for androgen ablation have occurred and may be in part due to the institution of medical and intermittent or somewhat reversible castration with LHRH agonists and antagonists in the 1980s, as well as nonsteroidal anti-hormonal therapies leading to better acceptance by patients.

The Medical Research Council randomized men with advanced or metastatic disease to immediate or delayed hormonal therapy and concluded immediate therapy prolonged survival in men with advanced disease, but for men with metastatic disease immediate therapy delayed disease related complications, but did not change survival (The Medical Research Council Prostate Cancer Working Party Investigators Group, 1997). The Eastern Cooperative Oncology Group reported upon a trial where men undergoing radical prostatectomy with pelvic lymph node dissection for presumed localized disease but found to positive lymph nodes were randomized to either immediate hormonal deprivation versus observation with initiation of treatment based if PSA recurred. They found a survival advantage for men undergoing immediate therapy. At 7 year follow up, 7 of 47 men undergoing immediate therapy died compared to 18 of 51 men in the delayed therapy arm (Messing et al., 1999). The radiation treatment protocols have also demonstrated a survival advantage for localized disease with hormonal therapy (Speight and Roach, 2005). Thus expansion into areas not well researched such as PSA recurrence after primary treatment, or high risk features such as seminal vesicle involvement at time of surgery are being instituted based on these previous studies showing benefit to asymptomatic men with advanced local or nodal disease. According to large population data bases men are choosing primary androgen deprivation even with low and moderate risk localized disease, advantages which have not been well studied (Cooperberg et al., 2005).

Life expectancy is difficult to gage with advanced disease as there has been a shift in earlier metastatic disease with the introduction of PSA. A recent report from MD Anderson looking at 4141 men registered with prostate cancer between 1982 and 2001 and found median survival for lymph node involvement was 134 months and for distant metastasis was 42 months (Taylor et al., 2005).

Newer therapies for prostate cancer are needed. The deaths from prostate cancer occur when the cells become androgen insensitive. Chemotherapy historically has not been effective in improving survival- just improving pain control. Recently docetaxel did show a survival advantage over standard therapy in hormone refractory disease increasing survival from 16 to 18 months in two trials, and has been approved for use (Ryan and Eisenberger, 2005). Further work is necessary.

In conclusion prostate cancer is a multifaceted disease which increases with aging- the clinical course of prostate cancer is impacted by genetic and environmental interactions. Much effort is aimed at understanding the etiology of prostate cancer so preventive efforts will be effective. As it is a disease of the old, delaying progression with diet, medications and lifestyle modifications, if possible, has tremendous implications allowing men to die of other diseases. The recent improvements in cardiovascular care has caused cancer to be the number one cause of death in those less than 85 in the U.S. This will impact the number of men living long enough to be affected by prostate cancer (Smith et al., 2005). The art of medicine for physicians involved with aging men is to determine when PSA screening and treatment need to be applied and when they do not. The exact etiology of prostate cancer is unknown, but the evidence that a healthy diet is associated with improved cancer mortality is growing, and should be encouraged for all throughout their lifetime.

REFERENCES

Albanes, D., et al. (1996) Alpha-Tocopherol and beta-carotene supplements and lung cancer incidence in the alpha-tocopherol, beta-carotene cancer prevention study: effects of base-line characteristics and study compliance. J Natl Cancer Inst, 88(21): 1560–70.

Albertsen, P.C., et al. (1998) Competing risk analysis of men aged 55 to 74 years at diagnosis managed conservatively for clinically localized prostate cancer. Jama, 280(11): 975–80.

Albertsen, P.C., Hanley, J.A. and Fine, J. (2005) 20-year outcomes following conservative management of clinically localized prostate cancer. Jama, 293(17): 2095–101.

Allen, E.A., Kahane, H. and Epstein, J.I. (1998) Repeat biopsy strategies for men with atypical diagnoses on initial prostate needle biopsy. Urology, 52(5): 803–7.

Allen, D.J., et al. (2004) Does body-coil magnetic-resonance imaging have a role in the preoperative staging of patients with clinically localized prostate cancer? BJU Int, 94(4): 534–8.

Andriole, G., et al. (2004) Chemoprevention of prostate cancer in men at high risk: rationale and design of the reduction by dutasteride of prostate cancer events (REDUCE) trial. J Urol, 172(4 Pt 1): 1314–7.

Andriole, G., et al. (2005) The effects of 5alpha-reductase inhibitors on the natural history, detection and grading of prostate cancer: current state of knowledge. J Urol, 174(6): 2098–104.

Andriole, G.L., et al. (2005) Prostate Cancer Screening in the Prostate, Lung, Colorectal and Ovarian (PLCO) Cancer Screening Trial: findings from the initial screening round of a randomized trial. J Natl Cancer Inst, 97(6): 433–8.

Angwafo, F.F., 3rd, et al. (2003) High-grade intra-epithelial neoplasia and prostate cancer in Dibombari, Cameroon. Prostate Cancer Prostatic Dis, 6(1): 34–8.

Attar-Bashi, N.M., Frauman, A.G. and Sinclair, A.J. (2004) Alpha-linolenic acid and the risk of prostate cancer. What is the evidence? J Urol, 171(4): 1402–7.

Baade, P.D., et al. (2005) Communicating prostate cancer risk: what should we be telling our patients? Med J Aust, 182(9): 472–5.

Bartsch, G., et al. (2001) Prostate cancer mortality after introduction of prostate-specific antigen mass screening in the Federal State of Tyrol, Austria. Urology, 58(3): 417–24.

Bianco, F.J., Jr., et al. (2002) Prostate cancer stage shift has eliminated the gap in disease-free survival in black and white American men after radical prostatectomy. J Urol, 168(2): 479–82.

Bill-Axelson, A., et al. (2005) Radical prostatectomy versus watchful waiting in early prostate cancer. N Engl J Med, 352(19): 1977–84.

Borboroglu, P.G., et al. (2001) Repeat biopsy strategy in patients with atypical small acinar proliferation or high grade prostatic intraepithelial neoplasia on initial prostate needle biopsy. J Urol, 166(3): 866–70.

Borgono, C.A. and Diamandis, E.P. (2004) The emerging roles of human tissue kallikreins in cancer. Nat Rev Cancer, 4(11): 876–90.

Boring, C.C., Squires, T.S. and Tong, T. (1992) Cancer statistics, 1992. CA Cancer J Clin, 42(1): 19–38.

Bostwick, D.G., et al. (2004) Does finasteride alter the pathology of the prostate and cancer grading? Clin Prostate Cance, 2(4): 228–35.

Bostwick, D.G. (1988) Premalignant lesions of the prostate. Semin Diagn Pathol, 5(3): 240–53.

Bostwick, D.G. (1996) Prospective origins of prostate carcinoma. Prostatic intraepithelial neoplasia and atypical adenomatous hyperplasia. Cancer, 78(2): 330–6.

Bostwick, D.G. and Brawer, M.K. (1987) Prostatic intra-epithelial neoplasia and early invasion in prostate cancer. Cancer, 59(4): 788–94.

Brausi, M., et al. (2004) Immediate radical prostatectomy in patients with atypical small acinar proliferation. Over treatment? J Urol, 172(3): 906–8; discussion 908–9.

Brawer, M.K., et al. (1991) Significance of prostatic intraepithelial neoplasia on prostate needle biopsy. Urology, 38(2): 103–7.

Breslow, N., et al. (1977) Latent carcinoma of prostate at autopsy in seven areas. The International Agency for Research on Cancer, Lyons, France. Int J Cancer, 20(5): 680–8.

Carpten, J., et al. (2002) Germline mutations in the ribonuclease L gene in families showing linkage with HPC1. Nat Genet, 30(2): 181–4.

Carroll, P., et al. (2001) Prostate-specific antigen best practice policy–part II: prostate cancer staging and post-treatment follow-up. Urology, 57(2): 225–9.

Carter, B.S., et al. (1993) Hereditary prostate cancer: epidemiologic and clinical features. J Urol, 150(3): 797–802.

Carver, B.S., et al. (2005) Gleason grade remains an important prognostic predictor in men diagnosed with prostate cancer while on finasteride therapy. BJU Int, 95(4): 509–12.

Catalona, W.J., et al. (1994) Comparison of prostate specific antigen concentration versus prostate specific antigen density in the early detection of prostate cancer: receiver operating characteristic curves. J Urol, 152(6 Pt 1): 2031–6.

Catalona, W.J., et al. (2004) Serum pro-prostate specific antigen preferentially detects aggressive prostate cancers in men with 2 to 4 ng/ml prostate specific antigen. J Urol, 171(6 Pt 1): 2239–44.

Chan, T.Y. and Epstein, J.I. (1999) Follow-up of atypical prostate needle biopsies suspicious for cancer. Urology, 53(2): 351–5.

Chan, D.W., et al. (1987) Prostate-specific antigen as a marker for prostatic cancer: a monoclonal and a polyclonal immunoassay compared. Clin Chem, 33(10): 1916–20.

Chan, J.M., et al. (2001) Dairy products, calcium, and prostate cancer risk in the Physicians' Health Study. Am J Clin Nutr, 74(4): 549–54.

Chan, J.M., Gann, P.H. and Giovannucci, E.L. (2005) Role of diet in prostate cancer development and progression. J Clin Oncol, 23(32): 8152–60.

Cheville, J.C., Reznicek, M.J. and Bostwick, D.G. (1997) The focus of "atypical glands, suspicious for malignancy" in prostatic needle biopsy specimens: incidence, histologic features, and clinical follow-up of cases diagnosed in a community practice. Am J Clin Pathol, 108(6): 633–40.

Christen, W.G., Gaziano, J.M. and Hennekens, C.H. (2000) Design of Physicians' Health Study II–a randomized trial of beta-carotene, vitamins E and C, and multivitamins, in prevention of cancer, cardiovascular disease, and eye disease, and review of results of completed trials. Ann Epidemiol, 10(2): 125–34.

Clark, L.C., et al. (1996) Effects of selenium supplementation for cancer prevention in patients with carcinoma of the skin. A randomized controlled trial. Nutritional Prevention of Cancer Study Group. Jama, 276(24): 1957–63.

Clark, L.C., et al. (1998) Decreased incidence of prostate cancer with selenium supplementation: results of a double-blind cancer prevention trial. Br J Urol, 81(5): 730–4.

Cooperberg, M.R., Moul, J.W. and Carroll, P.R. (2005) The changing face of prostate cancer. J Clin Oncol, 23(32): 8146–51.

D'Amico, A.V., et al. (2004) Prostate specific antigen doubling time as a surrogate end point for prostate cancer specific mortality following radical prostatectomy or radiation therapy. J Urol, 172(5 Pt 2): S42–6; discussion S46–7.

D'Amico, A.V., et al. (2004) Preoperative PSA velocity and the risk of death from prostate cancer after radical prostatectomy. N Engl J Med, 351(2): 125–35.

Dagnelie, P.C., et al. (2004) Diet, anthropometric measures and prostate cancer risk: a review of prospective cohort and intervention studies. BJU Int, 93(8): 1139–50.

Danley, K.L., et al. (1995) Prostate cancer: trends in mortality and stage-specific incidence rates by racial/ethnic group in Los Angeles County, California (United States). Cancer Causes Control, 6(6): 492–8.

Davidson, D., et al. (1995) Prostatic intraepithelial neoplasia is a risk factor for adenocarcinoma: predictive accuracy in needle biopsies. J Urol, 154(4): 1295–9.

Dhom, G. (1983) Epidemiologic aspects of latent and clinically manifest carcinoma of the prostate. J Cancer Res Clin Oncol, 106(3): 210–8.

Diblasio, C.J. and Kattan, M.W. (2003) Use of nomograms to predict the risk of disease recurrence after definitive local therapy for prostate cancer. Urology, 62 Suppl 1: 9–18.

Donn, A.S. and Muir, C.S. (1985) Prostatic cancer: some epidemiological features. Bull Cancer, 72(5): 381–90.

Drago, J.R., Mostofi F.K. and Lee Fred. (1989) Introductory Remarks and Workshop Summary. Urology, 34(6): 2–3.

Edwards, S.M. and Eeles, R.A. (2004) Unravelling the genetics of prostate cancer. Am J Med Genet C Semin Med Genet, 129(1): 65–73.

Feneley, M.R., et al. (1997) Prevalence of prostatic intra-epithelial neoplasia (PIN) in biopsies from hospital practice and pilot screening: clinical implications. Prostate Cancer Prostatic Dis, 1(2): 79–83.

Ferlay, J., B.F., Pisani P. and Parkin D.M. (2004) Cancer Incidence, Mortality and Prevalence Worldwide. IARC CancerBase. No. 5(version 2.0).

Fowler, F.J., Jr., et al. (2000) Comparison of recommendations by urologists and radiation oncologists for treatment of clinically localized prostate cancer. Jama, 283(24): 3217–22.

Fowler, F.J., Jr., et al. (1995) Effect of radical prostatectomy for prostate cancer on patient quality of life: results from a Medicare survey. Urology, 45(6): 1007–13; discussion 1013–5.

Fowler, F.J., Jr., et al. (1996) Outcomes of external-beam radiation therapy for prostate cancer: a study of Medicare beneficiaries in three surveillance, epidemiology, and end results areas. J Clin Oncol, 14(8): 2258–65.

Fowler, J.E., Jr., et al. (2000) Predictors of first repeat biopsy cancer detection with suspected local stage prostate cancer. J Urol, 163(3): 813–8.

Giovannucci, E.L., et al. (2005) A prospective study of physical activity and incident and fatal prostate cancer. Arch Intern Med, 165(9): 1005–10.

Giovannucci, E. (2005) The epidemiology of vitamin D and cancer incidence and mortality: a review (United States). Cancer Causes Control, 16(2): 83–95.

Gleason, D.F. (1966) Classification of prostatic carcinomas. Cancer Chemother Rep, 50(3): 125–8.

Glover, F.E., Jr., et al. (1998) The epidemiology of prostate cancer in Jamaica. J Urol, 159(6): 1984–6; discussion 1986–7.

Gokden, N., et al. (2005) High-grade prostatic intraepithelial neoplasia in needle biopsy as risk factor for detection of adenocarcinoma: current level of risk in screening population. Urology, 65(3): 538–42.

Goode, E.L., Ulrich, C.M. and Potter, J.D. (2002) Polymorphisms in DNA repair genes and associations with cancer risk. Cancer Epidemiol Biomarkers Prev, 11(12): 1513–30.

Grignon, D.J. and Sakr, W.A. (1996) Atypical adenomatous hyperplasia of the prostate: a critical review. Eur Urol, 30(2): 206–11.

Gsur, A., Feik, E. and Madersbacher, S. (2004) Genetic polymorphisms and prostate cancer risk. World J Urol, 21(6): 414–23.

Habel, L.A., Zhao, W. and Stanford, J.L. (2002) Daily aspirin use and prostate cancer risk in a large, multiracial cohort in the US. Cancer Causes Control, 13(5): 427–34.

Han, M., et al. (2001) Long-term biochemical disease-free and cancer-specific survival following anatomic radical retropubic prostatectomy. The 15-year Johns Hopkins experience. Urol Clin North Am, 28(3): 555–65.

Hebert, J.R., et al. (1998) Nutritional and socioeconomic factors in relation to prostate cancer mortality: a cross-national study. J Natl Cancer Inst, 90(21): 1637–47.

Helpap, B.G., Bostwick, D.G. and Montironi, R. (1995) The significance of atypical adenomatous hyperplasia and prostatic intraepithelial neoplasia for the development of prostate carcinoma. An update. Virchows Arch, 426(5): 425–34.

Hoedemaeker, R.F., et al. (1997) Comparison of pathologic characteristics of T1c and non-T1c cancers detected in a population-based screening study, the European Randomized Study of Screening for Prostate Cancer. World J Urol, 15(6): 339–45.

Holmberg, L., et al. (2002) A randomized trial comparing radical prostatectomy with watchful waiting in early prostate cancer. N Engl J Med, 347(11): 781–9.

Horninger, W., et al. (2005) Screening for prostate cancer: updated experience from the Tyrol study. Can J Urol, 12 Suppl 1: 7–13; discussion 92–3.

Hugosson, J., Aus, G. and Norlen, L. (1996) Surveillance is not a viable and appropriate treatment option in the management of localized prostate cancer. Urol Clin North Am, 23(4): 557–73.

Iczkowski, K.A., MacLennan, G.T. and Bostwick, D.G. (1997) Atypical small acinar proliferation suspicious for malignancy in prostate needle biopsies: clinical significance in 33 cases. Am J Surg Pathol, 21(12): 1489–95.

Iczkowski, K.A., et al. (1998) Diagnosis of "suspicious for malignancy" in prostate biopsies: predictive value for cancer. Urology, 51(5): 749–57; discussion 757–8.

Jackson, M.A., et al. (1980) Characterization of prostatic carcinoma among blacks: a comparison between a low-incidence area, Ibadan, Nigeria, and a high-incidence area, Washington, DC. Prostate, 1(2): 185–205.

Jackson, M.A., et al. (1981) Factors involved in the high incidence of prostatic cancer among American blacks. Prog Clin Biol Res, 53: 111–32.

Jemal, A., et al. (2005) Cancer statistics, 2005. CA Cancer J Clin, 55(1): 10–30.

Johns, L.E. and Houlston, R.S. (2003) A systematic review and meta-analysis of familial prostate cancer risk. BJU Int, 91(9): 789–94.

Kattan, M.W., et al. (2003) Counseling men with prostate cancer: a nomogram for predicting the presence of small, moderately differentiated, confined tumors. J Urol, 170(5): 1792–7.

Khan, M.A., et al. (2003) Long-term cancer control of radical prostatectomy in men younger than 50 years of age: update 2003. Urology, 62(1): 86–91; discussion 91–2.

Klotz, L. (2005) Active surveillance for prostate cancer: for whom? J Clin Oncol, 23(32): 8165–9.

Kristal, A.R., et al. (2002) Associations of energy, fat, calcium, and vitamin D with prostate cancer risk. Cancer Epidemiol Biomarkers Prev, 11(8): 719–25.

Kronz, J.D., et al. (2001) Predicting cancer following a diagnosis of high-grade prostatic intraepithelial neoplasia on needle biopsy: data on men with more than one follow-up biopsy. Am J Surg Pathol, 25(8): 1079–85.

Kuller, L.H., et al. (2004) Elevated prostate-specific antigen levels up to 25 years prior to death from prostate cancer. Cancer Epidemiol Biomarkers Prev, 13(3): 373–7.

Kundu, S.D., et al. (2004) Potency, Continence And Complications In 3,477 Consecutive Radical Retropubic Prostatectomies. J Urol, 172(6, Part 1 of 2): 2227–2231.

Labrie, F., et al. (1999) Screening decreases prostate cancer death: first analysis of the 1988 Quebec prospective randomized controlled trial. Prostate, 38(2): 83–91.

Labrie, F., et al. (2004) Screening decreases prostate cancer mortality: 11-year follow-up of the 1988 Quebec prospective randomized controlled trial. Prostate, 59(3): 311–8.

Latil, A.G., et al. (2001) Prostate carcinoma risk and allelic variants of genes involved in androgen biosynthesis and metabolism pathways. Cancer, 92(5): 1130–7.

Lee, M.M., et al. (1998) Case-control study of diet and prostate cancer in China. Cancer Causes Control, 9(6): 545–52.

Lefkowitz, G.K., et al. (2001) Is repeat prostate biopsy for high-grade prostatic intraepithelial neoplasia necessary after routine 12-core sampling? Urology, 58(6): 999–1003.

Lefkowitz, G.K., et al. (2002) Followup interval prostate biopsy 3 years after diagnosis of high grade prostatic intraepithelial neoplasia is associated with high likelihood of prostate cancer, independent of change in prostate specific antigen levels. J Urol, 168(4 Pt 1): 1415–8.

Leitzmann, M.F., et al. (2004) Dietary intake of n-3 and n-6 fatty acids and the risk of prostate cancer. Am J Clin Nutr, 80(1): 204–16.

Li, H., et al. (2004) A prospective study of plasma selenium levels and prostate cancer risk. J Natl Cancer Inst, 96(9): 696–703.

Li, H., et al. (2005) Manganese superoxide dismutase polymorphism, prediagnostic antioxidant status, and risk of clinical significant prostate cancer. Cancer Res, 65(6): 2498–504.

Lin, D.D., et al. (2005) Predictors of short postoperative prostate-specific antigen doubling time for patients diagnosed during PSA era. Urology, 65(3): 528–32.

Lippman, S.M., et al. (2005) Designing the Selenium and Vitamin E Cancer Prevention Trial (SELECT). J Natl Cancer Inst, 97(2): 94–102.

Luo, J., et al. (2003) Decreased gene expression of steroid 5 alpha-reductase 2 in human prostate cancer: implications for finasteride therapy of prostate carcinoma. Prostate, 57(2): 134–9.

Lu-Yao, G.L. and Yao, S.L. (1997) Population-based study of long-term survival in patients with clinically localised prostate cancer. Lancet, 349(9056): 906–10.

Mallick, S., Blanchet, P. and Multigner, L. (2005) Prostate cancer incidence in guadeloupe, a French Caribbean archipelago. Eur Urol, 47(6): 769–72.

Mann, T. (1974) Secretory function of the prostate, seminal vesicle and other male accessory organs of reproduction. J Reprod Fertil, 37(1): 179–88.

Marker, P.C., et al. (2003) Hormonal, cellular, and molecular control of prostatic development. Dev Biol, 253(2): 165–74.

McNeal, J.E. (1981) The zonal anatomy of the prostate. Prostate, 2(1): 35–49.

McNeal, J.E. and Bostwick, D.G. (1986) Intraductal dysplasia: a premalignant lesion of the prostate. Hum Pathol, 17(1): 64–71.

Merck VIOXX Timeline, in http://media.corporateir.net/media_files/irol/73/73184/VIOXX_Timeline.pdf. (2004).

Messing, E.M., et al. (1999) Immediate hormonal therapy compared with observation after radical prostatectomy and pelvic lymphadenectomy in men with node-positive prostate cancer. N Engl J Med, 341(24): 1781–8.

Messing, E. (2003) The timing of hormone therapy for men with asymptomatic advanced prostate cancer. Urol Oncol, 21(4): 245–54.

Mian, B.M., et al. (2002) Predictors of cancer in repeat extended multisite prostate biopsy in men with previous negative extended multisite biopsy. Urology, 60(5): 836–40.

Mikolajczyk, S.D. and Rittenhouse, H.G. (2004) Tumor-associated forms of prostate specific antigen improve the discrimination of prostate cancer from benign disease. Rinsho Byori, 52(3): 223–30.

Montironi, R., et al. (2004) Karyometry detects subvisual differences in chromatin organization state between cribriform and flat high-grade prostatic intraepithelial neoplasia. Mod Pathol, 17(8): 928–37.

Montironi, R., et al. (2005) Incidentally detected prostate cancer in cystoprostatectomies: pathological and morphometric comparison with clinically detected cancer in totally embedded specimens. Hum Pathol, 36(6): 646–54.
Moore, C.K., et al. (2005) Prognostic significance of high grade prostatic intraepithelial neoplasia and atypical small acinar proliferation in the contemporary era. J Urol, 173(1): 70–2.
Muir, C.S., Nectoux, J. and Staszewski, J. (1991) The epidemiology of prostatic cancer. Geographical distribution and time-trends. Acta Oncol, 30(2): 133–40.
Nakayama, M., et al. (2004) GSTP1 CpG island hypermethylation as a molecular biomarker for prostate cancer. J Cell Biochem, 91(3): 540–52.
Narain, V., et al. (2001) How accurately does prostate biopsy Gleason score predict pathologic findings and disease free survival? Prostate, 49(3): 185–90.
Nelson, W.G., et al. (2004) The role of inflammation in the pathogenesis of prostate cancer. J Urol, 172(5 Pt 2): S6–11; discussion S11–2.
Nguyen, E.V. (2003) Cancer in Asian American males: epidemiology, causes, prevention, and early detection. Asian Am Pac Isl J Health, 10(2): 86–99.
Orozco, R., et al. (1998) Observations on pathology trends in 62,537 prostate biopsies obtained from urology private practices in the United States. Urology, 51(2): 186–95.
Ostrander, E.A., Markianos, K. and Stanford, J.L. (2004) Finding prostate cancer susceptibility genes. Annu Rev Genomics Hum Genet, 5: 151–75.
Palapattu, G.S., et al. (2005) Prostate carcinogenesis and inflammation: emerging insights. Carcinogenesis, 26(7): 1170–81.
Pan, C.X., et al. (2004) PC cell-derived growth factor expression in prostatic intraepithelial neoplasia and prostatic adenocarcinoma. Clin Cancer Res, 10(4): 1333–7.
Pareek, G. and Nakada, S.Y. (2005) The current role of cryotherapy for renal and prostate tumors. Urol Oncol, 23(5): 361–6.
Partin, A.W. and Carter, H.B. (1996) The use of prostate-specific antigen and free/total prostate-specific antigen in the diagnosis of localized prostate cancer. Urol Clin North Am, 23(4): 531–40.
Partin, A.W., et al. (1993) The use of prostate specific antigen, clinical stage and Gleason score to predict pathological stage in men with localized prostate cancer. J Urol, 150(1): 110–4.
Partin, A.W., et al. (1997) Combination of prostate-specific antigen, clinical stage, and Gleason score to predict pathological stage of localized prostate cancer. A multi- institutional update. Jama, 277(18): 1445–51.
Partin, A.W., et al. (2001) Contemporary update of prostate cancer staging nomograms (Partin Tables) for the new millennium. Urology, 58(6): 843–8.
Petros, J.A., et al. (2005) mtDNA mutations increase tumorigenicity in prostate cancer. Proc Natl Acad Sci USA, 102(3): 719–24.
Plaskon, L.A., et al. (2003) Cigarette smoking and risk of prostate cancer in middle-aged men. Cancer Epidemiol Biomarkers Prev, 12(7): 604–9.
Platz, E.A., De Marzo, A.M. and Giovannucci, E. (2004) Prostate cancer association studies: pitfalls and solutions to cancer misclassification in the PSA era. J Cell Biochem, 91(3): 553–71.
Platz, E.A., et al. (2005) Nonsteroidal anti-inflammatory drugs and risk of prostate cancer in the Baltimore Longitudinal Study of Aging. Cancer Epidemiol Biomarkers Prev, 14(2): 390–6.
Postma, R., et al. (2004) Lesions predictive for prostate cancer in a screened population: first and second screening round findings. Prostate, 61(3): 260–6.
Potosky, A.L., et al. (2004) Five-year outcomes after prostatectomy or radiotherapy for prostate cancer: the prostate cancer outcomes study. J Natl Cancer Inst, 96(18): 1358–67.
Pound, C.R., et al. (1997) Prostate-specific antigen after anatomic radical retropubic prostatectomy. Patterns of recurrence and cancer control. Urol Clin North Am, 24(2): 395–406.
Pound, C.R., et al. (1999) Natural history of progression after PSA elevation following radical prostatectomy. Jama, 281(17): 1591–7.
Prentice, R.L. and Sheppard, L. (1990) Dietary fat and cancer: consistency of the epidemiologic data, and disease prevention that may follow from a practical reduction in fat consumption. Cancer Causes Control, 1(1): 81–97; discussion 99–109.

Pu, Y.S., et al. (2004) Changing trends of prostate cancer in Asia. Aging Male, 7(2): 120–32.

Qian, J., Jenkins, R.B. and Bostwick, D.G. (1998) Determination of gene and chromosome dosage in prostatic intraepithelial neoplasia and carcinoma. Anal Quant Cytol Histol, 20(5): 373–80.

Qian, J., Jenkins, R.B. and Bostwick, D.G. (1999) Genetic and chromosomal alterations in prostatic intraepithelial neoplasia and carcinoma detected by fluorescence in situ hybridization. Eur Urol, 35(5–6): 479–83.

Raviv, G., et al. (1996) Prostatic intraepithelial neoplasia: influence of clinical and pathological data on the detection of prostate cancer. J Urol, 156(3): 1050–4; discussion 1054–5.

Rebbeck, T.R., et al. (1998) Modification of clinical presentation of prostate tumors by a novel genetic variant in CYP3A4. J Natl Cancer Inst, 90(16): 1225–9.

Rishi, I., et al. (2003) Prostate cancer in African American men is associated with downregulation of zinc transporters. Appl Immunohistochem Mol Morphol, 11(3): 253–60.

Roehl, K.A., et al. (2004) Cancer progression and survival rates following anatomical radical retropubic prostatectomy in 3,478 consecutive patients: long-term results. J Urol, 172(3): 910–4.

Rose, D.P., Boyar, A.P. and Wynder, E.L. (1986) International comparisons of mortality rates for cancer of the breast, ovary, prostate, and colon, and per capita food consumption. Cancer, 58(11): 2363–71.

Rubin, M.A., et al. (2005) Effects of long-term finasteride treatment on prostate cancer morphology and clinical outcome. Urology, 66(5): 930–4.

Ryan, C.J. and Eisenberger, M. (2005) Chemotherapy for hormone-refractory prostate cancer: now it's a question of "when?". J Clin Oncol, 23(32): 8242–6.

Sakr, W.A. and Grignon, D.J. (1998) Prostatic intraepithelial neoplasia and atypical adenomatous hyperplasia. Relationship to pathologic parameters, volume and spatial distribution of carcinoma of the prostate. Anal Quant Cytol Histol, 20(5): 417–23.

Sakr, W.A. and Partin, A.W. (2001) Histological markers of risk and the role of high-grade prostatic intraepithelial neoplasia. Urology, 57(4 Suppl 1): 115–20.

Sakr, W.A., et al. (1993) The frequency of carcinoma and intraepithelial neoplasia of the prostate in young male patients. J Urol, 150(2 Pt 1): 379–85.

Sakr, W.A., et al. (1994) High grade prostatic intraepithelial neoplasia (HGPIN) and prostatic adenocarcinoma between the ages of 20–69: an autopsy study of 249 cases. In Vivo, 8(3): 439–43.

Sakr, W.A., et al. (1996) Age and racial distribution of prostatic intraepithelial neoplasia. Eur Urol, 30(2): 138–44.

Sakr, W.A., et al. (2000) Epidemiology and molecular biology of early prostatic neoplasia. Mol Urol, 4(3): 109–13; discussion 115.

Sakr, W.A. (1999) Prostatic intraepithelial neoplasia: A marker for high-risk groups and a potential target for chemoprevention. Eur Urol, 35(5–6): 474–8.

Sakr, W.A. (2004) Epidemiology of prostate cancer and its precursors. Mod Pathol.

San Francisco, I.F., et al. (2003) Clinical management of prostatic intraepithelial neoplasia as diagnosed by extended needle biopsies. BJU Int, 91(4): 350–4.

Sanchez-Chapado, M., et al. (2003) Prevalence of prostate cancer and prostatic intraepithelial neoplasia in Caucasian Mediterranean males: an autopsy study. Prostate, 54(3): 238–47.

Sandblom, G., et al. (2004) Clinical consequences of screening for prostate cancer: 15 years follow-up of a randomised controlled trial in Sweden. Eur Urol, 46(6): 717–23; discussion 724.

Schoonen, W.M., et al. (2005) Alcohol consumption and risk of prostate cancer in middle-aged men. Int J Cancer, 113(1): 133–40.

Sim, H.G. and Cheng C.W. (2005) Changing demography of prostate cancer in Asia. Eur J Cancer, 41(6): 834–45.

Sinha, A.A., et al. (2004) Microvessel density as a molecular marker for identifying high-grade prostatic intraepithelial neoplasia precursors to prostate cancer. Exp Mol Pathol, 77(2): 153–9.

Smith, R.A., Cokkinides, V. and Eyre, H.J. (2005) American Cancer Society Guidelines for the Early Detection of Cancer. CA Cancer J Clin, 2005, 55(1): 31–44; quiz 55–6.

Smith, J.A., Jr. and Herrell, S.D. (2005) Robotic-assisted laparoscopic prostatectomy: do minimally invasive approaches offer significant advantages? J Clin Oncol, 23(32): 8170–5.

Speight, J.L. and Roach, M. (2005) 3rd, Radiotherapy in the management of clinically localized prostate cancer: evolving standards, consensus, controversies and new directions. J Clin Oncol, 23(32): 8176–85.

Steiner, M.S. and Pound, C.R. (2003) Phase IIA clinical trial to test the efficacy and safety of Toremifene in men with high-grade prostatic intraepithelial neoplasia. Clin Prostate Cancer, 2(1): 24–31.

Steiner, M.S., B.R., Barnette G., et al. (2004) Evaluation of Acapodene in reducing prostate cancer incidence in high risk men. in Third Annual AACR Frontiers in Cancer Prevention Research. Seattle, WA.

Steiner, M.S. (2003) High-grade prostatic intraepithelial neoplasia and prostate cancer risk reduction. World J Urol, 21(1): 15–20.

Stratton, M.S., et al. (2003) Selenium and prevention of prostate cancer in high-risk men: the Negative Biopsy Study. Anticancer Drugs, 14(8): 589–94.

Taylor, S.H., et al. (2005) Inadequacies of the current American Joint Committee on cancer staging system for prostate cancer. Cancer.

The Medical Research Council Prostate Cancer Working Party Investigators Group. (1997) Immediate versus deferred treatment for advanced prostatic cancer: initial results of the Medical Research Council Trial. Br J Urol, 79(2): 235–46.

Thomas, L.N., et al. (2003) 5alpha-reductase type 1 immunostaining is enhanced in some prostate cancers compared with benign prostatic hyperplasia epithelium. J Urol, 170(5): 2019–25.

Thompson, I.M., et al. (2004) Prevalence of prostate cancer among men with a prostate-specific antigen level < or = 4.0 ng per milliliter. N Engl J Med, 350(22): 2239–46.

van der Cruijsen-Koeter, I.W., et al. (2005) Comparison of screen detected and clinically diagnosed prostate cancer in the European randomized study of screening for prostate cancer, section rotterdam. J Urol, 174(1): 121–5.

van der Kwast, T.H., et al. (2003) Guidelines for processing and reporting of prostatic needle biopsies. J Clin Pathol, 56(5): 336–40.

Vij, U. and Kumar, A. (2004) Phyto-oestrogens and prostatic growth. Natl Med J India, 17(1): 22–6.

Virtamo, J., et al. (2003) Incidence of cancer and mortality following alpha-tocopherol and beta-carotene supplementation: a postintervention follow-up. Jama, 290(4): 476–85.

Visvanathan, K., et al. (2004) Association among an ornithine decarboxylase polymorphism, androgen receptor gene (CAG) repeat length and prostate cancer risk. J Urol, 171(2 Pt 1): 652–5.

Walsh, P.C., Partin, A.W. and Epstein, J.I. (1994) Cancer control and quality of life following anatomical radical retropubic prostatectomy: results at 10 years. J Urol, 152(5 Pt 2): 1831–6.

Wei, J.T., et al. (2002) Comprehensive comparison of health-related quality of life after contemporary therapies for localized prostate cancer. J Clin Oncol, 20(2): 557–66.

Weinstein, M.H. and Epstein, J.I. (1993) Significance of high-grade prostatic intraepithelial neoplasia on needle biopsy. Hum Pathol, 24(6): 624–9.

Whittemore, A.S., et al. (1995) Prostate cancer in relation to diet, physical activity, and body size in blacks, whites, and Asians in the United States and Canada. J Natl Cancer Inst, 87(9): 652–61.

Wills, M.L., et al. (1997) Incidence of high-grade prostatic intraepithelial neoplasia in sextant needle biopsy specimens. Urology, 49(3): 367–73.

Wilt, T.J. and Brawer, M.K. (1997) The Prostate Cancer Intervention Versus Observation Trial (PIVOT). Oncology (Williston Park), 11(8): 1133–9; discussion 1139–40, 1143.

Wolf, J.S., Jr., et al. (1995) The use and accuracy of cross-sectional imaging and fine needle aspiration cytology for detection of pelvic lymph node metastases before radical prostatectomy. J Urol, 153 (3 Pt 2): 993–9.

Wu, C.L., et al. (2004) Analysis of alpha-methylacyl-CoA racemase (P504S) expression in high-grade prostatic intraepithelial neoplasia. Hum Pathol, 35(8): 1008–13.

Xu, J., et al. (2002) Germline mutations and sequence variants of the macrophage scavenger receptor 1 gene are associated with prostate cancer risk. Nat Genet, 32(2): 321–5.

Yatani, R., et al. (1982) Geographic pathology of latent prostatic carcinoma. Int J Cancer, 29(6): 611–6.

Yoshizawa, K., et al. (1998) Study of prediagnostic selenium level in toenails and the risk of advanced prostate cancer. J Natl Cancer Inst, 90(16): 1219–24.

Zaridze, D.G., Boyle, P. and Smans, M. (1984) International trends in prostatic cancer. Int J Cancer, 33(2): 223–30.

CHAPTER 14

HUMAN PREMATURE AGING DISEASES

Molecular biology to clinical diagnosis

DAI-DI GAN[1], MOHAMMAD HEDAYATI[1], TINNA STEVNSNER[2] AND VILHELM A. BOHR[1]

[1] *Laboratory of Molecular Gerontology, National Institute on Aging, NIH, Baltimore, USA*

[2] *Department of Molecular Biology, Aarhus University, Denmark*

Abstract: A number of rare human disorders are associated with distinct clinical features that resemble the aging process at an early stage in life. The study of these conditions has greatly advanced our insight into the aging process. Here, we discuss the clinical and molecular characteristics as well as the recent advances in our insight into the mechanisms of dysfunction in these diseases

Keywords: Aging, DNA repair, Werner syndrome, premature aging

Abbreviations: WS, Werner syndrome; WRN, Werner syndrome protein; RTS, Rothmund-Thomson syndrome; HGPS, Hutchinson-Gilford progeria; CS, Cockayne syndrome; BER, base excision repair; HR, homologous recombination repair; NHEJ, non-homologous end joining; TCR, transcription coupled repair; ROS, reactive oxygen species; 8-oxoG, 7,8-dihydroxyguanine, DSB, double strand breaks; Potl, Protection of telomere factor 1

1. INTRODUCTION

We are discussing human premature aging diseases, or segmental progerias. The latter name indicates that these conditions do not reflect all of the features of the normal aging process, but only a subset. Here, we describe clinical and molecular features of some of the prominent segmental progerias (Table 1), and we discuss the progress in this field and the challenges and complications of trying to understand the underlying molecular mechanism and in establishing the full clinical picture. These conditions are all very rare in the population, and thus in many cases there is not enough individuals to establish statistical significance.

S. Rattan and M. Kassem (eds.), Prevention and Treatment of Age-related Diseases, 271–295.

Table 1. Premature aging syndromes and their associated aging features

Aging features	WS	RTS	HGPS	CS	XP	TTD
Cataract	yes	yes	no	yes	yes	yes
Hair loss/Graying hair	yes/yes	yes/yes	yes/?	no	no	no
Skin aging	yes	yes	yes	yes	yes	yes
Osteoporosis	yes	?	yes	?	?	?
Cardiovascular diseases	yes	?	yes	yes	?	?
Diabetes	yes	no	yes	?	?	?
Hypogonadism	yes	yes	yes	yes	no	yes
Cancer	osteosarcoma	osteosarcoma	no	no	skin cancer	no

2. WERNER SYNDROME

2.1 Clinical features

The major clinical symptoms of Werner syndrome (WS; www.wernersyndrome.org/registry/registry.html) are: (1) appearance: short stature (dwarfism; usually less than 160 cm), thin extremities with smaller hands and shorter/deformed fingers and a pinched or beaked nose; (2) hair: graying and loss of hair, scanty eyebrows, absence of eyelashes; (3) eyes: cataracts, protuberant eyes; (4) pitch voice; (5) skin/muscle: scleroderma and wrinkled skin, ankle ulcers, muscle atrophy, soft tissue calcification, newly synthesized hyalinized collagen replaces the subcutaneous fat; (6) metabolism: type II diabetes mellitus, parathyroid glands disorder related osteoporosis, calcification of blood vessels (atherosclerosis/arteriosclerosis), hyperlipidemia, hypogonadism (appeared as: poorly developed genitalia/breasts and menstrual disorders), pituitary dysfunction; and (7) malignancies: osteosarcoma and soft tissue sarcoma (around 1 of 400 Japanese WS patients developed osteosarcoma and 70% of WS associated osteosarcoma are formed in the ankle or foot), melanomas, myeloid leukemia and myelodysplastic syndrome, epithelial neoplasm, carcinomas of the thyroid, and meningiomas. Cancers in lung, colon, and prostate, which are very often formed on elders, are rarely seen in WS patients. The average life span of WS patients is about 47 years. The principal causes of death are myocardial or cerebrovascular accidents and malignancy (Martin et al., 1999; Epstein et al., 1966; Goto et al., 1996; Ishikawa et al., 2000; Leone et al., 2005; Yamamoto et al., 2003).

These clinical symptoms will not appear until patients reach their puberty (about age 20). They then develop fully when the patients reach age around 30 to 40 (Martin et al., 1999; Epstein et al., 1966). The mechanism which associates with this delay in clinical phenotype development is still under investigation. Western blot and RT-PCR studies of human fetal and adult aorta samples showed that WRN mRNA were expressed at similar levels in all of the different age samples from normal individuals (Wang et al., 1999). Thus, these data support the hypothesis that WRN is expressed at all ages, whereas the WS phenotype becomes apparent after puberty. Another study, however, using immunohistochemistry and Western blot studies, performed with human pancreas, testis, and cortex of adrenal gland samples from normal individuals at the age of 11 to 32 years showed that WRN was only present after the age of 32 (Motonaga et al., 2002).

WS is an autosomal recessive disorder, which means the gene is located on not-sex related chromosomes and that the mutated gene functions as recessive. The Japanese population is more susceptible to WS (Martin et al., 1999; Epstein et al., 1966).

2.2 Gene

WS patients have mutations in the WRN (RECQL2) (RecQ3) (WRN) gene (Yu et al., 1996), which is located on chromosome 8p12-p11.2 and encodes a human homolog of the *Escherichia coli* (E. coli) RecQ DNA helicase named "Werner syndrome protein (WRN)". By multiple-tissue Northern blot analysis, the WRN gene is more highly expressed in pancreas, testis, ovary, muscle, placenta, and heart than in lung, brain, kidney, liver, and leukocytes (Yu et al., 1996; Furuichi, 2001). Since this tissue type specific gene expression pattern of pancreas, testis, and ovary correlates with WS clinical features of diabetes mellitus and early hypogonadism in both males and females, it has been suggested that WS is a helicase associated tissue-specific genomic instability disease (Furuichi, 2001).

WRN is a 1432 amino acids protein. It contains at least 7 domains: (1) an exonuclease domain which spans from amino acid 70 to 240, WRN has 3' to 5' exonuclease activity that can facilitate its helicase activity during in vitro DNA unwinding; (2) an acidic domain (also called "direct repeat domain) from amino acid 424 to 477, which functions in transcription activation with RNA polymerase II (Balajee et al., 1999; Ye et al., 1998); (3) a helicase motif which is located from amino acid 500 to 946, and contains a 3' to 5' helicase activity and an ATP-ase activity; (4) a RQC domain from amino acid 949 to 1092, which functions in protein-protein/protein-DNA interactions; (5) a HRDC domain which locates at amino acid 1072 to 1236 and has DNA binding activity; (6) a Nuclear localization signal sequence (NLS) from amino acid 1370 to 1375; the NLS functions as the recognition signal for the cytoplasmatically translated WRN to enter the nucleus; (7) Nucleolar targeting sequence (NTS): two NTS domains have been found to locate at amino acids 949 to 1092 (von Kobbe and Bohr, 2002) and 1403 to 1404 (Suzuki et al., 2001), which function as the entering signal to nucleolus. WRN activity can be regulated through protein-protein interactions, post-transcriptional modifications, and the presence of different metal cofactors (von Kobbe et al., 2003; Opresko et al., 2004; Choudhary et al., 2004; Lee et al., 2005) (for a WRN domain map, please see Figure 1).

Most of the known RECQL2 mutations result in a NLS deletion, or a C-terminal (which contains the NLS) truncated WRN mutant. These mutations have been termed "mutation 1 to 10" (for review, see Yu et al., 1996; Matsumoto et al., 1997). Northern blot studies, performed with WS patients' mRNA samples, showed

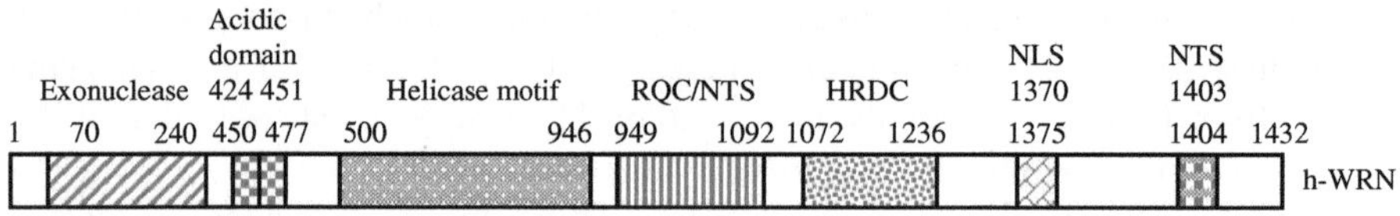

Figure 1. Schematic representation of human WRN protein. Each domain is highlighted with different patterns

reduced WRN mRNA intensity when WS patients are homozygous in 4/4, 6/6, and heterozygous in 1/4 RECQL2 mutations (Yamabe et al., 1997). These data may suggest that these mutations could result in degradation of mutated WRN mRNAs (Yamabe et al., 1997). However, the Northern blot probe that was used is within the 3' part of the WRN mRNA (Yamabe et al., 1997). Thus, it would be interesting to re-examine the presence of all types of mutant WRN mRNAs with a probe that recognizes the 5' part of the mRNA. Use of anti-N terminal and anti-C terminal WRN monoclonal antibodies showed that anti-N terminal WRN antibody could detect 1/1, 5/5, and 8/8 WRN mutant proteins but not 4/4, 6/6, and others (Goto et al., 1999). Thus, these data suggested the presence of 1/1, 5/5, and 8/8 WRN mutants in cells.

Due to lack of samples and sensitive methods, it is still not clear whether neurological abnormalities are part of WS clinical features (Postiglione et al., 1996; Leverenz et al., 1998; Mori et al., 2003; Payao et al., 2004; De Stefano et al., 2003). Sensitive immunohistochemistry methods showed that two female WS patients had increased amyloid deposits and plaque counts in the frontal cortex, parahippocampal gyrus, and hippocampus (Leverenz et al., 1998). Moreover, in the same studies, the number of amyloid deposits and plaque counts (counted from the parahippocampal gyrus and hippocampus) of a 57 year old WS patient were similar to 74 year old patients who had no WS but had sporadic Alzheimer's disease (Leverenz et al., 1998). Thus, these studies suggested that WS patients might have accelerated aging of the central nervous system. Yet, Alzheimer's symptoms are rarely reported with WS. These two female WS patients had a homozygous splice junction mutation, which results in a single exon deletion of amino acids 1047 to 1077, a deletion that is located in the RQC domain (Leverenz et al., 1998). This specific mutation also results in a premature stop codon at amino acid 1092, thus this particular WRN mutant protein contains not only a truncated RQC domain, but also a truncated HRDC domain, with no NLS sequence (Yu et al., 1996). Whether it has an intact NTS is unknown. Interestingly, in a different WS patient, a male at age 55 and with the same homozygous splice junction mutation, researchers reported no amyloid plaque formation in his central nervous system (Mori et al., 2003). Case studies showed that WS males are more likely to develop meningioma than females (Goto et al., 1996; Nakamura et al., 2005). Thus, WRN may function differently in brains from different sexes. Since WS syndrome does not develop until after puberty, which is also the time of human sexual development, it will be very interesting to investigate the association of sex hormones to WRN gene expression, functions, and WS clinical outcome.

There is some dispute as to whether polymorphisms in the amino acid 1367 cysteine polymorphic form of WRN (located between the HRDC and NLS domains) have a higher risk of vascular diseases. In a study in the Japanese population there was an association to myocardial infarction (Ye et al., 1997), but this was not the case in a Brazilian study of cardiovascular disease (Smith et al., 2005), nor in a North American study from the University of Washington of coronary artery disease (Castro et al., 2000), nor in male Caucasian patients from the Baltimore Longitudinal Study of Aging (BLSA) of coronary artery disease (Bohr et al., 2004). Different races may have different sensitivities toward polymorphism of amino acid 1367 on WRN and its specificity in vascular disease.

When comparing this Japanese 1367 polymorphism study (Ye et al., 1997) to the splice junction mutation studies (Leverenz et al., 1998; Mori et al., 2003), the data imply that different domains and regions of WRN may interact with different proteins in different tissues and have different functions in those tissues. Thus, it will be interesting to study different WRN polymorphism (Passarino et al., 2001)/mutation (Oshima, 2000) associated diseases to reveal functions of WRN in different tissues.

One of the particularly interesting WRN polymorphisms is R834C, which is located in the helicase motif (Kamath-Loeb et al., 2004). In in vitro assays, WRN R834C mutant protein showed a reduced helicase activity, reduced exonuclease activity, and reduced ATPase activity (Kamath-Loeb et al., 2004). This WRN R834C polymorphism is preferentially present in Spanish individuals (Kamath-Loeb et al., 2004). DNA sequencing studies showed that within 459 Spanish DNA samples, 6 of them are heterozygotes and 1 is a homozygote of the WRN R834C polymorphism (Kamath-Loeb et al., 2004). Thus, it will be very important to study the clinical reports and the data of physical examination of individuals who are WRN R834C homozygotic and compare their clinical data with those of WS patients. This approach will provide insight into WRN function in humans, its effect in different developmental stages.

2.3 Function

Studies in vitro and in vivo suggest that WRN is involved in DNA replication, DNA repair, telomere maintenance, and more. We will focus on some of these topics and discuss them below. However, this is only a limited overview since we have recently reviewed Werner functions thoroughly (Opresko et al., 2004; Opresko et al., 2005; Lee et al., 2005).

2.3.1 WRN is involved in DNA replication

Several observations suggest that WRN has potential activity in DNA replication. First, cells derived from WS patients have reduced frequency of replication initiation sites (Takeuchi et al., 1982) with extended S-phase (Takeuchi et al., 1982; Poot et al., 1992), and undergo premature replicative senescence in cell culture

(Salk et al., 1985; Martin et al., 1970). Second, in vitro helicase assays suggest that WRN can unwind a DNA substrate with replication fork structure (von Kobbe et al., 2003b). Moreover, WRN displays protein-protein interplay with lots of DNA replication related proteins, such as: RPA (Brosh Jr et al., 1999; Shen et al., 1998; Constantinou et al., 2000); proliferating cell nuclear antigen (PCNA) (Huang et al., 2000; Lebel et al., 1999), topoisomerase I (topo I) (Lebel et al., 1999), flap endonuclease 1 (FEN-1) (Brosh et al., 2001; Zheng et al., 2005), DNA polyermase delta (pol δ) (Szekely et al., 2000; Kamath-Loeb et al., 2000; 2001), and DNA polymerase beta (pol β) (Harrigan et al., 2003). In vitro studies have shown that RPA enhances WRN helicase activity (Brosh Jr et al., 1999). In vivo, WRN and RPA co-localize upon hydroxyurea induced replication fork arrest (Constantinou et al., 2000). Since WS cells are sensitive to the topoisomerase I inhibitor camptothecin (Laine et al., 2003) (CPT; blocks the replication fork by stabilizing the DNA topoisomerasei complex causing DNA double strand breaks), WRN may be involved in replication block resolution with topoisomerase I.

2.3.2 WRN is involved in DNA base excision repair (BER)

WRN has been found to interplay with several proteins that are involved in BER (Fan and Wilson III, 2005), one of them being poly(ADP-ribose)polymerase-1 (PARP-1) (von Kobbe et al., 2003a). PARP-1 suppresses WRN helicase activity and exonuclease activity (von Kobbe et al., 2004a). Furthermore, WS primary fibroblasts are deficient in poly(ADP-ribosyl)ation after hydrogen peroxide (H_2O_2) treatment (von Kobbe et al., 2003a) (which generates oxidative DNA lesion). WS cells have lower activity of 5-hydroxymethyluracil (HMU) glycosylase activity (Ganguly et al., 1992), which might result in an inefficiency in removing HMU. GST-pull down assay, ELISA, and dot blot methods have shown that WRN binds to APE1 (Ahn et al., 2004). Also, WRN and APE1 are co-localized in the nucleus (Ahn et al., 2004). Helicase assays showed that APE1 suppresses WRN helicase activity in a DNA substrate structure specific manner, in which BER DNA substrates would be affected by the presence of APE1 (Ahn et al., 2004). Interestingly, the presence of pol β could release the inhibitory activity of APE1 on WRN in the in vitro helicase assay (Ahn et al., 2004) suggesting an interplay of these three proteins. In addition, WRN binds and stimulates pol β strand displacement DNA synthesis at a nick on a BER substrate (Harrigan et al., 2003). In vivo FRET analysis showed that WRN interacts with FEN-1 in 4-nitroquinoline-1-oxide (4-NQO; DNA damaging agent which generates adducts in DNA that require NER/BER repair) treated cells and WRN stimulates the flap cleavage activity of FEN-1 in vitro (Sharma et al., 2004). These in vivo and in vitro studies of the interplays between WRN/FEN-1 imply that WRN is mainly involved in the long patch BER subpathway.

2.3.3 Homologous recombination repair (HR)

DNA damaging agents, ionizing radiation (IR), DNA cross-linking agents, enzymatic activity, and DNA replication errors during proliferation can induce DNA double strand breaks (DSB) in vivo. In the presence of a DNA template with certain homologous sequences, cells can repair DNA DSB by HR. Rad51 is the major protein in HR, which binds and stabilizes single-strand DNA for strand exchange during HR (West, 2003).

Cells derived from WS patients are sensitive to the DNA cross-linkers Mitomycin C (MMC) (Sharma et al., 2004) and ionizing radiation (IR) (Saintigny et al., 2002). In vivo studies suggested that WRN and Rad51 co-localize in Rad51 foci (Sakamoto et al., 2001). In addition, exogenously expressed dominant-negative Rad51 could reverse the DNA damage sensitivity of WS cells (Saintigny et al., 2002). These studies suggest that WRN plays an important role in Rad51 associated HR.

The Mre11/Rad50/Nbs1 complex is the other important player in HR. It has specificity to IR induced DNA damage HR repair, in which Mre11/Rad50/Nbs1 complex interplay with the ataxia-telangiectasia mutated (ATM) kinase in the detection of DNA DSB. The Mre11/Rad50/Nbs1 complex binds directly to telomere repeat binding factor 2 (TRF2; a regulator of telomere function) and telomeres specifically during S-phase. The Mre11/Rad50/Nbs1 complex also functions in telomere maintenance. Mutations in Nbs1 have been linked to Nijmegen breakage syndrome (NBS) with the clinical features of neuronal abnormality, neuronal degeneration, microcephaly, cancer predisposition, and immunodeficiency. Studies of NBS patient cells showed that NBS cells have shorter telomeres, which echo the TRF2-Mre11/Rad50/Nbs1 complex studies. Mutations in Mre11 can result in ataxia-telangiectasia-like disease (ATLD), which has a clinical phenotype similar to ataxia-telangiectasia (AT) and NBS. ATLD patients are also predisposed to cancer (D'Amours and Jackson, 2002; Lavin, 2004; Assenmacher and Hopfner, 2004; Kobayashi et al., 2004; Stracker et al., 2004).

Extensive interplay is going on between WRN and the Mre11/Rad50/Nbs1 complex (Cheng et al., 2004; Cheng et al., 2005; Franchitto and Pichierri, 2004; Franchitto and Pichierri, 2002). WRN and Mre11 co-localize in the nucleus (Franchitto and Pichierri, 2004). Furthermore, WRN binds to the Mre11/Rad50/Nbs1 complex and has a specific interaction with Nbs1 (Cheng et al., 2004). Also, WRN co-localizes with Nbs1 after IR treatment, and the Mre11/Rad50/Nbs1 complex enhances WRN helicase activity in vitro (Cheng et al., 2004).

2.3.4 *Telomere maintenance*

The telomere is the end of the eukaryotic linear chromosome with a specific DNA sequence, structure and associated proteins. Human telomeres consist of 5–15 kb of TTAGGG tandem repeats and end in a 3' single strand G-rich tail. This G-rich tail loops back and invades the telomeric duplex, which forms the intra-telomeric D-loop and a large lasso-like t-loop structure. Several proteins have been found to bind to the telomere, including telomere repeat binding factors 1 and 2 (TRF1 and TRF2) that bind to duplex (TTAGGG)n DNA and participate in the regulation of telomere length. Human protection of telomeres-1 (POT 1) protein binds specifically to telomeric single stranded DNA. These specific telomeric DNA sequence/structure/proteins complexes protect the end of the linear chromosome and prevent telomere dysfunction (Griffith et al., 1999; Kuimov, 2004; Baumann and Cech, 2001; Lei et al., 2004; Loayza et al., 2004).

Several in vitro and in vivo studies have suggested that WRN is involved in telomere maintenance: (1) TRF2 binds to WRN and stimulates the helicase activity of WRN when incubated with telomeric duplex substrate (Opresko et al., 2002; Machwe et al., 2004); (2) WRN associates to telomeres in human alternative lengthening of telomeres (ALT) cell lines (Opresko et al., 2004); (3) WRN associates to telomeres in S-phase of human primary fibroblasts (Crabbe et al., 2004); (4) WRN associates with the telomere lagging strand synthesis (Crabbe et al., 2004); (5) in vitro, POT 1 binds and interplays with WRN in the unwinding of telomeric forked duplexes and D-loop structure substrates with specificity towards telomere sequence and native D-loop structure (Opresko et al., 2005).

Studies of telomere erosion rates from primary WS fibroblasts suggest that WS cells have normal telomere erosion rate (Baird et al., 2004). Thus, these studies imply that WS cells associated in vitro pre-mature replicative senescence may be affected not only by telomeres but also by additional factors – perhaps, through the interplay with a p53 associated reactive oxygen species associated cellular senescence (for discussion please see below).

2.3.5 *p53/p21$^{Waf1/Cip1/Sdi1}$/reactive oxygen species (ROS) associated cellular senescence*

p53 is a transcription factor that is involved in DNA repair, DNA check point regulation, apoptosis, and cellular senescence (Kulju and Lehman, 1995; Sugrue et al., 1997; Gomez-Lazaro et al., 2004). Studies from both tumor cell lines and primary cells have shown that: (1) near-senescent human primary diploid fibroblast cultures have a higher protein level of p53 (Kulju and Lehman, 1995; Sugrue et al., 1997); (2) over-expressed p53 can result in the accumulation of ROS (Polyak et al., 1997); (3) p53 can induce the expression of cyclin-dependent kinase (CDK) inhibitor p21$^{Waf1/Cip1/Sdi1}$ (el-Deiry et al., 1993); (4) the expression of p21$^{Waf1/Cip1/Sdi1}$ can result in the accumulation of ROS in normal human fibroblasts and induce the ROS associated senescence (Macip et al., 2002).

Interestingly, several studies have shown that p53 interplays with WRN: (1) p53 down-regulates the gene expression of WRN (Yamabe et al., 1998); (2) p53 binds to WRN (Sommers et al., 2005); (3) WRN helicase activity is suppressed in the presence of p53 (Sommers et al., 2005). Moreover, senescent primary WS fibroblasts have a higher protein level of p21$^{Waf1/Cip1/Sdi1}$ (Davis et al., 2003). These senescent cells can re-enter the cell cycle by microinjection of a p53-neutralizing antibody (Davis et al., 2003). Experiments have also shown that ascorbic acid (an antioxygenic reagent) could delay cellular senescence in cultured normal human embryonic cells, human adult skin fibroblasts, and in a WS cell strain (Kashino et al., 2003). Interestingly, hydrogen peroxide (H_2O_2; a common ROS intermediate; when cells treated with H_2O_2 and generated/accumulated high amount of DNA damage, cells would enter irreversible proliferation arrest and premature senescence) treated WS primary cells and WRN depleted normal diploid fibroblasts showed an escape of the H_2O_2-induced cell proliferation arrest with a lack of the features of p53 and $p21^{Waf1/Cip1/Sdi1}$ accumulation in H_2O_2 treated normal diploid fibroblasts (von Kobbe et al., 2004b). These data could imply that WS cells have specific deficiency in sensing H_2O_2/ROS.

3. ROTHMUND-THOMSON SYNDROME

In 1868, August von Rothmund, Jr. discovered a familial syndrome characterized by cataracts, a depressed nasal bridge, and skin hypertrophy. Later on, in 1923, Matthew Sydney Thomson reported a similar disorder but without any cataracts. In 1957, William Taylor recognized that these were similar disorders and called it "Rothmund-Thomson syndrome (RTS)". There are several good electronic databases for RTS, such as: www.infobiogen.fr/services/chromcancer/Kprones/RothmundID10021.html.

3.1 Clinical Features

RTS is a rare, autosomal recessive disorder and to date there are only about 300 reported cases (www.geneclinics.org/profiles/rts/details.html#gcID1619). Patients have features of congenital poikiloderma which includes photosensitivity. Skin rash begins on the face and the cheeks with erythema, swelling and bullae. These symptoms usually appear around 3 to 6 months of age, but in some patients the symptoms may appear earlier just after birth or later around age 2. The skin rash can spread to the buttocks and flexural areas of the extremities. The rash then enters a chronic phase with the features of punctuate skin atrophy, telangiectasia, and hypo- or hyperpigmentation which persist throughout life. Other clinical features are: congenital cataracts, saddle nose, disturbances of hair growth, early graying and hair loss (partial to total alopecia), defective nails and teeth, short stature, and skeletal defects such as radial ray hypoplasia and absent thumbs. Many of these characteristics are consistent with premature aging. About one third of the RTS patients develop osteosarcoma with the median age of onset of about 9 to 11 years. Some RTS patients may have infertility problems, yet normal pregnancies have been reported in some cases. Immunological functions and intelligence appear normal in most RTS patients. Life span, in the absence of malignancy, is probably normal but more follow-up studies are needed to confirm this (http://www.ncbi.nlm.nih.gov/books/bv.fcgi?rid=gene.chapter.rts) (Wang et al., 2003; Spurney et al., 1998; Wang et al., 2001; Gelaw et al., 2004; Roth et al., 1989; Sim et al., 1992).

3.2 Gene

The RTS mutation was mapped to a gene called RecQL4, which is located on chromosome 8q24.3 (Kitao et al., 1999). This gene is another member of the RecQ family of helicases. It encodes a 1208 amino acid protein called RECQ4, which has been shown to be localized mainly in the nucleus (Kitao et al., 1999; Kitao et al., 1998). As with other RECQ helicases, RecQL4 has a central helicase domain, containing the seven conserved helicase motifs. The N- and C-terminal regions do not show any striking similarity with the other RecQ helicases. In RTS patients, RecQL4 mutations have been shown to be located at exons 5,8,10,11,12,13,14,15,19, and 21 (Wang et al., 2003; Kellermayer et al., 2005). Many of these mutations are mapped to the helicase domain (exons 8–14). Interestingly, mutations in the RecQL4 gene have also been shown to be associated with The RAPADILINO (Siitonen et al., 2003) and Baller-Gerold syndromes (Van Maldergem et al., 2005). These syndromes share some of the same characteristics with the RTS such as growth deficiency and radial ray defects but also display some distinct differences. In one study, osteosarcomas were observed only in RTS and RAPADILINO while cataracts were unique to RTS (Van Maldergem et al., 2005). RecQL4 mutation at a splice site which causes an in-frame skipping of exon 7 has been found in RAPADILINO syndrome (Siitonen et al., 2003). Mutations in exon 9 and exon 18 have been associated with Baller-Gerold syndromes (Van Maldergem et al., 2005). It is tempting to speculate that different mutation patterns of the RecQL4 gene in these three syndromes could in part explain the differences in the observed phenotypes. Further mapping of the mutations in RecQL4 gene and a better understanding of the RECQ4 function should provide useful insights into the etiology of these syndromes as well as mechanisms that lead to premature aging.

An interesting distinction between RTS and WS is that the symptoms of WS do not appear until after puberty, whereas the symptoms of Rothmund Thompson syndrome start in early childhood. Comparing the gene expression pattern and protein function of WRN and RECQ4 in vivo, could also enhance our understanding of the human aging process.

3.3 Function

RECQL4 mRNA is detected in most tissues in the body with higher expression levels in thymus, testis, and placenta; and moderate levels in heart, brain, small intestine, and colon. (Furuichi, 2001) (http://www.ncbi.nlm.nih.gov/unigene/clust.cgi?org=hs&cid=31442). RECQ4 may have an important function in osteoblasts since many of the RTS patients have joint and skeletal defects and a high incidence of osteosarcoma. Lymphoblasts and fibroblasts, obtained from some RTS patients, show a normal karyotype. However, Some RTS cells show genomic instability with high frequency of chromosomal rearrangements. This suggests that RECQ4 may play an important role in maintaining genomic stability (http://www.ncbi.nlm.nih.gov/books/bv.fcgi?rid=gene.chapter.rts) (Lindor et al. 2000; Beghini et al., 2003).

Recently, it was shown that the RECQ4 expression could be regulated by p53. Upon camptothecin treatment, the p53/Sp1 complex leaves the RECQL4 promoter leading to the down-regulation of the RecQL4 expression (Sengupta et al., 2005).

Interestingly, in the same study, the RECQ4 protein level was also down-regulated when the cell cycle was arrested by contact inhibition (Sengupta et al., 2005). The contact inhibition induced down-regulation of RECQ4 was not affected by the expression of human papillomavirus 16 E6 protein (Sengupta et al., 2005) (E6 degrades p53). In addition, the RECQ4 protein level was up-regulated faster in cells that contained E6 when released from contact inhibition induced cell cycle arrest (Sengupta et al., 2005). These data suggest that cells also have a contact inhibition inducible, p53 independent down-regulation of RECQ4, implying the importance of RECQ4 in cell cycle progression and cell proliferation.

In primary skin fibroblasts, RECQ4 protein was shown to be localized to the nucleus and form distinct foci (Petkovic et al., 2005). In etoposide (a DNA damaging agent) treated HeLa cells, the RECQ4 protein was shown to co-localize with promyelotic leukaemia protein nuclear bodies (PML) and RAD51, but not with BRCA1 (Petkovic et al., 2005). These data suggest that RECQ4 may play an important role in the repair of double strand breaks, but in a manner which may be different from that of BRCA1.

Studies of transgenic mice show that the defect in RECQL4 could be associated with cancer formation and premature centromere separation. These studies suggest that RECQ4 may play an important role in both genomic stability and sister-chromatid cohesion (Mann et al., 2005). Interestingly, exon 9–13 disrupted transgenic mice had the phenotypes of bone abnormality, cancer predisposition (osteosarcoma and lymphomas in RecQL4 mutation background and macroade-nomas with the genetic background of both RecQL4 mutation and APC mutations) and poikiloderma, but without graying, hair loss, or growth defects (Mann et al., 2005). In contrast, exon 13 disrupted transgenic mice had the features of bone cell abnormality, graying/hair loss, growth defect, immune system defect, yet no UV/IR sensitivity, no poikiloderma, and no cancer formation (Hoki et al., 2003). The differences between these two strains of transgenic mice are interesting and may imply that exon 13 is associated with cell growth and exon 9–12 is related to DNA repair or apoptosis. These animal models may reveal explanations for the variation of RTS patient clinical features. Future studies of these animal models as well as studies on the phenotypic variations in the RTS patients should provide useful clues about the function of RECQ4.

4. HUTCHINSON-GILFORD PROGERIA

4.1 Clinical Features

Hutchinson-Gilford progeria (HGPS) is an autosomal dominant disease (Progeria Research Foundation's medical and research databases: www.progeriaresearch.org; www.genereviews.org). Different from WS patients, HGPS patients display their senile appearance and clinical features before puberty; some have the symptoms already from birth. Their clinical features are: hair loss (alopecia), growth retardation, skin aging appearance, disproportionately large head, pinched facial features, lipodystrophy, incomplete extension at the knees and elbows indicating stiffness of joints, bone deformations, osteoporosis, delayed dentition, hip dislocations, sclerodermatous areas, and atherosclerosis. In the final stage of the disease, patients will have hypertension, angina, and atherosclerotic heart disease. Usually, patients die by coronary artery disease (such as myocardial infarction or stroke) at an average age of 13 (http://www.ncbi.nlm.nih.gov/books/bv.fcgi?rid=gene.chapter.hgps).

4.2 Gene

The HGPS gene had not been identified until 2003 (Eriksson et al., 2003). It was reported that LMNA, which is located on chromosome 1q21.2, is the HGPS gene (Eriksson et al., 2003). The LMNA gene produces 4 different proteins by alternative splicing: lamin A and lamin C are the major products, lamin AΔ10 and lamin C2 are the minor products (Gruenbaum et al., 2005; Burke and Stewart, 2002). The expression of A-type lamins is low or absent in highly proliferating cells and cells with low degree of differentiation (Broers et al., 1997). Once LMNA transcribes/translates prelamin A protein, it will go through a posttranslational farnesylation modification process before targeting to the nuclear envelope. Then, the farnesylated prelamin A is methylated before its last 15 amino acids are clipped off by ZMPSTE24 (a metalloproteinase) yielding mature lamin A that can be incorporated into lamina. Lamin A is a type V intermediate filament protein which has an N-terminal "head" domain, an alpha-helical "central rod" domain, and a globular tail domain. Lamins form dimers through parallel and in-register coil-coil interaction between central rod domains. These dimers associate in a head-to-tail fashion to form protofilaments that associate to form higher-order structures in nuclear lamina (the meshwork of filaments which is located at the inner layer of the nuclear membrane) (Gruenbaum et al., 2005; Burke and Stewart, 2002).

4.3 Function

Although it was thought that lamin A functions as the structure supporter protein in lamina, it has been observed that lamin A distributes throughout the nucleoplasm, binds to the retinoblastoma protein (Mancini et al., 1994), interacts with the RNA polymerase II transcription complex, and may be involved in gene transcription (Csoka et al., 2004; Kumaran et al., 2002). DNA repair studies (from both patients and mouse cell strains) suggested that lamin A may be involved in Rad51 associated homologous recombination DNA repair but not in non-homologous end-joining DNA repair (Johnson et al., 2004; Liu et al., 2005). Telomere length studies have shown that cells from HGPS patients have shorter telomeres than age matched controls (Allsopp et al., 1992), potentially suggesting that lamin A may function in telomere maintenance.

Most of the known HGPS cases contain a mutation in exon 11 of LMNA (Eriksson et al., 2003; Goldman et al., 2004). This common mutation, C1824T in LMNA cDNA (G608G on lamin A peptide), partially activates a cryptic splice site that can delete 150 nt in exon 11 and results in a deletion mutant of lamin A that lacks 50 amino acids at the C-terminal which includes the endoproteolytic site for ZMPSTE24 but retains the farnesylation site (Eriksson et al., 2003; Goldman et al., 2004). Thus, the lamin A mutant protein (called "progerin") can be farnesylated and targeted to the nuclear envelope, but cannot be endoproteolytically cleaved by ZMPSTE24. It has been observed that progerin accumulated on the nuclear envelope and associated with blebs, an abnormal nuclear envelope structure (Goldman et al., 2004). Besides the bleb formation, HGPS patient cells also show visible abnormalities of the interrupted nuclear membrane, lobulation of the nuclear envelope, thickening of the nuclear lamina, altered nuclear sizes and shapes, clustering of nuclear pores, loss of peripheral heterochromatin, and chromatin extrusion (Goldman et al., 2004). HGPS derived cells have abnormalities in cell proliferation and the degree of apoptosis (Bridger and Kill, 2004), yet lack of abnormality in protein synthetic errors (Harley et al., 1980).

Interestingly, mutations of G608G, E145K, R471C, and R527C can result in HGPS, however, a R527H mutation, which is the same site as R527C of HGPS, gives the clinical features of autosomal recessive mandibuloacral dysplasia (MADA): white fat atrophy, insulin resistance, skeletal malformations, and alopecia. MADA patients have no life span shortening (Burke and Stewart, 2002). Different mutations on LMNA can result in different diseases, grouped into 3 classes: (1) with adipose tissues features: MADA and familial partial lipodystrophy (FPLD; loss of adipose tissue, insulin resistance, and diabetes mellitus); (2) striated muscle disorders: an autosomal form of Emery-Dreifuss muscular dystrophy (EDMD), limb girdle muscular dystrophy (LGMD1B), and dilated cardiomyopathy (DCM); (3) with peripheral neuropathy: Charcot-Marie-Tooth disorder type 2 (CMT2; axonal neuropathy) (Burke and Stewart, 2002; Sullivan et al., 1999; De Sandre-Giovannoli et al., 2003). Mutations A57P, R133L, and L140R, which are located on the N-terminal of the lamin A peptide, give the clinical features of atypical Werner syndrome (Chen et al., 2003). Mutations on T10I, E578V, and R644C result in atypical progeroid syndromes (Csoka et al., 2004) (for all mutation sites, see Burke and Stewart, 2002). Cells from most of the patients with these diseases share the feature of nuclear shape abnormality. Most lamin A mutations that are associated with premature aging syndromes affect function/development of skin, hair, fat, muscle, bone and the cardiovascular system.

The observation that mutations in LMNA can result in different disease outcomes, implies that different mutations can disrupt different functions of lamin A. Perhaps one of its functions can be impaired without affecting others, or generate a gain-of-function-mutant in certain situations. The 3-D structure analysis of lamin A has shown that FPLD associated lamin A mutation sites are mainly located on a discrete solvent-exposed path on the surface of the domain. In contrast, EDMD associated lamin A mutation sites are mainly located within the hydrophobic core or participate in the salt bridge formation (Burke and Stewart, 2002) (for view the 3D structure of lamin A, see Burke and Stewart, 2002).

5. COCKAYNE SYNDROME (CS)

5.1 Clinical Features

Cockayne syndrome is a rare autosomal recessive disease, originally described by Dr. Edward Alfred Cockayne in the 1930s. Approximately 180 cases of CS have been reported from different parts of the world, with no apparent overrepresentation in any specific population (reviewed in Licht et al., 2003). CS patients have major clinical features including slow growth and dwarfism due to defects in bone formation, a premature aging phenotype of progeriod appearance, cutaneous photosensitivity, microcephaly (defects in brain/CNS cells development), neuronal demyelination (leukodystrophy, the loss of oligodendrocytes), and mental retardation. CS patients also express a variety of other clinical phenotypes such as: long limbs, large hands and feet, flexion contractures of joints, dry hair, deafness, retinal degeneration, atherosclerosis, dental caries, hypertension, renal disease, and decreased thyroid hormone. Although CS patients have cutaneous photosensitivity, there are no reports of increased skin cancer formation (Stefanini et al., 1996; Tan et al., 2005; Mahmoud et al., 2002; Mizuguchi and Itoh, 2005; Komatsu et al., 2004). However, due to the increased apoptosis propensity, mutated cells may be eliminated before they become cancer cells (Licht et al., 2003; D'Errico et al., 2005). The clinical symptoms of CS appear just after birth and the mean age of CS patients is around 12.5 years. Most CS patients die with the disease of atherosclerosis, but the most common cause of death is pneumonia (reviewed in Licht et al., 2003).

5.2 Gene

CS has two complementation groups: 20% of the CS patients are type A (CSA), who have mutations in the 46 kDa CSA protein (a WD-repeat family protein; 396 amino acid protein; functions as a ubiquitylation E3 ligase); 80% of the CS patients are type B (CSB) patients who have mutations in the 168 kDa CSB protein (1493 amino acids protein), a Swi/Snf-like DNA-dependent ATPase that belongs to SNF2 protein family (Licht et al., 2003; Cleaver et al., 1999; Mallery et al., 1998; Christiansen et al., 2003; Eisen et al., 1995; Ren et al., 2003; Cao et al., 2004; Ridley et al., 2005). The different mutations in the *CSB* gene is not only linked to CS, but also to i) the cerebro-oculo-facio-skeletal syndrome (OMIM#214150; Meira et al., 2000) (COFS; with the specific clinical features of little or no postnatal neurological development (microcephaly), growth retardation, early postnatal contractures of the spine and joints, some patients may have cataracts or other types of eye structural defects); ii) the DeSanctis-Cacchione severe neurological form of XP (Colella et al., 2000; OMIM#278800) (DS-C; with the specific clinical features of xeroderma pigmentosum, mental deficiency, progressive neurologic deterioration, dwarfism, and gonadal hypoplasia); and iii) UV-sensitive syndrome (Spivak, 2004; Horibata et al., 2004) (UV^sS; with the clinical features of photosensitivity and mild freckling but without neurological abnormalities or skin cancer predisposition). The mutation sites of CSB are mainly located to the central part of the CSB protein; the mutation sites which link to COFS are also located at the central part; the DS-C patients have mutation sites at the C-terminal part of the CSB protein; and the mutation site of UV^sS is located at the N-terminal of the CSB protein (Spivak, 2004) (for all mutation sites on the CSB protein, Spivak, 2004). The UV^sS-CSB mutant is a 76 amino acids protein (Horibata et al., 2004), thus, the protein has a molecular weight of about 8 kDa. It is interesting to speculate how this UV^sS-CSB mutated gene can be transcribed/translated into a protein and only cause such mild clinical symptoms.

5.3 Function

CS patients have major defects in brain, skin, and bone, but the CSB cDNA can be found in many organs and tissues. CSB protein can interact several other proteins, implying that it functions in a number of different pathways (Licht et al., 2003).

Cells derived from CS patients have increased sensitivity to UV irradiation, defects in recovery of RNA synthesis after UV irradiation, loss of the preferential repair of active genes but normal ability to repair the overall genome DNA after damage induced by e.g. UV-light (Licht et al., 2003). Kyng et al. have demonstrated that the transcriptional response after oxidative stress is defective in CSB deficient cells (Kyng et al., 2003), and furthermore, it has been demonstrated that repair of certain oxidative lesions, such as 7,8-dihydroxyguanine (8-oxoG) and 7,8-dihydroxyadenine is decreased in these cells (Dianov et al., 1999; Sunesen et al., 2002). We have also demonstrated that mitochondrial repair of 8-oxoG is deficient in CSB deficient cells (Stevnsner et al., 2002). Furthermore, CSB is implicated in the PARP-1 poly(ADP-ribosyl)ation response after oxidative stress (Thorslund et al., 2005). As patients with CS suffer from dramatic neurodegeneration and a variety of clinical features associated with progeria, we speculate that the reduced capability to respond to oxidative damage in the absence of CSB may contribute to these CS phenotypes.

Members of the SNF2-like family of DNA dependent ATPases, including CSB, contain seven characteristic motifs, I, Ia and II–VI, that are also present in DNA and RNA helicases (Eisen et al., 1995). However, helicase activity has not been demonstrated for any members of the SNF2-like family of DNA dependent ATPases. Mutations in motifs I and II of CSB eliminate the ATPase activity (reviewed in Licht et al., 2003), but interestingly an E646Q mutation in motif II, which eliminates the ATPase activity of CSB, does not affect the mitochondrial repair of 8-oxoG (Stevnsner et al., 2002; Selzer et al., 2002). This suggests that CSB may have separate roles in transcription coupled repair of e.g. UV lesions and in the repair of oxidative lesions such as 8-oxoG.

Under normal conditions the CSB protein is phosphorylated, but UV irradiation of cells leads to its dephosphorylation (Christiansen et al., 2003). The dephosphorylation of CSB in vitro results in increased ATPase activity of the protein, suggesting that the activity of CSB is subject to phosphorylation control in vivo. Very recently, we demonstrated that CSB forms homodimers in vitro and in vivo, and that the ATPase activity of CSB elutes as a dimer when gel filtration chromatography analysis is performed (Christiansen et al., 2005). Beerens et al. reported that CSB wraps DNA around its surface and ATP hydrolysis leads to unwrapping (Beerens et al., 2005). Size analysis of scanning force microscopy pictures indicated that the DNA was wrapped around two CSB molecules at a time (Beerens et al., 2005). These observations have important implications for the mechanism of action of CSB.

The CSB protein has been suggested to promote RNA Pol II, as well as RNA Pol I and RNA pol III transcription (reviewed in Licht et al., 2003). The protein has also been suggested to remodel the DNA-pol II interface to allow DNA repair of some types of damage. Finally, CSA and CSB seem to be implicated in the ubiquitination of RNA pol II after treatment of cells with UV (Bregman et al., 1996). Thus, CSB is suggested to stimulate elongation when an RNA polymerase is

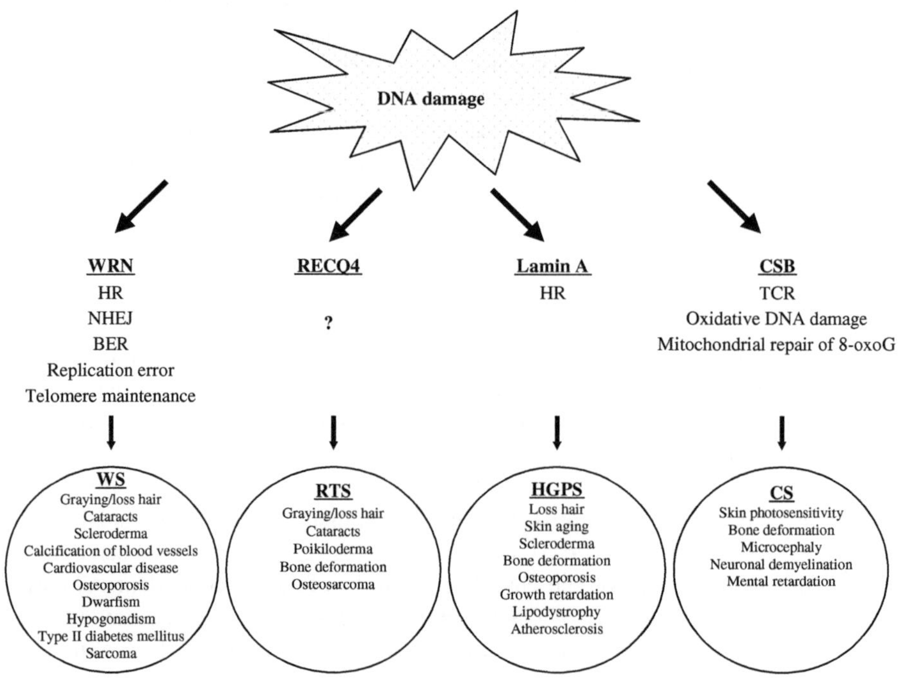

Figure 2. Association of premature aging syndromes and DNA repair pathways

paused at a natural pause site or strong RNA secondary structures. A part of the role in transcription coupled repair is likely to be the removal of RNA polymerase II by ubiquitination and proteosomal degradation of the large subunit. This degradation may be necessary for the cellular recovery of RNA synthesis after polymerase blocking damage to DNA (reviewed in Licht et al., 2003). Finally, Citterio et al. found that CSB has chromatin remodeling activity *in vitro*, and this activity is dependent on a functional motif I (Citterio et al., 1998; Citterio et al., 2000). The chromatin remodeling activity may have implications for the role of CSB in the DNA repair processes.

6. CONCLUDING REMARKS

A central hypothesis of aging suggests that the genomic instability and other molecular and clinical aspects of the aging phenotype is associated with accumulated DNA damage and maybe also with damage accumulated in other macromolecules. Here, we have discussed the molecular and clinical features associated with some of the significant human progerias. It is evident that these conditions involve defects in the DNA repair mechanisms at the molecular level, and thus this supports the possibility that DNA damage accumulates with age in those patients more than it does in normals. This notion is also illustrated in Figure 2, where we have indicated deficiencies in these pathways. Future studies need to be directed at further establishing these connections and development of therapeutic strategies to help these patients.

REFERENCES

Assenmacher, N., Hopfner, K.P. (2004) MRE11/RAD50/NBS1: complex activities. Chromosoma., 113: 157–66.

Ahn, B., Harrigan, J.A., Indig, F.E., Wilson, DM III and Bohr, V.A. (2004) Regulation of WRN helicase activity in human base excision repair. J Biol Chem., 279: 53465–74.

Allsopp, R.C., Vaziri, H., Patterson, C., Goldstein, S., Younglai, E.V., Futcher, A.B., Greider, C.W. and Harley, C.B. (1992) Telomere length predicts replicative capacity of human fibroblasts. Proc Natl Acad Sci USA., 89: 10114–8.

Baird, D.M., Davis, T., Rowson, J., Jones, C.J. and Kipling, D. (2004) Normal telomere erosion rates at the single cell level in Werner syndrome fibroblast cells. Hum Mol Genet., 13: 1515–24.

Balajee, A.S., Machwe, A., May, A., Gray, M.D., Oshima, J., Martin, G.M., Nehlin, J.O., Brosh, R., Orren, D.K. and Bohr, V.A. (1999) The Werner syndrome protein is involved in RNA polymerase II transcription. Mol Biol Cell., 10: 2655–68.

Baumann, P. and Cech, T.R. (2001) Pot1, the putative telomere end-binding protein in fission yeast and humans. Science., 292: 1171–5.

Beerens, N., Hoeijmakers, J.H., Kanaar, R., Vermeulen, W. and Wyman, C. (2005) The CSB protein actively wraps DNA. J Biol Chem., 280: 4722–9.

Beghini, A., Castorina, P., Roversi, G., Modiano, P. and Larizza, L. (2003) RNA processing defects of the helicase gene RECQL4 in a compound heterozygous Rothmund-Thomson patient. Am J Med Genet A., 120: 395–9.

Bohr, V.A., Metter, E.J., Harrigan, J.A., von Kobbe, C., Liu, J.L., Gray, M.D., Majumdar, A., Wilson, D.M., III and Seidman, M.M. (2004) Werner syndrome protein 1367 variants and disposition towards coronary artery disease in Caucasian patients. Mech Ageing Dev., 125: 491–6.

Bregman, D.B., Halaban, R., van Gool, A.J., Henning, K.A., Friedberg, E.C. and Warren, S.L. (1996) UV-induced ubiquitination of RNA polymerase II: a novel modification deficient in Cockayne syndrome cells. Proc Natl Acad Sci USA., 93: 11586–90.

Broers, J.L., Machiels, B.M., Kuijpers, H.J., Smedts, F., van den Kieboom, R., Raymond, Y. and Ramaekers, F.C. (1997) A- and B-type lamins are differentially expressed in normal human tissues. Histochem Cell Biol., 107: 505–17.

Brosh, R.M., Jr., von Kobbe, C., Sommers, J.A., Karmakar, P., Opresko, P.L., Piotrowski, J., Dianova, I., Dianov ,G.L. and Bohr, V.A. (2001) Werner syndrome protein interacts with human flap endonuclease 1 and stimulates its cleavage activity. EMBO J., 20: 5791–801.

Brosh, R.M., Jr. Orren, D.K., Nehlin, J.O., Ravn, P.H., Kenny, M.K., Machwe, A. and Bohr, V.A. (1999) Functional and physical interaction between WRN helicase and human replication protein A. J Biol Chem., 274: 18341–50.

Bridger, J.M. and Kill, I.R. (2004) Aging of Hutchinson-Gilford progeria syndrome fibroblasts is characterised by hyperproliferation and increased apoptosis. Exp Gerontol., 39: 717–24.

Burke, B., Stewart, C.L. (2002) Life at the edge: the nuclear envelope and human disease. Nat Rev Mol Cell Biol., 3: 575–85.

Castro, E., Edland, S.D., Lee, L., Ogburn, C.E., Deeb, S.S., Brown, G., Panduro, A., Riestra, R., Tilvis, R., Louhija, J., Penttinen, R., Erkkola, R., Wang, L., Martin, G.M. and Oshima, J. (2000) Polymorphisms at the Werner locus: II. 1074Leu/Phe, 1367Cys/Arg, longevity and atherosclerosis. Am J Med Genet., 95: 374–80.

Cao, H., Williams, C., Carter, M. and Hegele, R.A. (2004) CKN1 (MIM 216400): mutations in Cockayne syndrome type A and a new common polymorphism. J Hum Genet., 49: 61–3.

Chen, L., Lee, L., Kudlow, B.A., Dos Santos, H.G., Sletvold, O., Shafeghati, Y., Botha, E.G., Garg, A., Hanson, N.B., Martin, G.M., Mian, I.S., Kennedy, B.K. and Oshima, J. (2003) LMNA mutations in atypical Werner's syndrome. Lancet., 362: 440–5.

Cheng, W.H., von Kobbe, C., Opresko, P.L., Arthur, L.M., Komatsu, K., Seidman, M.M., Carney, J.P. and Bohr, V.A. (2004) Linkage between Werner syndrome protein and the Mre11 complex via Nbs1. J Biol Chem., 279: 21169–76.

Cheng, W.H., Sakamoto, S., Fox, J.T., Komatsu, K., Carney, J. and Bohr, V.A. (2005) Werner syndrome protein associates with gamma H2AX in a manner that depends upon Nbs1. FEBS Lett., 579: 1350–6.

Choudhary, S., Sommers, J.A. and Brosh, R.M., Jr. (2004) Biochemical and kinetic characterization of the DNA helicase and exonuclease activities of werner syndrome protein. J Biol Chem., 279: 34603–13.

Christiansen, M., Stevnsner, T., Modin, C., Martensen, P.M., Brosh, R.M., Jr. and Bohr, V.A. (2003) Functional consequences of mutations in the conserved SF2 motifs and post-translational phosphorylation of the CSB protein. Nucleic Acids Res., 31: 963–73.

Christiansen, M., Thorslund, T., Jochimsen, B., Bohr, V.A. and Stevnsner, T. (2005) The Cockayne syndrome group B protein is a functional dimer. FEBS J., 272: 4306–14.

Citterio, E., Rademakers, S., van der Horst, G.T., van Gool, A.J., Hoeijmakers, J.H. and Vermeulen, W. (1998) Biochemical and biological characterization of wild-type and ATPase-deficient Cockayne syndrome B repair protein. J Biol Chem., 273: 11844–51.

Citterio, E., Van Den Boom, V., Schnitzler, G., Kanaar, R., Bonte, E., Kingston, R.E., Hoeijmakers, J.H. and Vermeulen, W. (2000) ATP-dependent chromatin remodeling by the Cockayne syndrome B DNA repair-transcription-coupling factor. Mol Cell Biol., 20: 7643–53.

Cleaver, J.E., Thompson, L.H., Richardson, A.S. and States, J.C. (1999) A summary of mutations in the UV-sensitive disorders: xeroderma pigmentosum, Cockayne syndrome and trichothiodystrophy. Hum Mutat., 14: 9–22.

Colella, S., Nardo, T., Botta, E., Lehmann, A.R. and Stefanini, M. (2000) Identical mutations in the CSB gene associated with either Cockayne syndrome or the DeSanctis-cacchione variant of xeroderma pigmentosum. Hum Mol Genet., 9: 1171–5.

Constantinou, A., Tarsounas, M., Karow, J.K., Brosh, R.M., Bohr, V.A., Hickson, I.D. and West, S.C. (2000) Werner's syndrome protein (WRN) migrates Holliday junctions and co-localizes with RPA upon replication arrest. EMBO Rep., 1: 80–4.

Crabbe, L., Verdun, R.E., Haggblom, C.I. and Karlseder, J. (2004) Defective telomere lagging strand synthesis in cells lacking WRN helicase activity. Science., 306: 1951–3.

Csoka, A.B., Cao, H., Sammak, P.J., Constantinescu, D., Schatten, G.P. and Hegele, R.A. (2004) Novel lamin A/C gene (LMNA) mutations in atypical progeroid syndromes. J Med Genet., 41: 304–8.

Csoka, A.B., English, S.B., Simkevich, C.P., Ginzinger, D.G., Butte, A.J., Schatten, G.P., Rothman, F.G. and Sedivy, J.M. (2004) Genome-scale expression profiling of Hutchinson-Gilford progeria syndrome reveals widespread transcriptional misregulation leading to mesodermal/mesenchymal defects and accelerated atherosclerosis. Aging Cell., 3: 235–43.

D'Amours, D. and Jackson, S.P. (2002) The Mre11 complex: at the crossroads of dna repair and checkpoint signalling. Nat Rev Mol Cell Biol., 3: 317–27.

D'Errico, M., Teson, M., Calcagnile, A., Nardo, T., De Luca, N., Lazzari, C., Soddu, S., Zambruno, G., Stefanini, M. and Dogliotti, E. (2005) Differential role of transcription-coupled repair in UVB-induced response of human fibroblasts and keratinocytes. Cancer Res., 65: 432–8.

Davis, T., Singhrao, S.K., Wyllie, F.S., Haughton, M.F., Smith, P.J., Wiltshire, M., Wynford-Thomas, D., Jones, C.J., Faragher, R.G. and Kipling, D. (2003) Telomere-based proliferative lifespan barriers in Werner-syndrome fibroblasts involve both p53-dependent and p53-independent mechanisms. J Cell Sci., 116(Pt 7):1349–57.

De Sandre-Giovannoli, A., Bernard, R., Cau, P., Navarro, C., Amiel, J., Boccaccio, I., Lyonnet, S., Stewart, C.L., Munnich, A., Le Merrer, M. and Levy, N. (2003) Lamin a truncation in Hutchinson-Gilford progeria. Science., 300: 2055.

De Stefano, N., Dotti, M.T., Battisti, C., Sicurelli, F., Stromillo, M.L., Mortilla, M. and Federico, A. (2003) MR evidence of structural and metabolic changes in brains of patients with Werner's syndrome. J Neurol., 250: 1169–73.

Dianov, G., Bischoff, C., Sunesen, M. and Bohr, V.A. (1999) Repair of 8-oxoguanine in DNA is deficient in Cockayne syndrome group B cells. Nucleic Acids Res., 27: 1365–8.

el-Deiry, W.S., Tokino, T., Velculescu, V.E., Levy, D.B., Parsons, R., Trent, J.M., Lin, D., Mercer, W.E., Kinzler, K.W. and Vogelstein, B. (1993) WAF1, a potential mediator of p53 tumor suppression. Cell., 75: 817–25.

Eisen, J.A., Sweder, K.S. and Hanawalt, P.C. (1995) Evolution of the SNF2 family of proteins: subfamilies with distinct sequences and functions. Nucleic Acids Res., 23: 2715–23.

Epstein, C.J., Martin, G.M., Schultz, A.L. and Motulsky, A.G. (1966) Werner's syndrome a review of its symptomatology, natural history, pathologic features, genetics and relationship to the natural aging process. Medicine (Baltimore)., 45: 177–221.

Eriksson, M., Brown, W.T., Gordon, L.B., Glynn, M.W., Singer, J., Scott, L., Erdos, M.R., Robbins, C.M., Moses, T.Y., Berglund, P., Dutra, A., Pak, E., Durkin, S., Csoka, A.B., Boehnke, M., Glover, T.W. and Collins, F.S. (2003) Recurrent de novo point mutations in lamin A cause Hutchinson-Gilford progeria syndrome. Nature., 423: 293–8.

Fan, J. and Wilson, D.M., III. (2005) Protein–protein interactions and posttranslational modifications in mammalian base excision repair. Free Radic Biol Med., 38: 1121–38.

Franchitto, A. and Pichierri, P. (2002) Protecting genomic integrity during DNA replication: correlation between Werner's and Bloom's syndrome gene products and the MRE11 complex. Hum Mol Genet., 11: 2447–53.

Franchitto, A. and Pichierri, P. (2004) Werner syndrome protein and the MRE11 complex are involved in a common pathway of replication fork recovery. Cell Cycle., 3: 1331–9.

Furuichi, Y. (2001) Premature aging and predisposition to cancers caused by mutations in RecQ family helicases. Ann N Y Acad Sci., 928: 121–31.

Ganguly, T. and Duker, N.J. (1992) Reduced 5-hydroxymethyluracil-DNA glycosylase activity in Werner's syndrome cells. Mutat Res., 275: 87–96.

Gelaw, B., Ali, S. and Becker, J. (2004) Rothmund-Thomson syndrome, Klippel-Feil syndrome and osteosarcoma. Skeletal Radiol., 33: 613–5.

Goldman, R.D., Shumaker, D.K., Erdos, M.R., Eriksson, M., Goldman, A.E., Gordon, L.B., Gruenbaum, Y., Khuon, S., Mendez, M., Varga, R. and Collins, F.S. (2004) Accumulation of mutant lamin A causes progressive changes in nuclear architecture in Hutchinson-Gilford progeria syndrome. Proc Natl Acad Sci USA., 101: 8963–8.

Gomez-Lazaro, M., Fernandez-Gomez, F.J. and Jordan, J. (2004) p53: twenty five years understanding the mechanism of genome protection. J Physiol Biochem., 60: 287–307.

Goto, M., Miller, R.W., Ishikawa, Y. and Sugano, H. (1996) Excess of rare cancers in Werner syndrome (adult progeria). Cancer Epidemiol Biomarkers Prev., 5: 239–46.

Goto, M., Yamabe, Y., Shiratori, M., Okada, M., Kawabe, T., Matsumoto, T., Sugimoto, M. and Furuichi, Y. (1999) Immunological diagnosis of Werner syndrome by down-regulated and truncated gene products. Hum Genet., 105: 301–7.

Griffith, J.D., Comeau, L., Rosenfield, S., Stansel, R.M., Bianchi, A., Moss, H. and de Lange, T. (1999) Mammalian telomeres end in a large duplex loop. Cell., 97: 503–14.

Gruenbaum, Y., Margalit, A., Goldman, R.D., Shumaker, D.K. and Wilson, K.L. (2005) The nuclear lamina comes of age. Nat Rev Mol Cell Biol., 6: 21–31.

http://www.ncbi.nlm.nih.gov/books/bv.fcgi?rid=gene.chapter.hgps

http://www.ncbi.nlm.nih.gov/books/bv.fcgi?rid=gene.chapter.rts

Harley, C.B., Pollard, J.W., Chamberlain, J.W., Stanners, C.P. and Goldstein, S. (1980) Protein synthetic errors do not increase during aging of cultured human fibroblasts. Proc Natl Acad Sci USA., 77: 1885–9.

Harrigan, J.A, Opresko, P.L., von Kobbe, C., Kedar, P.S., Prasad, R., Wilson, S.H. and Bohr, V.A. (2003) The Werner syndrome protein stimulates DNA polymerase beta strand displacement synthesis via its helicase activity. J Biol Chem., 278: 22686–95.

Hoki, Y., Araki, R., Fujimori, A., Ohhata, T., Koseki, H., Fukumura, R., Nakamura, M., Takahashi, H., Noda Y., Kito, S. and Abe, M. (2003) Growth retardation and skin abnormalities of the Recql4-deficient mouse. Hum Mol Genet., 12: 2293–9.

Horibata, K., Iwamoto, Y., Kuraoka, I., Jaspers, N.G., Kurimasa, A., Oshimura, M., Ichihashi, M. and Tanaka, K. (2004) Complete absence of Cockayne syndrome group B gene product gives rise to UV-sensitive syndrome but not Cockayne syndrome. Proc Natl Acad Sci USA., 101: 15410–5.

Huang, S., Beresten, S., Li B., Oshima, J., Ellis, NA. and Campisi, J. (2000) Characterization of the human and mouse WRN 3'–>5' exonuclease. Nucleic Acids Res., 28: 2396–405.

Ishikawa, Y., Miller, R.W., Machinami, R., Sugano, H. and Goto, M. (2000) Atypical osteosarcomas in Werner Syndrome (adult progeria). Jpn J Cancer Res., 91: 1345–9.

Johnson, B.R., Nitta, R.T., Frock, R.L., Mounkes, L., Barbie, D.A., Stewart, C.L., Harlow, E. and Kennedy, B.K. (2004) A-type lamins regulate retinoblastoma protein function by promoting subnuclear localization and preventing proteasomal degradation. Proc Natl Acad Sci USA., 101: 9677–82.

Kamath-Loeb, A.S., Johansson, E., Burgers, P.M. and Loeb, L.A. (2000) Functional interaction between the Werner Syndrome protein and DNA polymerase delta. Proc Natl Acad Sci USA., 97: 4603–8.

Kamath-Loeb, A.S., Loeb, L.A., Johansson, E., Burgers, P.M. and Fry, M. (2001) Interactions between the Werner syndrome helicase and DNA polymerase delta specifically facilitate copying of tetraplex and hairpin structures of the d(CGG)n trinucleotide repeat sequence. J Biol Chem., 276: 16439–46.

Kamath-Loeb, A.S., Welcsh, P., Waite, M., Adman, E.T. and Loeb, L.A. (2004) The enzymatic activities of the Werner syndrome protein are disabled by the amino acid polymorphism R834C. J Biol Chem., 279: 55499–505.

Kashino, G., Kodama, S., Nakayama, Y., Suzuki, K., Fukase, K., Goto, M. and Watanabe, M. (2003) Relief of oxidative stress by ascorbic acid delays cellular senescence of normal human and Werner syndrome fibroblast cells. Free Radic Biol Med., 35: 438–43.

Kellermayer, R., Siitonen, H.A., Hadzsiev, K., Kestila, M. and Kosztolanyi, G. (2005) A patient with Rothmund-Thomson syndrome and all features of RAPADILINO. Arch Dermatol., 141: 617–20.

Kitao, S., Ohsugi, I., Ichikawa, K., Goto, M., Furuichi, Y. and Shimamoto, A. (1998) Cloning of two new human helicase genes of the RecQ family: biological significance of multiple species in higher eukaryotes. Genomics., 54: 443–52.

Kitao, S., Lindor, N.M., Shiratori, M., Furuichi, Y. and Shimamoto, A. (1999) Rothmund-thomson syndrome responsible gene, RECQL4: genomic structure and products. Genomics., 61: 268–76.

Kitao, S., Shimamoto, A., Goto, M., Miller, R.W., Smithson, W.A., Lindor, N.M. and Furuichi, Y. (1999) Mutations in RECQL4 cause a subset of cases of Rothmund-Thomson syndrome. Nat Genet., 22: 82–4.

Kobayashi, J., Antoccia, A., Tauchi, H., Matsuura, S. and Komatsu, K. (2004) NBS1 and its functional role in the DNA damage response. DNA Repair (Amst)., 3: 855–61.

Komatsu, A., Suzuki, S., Inagaki, T., Yamashita, K. and Hashizume, K. (2004) A kindred with Cockayne syndrome caused by multiple splicing variants of the CSA gene. Am J Med Genet A., 128: 67–71.

Kuimov, A.N. (2004) Polypeptide components of telomere nucleoprotein complex. Biochemistry (Mosc)., 69: 117–29.

Kulju, K.S. and Lehman, J.M. (1995) Increased p53 protein associated with aging in human diploid fibroblasts. Exp Cell Res., 217: 336–45.

Kumaran, R.I., Muralikrishna, B. and Parnaik, V.K. (2002) Lamin A/C speckles mediate spatial organization of splicing factor compartments and RNA polymerase II transcription. J Cell Biol., 159: 783–93.

Kyng, K.J., May, A., Brosh, R.M., Jr, Cheng, W.H., Chen, C., Becker, K.G. and Bohr, V.A. (2003) The transcriptional response after oxidative stress is defective in Cockayne syndrome group B cells. Oncogene., 22:1135–49.

Laine, J.P., Opresko, P.L., Indig, F.E., Harrigan, J.A., von Kobbe, C. and Bohr, VA. (2003) Werner protein stimulates topoisomerase I DNA relaxation activity. Cancer Res., 63: 7136–46.

Lavin, M.F. (2004) The Mre11 complex and ATM: a two-way functional interaction in recognising and signaling DNA double strand breaks. DNA Repair (Amst)., 3: 1515–20.

Lebel, M., Spillare, E.A., Harris, C.C. and Leder, P. (1999) The Werner syndrome gene product copurifies with the DNA replication complex and interacts with PCNA and topoisomerase I. J Biol Chem., 274: 37795–9.

Lee, J.W., Harrigan, J., Opresko, P.L. and Bohr, V.A. (2005) Pathways and functions of the Werner syndrome protein. Mech Ageing Dev., 126: 79–86.

Lee, J.W., Kusumoto, R., Doherty, K.M., Lin, G.X., Zeng, W., Cheng, W.H., von Kobbe, C., Brosh, R.M., Jr., Hu J.S. and Bohr V.A. (2005) Modulation of Werner syndrome protein function by a single mutation in the conserved RQC domain. J Biol Chem., 280: 38627–36.

Lei, M., Podell, E.R. and Cech, T.R. (2004) Structure of human POT1 bound to telomeric single-stranded DNA provides a model for chromosome end-protection. Nat Struct Mol Biol., 11: 1223–9.

Leone, A., Costantini, A.M., Brigida, R., Antoniol, O.M., Antonelli-Incalzi, R. and Bonomo, L. (2005) Soft-tissue mineralization in Werner syndrome. Skeletal Radiol., 34: 47–51.

Leverenz, J.B., Yu, C.E. and Schellenberg, G.D. (1998) Aging-associated neuropathology in Werner syndrome. Acta Neuropathol (Berl)., 96: 421–4.

Licht, C.L., Stevnsner, T. and Bohr, V.A. (2003) Cockayne syndrome group B cellular and biochemical functions. Am J Hum Genet., 73: 1217–39.

Lindor, N.M., Furuichi, Y., Kitao, S., Shimamoto, A., Arndt, C. and Jalal, S. (2000) Rothmund-Thomson syndrome due to RECQ4 helicase mutations: report and clinical and molecular comparisons with Bloom syndrome and Werner syndrome. Am J Med Genet., 90: 223–8.

Liu, B., Wang, J., Chan, K.M., Tjia, W.M., Deng, W., Guan, X., Huang, J.D., Li, K.M., Chau, P.Y., Chen, D.J., Pei, D., Pendas, A.M., Cadinanos, J., Lopez-Otin, C., Tse, H.F., Hutchison, C., Chen, J., Cao, Y., Cheah, K.S., Tryggvason K,. and Zhou, Z. (2005) Genomic instability in laminopathy-based premature aging. Nat Med., 11: 780–5.

Loayza, D., Parsons, H., Donigian, J., Hoke, K. and de Lange, T. (2004) DNA binding features of human POT1: a nonamer 5′-TAGGGTTAG-3′ minimal binding site, sequence specificity and internal binding to multimeric sites. J Biol Chem., 279: 13241–8.

Machwe, A., Xiao, L. and Orren, D.K. (2004) TRF2 recruits the Werner syndrome (WRN) exonuclease for processing of telomeric DNA. Oncogene., 23: 149–56.

Macip, S., Igarashi, M., Fang, L., Chen, A., Pan, Z.Q., Lee, S.W. and Aaronson, S.A. (2002) Inhibition of p21-mediated ROS accumulation can rescue p21-induced senescence. EMBO J., 21: 2180–8.

Mahmoud, A.A., Yousef, G.M., Al-Hifzi, I. and Diamandis, E.P. (2002) Cockayne syndrome in three sisters with varying clinical presentation. Am J Med Genet., 111: 81–5.

Mallery, D.L., Tanganelli, B., Colella, S., Steingrimsdottir, H., van Gool, A.J., Troelstra, C., Stefanini, M. and Lehmann, A.R. (1998) Molecular analysis of mutations in the CSB (ERCC6) gene in patients with Cockayne syndrome. Am J Hum Genet., 62: 77–85.

Mancini, M.A., Shan, B., Nickerson, J.A., Penman, S. and Lee, W.H. (1994) The retinoblastoma gene product is a cell cycle-dependent, nuclear matrix-associated protein. Proc Natl Acad Sci USA., 91: 418–22.

Mann, M.B., Hodges, C.A., Barnes, E., Vogel, H., Hassold, T.J. and Luo, G. (2005) Defective sister-chromatid cohesion, aneuploidy and cancer predisposition in a mouse model of type II Rothmund-Thomson syndrome. Hum Mol Genet., 14: 813–25.

Martin, G.M., Sprague, C.A. and Epstein, C.J. (1970) Replicative life-span of cultivated human cells. Effects of donor's age, tissue, and genotype. Lab Invest., 23: 86–92.

Martin, G.M., Oshima, J., Gray, M.D. and Poot, M. (1999) What geriatricians should know about the Werner syndrome. J Am Geriatr Soc., 47: 1136–44.

Matsumoto, T., Imamura, O., Yamabe, Y., Kuromitsu, J., Tokutake, Y., Shimamoto, A., Suzuki, N., Satoh, M., Kitao, S., Ichikawa, K., Kataoka, H., Sugawara, K., Thomas, W., Mason, B., Tsuchihashi, Z., Drayna, D., Sugawara, M., Sugimoto, M., Furuichi, Y. and Goto, M. (1997) Mutation and haplotype analyses of the Werner's syndrome gene based on its genomic structure: genetic epidemiology in the Japanese population. Hum Genet., 100: 123–30.

Meira, L.B., Graham, J.M., Jr, Greenberg, C.R., Busch, D.B., Doughty, A.T., Ziffer, D.W., Coleman, D.M., Savre-Train I and Friedberg, E.C. (2000) Manitoba aboriginal kindred with original cerebro-oculo-facio-skeletal syndrome has a mutation in the Cockayne syndrome group B (CSB) gene. Am J Hum Genet., 66: 1221–8.

Mizuguchi, M. and Itoh, M. (2005) A 35-year-old female with growth and developmental retardation, progressive ataxia, dementia and visual loss. Neuropathology., 25: 103–6.

Mori, H., Tomiyama, T., Maeda, N., Ozawa, K. and Wakasa, K. (2003) Lack of amyloid plaque formation in the central nervous system of a patient with Werner syndrome. Neuropathology., 23: 51–6.

Motonaga, K., Itoh, M., Hachiya, Y., Endo, A., Kato, K., Ishikura, H., Saito, Y., Mori, S., Takashima, S. and Goto, Y. (2002) Age related expression of Werner's syndrome protein in selected tissues and coexpression of transcription factors. J Clin Pathol., 55: 195–9.

Nakamura, Y., Shimizu, T., Ohigashi, Y., Itou, N. and Ishikawa, Y. (2005) Meningioma arising in Werner syndrome confirmed by mutation analysis. J Clin Neurosci., 12: 503–6.

OMIM#214150.

OMIM#278800.

Opresko, P.L., von Kobbe, C., Laine, J.P., Harrigan, J., Hickson, I.D. and Bohr, V.A. (2002) Telomere-binding protein TRF2 binds to and stimulates the Werner and Bloom syndrome helicases. J Biol Chem., 277: 41110–9.

Opresko, P.L., Cheng, W.H., von Kobbe, C., Harrigan, J.A. and Bohr, V.A. (2003) Werner syndrome and the function of the Werner protein; what they can teach us about the molecular aging process. Carcinogenesis., 24: 791–802.

Opresko, P.L., Cheng, W.H. and Bohr, V.A. (2004) Junction of RecQ helicase biochemistry and human disease. J Biol Chem., 279: 18099–102.

Opresko, P.L., Otterlei, M., Graakjaer, J., Bruheim, P., Dawut, L., Kolvraa, S., May, A., Seidman, M.M. and Bohr, V.A. (2004) The Werner syndrome helicase and exonuclease cooperate to resolve telomeric D loops in a manner regulated by TRF1 and TRF2. Mol Cell., 14: 763–74.

Opresko, P.L., Mason, P.A., Podell, E.R., Lei, M., Hickson, I.D., Cech, T.R. and Bohr, VA. (2005) POT1 stimulates RecQ helicases WRN and BLM to unwind telomeric DNA substrates. J Biol Chem., 280: 32069–80.

Oshima, J. (2000) Comparative aspects of the Werner syndrome gene. In Vivo., 14: 165–72.

Passarino, G., Shen, P., Van Kirk, J.B., Lin, A.A., De Benedictis, G., Cavalli Sforza, L.L., Oefner, P.J. and Underhill, P.A. (2001) The Werner syndrome gene and global sequence variation. Genomics., 71: 118–22.

Payao, S.L., de Labio, R.W., Gatti, L.L., Rigolin, V.O., Bertolucci, P.H. and Smith, Mde, A. (2004) Werner helicase polymorphism is not associated with Alzheimer's disease. J Alzheimers Dis., 6: 591–4; discussion 673–81.

Petkovic, M., Dietschy, T., Freire, R., Jiao, R. and Stagljar, I. (2005) The human Rothmund-Thomson syndrome gene product, RECQL4, localizes to distinct nuclear foci that coincide with proteins involved in the maintenance of genome stability. J Cell Sci., 118(Pt 18): 4261–9.

Polyak, K., Xia, Y., Zweier, J.L., Kinzler, K.W. and Vogelstein, B. (1997) A model for p53-induced apoptosis. Nature., 389: 300–5.

Poot, M., Hoehn, H., Runger, T.M., Martin, G.M. (1992) Impaired S-phase transit of Werner syndrome cells expressed in lymphoblastoid cell lines. Exp Cell Res., 202: 267–73.

Postiglione, A., Soricelli, A., Covelli, E.M., Iazzetta, N., Ruocco, A., Milan, G., Santoro, L., Alfano, B. and Brunetti, A. (1996) Premature aging in Werner's syndrome spares the central nervous system. Neurobiol Aging., 17: 325–30.

Ren, Y., Saijo, M., Nakatsu, Y., Nakai, H., Yamaizumi, M. and Tanakam K. (2003) Three novel mutations responsible for Cockayne syndrome group A. Genes Genet Syst., 78: 93–102.

Ridley, A.J., Colley, J., Wynford-Thomas, D. and Jones, C.J. (2005) Characterisation of novel mutations in Cockayne syndrome type A and xeroderma pigmentosum group C subjects. J Hum Genet., 50: 151–4.

Roth, D.E., Campisano, L.C., Callen, J.P., Hersh, J.H. and Yusk, J.W. (1989) Rothmund-Thomson syndrome: a case report. Pediatr Dermatol., 6: 321–4.

Saintigny, Y., Makienko, K., Swanson, C., Emond, M.J. and Monnat, R.J., Jr. (2002) Homologous recombination resolution defect in werner syndrome. Mol Cell Biol., 22: 6971–8.

Sakamoto, S., Nishikawa, K., Heo, S.J., Goto, M., Furuichi, Y. and Shimamoto, A. (2001) Werner helicase relocates into nuclear foci in response to DNA damaging agents and co-localizes with RPA and Rad51. Genes Cells., 6: 421–30.

Salk, D., Bryant, E., Hoehn, H., Johnston, P. and Martin, G.M. (1985) Growth characteristics of Werner syndrome cells in vitro. Adv Exp Med Biol., 190: 305–11.

Selzer, R.R., Nyaga, S., Tuo, J., May, A., Muftuoglu, M., Christiansen, M., Citterio, E., Brosh, R.M., Jr., and Bohr, V.A. (2002) Differential requirement for the ATPase domain of the Cockayne syndrome group B gene in the processing of UV-induced DNA damage and 8-oxoguanine lesions in human cells. Nucleic Acids Res., 30: 782–93.

Sengupta, S., Shimamoto, A., Koshiji, M., Pedeux, R., Rusin, M., Spillare, E.A., Shen, J.C., Huang, L.E., Lindor, N.M., Furuichi, Y. and Harris, C.C. (2005) Tumor suppressor p53 represses transcription of RECQ4 helicase. Oncogene., 24: 1738–48.

Sharma, S., Otterlei, M., Sommers, J.A., Driscoll, H.C., Dianov, G.L., Kao, H.I., Bambara, R.A. and Brosh, R.M., Jr. (2004) WRN helicase and FEN-1 form a complex upon replication arrest and together process branchmigrating DNA structures associated with the replication fork. Mol Biol Cell., 15: 734–50.

Shen, J.C., Gray, M.D., Oshima, J. and Loeb, L.A. (1998) Characterization of Werner syndrome protein DNA helicase activity: directionality, substrate dependence and stimulation by replication protein A. Nucleic Acids Res., 26: 2879–85.

Siitonen, H.A., Kopra, O., Kaariainen, H., Haravuori, H., Winter, R.M., Saamanen, A.M., Peltonen, L. and Kestila, M. (2003) Molecular defect of RAPADILINO syndrome expands the phenotype spectrum of RECQL diseases. Hum Mol Genet., 12: 2837–44.

Sim, F.H., DeVries, E.M., Miser, J.S. and Unni, K.K. (1992) Case report 760. Osteoblastic osteosarcoma (grade 4) with Rothmund-Thomson syndrome. Skeletal Radiol., 21: 543–5.

Smith, M.A., Silva, M.D., Araujo, L.Q., Ramos, L.R., Labio, R.W., Burbano, R.R., Peres, C.A., Andreoli, S.B., Payao, S.L. and Cendoroglo, M.S. (2005) Frequency of Werner helicase 1367 polymorphism and age-related morbidity in an elderly Brazilian population. Braz J Med Biol Res., 38: 1053–9.

Sommers, J.A., Sharma, S., Doherty, K.M., Karmakar, P., Yang, Q., Kenny, M.K., Harris, C.C. and Brosh, R.M., Jr. (2005) p53 modulates RPA-dependent and RPA-independent WRN helicase activity. Cancer Res., 65: 1223–33.

Spivak, G. (2004) The many faces of Cockayne syndrome. Proc Natl Acad Sci USA., 101: 15273–4.

Spurney, C., Gorlick, R., Meyers, P.A., Healey, J.H. and Huvos, A.G. (1998) Multicentric osteosarcoma, Rothmund-Thomson syndrome, and secondary nasopharyngeal non-Hodgkin's lymphoma: a case report and review of the literature. J Pediatr Hematol Oncol., 20: 494–7.

Stefanini, M., Fawcett, H., Botta, E., Nardo, T. and Lehmann, A.R. (1996) Genetic analysis of twenty-two patients with Cockayne syndrome. Hum Genet., 97: 418–23.

Stevnsner, T., Nyaga, S., de Souza-Pinto N.C., van der Horst, G.T., Gorgels, T.G., Hogue, B.A., Thorslund, T. and Bohr, V.A. (2002) Mitochondrial repair of 8-oxoguanine is deficient in Cockayne syndrome group B. Oncogene., 21: 8675–82.

Stracker, T.H., Theunissen, J.W., Morales, M. and Petrini, J.H. (2004) The Mre11 complex and the metabolism of chromosome breaks: the importance of communicating and holding things together. DNA Repair (Amst)., 3: 845–54.

Sugrue, M.M., Shin, D.Y., Lee, S.W. and Aaronson, S.A. (1997) Wild-type p53 triggers a rapid senescence program in human tumor cells lacking functional p53. Proc Natl Acad Sci USA., 94: 9648–53.

Sullivan, T., Escalante-Alcalde, D., Bhatt, H., Anver, M., Bhat, N., Nagashima, K., Stewart, C.L. and Burke, B. (1999) Loss of A-type lamin expression compromises nuclear envelope integrity leading to muscular dystrophy. J Cell Biol., 147: 913–20.

Sunesen, M., Stevnsner, T., Brosh, Jr, R.M., Dianov G.L. and Bohr, V.A. (2002) Global genome repair of 8-oxoG in hamster cells requires a functional CSB gene product. Oncogene., 21: 3571–8.

Suzuki, T., Shiratori, M., Furuichi, Y. and Matsumoto, T. (2001) Diverged nuclear localization of Werner helicase in human and mouse cells. Oncogene., 20: 2551–8.

Szekely, A.M., Chen, Y.H., Zhang, C., Oshima, J. and Weissman, S.M. (2000) Werner protein recruits DNA polymerase delta to the nucleolus. Proc Natl Acad Sci USA., 97: 11365–70.

Takeuchi, F., Hanaoka, F., Goto, M., Akaoka, I., Hori, T., Yamada, M. and Miyamoto, T. (1982a) Altered frequency of initiation sites of DNA replication in Werner's syndrome cells. Hum Genet., 60: 365–8.

Takeuchi, F., Hanaoka, F., Goto, M., Yamada, M. and Miyamoto, T. (1982b) Prolongation of S phase and whole cell cycle in Werner's syndrome fibroblasts. Exp Gerontol., 17: 473–80.

Tan, W.H., Baris, H., Robson, C.D. and Kimonis, V.E. (2005) Cockayne syndrome: the developing phenotype. Am J Med Genet A., 135: 214–6.

Thorslund, T., von Kobbe, C., Harrigan, J.A., Indig, F.E., Christiansen, M., Stevnsner, T. and Bohr VA. (2005) Cooperation of the Cockayne syndrome group B protein and poly(ADP-ribose) polymerase 1 in the response to oxidative stress. Mol Cell Biol., 25: 7625–36.

UniGene Hs.31442 Homo sapiens RECQL4. http://www.ncbi.nlm.nih.gov/unigene/clust.cgi?org=hs&cid=31442

von Kobbe, C. and Bohr, V.A. (2002) A nucleolar targeting sequence in the Werner syndrome protein resides within residues 949–1092. J Cell Sci., 115(Pt 20): 3901–7.

von Kobbe, C., Harrigan, J.A., May, A., Opresko, P.L., Dawut, L., Cheng, W.H. and Bohr, V.A. (2003a) Central role for the Werner syndrome protein/poly(ADP-ribose) polymerase 1 complex in the poly(ADP-ribosyl)ation pathway after DNA damage. Mol Cell Biol., 23: 8601–13.

von Kobbe, C., Thoma, N.H., Czyzewski, B.K., Pavletich, N.P. and Bohr, V.A. (2003b) Werner syndrome protein contains three structure-specific DNA binding domains. J Biol Chem., 278: 52997–3006.

von Kobbe, C., Harrigan, J.A., Schreiber, V., Stiegler, P., Piotrowski, J., Dawut, L. and Bohr, V.A. (2004a) Poly(ADP-ribose) polymerase 1 regulates both the exonuclease and helicase activities of the Werner syndrome protein. Nucleic Acids Res., 32: 4003–14.

Von Kobbe, C., May, A., Grandori, C. and Bohr, V.A. (2004b) Werner syndrome cells escape hydrogen peroxide-induced cell proliferation arrest. FASEB J., 18: 1970–2.

Van Maldergem, L., Siitonen, H.A., Jalkh, N., Chouery, E., De Roy, M., Delague, V., Muenke, M., Jabs, E.W., Cai, J., Wang, L.L., Plon, S.E., Fourneau, C., Kestila, M., Gillerot, Y., Megarbane, A. and Verloes, A. (2006) Revisiting the craniosynostosis-radial ray hypoplasia association : Baller-Gerold syndrome caused by mutations in RECQL4 gene. J Med Genet., 43: 148–52.

Wang, L., Evans, A.E., Ogburn, C.E., Youssoufian, H., Martin, G.M. and Oshima, J. (1999) Werner helicase expression in human fetal and adult aortas. Exp Gerontol., 34: 935–41.

Wang, L.L., Levy, M.L., Lewis, R.A., Chintagumpala, M.M., Lev, D., Rogers, M. and Plon, S.E. (2001) Clinical manifestations in a cohort of 41 Rothmund-Thomson syndrome patients. Am J Med Genet., 102: 11–7.

Wang, L.L., Gannavarapu, A., Kozinetz, C.A., Levy, M.L., Lewis, R.A., Chintagumpala, M.M., Ruiz-Maldanado, R., Contreras-Ruiz, J., Cunniff, C., Erickson, R.P., Lev, D., Rogers, M., Zackai, E.H. and Plon, S.E. (2003) Association between osteosarcoma and deleterious mutations in the RECQL4 gene in Rothmund-Thomson syndrome. J Natl Cancer Inst., 95: 669–74.

West, S.C. (2003) Molecular views of recombination proteins and their control. Nat Rev Mol Cell Biol., 4: 435–45.

Yamabe, Y., Sugimoto, M., Satoh M., Suzuki, N., Sugawara, M., Goto, M. and Furuichi, Y. (1997) Down-regulation of the defective transcripts of the Werner's syndrome gene in the cells of patients. Biochem Biophys Res Commun., 236: 151–4.

Yamabe, Y., Shimamoto, A., Goto, M., Yokota, J., Sugawara, M. and Furuichi, Y. (1998) Sp1-mediated transcription of the Werner helicase gene is modulated by Rb and p53. Mol Cell Biol., 18: 6191–200.

Yamamoto, K., Imakiire, A., Miyagawa, N. and Kasahara, T. (2003) A report of two cases of Werner's syndrome and review of the literature. J Orthop Surg (Hong Kong)., 11: 224–33.

Ye, L., Miki, T., Nakura, J., Oshima, J., Kamino, K., Rakugi, H., Ikegami, H., Higaki, J., Edland, S.D., Martin, G.M. and Ogihara, T. (1997) Association of a polymorphic variant of the Werner helicase gene with myocardial infarction in a Japanese population. Am J Med Genet., 68: 494–8.

Ye, L., Nakura, J., Morishima, A. and Miki, T. (1998) Transcriptional activation by the Werner syndrome gene product in yeast. Exp Gerontol., 33: 805–12.

Yu, C.E., Oshima, J., Fu, Y.H., Wijsman, E.M., Hisama, F., Alisch, R., Matthews, S., Nakura, J., Miki, T., Ouais, S., Martin, G.M., Mulligan, J. and Schellenberg, G.D. (1996) Positional cloning of the Werner's syndrome gene. Science., 272: 258–62.

Zheng, L., Zhou, M., Chai, Q., Parrish, J., Xue, D., Patrick, S.M., Turchi, J.J., Yannone, S.M., Chen, D. and Shen, B. (2005) Novel function of the flap endonuclease 1 complex in processing stalled DNA replication forks. EMBO Rep., 6: 83–9.

CHAPTER 15

PROTEIN AGGREGATION IN AGING AND AGE-RELATED NEURODEGENERATIVE DISORDERS

JEFFREY N. KELLER[1,2] AND QUNXING DING[1]

[1] *Anatomy and Neurobiology*

[2] *Sanders-Brown Center on Aging, University of Kentucky, Lexington KY, 40536–0203*

Abstract: The purpose of this chapter is to provide a background on the effects of aging on proteolytic pathways and protein aggregation, and to discuss the contribution of altered protease function and protein aggregation to brain function. Studies will focus on the proteasome proteolytic pathway. Lastly, these studies will also discuss the relationship between aging and age-related neurodegenerative disorders

Keywords: Aging; Alzheimer's disease; lysosome; neuron; oxidative stress; proteasome

1. INTRODUCTION

Recent studies indicate that proteasome inhibition likely occurs during, and may contribute to, multiple aspects of aging. In particular, studies now demonstrate that inhibition of proteasome function is sufficient to induce a variety of pathological events associated with aging. Specifically, alterations in the proteasome proteolytic pathway may contribute to the elevations in protein oxidation, protein aggregation, and neurodegeneration evident in the aging central nervous system (CNS). The focus of this chapter is to discuss what is presently known about the effects of aging on proteolysis, and to describe the possible role alterations in proteolysis may play in mediating protein aggregation in the CNS. Studies will focus on the role of the proteasome proteolytic pathway.

2. AGING ALTERS PROTEASOME ACTIVITY

Alterations in proteasome function during normal aging have been described in a wide range of species, including humans, and reported to occur in a wide variety of tissues. It is important to point out that even within individual organs a regional

S. Rattan and M. Kassem (eds.), Prevention and Treatment of Age-related Diseases, 297–312.

specificity, with regards to the severity of proteasome inhibition can occur. This is best illustrated in the CNS where there are clearly brain region susceptibilities with regards to age-related proteasome inhibition (Keller et al., 2000a; Ding and Keller, 2001; Goto et al., 2002; Gray et al., 2003). In addition to these *in vivo* examples of age-related proteasome inhibition, *in vitro* aging is also associated with declines in proteasome function, occurring in a diverse range of cell types. The proliferative state of cells also appears to be an important factor regulating age-related impairments in proteasome function. As an example, post-mitotic cells undergo more severe inhibition of proteasome activity as compared to mitotic cells (Sitte et al., 2000a,b,c,d; Chondrogianni et al., 2003).

2.1 Age related alterations in protease function in the brain

It has been demonstrated that post-mitotic cells exhibit a preferential loss of postglutamyl peptidase activity, while mitotic cells undergo a loss in trypsin-like, chymotrypsin-like, and postglutamyl peptidase activities of the proteasome (Sitte et al., 2000a,b,c,d; Chondrogianni et al., 2003). In the liver, there is a 50% reduction in proteasomal postglutamyl peptidase activity with no significant differences in either trypsin-like or chymotrypsin-like activity reported (Conconi et al., 1996). In rats there is a loss in chymotrypsin-like proteasome activity throughout the CNS during aging. Decreases in chymotrypsin-like activity are evident within the cortex, hippocampus, and spinal cord of 12-month-old rats (Keller et al., 2000a,b). In contrast, no impairment in chymotrypsin-like activity is evident in either the brain stem or cerebellum. Impairments in the chymotrypsin-like activity of the proteasome are also evident by 12-months of age in the heart, kidney, liver, but not the lung of these aged rats (Keller et al., 2000a,b).

2.2 Age related altrerations in protease function outside of the CNS

In addition to age-related alterations in basal proteasome activity, it is important to point out that aging has been demonstrated to impair the ability of the proteasome to respond to stress (Merker et al., 2001; Beedholm et al., 2004). The ability of the proteasome to up-regulate its activity in response to environmental or genetic stressors would be expected to play a pivotal role in determining whether a cell was able to survive the wide variety of stressors it is likely to encounter during aging. In this scenario, the lack of proteasome plasticity would result in an ineffective or inhibited proteasome, which could contribute to cell pathology and cytotoxicity. As mentioned previously, the expression of the proteasome in neural cells is dramatically altered in response to oxidative stress and the expression of proteins with an increased propensity to aggregate (Ding et al., 2002; Pacifici et al., 1993). Together; these studies show an apparent increase in immunoproteasome complex formation. Interestingly, studies in neural cells expressing polyglutamine containing proteins suggest that the immunoproteasome is not capable of increasing activity in response to subsequent stressors (Ding et al., 2002), and may ultimately be deleterious towards long-term viability.

3. BASIS FOR AGE-RELATED CHANGES IN THE PROTEASOME

At the present time it is believed that age-related impairments in proteasome-mediated protein degradation can occur as the result of alterations in protein targeting, excessive cross linking proteasome substrates, compromises in heat shock protein (HSP) capacity, alterations in the intracellular localization of proteasome complexes, alterations in proteasome composition, impairments in proteasome plasticity, and increased oxidative damage to the proteasome complex. Each of these events is discussed in detail below.

Increases in protein hydrophobicity appear to be central mechanism for targeting proteins to be degraded by the 20S or 26S proteasome. In order to efficiently degrade these "marked" proteins they must be rapidly identified, and upon identification be brought together with the proteasome complex in a timely and efficient manner. In most aging tissues it is likely that there may be an overwhelming amount of proteins targeted to the proteasome. Oxidized, misfolded, and damaged proteins are all proteasome substrates, and increases in their formation undoubtedly occur in aging cells. This increase in substrates may override the targeting systems, contributing to inefficiency in proteasome-mediated protein degradation, as some proteins are unable to reach a proteasome complex. The ubiquitin-pathway is known to be negatively affected by oxidative stress (Obin et al., 1998), may be deleteriously affected by aging. Inefficiencies in the ubiquitin system would also be expected to negatively affect proteasome-mediated protein degradation. Each of these manifestations may lead to a specialized form of proteasome inhibition, namely the inhibition of protein turnover by failure to deliver proteins to the proteasome.

3.1 Proteolysis and oxidative stress

While the mild oxidation of proteins is known to serve as a potent inducer of proteasome mediated proteolysis (Grune et al., 1998; Davies, 2001; Squier, 2001; Sohal and Weindruch, 1996), excessive oxidation is known to mediate inhibition of the proteasome. Impairment of proteasome-mediated protein degradation by excessively cross linked proteins is believed to be mediated by the blockage that occurs at the entrance of proteasome complex. This obstruction at the openings between the α- and β-subunits is sufficient to block the entrance of subsequent protein substrates into the proteasome. Cross linking may be achieved by oxidants (ROS) (Squier, 2001; Sohal and Weindruch, 1996), or as the result of lipid peroxidation products such as 4-hydroxynonenal (HNE) (Friguet and Szweda, 1997). Increased oxidative damage to proteins, including increased levels of protein cross linking, is known to occur during normal aging. These data are consistent with a role for increased protein cross linking mediating inhibition of the proteasome during normal aging. Cross linking of proteins is also likely to impair the unfolding of proteins, which is required for their degradation by the proteasome (Benaroudj et al., 2001). Inhibition of this process could also provide an additional mechanism for impairment of proteasome mediated protein degradation.

Increasing evidence suggests that oxidative damage to the proteasome complex may be a mediator of at least some forms of proteasome inhibition in the CNS. Studies from our laboratory demonstrate that dopamine may support ROS-induced impairment of proteasome function in the CNS (Keller et al., 2000c). Several features of the CNS presumably make it very vulnerable to oxidative stress including the fact that the CNS has a high metabolic rate that may produce a higher level of mitochondrial derived ROS, may undergo age-related decreases in antioxidant levels, and has a high content of readily oxidized lipids that are capable of promoting oxidative stress. Post mitotic cells in the CNS, which survive for decades, are particularly susceptible to an age-related accrual and elevation in oxidative damage. Proteasomes can undergo direct oxidative modification by a variety of mechanisms. For example, peroxynitrite and HNE can be generated in the intraceullular environment and directly interact with the proteasome and inhibit its function (Keller et al., 2000a; Esterbauer et al., 1991; Glockzin et al., 1999; Okada et al., 1999; Hyun et al., 2002; Amici et al., 2003; Uchida, 2003). This inhibition is mediated in part by changes in proteasome stability as well as potentially mediated by oxidative modification of the active enzymatic sites. However, because the proteolytic activities of the proteasome face the inner core of the proteasome, it is unlikely that much interaction between oxidants and the actual enzymatic sites occurs. Studies have now demonstrated that oxidative modification of the proteasome occurs in conditions where proteasome inhibition is present (Keller et al., 2000a,b; Okada et al., 1999). In particular, oxidation of the proteasome is observed during normal aging in the spinal cord and in experimental models of ischemia-reperfusion injury (Keller et al., 2000a,b). It is interesting to point out that within the spinal cord there are detectable levels of proteasome oxidation within 3-month-old rats, which are not detectable in other regions of the CNS, without any apparent loss of proteasome activity (Keller et al., 2000b). These data suggest that increased oxidation of the proteasome does not always result in proteasome inhibition.

The degradation of proteins by the proteasome requires that proteins be unfolded and inserted within the proteasome complex (Benaroudj et al., 2001). The unfolding of proteins must be mediated by HSP. Studies have demonstrated that increased HSP expression ameliorates oxidative stress-induced proteasome inhibition (Ding and Keller, 2001), consistent with HSP playing a critical role in preserving proteasome function during periods of oxidative stress. The identification of which HSP are most important in this process has not been elucidated. Age-related compromises in HSP capacity therefore provide a mechanism by which proteasome-mediated protein degradation may be inhibited, via failure to deliver and/or unfold proteasome substrates.

It is clear that the localization of proteasome complexes can be altered in response to specific stressors (Rivett, 1993; Noda et al., 2000; Ogiso et al., 2002; Adam et al., 2004). The localization of the proteasome to either nuclear or synaptic compartments may be particularly important for neuron function and neuron viability. It is important to point out that localized alterations in proteasome function, through decreases in the number of available of proteasome complexes or decreases in specific activity distinct proteasome populations, may not be readily evident when measuring proteasome function in brain homogenates. In neurons, the loss of proteasome function in the synapse could be particularly deleterious to neuronal signaling, excitotoxicity, and synaptic plasticity. Impairments in nuclear proteasome function could selectively affect the activity of transcription factors, histone function, and chromatin remodeling. Elucidating these localized alterations in proteasome function are critical to accurately understanding the contribution proteasome inhibition may play in aging and age-related disorders of the CNS.

3.2 Plasticity of the proteolytic system in the CNS

Continual generation of new proteasome complexes is presumably necessary to replace damaged and/or less efficient proteasome complexes. Additionally, a perpetual generation of proteasome complexes allows for the generation of proteasomes with altered composition, and the generation of proteasomes that are more efficient at degrading proteins under stressful conditions. In aging, and age-related disorders of the CNS, proteasome biogenesis may be altered and contribute to the loss of proteasome function. This impairment in biogenesis could result from a loss of proteoassemblin (Schmidt and Kloetzel, 1997; Griffin et al., 2000; Kruger et al., 2001), reduced levels of molecular chaperones that participate in proteasome biogenesis, alterations in proteasome subunit expression, oxidative modification of proteasome subunits, or oxidative attack on a developing proteasome complex. Additionally, polymorphisms in proteasome subunits may contribute to alterations in proteasome subunit expression. A number of studies now demonstrate a clear association between polymorphisms in proteasome subunits and Graves' disease, ankylosing spondylitis, and insulin-dependent diabetes mellitus (Heward et al., 1999; Maksymowych et al., 2000; Deng et al., 1995; Vinasco et al., 1998; Mishto et al., 2002). Studies have shown that LMP2 codon polymorphisms can alter age-related susceptibility to TNF-α induced apoptosis in peripheral blood mononuclear cells (Mishto et al., 2002). LMP2 polymorphisms may also be associated with AD (Mishto et al., 2006). Presumably, these polymorphisms in the LMP2 subunit promote deleterious alterations in proteasome function and may provide an additional means by which proteasome inhibition occurs in aging and age-related disorders of the CNS.

Changes in proteasome composition appear to be an important means by which proteasome function can be specialized in order to address a specific need. Changes in proteasome subunit expression occur in the aging of the retina, fibroblast, muscle, and liver (Chondrogianni et al., 2003; Louie et al., 2002; Friguet et al., 2002; Bulteau et al., 2002; Bulteau et al., 2000; Anselmi et al., 1998). Cytokine-induced expression of immunoproteasome has been reported in a variety of tissues and cell types that are not part of the immune system (Louie et al., 2002; Singh et al., 2002;

Piccinini et al., 2003). These data raise the possibility that immunoproteasomes may be generated as a means of increasing the turnover of specific proteins in aging, including the degradation of oxidized proteins. Additionally, studies have demonstrated that proteasome subunits exhibit a hierarchical susceptibility to HNE modification (Ferrington and Kapphahn, 2004), which may be important in determining the amount of HNE-induced inactivation that occurs following a variety of stressors. It is interesting to note that formation of immunoproteasome, while allowing for continued proteasome function, may impair the ability of the proteasome to respond to subsequent stressors (Ding et al., 2002). Aging and age-related diseases of the CNS may promote changes in proteasome composition that in the short term allow for maintenance of proteasome function, but in the long term promote proteasome inhibition or at least impair the ability of the proteasome to respond to subsequent stressors.

4. EFFECTS OF PROTEASOME INHIBITION WITHIN THE CNS

Numerous studies have now demonstrated that inhibition of the proteasome is sufficient to induce neuron death in primary neuronal cultures, as well as neural cell lines (Lopes et al., 1997; Keller and Markesbery, 2000; Pasquini et al., 2000; Qiu et al., 2000). A number of the 26S proteasome substrates are involved in the apoptotic pathway (Wojcik, 1999; Grimm and Osborne, 1999), with the best characterized of these substrates is p53. Normally a very short-lived protein, the expression of p53 is kept at a low level, and thus is unable to induce its pro-apoptotic effects. However, following inhibition of proteasome function the level of p53 would be expected to become elevated (Jesenberger and Jentsch, 2002; Dietrich et al., 2003; Williams and McConkey, 2003; Nakaso ct al., 2004), eventually elevating to the point that it is able to induce its pro-apoptotic pathways. Indeed, p53 has been demonstrated to play a causal role in the apoptosis induced by severe proteasome inhibition (Nakaso et al., 2004).

It is important to point out that proteasome inhibition does not appear to induce neuron death in all neuron populations or experimental paradigms. These data raise the possibility that proteasome inhibitor toxicity may be cell type specific, based on the function of the proteasome in a given cell. For example, the proteasome is responsible for some forms of NF*k*B activation, which can have pro-apoptotic or anti-apoptotic effects depending on cell type. As such, proteasome inhibition could have very different effects on cell survival based on the differential role of NF*k*B in these two cell populations. Alternatively, these data could indicate the inadequacy of some neuronal populations to utilize non-proteasomal proteolysis, in order to maintain neuronal homeostasis. In such a scenario, cells able to sufficiently up-regulate lysosomal activity would be expected to exhibit little toxicity in response to the application of proteasome inhibitors. Cell specific susceptibilities to proteasome inhibition may also be due in part to alterations in HSP capacity, with neurons possessing higher levels of HSP capacity being more resistant to proteasome inhibitor toxicity. It is important to keep in mind that the majority of *in vitro* studies are conducted in cultures established from embryonic tissue, or tissue from early postnatal brain. As such, one must take into account the possibility that embryonic tissue may have a different dependence on proteasome activity than established neurons within the mature and developed CNS.

4.1 Proteasome as a "secondary antioxidant"

The clearance of oxidized proteins is an important means by which cells are able to prevent the increase in oxidative damage (most notably increased protein oxidation), and thus proteasome-mediated protein degradation is an important "antioxidant" (Pacifici et al., 1989; Grune et al., 1997; Grune and Davies, 1997). In this capacity the proteasome aids in preventing the elevation in oxidative damage and induction of oxidative stress. This "antioxidant" feature of the 20S proteasome is not only important in the aging of the CNS, but also is likely important in numerous age-related disorders of the CNS.

Impairments in 20S proteasome function likely play an important role in the age-related increases in protein oxidation observed in a variety of tissues, including the CNS (Louie et al., 2002; Agarwal and Sohal, 1994; Radak et al., 2002; Viteri et al., 2004). It is important to note that during aging protein oxidation does not typically exhibit a gradual and progressive increase, rather during aging there is a very low level increase in protein oxidation that dramatically increases several fold in late age Squier, 2001; Beckman and Ames, 1998; Petropoulos et al., 2000; Barja, 2002; Hensley and Floyd, 2002; Keller et al., 2004). Proteasome inhibition may serve an important role as a trigger for the sudden and dramatic spike in protein oxidation observed in very late age. Therefore, early in the aging process there is likely a dynamic cellular environment that helps to prevent large increases in protein oxidation. For example, it is likely that proteasome plasticity and increases in stress response (present in young cells) prevent the accumulation of oxidative damage that could potentially occur as the result of cellular stressors. Over time the ability of these protective pathways to prevent increases in protein oxidation dramatically decrease, with inhibition of proteasome function serving as a mechanism for rapidly and profoundly elevating protein oxidation. Additionally, once the levels of oxidized proteins are increased to a deleterious stage, or allowed to persist in the intracellular space for prolonged periods of time, they may serve as potent inhibitors of proteasome function. In this model, excessively oxidized proteins inhibit the entry of other proteasome substrates, thus causing inhibition of proteasome-mediated protein degradation. Consistent with this model, studies from our laboratory have demonstrated that increased heat shock protein expression ameliorates oxidative stress-induced proteasome inhibition (Ding and Keller, 2001).

4.2 Proteasome and lipofuscin formation

Recent studies provide direct experimental evidence for proteasome inhibition serving as a mediator of lipofuscin-ceroid, which is one of the most common forms of oxidative damage observed in aged tissues. Interestingly, this increase in lipofuscin-ceroid may be related to impairment in mitochondria turnover and mitochondrial function (Sullivan et al., 2004). Because of the importance to mitochondria dysfunction to aging and age-related diseases of the CNS, these data indicate a novel mechanism by which proteasome inhibition may contribute to neuropathogenesis. Additionally, our laboratory has demonstrated that inhibition of proteasome function (low-level inhibition) is sufficient to increase autophagy (Ding et al., 2003), which are observed in the aging CNS as well as several age-related disorders of the CNS. The chronic activation of autophagy is likely deleterious towards neural homeostasis, based on the fact that rapid and large scale degradation of cytoplasmic complexes and organelles cannot be beneficial towards the long term cellular viability (Larsen and Sulzer, 2002). Therefore, induction of autophagy may serve as an additional mechanism by which proteasome inhibition contributes to cytotoxicity in the CNS. Lastly, inhibition of proteasome function in neural cells alters gene expression in a manner that is highly relevant to a variety of age-related disorders (Ding et al., 2004a), including modulating the genes involved in regulating beta amyloid metabolism.

A number of studies have suggested a link between DNA repair and the proteasome. For example, the degradation of oxidized histones is mediated by the proteasome (Ullrich et al., 1999; Ullrich and Grune, 2001), with additional studies showing that proteasome subunits may play a role in DNA repair (Walters et al., 2003; Elsasser et al., 2004). Data from our laboratory demonstrated that proteasome inhibition is sufficient to induce RNA and DNA oxidation in primary CNS cultures (Ding et al., 2004b). Interestingly, nucleic acid oxidation occurred in neurons and astrocytes, although it was much more severe in neurons as compared to astrocyte cultures. The oxidation of RNA was associated with an alteration in RNA processing (Ding et al., 2004b). These data suggest that there is potential crosstalk between proteasome-mediated protein degradation and the translation/protein synthesis processes. The proteasome is also capable of increasing ROS production (Ding and Keller, 2001; Sullivan et al., 2004; Fribley et al., 2004; Ling et al., 2003), which can increase oxidative stress. Studies have shown that both severe and moderate proteasome inhibition are capable of stimulating ROS generation in neural and non-neural cells. In at least 1 study the increase in mitochondrial derived ROS has been reported (Sullivan et al., 2004).

5. INTERPLAY BETWEEN PROTEIN DEGRADATION, PROTEIN SYNTHESIS, AND PROTEIN AGGREGATION

In addition to alterations in proteolysis, in a variety of age-related neurodegenerative conditions there is known to be increases in protein aggregation. As outlined above, it appears that protein aggregation may be a potential mediator of impairments of proteasome function. Conversely, inhibition of proteasome function has been reported to be sufficient to induce protein aggregation. These previous studies have been construed to indicate that failures in the lysosomal and proteasomal proteolytic pathways may contribute to elevations in protein aggregation and the

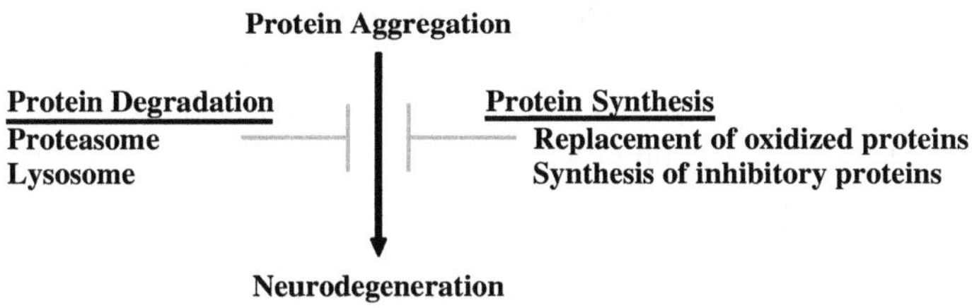

Figure 1. Protein degradation, protein synthesis, and protein aggregation. Preventing the toxicity of protein aggregation depends on maintaining a sufficient balance between proteolytic pathways and protein synthesis. In the face of impaired proteolytic pathways it is likely that cells induce compensatory impairments in protein synthesis, so as to maintain favorable steady state protein kinetics. While in the short term such changes may be beneficial, it is likely that long term impairments in protein degradation and protein synthesis are deleterious to cellular homeostasis

toxicity of protein aggregation, in a feed forward pathway that involves the ability of protein aggregates to inhibit both proteasomal and lysosomal proteolysis (Figure 1). However, it is important to point out that in addition to impairments in proteolysis, there is known to be impairments in protein synthesis in a variety of age related neurodegenerative conditions, as well as aging itself. If impairments in protein synthesis are occurring at the same time that impairments in proteolysis are occurring it is unlikely that gross alterations in steady state protein dynamics will occur (Figure 1). In this model, it is likely that decreasing the amount of substrate (proteins) and decreasing the amount of enzyme (proteolytic pathways) will balance out one another to aid in maintaining a favorable steady state dynamic for proteins inside of the cell (Figure 1). However, it is likely that long term impairments in protein synthesis and long term impairments in protein degradation are deleterious to cellular homeostasis, even though they may be beneficial in the short term. This is based on the fact that cells would not be expected to impair the ability of cells to successfully respond to the numerous environmental and genetic stressors over a long period of time, based on the fact that maintaining homeostasis requires rapid protein synthesis and rapid protein degradation. This is particularly true for the regulation of HSPs and transcription factors.

6. ROLE OF PROTEASOME INHIBITION AS MEDIATOR OF AGING

Proteasome inhibition occurs in the aging of most cell types and tissue, but does it play any role in mediating aging? Numerous studies suggest that proteasome inhibition may not only occur during normal aging, but may play a direct role in the aging process. As discussed previously, studies have demonstrated that proteasome inhibition is sufficient to induce multiple pathological alterations observed in aging including increased protein oxidation, nucleic acid oxidation, protein aggregation, increased lipofuscin/ceroid, induction of autophagy, and induction of mitochondrial dysfunction. The induction of cellular senescence is also tightly correlated with a loss of proteasome function (Sitte et al., 2000b,c,d; Caballero et al., 2004; Grune et al., 2001), with proteasome inhibition sufficient to induce multiple aspects of cellular senescence (Chondrogianni et al., 2003; Chondrogianni and Gonos, 2004). Such studies indicate that proteasome inhibition is not only a common feature of cellular and tissue aging, but demonstrate that proteasome inhibition is sufficient to induce age-related pathologies observed in a variety of tissues.

Caloric restriction (CR) is the only manipulation that consistently and reproducibly increases lifespan (average and maximal lifespan) in mammals (Sohal and Weindruch, 1996; Weindruch, 1996). Some studies suggest that CR may blunt age-related impairments in proteasome function (Merker et al., 2001; Anselmi et al., 1998), supporting a potential role for the preservation of proteasome function as a means by which CR increases lifespan. Interestingly, CR is also associated with an amelioration of oxidative damage (including protein oxidation) (Sohal and Weindruch, 1996; Weindruch, 1996; Forster et al., 2000), raising the possibility that the preservation of proteasome function contributes to the decreased levels of oxidative damage observed in CR tissues. Alternatively, it may be that the decrease in oxidative damage is what promotes the preservation of proteasome function in CR tissues. Clarification of this issue is essential and highlights the importance of determining whether proteasome inhibition necessary for aging. Perhaps even more importantly it remains to be elucidated whether the proteasome plays a role in regulating lifespan. Data from our laboratory demonstrate that the proteasome is essential for yeast aging (Chen et al., 2004), with decreases in proteasome function decrease lifespan, consistent with the proteasome playing a role in regulating lifespan.

At the present time we believe that the proteasome plays a direct role in regulating aging, with preservation of proteasome function slowing the rate of aging, and inhibition of proteasome function increasing the rate of aging. We believe that the ability of the proteasome to regulate aging is consistent with both the free radical theory or aging and the adaptation model of aging (Beckman and Ames, 1998; Harman, 2001; Mangel, 2001; Parsons, 2003). The free radical theory of aging proposes that aging is the result of cumulative oxidative damage inducing cellular aging, while the adaptation theory of aging suggest that lifespan is regulated by the ability to successfully adapt to stressors and that the accumulation of adaptations alters cellular function in a manner that ultimately causes aging. In this model the proteasome serves as the trigger for the majority of age-related alterations. In young healthy cells there is considerable proteasome plasticity, allowing the cells to rapidly respond to stressors, and the proteasome providing a barrier of safety from the deleterious effects of cellular stressors. Following exposure to stress, in young healthy cells the proteasome becomes inhibited for a brief period, with proteasome capacity rapidly brought back to basal levels through a host of events including antioxidants, heat shock proteins, and proteasome plasticity. With continual adaptation to stress revolving around the capacity of cells to maintain proteasome function. In aging cells, the ability of the proteasome to regain its full capacity is impaired, thus allowing for the persistence of proteasome inhibition.

Sustained proteasome impairment is the result of multiple factors including a decreased antioxidant defense system, reduced HSP capacity, and reduced proteasome plasticity. During the prolonged low-level proteasome inhibition a number of deleterious events occur, promoted by the presence of proteasome inhibition. For example, elevations in oxidative damage and pro-apoptotic pathways occur, thus promoting further inhibition of proteasome function. Once this process is set in motion, a catastrophic feed forward pathway is established, ultimately contributing to cellular aging. Proteasome inhibition thereby serves as a trigger for oxidative stress in the free radical theory of aging, and serves as the switch by which aging is promoted in the adaptation theory of aging. In this model the proteasome is not only affected by aging, but is a central mediator and regulator of aging.

7. THERAPEUTIC INTERVENTIONS FOR PREVENTING INCREASES IN PROTEIN AGGREGATION

While there remains some controversy as to whether protein aggregates are always deleterious to neuronal function, a number of studies have sought to elucidate pharmaceutical and environmental interventions which may suppress the formation of protein aggregates. For example, studies have demonstrated that addition of geldamycin and histone deacetylase inhibitors are sufficient to decrease protein aggregation in a variety of disorders (Corcoran et al., 2004; Ryu et al., 2005; Sittler et al., 2001; Auluck et al., 2005). Interestingly, both of these interventions may mediate their beneficial effects via the elevation of HSP pathways. Such an observation may highlight the potential for hormesis as a neuroprotective pathway (Rattan, 2004). In the model of hormesis a mild stress is activated in cells, which allows for a beneficial induction of HSP components and proteasome function (Breedholm et al., 2004), both of which may then allow for the degradation of potentially deleterious protein aggregates. Studies are needed in the future to explore the potential for pharmaceutical interventions, and environmental interventions, to activate beneficial hormesis in the aging CNS. Such interventions may be useful for delaying the development and progression of age-related disorders such as AD and PD.

ACKNOWLEDGEMENTS

The authors would like to thank Dr W.R. Markesbery for his support. This work was funded in part by a grant from the NIH (AG18437; J.N.K.).

REFERENCES

Adam, G., Gausz, J., Noselli, S., Kurucz, E., Ando, I. and Udvardy, A. (2004) Tissue- and developmental stage-specific changes in the subcellular localization of the 26S proteasome in the ovary of Drosophila melanogaster. Gene Expr Patterns, 4: 329–333.

Agarwal, S. and Sohal, R.S. (1994) Aging and proteolysis of oxidized proteins. Arch Biochem Biophys, 309: 24–28.

Amici, M., Lupidi, G., Angeletti, M., Fioretti, E. and Eleuteri, A.M. (2003) Peroxynitrite-induced oxidation and its effects on isolated proteasomal systems. Free Radic Biol Med, 34: 987–996.

Anselmi, B., Conconi, M., Veyrat-Durebex, C., Turlin, E., Biville, F., Alliot, J. and Friguet, B. (1998) Dietary self-selection can compensate an age-related decrease of rat liver 20 S proteasome activity observed with standard diet. J Gerontol A Biol Sci Med Sci, 53: B173–179.

Auluck, P.K., Meuleener, M.C. and Bonini, N.M. (2005) Mechanisms of suppression of a-synuclein neurotoxicity by geldamycin in Drosophila. J Biol Chem, 280: 2873–2878.

Barja, G. (2002) Rate of generation of oxidative stress-related damage and animal longevity. Free Radic Biol Med, 33: 1167–1172.

Beckman, K.B. and Ames, B.N. (1998) The free radical theory of aging matures. Physiol Rev, 78: 547–581.

Beedholm, R., Clark, B.F. and Rattan, S.I. (2004) Mild heat stress stimulates 20S proteasome and its 11S activator in human fibroblasts undergoing aging in vitro. Cell Stress Chaperones, 9: 49–57.

Benaroudj, N., Tarcsa, E., Cascio, P. and Goldberg, A.L. (2001) The unfolding of substrates and ubiquitin-independent protein degradation by proteasomes. Biochimie, 83: 311–318.

Breedholm, R., Clark, B.F. and Rattan, S.I. (2004) Mild heat stress stimulates 20S proteasome and its 11S activator in human fibroblasts undergoing aging in vitro. Cell Stress Chaperones, 9: 49–57.

Bulteau, A.L., Petropoulos, I. and Friguet, B. (2000) Age-related alterations of proteasome structure and function in aging epidermis. Exp Gerontol, 35: 767–777.

Bulteau, A.L., Szweda, L.I. and Friguet, B. (2002) Age-dependent declines in proteasome activity in the heart. Arch Biochem Biophys, 397: 298–304.

Caballero, M., Liton, P.B., Challa, P., Epstein, D.L. and Gonzalez, P. (2004) Effects of donor age on proteasome activity and senescence in trabecular meshwork cells. Biochem Biophys Res Commun, 323: 1048–1054.

Chen, Q., Thorpe, J., Ding, Q., El-Amouri, I.S. and Keller, J.N. (2004) Proteasome synthesis and assembly are required for survival during stationary phase. Free Radic Biol Med, 37: 859–868.

Chondrogianni, N. and Gonos, E.S. (2004) Proteasome inhibition induces a senescence-like phenotype in primary human fibroblasts cultures. Biogerontology, 5: 55–61.

Chondrogianni, N., Stratford, F.L., Trougakos, I.P., Friguet, B., Rivett, A.J. and Gonos, E.S. (2003) Central role of the proteasome in senescence and survival of human fibroblasts: induction of a senescence-like phenotype upon its inhibition and resistance to stress upon its activation. J Biol Chem, 278: 28026–28037.

Conconi, M., Szweda, L.I., Levine, R.L., Stadtman, E.R. and Friguet, B. (1996) Age-related decline of rat liver multicatalytic proteinase activity and protection from oxidative inactivation by heat-shock protein 90. Arch Biochem Biophys, 331: 232–240.

Corcoran, L.J., Mitchison, T.J. and Liu, Q. (2004) A novel action of histone deactylase inhibitors in a protein aggresome disease model. Curr Biol, 14: 488–492.

Davies, K.J. (2001) Degradation of oxidized proteins by the 20S proteasome. Biochimie, 83: 301–310.

Deng, G.Y., Muir, A., Maclaren, N.K. and She, J.X. (1995) Association of LMP2 and LMP7 genes within the major histocompatibility complex with insulin-dependent diabetes mellitus: population and family studies. Am J Hum Genet, 56: 528–534.

Dietrich, P., Rideout, H.J., Wang, Q. and Stefanis, L. (2003) Lack of p53 delays apoptosis, but increases ubiquitinated inclusions, in proteasomal inhibitor-treated cultured cortical neurons. Mol Cell Neurosci, 24: 430–441.

Ding, Q. and Keller, J.N. (2001) Proteasomes and proteasome inhibition in the central nervous system. Free Radic Biol Med, 31: 574–584.

Ding, Q. and Keller, J.N. (2001) Proteasome inhibition in oxidative stress neurotoxicity: implications for heat shock proteins. J Neurochem, 77: 1010–1017.

Ding, Q., Lewis, J.J., Strum, K.M., Dimayuga, E., Bruce-Keller, A.J., Dunn, J.C. and Keller, J.N. (2002) Polyglutamine expansion, protein aggregation, proteasome activity, and neural survival. J Biol Chem, 277: 13935–13942.

Ding, Q., Dimayuga, E., Martin, S., Bruce-Keller, A.J., Nukala, V., Cuervo, A.M. and Keller, J.N. (2003) Characterization of chronic low-level proteasome inhibition on neural homeostasis. J Neurochem, 86: 489–497.

Ding, Q., Bruce-Keller, A.J., Chen, Q. and Keller, J.N. (2004a) Analysis of gene expression in neural cells subject to chronic proteasome inhibition. Free Radic Biol Med, 36: 445–455.

Ding, Q., Dimayuga, E., Markesbery, W.R. and Keller, J.N. (2004b) Proteasome inhibition increases DNA and RNA oxidation in astrocyte and neuron cultures. J Neurochem, 91: 1211–1218.

Elsasser, S., Chandler-Militello, D., Muller, B., Hanna, J. and Finley, D. (2004) Rad23 and Rpn10 serve as alternative ubiquitin receptors for the proteasome. J Biol Chem, 279: 26817–26822.

Esterbauer, H., Schaur, R.J. and Zollner, H. (1991) Chemistry and biochemistry of 4-hydroxynonenal, malonaldehyde and related aldehydes. Free Radic Biol Med, 11: 81–128.

Ferrington, D.A. and Kapphahn, R.J. (2004) Catalytic site-specific inhibition of the 20S proteasome by 4-hydroxynonenal. FEBS Lett, 578: 217–223.

Forster, M.J., Sohal, B.H. and Sohal, R.S. (2000) Reversible effects of long-term caloric restriction on protein oxidative damage. J Gerontol A Biol Sci Med Sci, 55: B522–529.

Fribley, A., Zeng, Q. and Wang, C.Y. (2004) Proteasome inhibitor PS-341 induces apoptosis through induction of endoplasmic reticulum stress-reactive oxygen species in head and neck squamous cell carcinoma cells. Mol Cell Biol, 24: 9695–9704.

Friguet, B. and Szweda, L.I. (1997) Inhibition of the multicatalytic proteinase (proteasome) by 4-hydroxy-2-nonenal cross-linked protein. FEBS Lett, 405: 21–25.

Friguet, B., Bulteau, A.L., Conconi, M. and Petropoulos, I. (2002) Redox control of 20S proteasome. Methods Enzymol, 353: 253–262.

Glockzin, S., von Knethen, A., Scheffner, M. and Brune, B. (1999) Activation of the cell death program by nitric oxide involves inhibition of the proteasome. J Biol Chem, 274: 19581–19586.

Goto, S., Takahashi, R., Araki, S. and Nakamoto, H. (2002) Dietary restriction initiated in late adulthood can reverse age-related alterations of protein and protein metabolism. Ann N Y Acad Sci, 959: 50–56.

Gray, D.A., Tsirigotis, M. and Woulfe, J. (2003) Ubiquitin, proteasomes, and the aging brain. Sci Aging Knowledge Environ 2003, RE6.

Griffin, T.A., Slack, J.P., McCluskey, T.S., Monaco, J.J. and Colbert, R.A. (2000) Identification of proteassemblin, a mammalian homologue of the yeast protein, Ump1p, that is required for normal proteasome assembly. Mol Cell Biol Res Commun, 3: 212–217.

Grimm, L.M. and Osborne, B.A. (1999) Apoptosis and the proteasome. Results Probl Cell Differ, 23: 209–228.

Grune, T. and Davies, K.J. (1997) Breakdown of oxidized proteins as a part of secondary antioxidant defenses in mammalian cells. Biofactors, 6: 165–172.

Grune, T., Reinheckel, T. and Davies, K.J. (1997) Degradation of oxidized proteins in mammalian cells. Faseb J, 11: 526–534.

Grune, T., Blasig, I.E., Sitte, N., Roloff, B., Haseloff, R. and Davies, K.J. (1998) Peroxynitrite increases the degradation of aconitase and other cellular proteins by proteasome. J Biol Chem, 273: 10857–10862.

Grune, T., Shringarpure, R., Sitte, N. and Davies, K. (2001) Age-related changes in protein oxidation and proteolysis in mammalian cells. J Gerontol A Biol Sci Med Sci, 56: B459–467.

Harman, D. (2001) Aging: overview. Ann N Y Acad Sci, 928: 1–21.

Hensley, K. and Floyd, R.A. (2002) Reactive oxygen species and protein oxidation in aging: a look back, a look ahead. Arch Biochem Biophys, 397: 377–383.

Heward, J.M., Allahabadia, A., Sheppard, M.C., Barnett, A.H., Franklyn, J.A. and Gough, S.C. (1999) Association of the large multifunctional proteasome (LMP2) gene with Graves' disease is a result of linkage disequilibrium with the HLA haplotype DRB1*0304-DQB1*02-DQA1*0501. Clin Endocrinol (Oxf), 51: 115–118.

Hyun, D.H., Lee, M.H., Halliwell, B. and Jenner, P. (2002) Proteasomal dysfunction induced by 4-hydroxy-2,3-trans-nonenal, an end-product of lipid peroxidation: a mechanism contributing to neurodegeneration? J Neurochem, 83: 360–370.

Jesenberger, V. and Jentsch, S. (2002) Deadly encounter: ubiquitin meets apoptosis. Nat Rev Mol Cell Biol, 3: 112–121.
Keller, J.N. and Markesbery, W.R. (2000) Proteasome inhibition results in increased poly-ADP-ribosylation: implications for neuron death. J Neurosci Res, 61: 436–442.
Keller, J.N., Hanni, K.B. and Markesbery, W.R. (2000a) Possible involvement of proteasome inhibition in aging: implications for oxidative stress. Mech Ageing Dev, 113: 61–70.
Keller, J.N., Huang, F.F. and Markesbery, W.R. (2000b) Decreased levels of proteasome activity and proteasome expression in aging spinal cord. Neuroscience, 98: 149–156.
Keller, J.N., Huang, F.F., Dimayuga, E.R. and Maragos, W.F. (2000c) Dopamine induces proteasome inhibition in neural PC12 cell line. Free Radic Biol Med, 29: 1037–1042.
Keller, J.N., Dimayuga, E., Chen, Q., Thorpe, J., Gee, J. and Ding, Q. (2004) Autophagy, proteasomes, lipofuscin, and oxidative stress in the aging brain. Int J Biochem Cell Biol, 36: 2376–2391.
Kruger, E., Kloetzel, P.M. and Enenkel, C. (2001) 20S proteasome biogenesis. Biochimie, 83: 289–293.
Larsen, K.E. and Sulzer, D. (2002) Autophagy in neurons: a review. Histol Histopathol, 17: 897–908.
Ling, Y.H., Liebes, L., Zou, Y. and Perez-Soler, R. (2003) Reactive oxygen species generation and mitochondrial dysfunction in the apoptotic response to Bortezomib, a novel proteasome inhibitor, in human H460 non-small cell lung cancer cells. J Biol Chem, 278: 33714–33723.
Lopes, U.G., Erhardt, P., Yao, R. and Cooper, G.M. (1997) p53-dependent induction of apoptosis by proteasome inhibitors. J Biol Chem, 272: 12893–12896.
Louie, J.L., Kapphahn, R.J. and Ferrington, D.A. (2002) Proteasome function and protein oxidation in the aged retina. Exp Eye Res, 75: 271–284.
Maksymowych, W.P., Tao, S., Vaile, J., Suarez-Almazor, M., Ramos-Remus, C. and Russell, A.S. (2000) LMP2 polymorphism is associated with extraspinal disease in HLA-B27 negative Caucasian and Mexican Mestizo patients with ankylosing spondylitis. J Rheumatol, 27: 183–189.
Mangel, M. (2001) Complex adaptive systems, aging and longevity. J Theor Biol, 213: 559–571.
Merker, K., Stolzing, A. and Grune, T. (2001) Proteolysis, caloric restriction and aging. Mech Ageing Dev, 122: 595–615.
Mishto, M., Bonafe, M., Salvioli, S., Olivieri, F. and Franceschi, C. (2002) Age dependent impact of LMP polymorphisms on TNFalpha-induced apoptosis in human peripheral blood mononuclear cells. Exp Gerontol, 37: 301–308.
Mishto, M., Bellavista, E., Santoro, A., Stolzing, A., Ligorio, C., Nacmias, B., Spazzafumo, L., Chiappelli, M., Licastro, F., Sorbi, S., Pession, A., Ohm, T., Grune, T. and Franceschi, C. (2006) Immunoproteasome and LMP2 polymorphism in aged and Alzheimer's disease brains. Neurobiol Aging, 27: 54–66.
Nakaso, K., Yoshimoto, Y., Yano, H., Takeshima, T. and Nakashima, K. (2004) p53-mediated mitochondrial dysfunction by proteasome inhibition in dopaminergic SH-SY5Y cells. Neurosci Lett, 354: 213–216.
Noda, C., Tanahashi, N., Shimbara, N., Hendil, K.B. and Tanaka, K. (2000) Tissue distribution of constitutive proteasomes, immunoproteasomes, and PA28 in rats. Biochem Biophys Res Commun, 277: 348–354.
Obin, M., Shang, F., Gong, X., Handelman, G., Blumberg, J. and Taylor, A. (1998) Redox regulation of ubiquitin-conjugating enzymes: mechanistic insights using the thiol-specific oxidant diamide. Faseb J, 12: 561–569.
Ogiso, Y., Tomida, A. and Tsuruo, T. (2002) Nuclear localization of proteasomes participates in stress-inducible resistance of solid tumor cells to topoisomerase II-directed drugs. Cancer Res, 62: 5008–5012.
Okada, K., Wangpoengtrakul, C., Osawa, T., Toyokuni, S., Tanaka, K. and Uchida, K. (1999) 4-Hydroxy-2-nonenal-mediated impairment of intracellular proteolysis during oxidative stress. Identification of proteasomes as target molecules. J Biol Chem, 274: 23787–23793.
Pacifici, R.E., Salo, D.C. and Davies, K.J. (1989) Macroxyproteinase (M.O.P.): a 670 kDa proteinase complex that degrades oxidatively denatured proteins in red blood cells. Free Radic Biol Med, 7: 521–536.

Pacifici, R.E., Kono, Y. and Davies, K.J. (1993) Hydrophobicity as the signal for selective degradation of hydroxyl radical-modified hemoglobin by the multicatalytic proteinase complex, proteasome. J Biol Chem, 268: 15405–15411.
Parsons, P.A. (2003) From the stress theory of aging to energetic and evolutionary expectations for longevity. Biogerontology, 4: 63–73.
Pasquini, L.A., Besio Moreno, M., Adamo, A.M., Pasquini, J.M. and Soto, E.F. (2000) Lactacystin, a specific inhibitor of the proteasome, induces apoptosis and activates caspase-3 in cultured cerebellar granule cells. J Neurosci Res, 59: 601–611.
Petropoulos, I., Conconi, M., Wang, X., Hoenel, B., Bregegere, F., Milner, Y. and Friguet, B. (2000) Increase of oxidatively modified protein is associated with a decrease of proteasome activity and content in aging epidermal cells. J Gerontol A Biol Sci Med Sci, 55: B220–227.
Piccinini, M., Mostert, M., Croce, S., Baldovino, S., Papotti, M. and Rinaudo, M.T. (2003) Interferon-gamma-inducible subunits are incorporated in human brain 20S proteasome. J Neuroimmunol, 135: 135–140.
Qiu, J.H., Asai, A., Chi, S., Saito, N., Hamada, H. and Kirino, T. (2000) Proteasome inhibitors induce cytochrome c-caspase-3-like protease-mediated apoptosis in cultured cortical neurons. J Neurosci, 20: 259–265.
Radak, Z., Takahashi, R., Kumiyama, A., Nakamoto, H., Ohno, H., Ookawara, T. and Goto, S. (2002) Effect of aging and late onset dietary restriction on antioxidant enzymes and proteasome activities, and protein carbonylation of rat skeletal muscle and tendon. Exp Gerontol, 37: 1423–1430.
Rattan, S.I. (2004) Hormetic mechanisms of anti-aging and rejuvenating effects of repeated mild heat stress on human fibroblasts in vitro. Rejuvenation Res, 7: 40–48.
Rivett, A.J. (1993) Proteasomes: multicatalytic proteinase complexes. Biochem J, 291 (Pt 1), 1–10.
Ryu, H., Smith, K., Camelo, S.I., Carreras, I., Lee, J., Iglesias, A.H., Dandond, F., Cormier, K.A., Cudkowicz, M.F., Brown, R.H. and Ferrante, R.J. (2005) Sodium phenylbutyrate prolongs survival and regulates expression of anti-apoptotic genes in transgenic amyotrophic lateral sclerosis mice. J Neurochem, 93: 1087–1098.
Schmidt, M. and Kloetzel, P.M. (1997) Biogenesis of eukaryotic 20S proteasomes: the complex maturation pathway of a complex enzyme. Faseb J, 11: 1235–1243.
Singh, S., Awasthi, N., Egwuagu, C.E. and Wagner, B.J. (2002) Immunoproteasome expression in a nonimmune tissue, the ocular lens. Arch Biochem Biophys, 405: 147–153.
Sitte, N., Merker, K., von Zglinicki, T. and Grune, T. (2000a) Protein oxidation and degradation during proliferative senescence of human MRC-5 fibroblasts. Free Radic Biol Med, 28: 701–708.
Sitte, N., Huber, M., Grune, T., Ladhoff, A., Doecke, W.D., Von Zglinicki, T. and Davies, K.J. (2000b) Proteasome inhibition by lipofuscin/ceroid during postmitotic aging of fibroblasts. Faseb J, 14: 1490–1498.
Sitte, N., Merker, K., Von Zglinicki, T., Davies, K.J. and Grune, T. (2000c) Protein oxidation and degradation during cellular senescence of human BJ fibroblasts: part II–aging of nondividing cells. Faseb J, 14: 2503–2510.
Sitte, N., Merker, K., Von Zglinicki, T., Grune, T. and Davies, K.J. (2000d) Protein oxidation and degradation during cellular senescence of human BJ fibroblasts: part I–effects of proliferative senescence. Faseb J, 14: 2495–2502.
Sittler, A., Lurz, R., Lueder, G., Priller, J., Leharch, H., Hayer-Hartl, M.K., Hartl, F.U. and Wanker, E.E (2001) Geldamycin activates a heat shock response and inhibits huntingtin aggregation in a cell culture model of Huntington's disease. Hum Mol Genet, 10: 1307–1315.
Sohal, R.S. and Weindruch, R. (1996) Oxidative stress, caloric restriction, and aging. Science, 273: 59–63.
Squier, T.C. (2001) Oxidative stress and protein aggregation during biological aging. Exp Gerontol, 36: 1539–1550.
Sullivan, P.G., Dragicevic, N.B., Deng, J.H., Bai, Y., Dimayuga, E., Ding, Q., Chen, Q., Bruce-Keller, A.J. and Keller, J.N. (2004) Proteasome inhibition alters neural mitochondrial homeostasis and mitochondria turnover. J Biol Chem, 279: 20699–20707.

Uchida, K. (2003) 4-Hydroxy-2-nonenal: a product and mediator of oxidative stress. Prog Lipid Res, 42: 318–343.

Ullrich, O. and Grune, T. (2001) Proteasomal degradation of oxidatively damaged endogenous histones in K562 human leukemic cells. Free Radic Biol Med, 31: 887—893.

Ullrich, O., Reinheckel, T., Sitte, N., Hass, R., Grune, T. and Davies, K.J. (1999) Poly-ADP ribose polymerase activates nuclear proteasome to degrade oxidatively damaged histones. Proc Natl Acad Sci USA, 96: 6223–6228.

Vinasco, J., Fraile, A., Nieto, A., Beraun, Y., Pareja, E., Mataran, L. and Martin, J. (1998) Analysis of LMP and TAP polymorphisms by polymerase chain reaction-restriction fragment length polymorphism in rheumatoid arthritis. Ann Rheum Dis, 57: 33–37.

Viteri, G., Carrard, G., Birlouez-Aragon, I., Silva, E. and Friguet, B. (2004) Age-dependent protein modifications and declining proteasome activity in the human lens. Arch Biochem Biophys, 427: 197–203.

Walters, K.J., Lech, P.J., Goh, A.M., Wang, Q. and Howley, P.M. (2003) DNA-repair protein hHR23a alters its protein structure upon binding proteasomal subunit S5a. Proc Natl Acad Sci USA, 100: 12694–12699.

Weindruch, R. (1996) The retardation of aging by caloric restriction: studies in rodents and primates. Toxicol Pathol, 24: 742–745.

Williams, S.A. and McConkey, D.J. (2003) The proteasome inhibitor bortezomib stabilizes a novel active form of p53 in human LNCaP-Pro5 prostate cancer cells. Cancer Res, 63: 7338–7344.

Wojcik, C. (1999) Proteasomes in apoptosis: villains or guardians? Cell Mol Life Sci, 56: 908–917.

CHAPTER 16

NUTRITIONAL DEFICIENCY AND ITS MODULATION IN OLD AGE

CARLOS K.B. FERRARI

Postgraduate Program of Functional Foods & Nutraceuticals – Racine Institute (IR); R. Padre Chico, 93, São Paulo (SP), Brazil, 05008-10. Postgraduate Program of Functional Nutrition and Quality of Life – Centro Universitário Adventista (Unasp).

Abstract: This chapter covers physiological roles, dietary requirements and deficiencies associated to macronutrients and micronutrients in older people. It has been postulated that mitochondrial failure and oxidative stress are major events in cell aging and death. Nutritional modulation of mitochondria and antioxidant activities of nutrients and other bioactive compounds from functional foods help to reduce the risk of chronic diseases of aging

Keywords: aging; nutritional deficiencies; vitamins; minerals; mitochondrial nutrition; apoptosis; cancer; coenzyme Q10; lipoic acid; polyphenols; antioxidants

During aging and senescence, a variety of complex biochemical, cellular, and physiological changes are processed in the human body (Rattan, 2003a). Some of these changes can seriously affect the nutritional status of humans, although others cannot. Geriatrics and gerontology specialists have considered the most important age-related health problems as following below (Pahor and Applegate, 1997; Mecocci et al., 2000; Morley, 2004; Lyle et al., 1999).

- Cardiovascular diseases (atherosclerosis and other heart diseases);
- Hypertension;
- Immunological system disorders, including inflammation and immunosenescence;
- Cognitive decline;
- Oxidative stress;
- Diabetes, obesity, and osteoporosis; and
- Cataract and eye diseases.

S. Rattan and M. Kassem (eds.), Prevention and Treatment of Age-related Diseases, 313–334.

Nutritional deficiency in the elderly is very common and can seriously increase morbidity and mortality, compromising human life span. Clinical signs of nutritional deficiencies are presented in Table 1.

Many epidemiological studies with healthy centenarians have revealed that the most important determinants of human longevity are (Mecocci et al., 2000; Morley, 2004; Perls and Terry, 2003; Zyczkowska et al., 2004).

- Better glucose control and reduced risk of diabetes;
- Reduced risk of hypertension;
- Maintenance of higher levels of blood and body antioxidants (vitamins A and E); and
- Decreased risk of heart diseases.

Table 1. Clinical signs of possible nutritional deficiency

System	Sign or symptom	Nutritional deficiency
General	Wasted/thin	Calorie
	Stooped posture	Calcium, vitamin D
Skin	Dry scaly skin	Zinc/essential fatty acids
	Follicular hyperkeratosis	Vitamins A and C
	Petechia	Vitamins C and K
	Poor wound healing	Zinc/vitamin C
	Scrotal dermatosis	Riboflavin
Hair	Thin/despigmented	Protein
	Easy plukability	Protein/zinc
Nails	Transverse depigmentation	Albumin
	Spooning	Iron
Eyes	Night blindness	Vitamin A, zinc
	Conjunctival inflammation	Riboflavin
	Keratomalacia	Vitamin A
Mouth	Bleeding gums	Vitamin C, riboflavin
	Glossitis	Niacin, piridoxin, riboflavin
	Atrophic papillae	Iron
	Hypogeusia	Zinc/vitamin A
Neck	Thyroid enlargement	Iodine
	Parotid enlargement	Protein
Abdomen	Diarrhea	Niacin, folate, vitamin B_{12}
	Hepatomegaly	Protein
Extremities	Bone tenderness	Vitamin D
	Joint pain	Vitamin C
	Muscle tenderness	Thiamin
	Muscle wasting	Protein-calorie, selenium, vitamin D
	Edema	Protein
Neurologic	Tetany	Calcium, magnesium
	Paresthesia	Thiamin, vitamin B_{12}
	Ataxia	Vitamin B_{12}
	Dementia	Vitamin B_{12}, niacin
	Hyporeflexia	Thiamin

Source: Morley (2004).

Nutritional deficiency in old age and its nutritional modulation by macronutrients, micronutrients, and nutraceuticals, beyond the advanced studies regarding molecular nutritional physiology are presented in this chapter.

1. ENERGY BALANCE AND BODY SHAPE: KEYS FOR LONGEVITY?

Energy is necessary for almost all cellular and physiological events. Caloric needs should be measured in an exercise physiology laboratory. Generally, most elderly people need 1.600 Kcal every day (Wolf and Tanner, 2002). But in pathological conditions, such as infections and fever, energy metabolism is enhanced (ambulatory condition), increasing the caloric needs. There are predictive equations for caloric needs of older, presented in Table 2 (WHO, 1985). Even in health or disease, those energy predictions are not too much accurate, but can help the clinician to understand if those needs can influence normal body mass index ([BMI: weight (Kg)/(height)^2 (m)], ranging from 18,5 to 24.9) and adequate waist circumferences (women: <80 cm; men: <94 cm), leading to clinical malnutrition in the elderly, where these anthropometrical measures are under the normal values. Higher anthropometric values are associated with overweight (25 $\leq$BMI $\leq$30) or obesity (BMI $\geq$ 30), increasing the risk of diabetes, cardiovascular and other chronic non-communicable diseases, especially when associated with higher waist circumferences (women: >88 cm; men: >102 cm) (Wolf and Tanner, 2002).

Decreased energy expenditure (sedentarism) and increased caloric intake could lead the organism to overweight or obesity. Obesity has been associated with decreased life expectancy, especially when body mass index is higher than 45 (Fontaine et al., 2003). It is also associated to increased cancer mortality in men (52%) and women (62%) (Calle et al., 2003).

Aging is accelerated by with higher mitochondrial respiratory activities, with consequent superoxide ($O_2^{\cdot-}$) releasing, and simultaneous lower rates of antioxidant defenses. On the contrary, lower rates of superoxide releasing from mitochondria and higher levels of SOD, glutathion peroxidase (GPx) and GSH could be the major determinants of maximum life span (Ku and Sohal, 1993; Barja et al., 1994). This

Table 2. Estimation of basal and total metabolic rate for older people

Age/sex	Basal metabolic rate (Kcal/d)	Total metabolic rate (Kcal/d)
Women >60	10.5 × weight + 596	BMR X Activity factor (AF) Resting: 1,2 Ambulatory: 1,3 Normal activity: 1,5 Exercise: 2,0
Men >60	13.5 × weight + 487	BMR X AF

Source: WHO (1985).

explains why life span of some organisms could be increased by caloric restriction (Rattan, 2003a), an anti-aging intervention that decreases breast cancer risk in humans (Michels and Ekbom, 2004).

2. WATER, PROTEIN AND AMINOACIDS NEEDS OF OLDER

Dehydration in very common in the old people and it is caused by poor liquid and food ingestion, as well as gastrointestinal disorders, such as vomiting and diarrhea. Considering that water constitutes 60 to 70% of the human body cells, fluids, and tissues, people should pay attention in order to drink at least eight cups of water or liquids per day (1.500 mL) (Thaler et al., 1999). In a healthy-balanced diet, carbohydrates account for 60% of the total caloric intake, lipids for 20 to 30%, and 15 to 20% should be provided by protein sources. A gram of carbohydrates, proteins, lipids, and alcohol provides 4, 4, 9, and 7 kilocalories, respectively (Garrett and Grisham, 1995). Adults need about 1,0 gram of protein *per* kilogram of body weight and 6.5 g of the essential aminoacids (Somer, 2003).

Arginine is an important aminoacid with many physiological functions, such as (Zimmermann, 2001; Heys et al., 2004):

- Stimulation of prolactin, glucagon, and insulin secretions;
- Precursor of nitric oxide biosynthesis, a potent vasodilator agent;
- Increasing collagen synthesis and wound healing;
- Stimulation of cellular immunity (increase both NK cells populations, delayed-type hypersensitivity responses, and T-cell mitogenic responses; enhance cytokines' releasing).

Besides the same immunological actions executed by arginine, glutamine also plays an important role in B-cell differentiation and antibody biosynthesis, macrophage phagocytic functions, neutrophil activation, and improvement of gut mucosal barrier (Zimmermann, 2001; Heys et al., 2004).

3. VITAMINS AND AGING: CAN THEY MODULATE OUR HEALTH?

Originally designed as being "essential amines", the term vitamin has been used for decades. But today, it refers only to a group of many different organic compounds (not restricted to amines) not synthesized by human cells but necessary for cellular metabolic reactions and normal body development (Garrett and Grisham, 1995). Vitamins are essential coenzymes or co-factors in the intermediary metabolism and some of them can perform important antioxidant activities. Vegetables and fruits are good sources of vitamins, but all of us should also ingest legumes, seeds, eggs, milk, fish, poultry, and meat in order to get vitamins, essential aminoacids and heme-iron from meat. In general, contradicting the common sense, older adults do not require more vitamins and minerals than children or adolescents. In fact, if they need less energy, lower vitamin and mineral intake is also required. Nowadays scientists have demonstrated that higher intake of some vitamins, minerals, and

Table 3. Biochemico-physiological roles and reference intakes of vitamins for elderly

Name (chemical)	Biochemical actions	Physiological roles	Dietary reference intake (DRI)*
Water-soluble			
Thiamin (B_1) (Thiamin pyrophosphate)	Coenzyme in energy metabolism essential for acetyl-coenzyme A and succinyl-CoA biosynthesis	Essential for brain, Increases appetite, Can help in depression treatment	1.1 mg (♀) 1.2 mg (♂)
Riboflavin (B_2) (Flavin adenine mononucleotide [FMN] and Flavin adenine dinucleotide [FAD])	Coenzymes of the electron transport chain. Structural component of the flavoproteins (succinate dehydrogenase and Acyl-CoA), and cytochrome P_{450} hydroxilase	Essential for skin, eye, tongue and physical work capacity	1.1 mg (♀) 1.3 mg (♂)
Niacin (B_3) (Nicotinamide and nicotic acid)	Nicotinamide adenine dinucleotide (NAD^+) and its phosphorylated form ($NADP^+$) are respectively electron and proton acceptors of the respiratory chain. It participates in carbohydrate metabolism and lipid synthesis, inhibiting fatty acid releasing	Maintain skin, gut and nervous system	14 mg (♀) 16 mg (♂)
Biotin (B_4)	Coenzyme in carboxylation reactions of energy, protein and fat metabolism	Essential for skin, hair, muscle, and brain	30 μg
Pantotenic acid (B_5)	Co-factor of the coenzime A; Gluconeogenesis; energy metabolism; fatty acid metabolism; and acetilcholine synthesis	Important for appetite, work capacity, and brain	5 mg
Piridoxin (B_6) (Piridoxal-phosphate, piridoxin, and piridoxamine)	Participates in transamination, decarboxilation and isomerization reactions. Activates gluconeogenesis and glicogenolysis	Essential to the skin, tongue, nervous and muscle functions	1.5 mg (♀) 1.7 mg (♂)

(Continued)

Table 3. (Continued)

Name (chemical)	Biochemical actions	Physiological roles	Dietary reference intake (DRI)*
Vitamin B_{12} (Cianocobalamine)	B_{12} coenzyme Transmetylation reactions: conversion of homocysteine (Hcy) in methionine (Met); conversion of L-methyl-malonil-CoA into succinyl-CoA	Essential to erythropoiesis, DNA synthesis, cerebrovascular and cardiovascular systems	2.4 μg
Folic acid (Tetrahydrofolate)	Conversion of glycine to serine; synthesis of timidine-5-P (DNA precursor); conversion of homocysteine (Hcy) to Met (methionine)	Prevents neural tube defects; aids work capacity; prevents vascular impairment	400 μg
Vitamin C (Ascorbic acid)	Lisine hydroxilation to form proline (collagen synthesis); increases iron absorption in the gut; regulates folate, cholesterol, and aminoacid metabolism	Important to erithropoiesis, collagen synthesis, iron gut absorption, and adrenaline synthesis	75 mg (♀) 90 mg (♂)
Lipid-soluble			
Vitamin A (*cis*-Retinal, retinol, retinoic acid, all *trans*-retinal)	Form the visual eye pigment; some forms are free radical scavengers	Essential for the eye, skin, mucosae, bones, and immunity	700 μg** (♀) 900 μg** (♂)
Vitamin E (Calciferol)	Promotes calcium gut absorption; regulates phosphorus metabolism	Maintain skin and bones and is used in psoriasis treatment	51–70y: 10 μg >70 y: 15 μg
	Antioxidant that participates in the structure of cell membranes and scavenges free radicals	Inhibit LDL-cholesterol oxidation; decrease Alzheimer's and Parkinson's disease risk	15 mg
Vitamin K (Menadione, filoquinone, and menaquinone)	Blood clothing factor (promotes carboxylation of VII and X blood coagulation factors); participates in protein synthesis; increases osteocalcin function	Anti-hemorrhagic factor II	90 μg (♀) 120 μg (♂)

* RDI according to the Institute of Medicine, Food and Nutrition Board (www.nap.edu).

** RAEs = retinol activity equivalents; 1RAE = 1 μg retinol, 12 μg β-carotene, 24 μg α-carotene, or 24 μg β-cryptoxanthin.

Table 4. Food sources of vitamins

Vitamin	Foods
B_1	Ham, meats, grains, legumes, whole cereals and enriched breads, liver, fish, poultry, pasta, nuts and yeast
B_2	Milk and derivatives, meats (liver), grains, green leafy vegetables, beans, eggs, and yoghurt
B_3	Meats, fish, poultry, grains, beans, yeast, liver, legumes, milk, seeds, eggs
B_4	Produced by intestinal bacteria. Found in liver, egg yolk, peas, beans, and green leafy vegetables
B_5	Liver, beef, milk, eggs, legumes, grains and vegetables
B_6	Protein-rich foods; liver; slim meat; fish; poultry; green leafy vegetables; banana; and whole cereals
B_{12}	Foods of animal origin
Folate	Liver; green leafy vegetables; legumes; nuts and seeds; rice; cereals, and pasta
Vitamin C	Citric fruits and juices; cashew; acerola; green leafy vegetables; broccolis; red and sweet peppers; strawberry; potatoes
A	Buriti palm oil; liver; milk; cheese; carrots; green leafy vegetables; sweet potato; yellow-orange fruits (mango, peach)
D	UV sunlight induces skin synthesis of vitamin D; fish oils; margarine and enriched milk products
E	Vegetable oils and seeds; wheat germ; whole products; and egg yolk
K	Liver; eggs; spinach; cauli-flower; broccolis; microbial gut biosynthesis

phytochemicals usually has potent anti-aging effects on cell proliferation, injury, and death. This fact could be associated with increased cell survival and improved physiology of human body, offering better health outcomes during aging. Vitamins can be classified in water-soluble (B complex, C, folic acid, biotin, pantotenic acid, etc) and lipid soluble (A, D, E and K) compounds. Their functions and biochemical and physiological roles are summarized in Table 3, whereas their common food sources are listed in Table 4.

4. ESSENTIAL MINERALS AND OLD AGE

Important since the early life, minerals execute many biological functions. Calcium is important to bones and teeths and participates of nervous transmission, muscle contraction, and blood clothing (Somer, 2003). In recent years researchers have been discovered that calcium decreases hyperproliferation of colon cancer cells(Lipkin, 1999; Kelloff et al., 2000), inhibits ornithine decarboxilase activity, decreases the mutation rate of *ras* (a gene involved in proliferative responses), and promotes the formation of insoluble complexes with bile and fatty acids, decreasing proliferative and irritative effects on intestine (Lipkin, 1999), mechanisms that might result in decreased intestinal cancer risk.

Another important mineral is copper. It is essential constituent of at least 15 enzymes, including the antioxidant Cu,Zn-superoxide dismutase (Cu,Zn-SOD), and participates of collagen, melanin, and myelin synthesis maintaining the integrity of bones, cartilages, connective and nervous tissues (Somer, 2003; Richard and Roussel, 1999). Copper deficiency is common feature in diabetes mellitus patients, and its supplementation should be feasible (Pedrosa and Cozzolino, 1999).

Chromium participates of energy metabolism, estabilizes DNA and RNA against mutations, and improves glucose-mediated insulin sensitivity, helping diabetic patients (Lukaski, 2004).

Iodine and iron are essential minerals for any person, especially those aging subjects. Iodine is the active center of the thyroid hormones (T_3 and T_4) that regulates energy metabolism, physical and mental development and reproductive functions (Somer, 2003). Hypothyroidism and goitre are consequences of iodine deficiency (Ramalingaswami, 1992). Iron is the active center of hemeproteins such as hemoglobin (erythrocyte), myoglobin (muscle), and mitochondrial cytochromes (Richard and Roussel, 1999). Iron-deficiency anemia severily affects work capacity, aerobic and brain functions (Lukaski, 2004).

Dental caries are caused by acids produced in carbohydrate glucolysis. Fluoride inactivates some of these acids, reducing dental caries, and also helps bone mineralization.

Magnesium is a structural component of mitochondrial membrane and is involved in nervous transmission, protein catabolism, insulin synthesis, muscle relaxation, and estructure of protective teeths' enamel (Zimmermann, 2001; Lukaski, 2004).

Insulin activity, conjuntive tissues' anabolism, vitamin K cofactor, cholesterol and DNA biosynthesis, carbohydrate catabolism, and estructure of the Mn-SOD are the main functions of manganese (Somer, 2003; Lukaski, 2004).

Molibdenium constitutes dental enamel, decrease uric acid production, and is an active factor of enzymes that metabolizes carbohydrates, lipids, proteins, iron, sulfur aminoacids, and DNA (Zimmermann, 2001).

Cells use phosphorus to produce many membrane phospholipids. In bones, its co-deposition with calcium forms hydroxyapatite. Acid-base equilibrium of blood and fluids, muscle anabolism, and energy production are also performed by phosphorus.

Sodium chloride participates in the hydroelectrolytic equilibrium, but its excessive dietary intake is associated with hypertension. Potassium, another important electrolyte, controls nervous transmission and induces pos-contraction muscle relaxation, decreasing cardiac frequency (Thaler et al., 1999; Lukaski, 2004).

Selenium is a component of many different selenoproteins and enzymes, such as Se-glutathione-peroxidase (Se-GPx). It controls antibody production by B cells and phagocytic functions (Hughes, 2000). Lower levels of selenium increases the risk of cardiomyopathies, myositis, impair physical growth, and increase the risk of infections, especially that caused by viruses (Richard and Roussel, 1999; Levander, 2000). Higher selenium status is associated with decreased risk of prostate cancer, since selenium acts as a potent antioxidant, able to induce tumor cell death and inhibit new angiogenesis in tumoral tissues (Nève, 2002). It has been suggested that supranutritional doses of selenium (200 μg equivalent to 4-fold RDA) could protect against cancer (Nève, 2002). Administration of ebselen (10 mg/Kg, twice),

Table 5. Mineral requirements for older people and its food sources

Mineral	Requirements*	Foods
Calcium	1,200 mg	Milk and derivatives, fish with edible bones, dark green vegetables, fortified foods
Chromium	20 μg (♀) 30 μg (♂)	Wheat germ, brewer's yeast, cheese, and whole grains
Copper	900 μg	Cocoa powder, nuts and seeds, liver, seafoods
Fluoride	3 mg (♀) 4 mg (♂)	Tea, fluoridated water, ocean fish with edible bones
Iodine	150 μg	Iodized salt, seafood
Iron	8 mg	Meat, liver, egg yolk, dark green vegetables, and whole grains
Magnesium	320 mg (♀) 420 mg (♂)	Nuts, seeds, whole grains, wheat germ, bran, green vegetables, bananas
Manganese	1.8 mg (♀) 2.3 mg (♂)	Whole grains, fruits, vegetables, teas
Molybdenium	45 μg	Milk, beans, grains
Phosphorus	700 mg	Animal and high-protein vegetable products, whole grains
Selenium	55 μg	Seafood, meats, liver and kidney, onions, grains
Zinc	8 mg (♀) 11 mg (♂)	Meat, liver, eggs, seafoods, whole grains

* According to the Institute of Medicine, Food and Nutrition Board (www.nap.edu).

an organic-selenium, before ischemia and 12 hours after reperfusion, attenuates neuronal damage indicating a promising therapy for stroke (Namura et al., 2001).

Zinc is essential to imunological system function, physical growth, skin (cicatrization), and hair. Impairment of immunity and physical growth, nightly blindness, alopecia, hipogonadism, dermatitis, and increased mortality risk is associated with poor zinc status (Zimmermann, 2001; Lukaski, 2004). Zinc deficiency is also associated with decreased B and T cells functions, with impairment in cytokine responses and macrophage activation, and compromise of epithelials' physiology (Berger, 2002).

Mineral requirements and food sources are listed in Table 5.

Another recent functional aspects of mineral, vitamin and phytochemicals on healthy aging are further presented on this chapter.

5. BIOCHEMICAL AND PHARMACOLOGICAL TARGETS FOR NUTRITIONAL MODULATION OF AGING

Aging and its related disorders have been associated with specific cellular, molecular, and tissue changes during the life course. Nutritional modulation of these cell changes should consider the following biochemical-pharmacology approaches (or cell targets)(Ames et al., 1993; Mahoney et al., 2002; Rattan, 2003b; Ferrari, 2004):

- Mitochondrial function and structural stabilizers that avoids cell death, increasing its survival;
- Antioxidant activities (directly and indirectly free radical scavenging);

- Apoptotic inducers that avoid proliferation of aged-degenerated cells or neoplastic cells;
- Metal-chelators (Cuajungco et al., 2000);
- Immunological stimulators and inflammatory suppressors.

6. ANTIOXIDANTS: SCAVENGING OF FREE RADICALS AND PREVENTION OF CHRONIC DISEASES OF AGING

Free radicals (FR), produced during mitochondrial respiration and also released by peroxisomes, catalyze many redox reactions of various compounds in living tissues and cells (Halliwell, 1994; Gutteridge, 1995) and also in foods (Ferrari, 1998; Ferrari, 1999). FR is any atom, molecule or compound that present unpaired electrons and is able to receive or give them. The First is the oxidant and second is the reducer. Between reactive oxygen (ROS) or nitrogen species (RNS), there are FR [superoxide ($O_2^{\cdot-}$), hydroxyl ($^{\cdot}OH$), perhydroxyl ($HO_2^{\cdot-}$), peroxyl ($CO_2^{\cdot}$), nitric oxide ($NO^{\cdot}$), peroxynitrite ($ONOO^-$)] and reactive non-radical molecules, such as hydrogen peroxide (H_2O_2), singlet oxygen ($^1O_2$); although exist another FR without oxygen (tyil or CH_2S^-)(Gutteridge, 1995). FR catalysts promote oxidative stress yielding ROS could be (Ferrari, 1999; Möller et al., 1996).

1) High impact energy sources (thermal, microwave, radioactive);
2) Metals (cadmium, copper, iron, mercury, zinc, etc);
3) Enzymes (metalic or not), dispersed or grouped in cytoplasmic organelles (hepatic microsome, peroxisome, mitochondria, chloroplast and other plastids);
4) Mechanic action;
5) Toxic agents (alcohol, pesticides, cigarette, air pollutants, etc);
6) Physical and psychological stresses.

As discussed earlier, low free radicals releasing and high antioxidant protection, offered by cell antioxidants (SOD, GSH, GPx, catalase, etc) or by dietary intake decrease oxidation and increase cell and organism longevity (Mecocci et al., 2000; Ku and Sohal, 1993; Barja et al., 1994; Ames et al., 1993; Rattan, 2003b).

7. DIETARY ANTIOXIDANTS AND BRAIN PROTECTION

The brain of old rats had increased oxidative DNA and RNA markers both associated with temporal and special memory losses but acetyl-L-carnitine and lipoic acid reversed these adverse effects of aging (Liu et al., 2002). In this respect, it was observed an inverse consistent association between vitamin E levels and lower memory performance in NHANES III study (Perkins et al., 1999).

Higher intake of vegetables and fruits rich in vitamin C and carotenoids was positively associated with better cognitive function in the elderly (Berr, 2000). Besides contradictory results of epidemiological studies regarding aging-related dementia and intake of antioxidants (ascorbate, carotenoids, tocopherol), it has been postulated that a rich consumption of fruits and vegetables, plenty of antioxidants, can enhance cognition in the elderly (Bates et al., 2002; Engelhart et al., 2002).

High dietary intake of fruits and vegetables and the corresponding protective antioxidants (α-carotene, β-carotene, lycopene and vitamin C) were inversely associated with Alzheimer's disease risk (Smith et al., 1999). Other studies have been confirmed the protective effects of dietary antioxidants, including phytochemicals (flavonoids and phenolics), on the risk of neurodegenerative disorders (Ferrari, 2004).

8. ANTIOXIDANTS, HYPERTENSION AND CARDIOVASCULAR PROTECTION

Hypertension and cardiovascular diseases are common in the middle-aged and elderly people and successful aging is associated with better blood pressure control (Perls and Terry, 2003). Aging and hypertension is associated with sustained and intensive activation of NADP(H)-oxidase, potential amplification of superoxide release in aortic rings with massive degradation of nitric oxide by superoxide, leading to impaired vasodilation responses (Hamilton et al., 2001).

Vitamin C is a promising anti-hypertensive, once its plasmatic levels were inversely associated with arterial blood pressure (Block et al., 2001).

Scavenging free radicals produced during ischemic conditions constitute one of the most important cardiovascular benefits of antioxidant phytochemicals, vitamins and minerals in foods (Ferrari, 2004; Cui et al., 2002; Hu et al., 1998). Blood cholesterol lowering effects represents another important mechanism to protect against cardiovascular diseases (Ferrari, 2004; Ferrari and Torres, 2003).

Antioxidant vitamins, whole grains, and phytochemicals also protect vascular systems in heart and brain against homocysteine, and independent vascular risk factor (Broekmans et al., 2000). Antioxidant vitamins (E and C), carotenoids, soy, and green tea can inhibit LDL oxidation, protecting against atherosclerotic plaque formation (Ferrari, 2004; Ferrari, 2001). In a similar manner, antioxidant vitamins, whole grains, and phytochemicals also protect vascular systems in heart and brain against homocysteine, and independent vascular risk factor (Broekmans et al., 2000; McKeown et al., 2002). Previous treatment with vitamins E (800IU) and C (1,000mg) reversed deleterious effects of homocysteine (Nappo et al., 1999).

Intake of flavonoids, from apples and onion, has been associated with decreased cardiovascular mortality (Knekt et al., 1996). Within the cardiovascular protective mechanisms of flavonoids (from grapes and red wine), inhibition of platelet aggregation, increasing of nitric oxide synthesis and lowering of superoxide production seems to be important (Freedman et al., 2001).

Dietary intake of natural antioxidants (antioxidant vitamins, catechins, flavonol, and flavone) from diet has many benefits to older people, such as:

- Increase lung respiratory functions (Grievink et al., 1998; Tabak et al., 2001);
- Decrease coronary mortality (Knekt et al., 1996);
- Decreased Alzheimer's disease risk (Engelhart et al., 2002; Smith et al., 1999);
- Increase longevity (Trichopoulou and Vasilopoulou, 2000);

- Increase cognitive function in the elderly (Bates et al., 2002);
- Decrease cancer's risk (Kelloff et al., 2000; Ferrari, 2004);
- Decrease age-related macular degeneration and cataract's risk (Lyle et al., 1999; Cho et al., 2004).

9. MITOCHONDRIAL STABILITY: AN IMPORTANT TOOL FOR LONGEVITY AND HEALTH

9.1 Mitochondria is one of the most important targets for anti-aging interventions

Aging cells present mitochondrial dysfunction and failure characterized by impaired Mn-SOD synthesis, which compromises the dismutation of superoxide into hydrogen peroxide (H_2O_2), increasing cell oxidative stress. This mitochondrial SOD failure is associated with ovarian cancers, type I diabetes, neuronal degeneration, cardiac myocytes death, ischemic brain infarction, and normal aging (Lebovitz et al., 1996; Larsson and Luft, 1999; Ferrari, 2000) (Lebovitz et al., 1996; Larsson and Luft, 1999; Ferrari, 2000; Kim et al., 2002; Tornero et al., 2002; Viña et al., 2004).

Chronic imbalance of mitochondrial superoxide scavenging is associated with mitochondrial pore opening and intensification of ROS leakage, which induces mitochondrial and nuclear DNA mutations, cell aging and apoptosis (Linnane et al., 1989; Lenaz, 1998 (Linnane et al., 1989; Lenaz, 1998).

Five complexes are present in the mitochondrial respiratory chain: complex I (NADH-ubiquinone oxidoreductase), II (succinate-ubiquinone oxidoreductase), III (ubiquinol cytochrome c reductase), IV (cytochrome c oxidase/ATP synthase), and V (ATP-synthase) (Garrett and Grisham, 1995; Cardoso et al., 1999). Then, stability of respiratory chain requires adequate levels of iron and ubiquinone. Important mitochondrial disorders and associated disorders are listed in Table 6.

Heme iron deficiency impaired cytochrome c oxidase activity, impairing the control of mitochondrial respiratory functions and increase oxidative mtDNA damage (Atamna et al., 2001; Walter et al., 2002). However, excessive available iron has been also linked to increase mtDNA oxidation (Walter et al., 2002). Copper is very important for iron incorporation, since its deficiency has been linked to impaired incorporation of heme groups into cytochrome c oxidase molecules, decreasing IV complex and cytochrome c mitochondrial content (Rossi et al., 1998).

Coenzyme Q10 (ubiquinone), an electron acceptor of the complex I and II of the respiratory chain, when administered to a mice model of amyotrophic lateral sclerosis (ALS) reversed mitochondrial decay and decreased brain striatal damage induced by 3-nitropropionic acid, increasing animal life span (Matthews et al., 1998). Kelso et al. (Kelso et al., 2001) reported that a mitochondrial targeted ubiquinone compound had the ability to abrogate hydrogen peroxide-induced apoptosis, but not tumor necrosis factor-α induced cell death. Ubiquinone also improves mitochondrial respiration and enhances post-ischemic myocardial

Table 6. Disorders of the mitochondrial enzymatic complexes

Complex	Disease
I	Alzheimer's disease
	Cardiomiopathies
	Leber's disease
	Leigh's disease
	Miopathies
	Parkinson's disease
II	Leigh's disease
	Miopathies
	Ganglyome
	Pheochrocytome
III	Cardiomyopathies
	Leigh's disease
	Miopathies
IV	Alper's disease
	Ataxia
	Leber's disease
	Leigh's disease
	Miopathies
	Rhabdomyolisis
V	Leber's disease
	Leigh's disease

Adapted from: Tornero et al. (2002).

contractile function and decreases myocardial damage (Rosenfeldt et al., 2002). Recently, it was verified that ALS patients had increased plasma concentrations of oxidized coenzymeQ10 (Sohmiya et al., 2005), suggesting a potential need for coenzyme Q10 nutritional replacement. Coenzyme Q10, but not vitamin E, had prolonged life span of *Caenorhabiditis elegans*, effect mediated by apoptosis inhibition and possibly *in situ* superoxide scavenging action (Ishii et al., 2004). But human observational and clinical studies on this regard are necessary.

Magnesium, another important mitochondrial stabilizer, is a structural component of the mitochondrial membrane. Its important in ATP synthesis and its deficiency is associated with hypertension, diabetes, hyperlipidemia, chronic cardiovascular diseases, ALS, neuromuscular disorders, dementia, and Parkinson's disease (Fox et al., 2001; Durlach et al., 2004). Its nutritional replacement is very important, since older people has increased risk of magnesium deficiency due to massive gastrointestinal and renal losses of this nutrient, increasing the risk of asthma, coronary syndromes, and ischemic brain injury (Tong and Rude, 2005).

Selenium deficiency impairs antioxidant defenses by decreasing glutathione peroxidase (GPx) synthesis, increasing the risk of influenza and coxsackievirus infections, and heart disease as classically found in Keshan's disease (Levander, 2000; Beck, 2001). Dietary selenium supplementation has been found to recover cardiac, mitochondrial and cytosolic GPx values in aged rats previously submitted

Table 7. Coenzyme Q10 content of some foods

Food	Ubiquinone content (mg/100g)
Soy oil	92
Colza seed oil	73
Mackerel fish	43
Sesame seed oil	32
Meat	32
Peanut	27
Pork meat	25
Fish filet	24
Chicken	21
Nuts	19

Source: Duthie (1993).

to ischemic-reperfusion injury (Tanguy et al., 2003). Selenium is also associated with decreased risk of cancers (Schrauzer, 2000).

Vitamin E should also be considered a mitochondrial stabilizer agent. It has been observed that vitamin E deficiency was associated with increased lipid peroxidation and partially impaired mitochondrial respiration, since NADH-CoQ10 reductase and cytochrome oxidase activities were diminished in skeletal muscle cells (Rafique et al., 2004). However, the same authors reported increased mitochondrial activities and lipid peroxidation in the liver. However, other authors have found mitochondrial failure during liver aging in vitamin E-deficient rats (Armeni et al., 2003). Far beyond its general protective effect on biological membranes (Brown et al., 1998), tocopherol can specifically abrogate the oxidative decay of respiratory complex III (Atamna et al., 2001).

Dietary omega-3 fatty acids have been recognized as protective agents of mitochondrial membrane lipids, decreasing calcium release, a potent cell death element, and pyruvate dehydrogenase activity (Pepe et al., 1999). Padma and Devi (Padma and Devi, 2002) had reported that fish oil reversed mitochondrial respiratory failure. These findings constitute the basis for cardiovascular protective effects of fish and nuts dietary intake (Hu et al., 1998; Sheard, 1998; Fraser, 1999). It is postulated that the recognized neurological benefits of docosahexaenoic acid, from fatty fish, can be explained also by its capacity to stabilize phospholipids in biological membranes (He et al., 2002; Horrocks and Farooqui, 2004).

L-carnitine is a mitochondrial membrane fatty acid transporter and stabilizer in aging cells and neurons (Hagen et al., 1998; Binienda, 2003; Virmani et al., 2003), enhancing strength and cardio and encephalomyopathies (Mahoney et al., 2002).

Lipoic acid supplementation decreased heart mitochondrial DNA oxidation (Suh et al., 2001), once it has many free radical scavenging activities (Pioro, 2000). Caffeine and nicotinamide also showed to protect mitochondria against oxidative stress and dysfunction in a rat model of radiation-induced oxidative damage (Kamat and Devasagayam, 2000). Nicotinamide could also decrease free radicals and extend life span (Driver, 2003). Arivazhagan et al. (2001) (Arivazhagan et al., 2001) reported that lipoic acid supplementation reversed aging-associated mitochondrial oxidative stress, decreasing lipid peroxidation, but enhancing GSH, ascorbate and tocopherol content.

Many studies has been supported the concept that lower respiratory activity and higher mitochondrial DNA repair capacity are both associated with increased life span, as discussed by Barja (Barja, 1998).

10. DIABETES AND LONGEVITY

Healthy centenarians have lower rates of insulin resistance coupled to better glycemic control have been found (Morley, 2004; Perls and Terry, 2003; Paolisso et al., 2001). There is no "magic" dietary supplement that decreases diabetes risk. However, a diet rich in fruits and vegetables (rich in fibers and antioxidants) with moderate intake of meat foods, and lower intake of refined carbohydrates and fats, associated with a healthy life style incorporating limitation or avoidance of alcohol drinking and regular practice of physical activities is inversely associated with diabetes risk (Ford and Mokdad, 2001; Sato, 2000).

11. NUTRITIONAL MODULATION AGAINST CANCER

Many nutrients and non-nutrients can modulate cell and molecular targets providing efficient protection against cancer. Nutrients and recently discovered phytochemicals can promote (Kelloff et al., 2000; Ferrari and Torres, 2003; Ferrari and Torres, 2002; Heber, 2004):

- Apoptosis of cancer cells, killing neoplastic cells and decreasing tumor mass formation;
- Antioxidant protection of DNA, avoiding oxidative DNA and RNA mutations;
- Decreasing of oxidative stress and inflammation, avoiding genetic and cell damages;
- Antiangiogenesis, related to inhibition of neovascularization that is essential for tumor metastasis.

Table 8 summarizes the essential anticancer bioactive compounds in foods.

12. NUTRITIONAL MODULATION OF IMMUNITY AND INFLAMMATORY REACTIONS

Immunosenescence is a recognized pattern of immunological system in the elderly. It is characterized by suppression of T cell maturation, with thymic atrophy, disruption of normal immune activation, and presence of circulatory aged T-cells (Pawelec et al., 2002). Inflammation is also frequently positively associated with aging (Franceschi and Bonafé, 2003). In order to avoid increased mortality due to infectious diseases in the elderly (Yoshikawa, 1997), nutritional deficiencies of macro and micronutrients should be adequately treated. Table 8 lists important

Table 8. Anti-aging mechanisms of bioactive compounds from foods and herbs

Mechanisms	Bioactive compounds	Food Source
Antioxidants	Flavonoids (apigenin,	Onion, garlic, tomato, fruits and vegetables
	kaempferol, luteolin,	Grapes (juices), wines, berries, apples, cocoa
	myricetin, quercetin,	and chocolate, eggplant, teas, etc
	lycopene)	Turmeric
	Polyphenols	Oils and seeds
	Curcuminoids	Soybean
	Monounsaturated fatty acids	Oils and seeds
	Phytosterols (genistein and daidzein)	
	Tocopherols	
Anti-apoptotic	Ascorbic acid	Citrus and other fruits
agents	Egb761 extract (quercetin,	*Ginkgo biloba*
	kaempferol, isorhamnetin and	Polyphenol rich foods
	bilobalide, a terpene lactone)	
	Gallic acid	
Metal chelators	Phenols, Polyphenols, and	Grapes and wine
	their acids (quercetin, rutin,	Green teas
	catechins, sesamol, caffeic, ferulic and tannic acids)	
Proapoptotic agents	Artellipin C	Brazilian propolis
	Butyrate	Vegetable fibers
	Catechins	Tea polyphenols
	Genistein	Soy
	Indol-3-carbinol	Cruciferae vegetables (brocolis)
	Isoprenoids, terpenoids and tocotrienols	Vegetable oils (olive oil), nuts (Brazil nuts, cashew nuts, almonds, etc) and seeds
	Isothiocyanates	Cruciferae
	Fish oils	Fish oil
	Retinoids (vitamin A-related)	Vitamin A rich foods (oils, dark green leaf
	Polyphenols	vegetables and fruits)
	Protopanaxadiol	Persimmon (*Diospyros kaki*), green teas,
	Organosulfur compounds	wine, berries, purple fruit, and bananas
		Metabolites from ginsenosides (Rb1/Rb2/Rc)
		Garlic and onion
Mitochondrial stabilizers	Carnosine (β-alanyl-L-histidine)	Muscle foods
	Coenzyme Q10 (ubiquinone)	Soy oil, colza seed oil, mackarel fish, sesame seed oil, meat, peanut, pork meat, fish filet, chicken, and nuts
	Melatonin	*Scutellaria biacalensis* (Huang-qin), *Hypericum perfuratum* (St. John's wort), fever few, mustard and fenugreek seeds
	Lipoic acid	Meat, liver and heart
	Nicotinamide	Meats, grains, beans, fish, milk, eggs, seeds, vegetables
	n-3 fatty acids	Fatty fish (tuna, mackerel, salmon), canola (rapeseed) and flaxseed oils, flaxseed and nuts
	Tocopherol	Oils (olive) and seeds

(Continued)

Table 8. (Continued)

Mechanisms	Bioactive compounds	Food Source
Anti-inflammatory agents	Tocopherol	Oils and seeds
	Omega-3 fatty acids	Fish and vegetable oils
	Lycopene	Tomato and its products
	Polyphenols	Green tea, pomegranate, grapes and wine
Imunostimulatory agents	Selenium	Seafood, meat and grains
	Zinc	Meat and grains
	Tocopherols and tocotrienols	Oils and seeds
	Ginsenosides	*Panax ginseng*
	Garlic aqueous extract	Garlic
	Proteoglycans and β-D-glucans	Shiitake and other medicinal mushrooms

anti-inflammatory, immunoestimulatory, anticarcinogenic, antioxidant and other protective nutrients and food compounds.

13. CONCLUSIONS

Avoid nutrient deficiencies, control mitochondrial functions, block excessive oxidative stress and apoptosis of target cells, induce of cancer cell apoptosis, and increase immunological system performance are key factors of nutritional modulation for healthy aging.

Adherence to a Mediterranean diet, a model for healthy nutrition, rich in fruits, vegetables, legumes, monounsaturated and polyunsaturated fatty acids, was significantly associated with mortality risk reduction by 8% in the EPIC-elderly prospective cohort study (Trichopoulou et al., 2005). Then, adequate nutrition is a fundamental environmental approach to increase longevity.

REFERENCES

Ames, B.N., Shigenaga, M.K. and Hagen, T.M. (1993) Oxidants, antioxidants, and the degenerative diseases of aging. Proc Natl Acad Sci., 90: 7915–7922.

Arivazhagan, P., Ramanathan, K. and Panneerselvam, C. (2001) Effect of DL-alpha-lipoic acid on mitochondrial enzymes of aged rats. Chem Biol Inter., 138: 189–198.

Armeni, T., Principato, G., Quiles, J.L., Pieri, C., Bompadre, S., Battino, M. (2003) Mitochondrial dysfunction during aging: vitamin E deficiency or caloric restriction-two different ways of modulating stress. J Bionerg Biomembr., 35: 181–91.

Atamna, H., Liu, J., Ames, B.N. (2001) Heme deficiency selectively interrupts assembly of mitochondrial complex IV in human fibroblasts. Relevance to aging. J Biol Chem., 276: 48410–48416.

Barja, G., Cadenas, S., Rojas, C., Perez-Campo, R. and Lopez-Torres, M. (1994) Low mitochondrial free radical production per unit O_2 consumption can explain the simultaneous presence of high longevity and high aerobic metabolic rate in birds. Free Rad Res., 21: 317–327.

Barja, G. (1998) Mitochondrial free radical production and aging in mammals and birds. Ann NY Acad Sci., 854: 224–238.

Bates, C.J., Benton, D., Biesalski, H.K., et al. (2002). Nutrition and aging: a consensus statement. J Nutr Health Aging, 6: 103–116.

Beck, M.A. (2001) Antioxidants and Viral Infections: Host Immune Response and Viral Pathogenicity. J Am Coll Nutr., 20(suppl): 384S–388S.

Berger, A. (2002) Science commentary: what does zinc do? BMJ, 325: 1062–1063.

Berr, C. (2000) Cognitive impairment and oxidative stress in the elderly: Results of epidemiological studies. BioFactors, 13: 205–209.

Binienda, Z.K. (2003) Neuroprotective effects of L-carnitine in induced mitochondrial dysfunction. Ann NY Acad Sci., 993: 289–295.

Block, G., Mangels, A.R., Norkus, E.P., Patterson, B.H., Levander, O.A. and Taylor P.R. (2001) Ascorbic acid status and subsequent diastolic and systolic blood pressure. Hypertension, 37: 261–267.

Broekmans, W.M.R., Klöpping-Ketelaars, I.A.A., Schuurman, C.R.W.C., et al. (2000) Fruits and vegetables increase plasma carotenoids and vitamins and decrease homocysteine in humans. J Nutr., 130: 1578–1583.

Brown, K.M., Morrice, P.C. and Duthie G.G. (1998) Erythrocyte membrane fatty acid composition of smokers and non-smokers: effects of vitamin E supplementation. Eur J Clin Nutr., 52: 145–150.

Calle, E.E., Rodriguez, C., Walker-Thurmond, K. and Thun, M.J. (2003) Overweight, obesity and mortality from cancer in a prospectively studied cohort of US adults. New Engl J Med., 348: 1625–1638.

Cardoso, S.M., Pereira, C. and Oliveira, C. (1999) Mitochondrial function is differentially affected upon oxidative stress. Free Rad Biol Med., 26: 3–13.

Cho, E., Seddon, J.M., Rosner, B., Willett, W.C. and Hankinson, S.E. (2004) Prospective study of intake of fruits, vegetables, vitamins, and carotenoids, and risk of age-related maculopathy. Arch Ophtalmol., 122: 883–892.

Cuajungco, M.P., Fagét, K.Y., Huang, X., Tanzi, R.E. and Bush, A.I. (2000) Metal chelation as a potential therapy for Alzheimer's disease. Ann NY Acad Sci., 920: 292–304.

Cui, J., Cordis, G.A., Tosaki, A. et al. (2002) Reduction of myocardial ischemia reperfusion injury with regular consumption of grapes. Ann NY Acad Sci, 957: 302–307.

Driver, C. (2003) Mitochondrial interventions in aging and longevity. In: Modulating aging and longevity. Biology of aging and its modulation v.5 (ed. Rattan, S.I.S.) Pages 205–217, Kluwer Academic Publishers, Dordrecht, The Netherlands.

Durlach, J., Pagès, N., Bac, P., Bara, M. and Guiet-Bara, A. (2004) Magnesium research: from begginings to today. Magnes Res., 17: 163–168.

Duthie, G.G. (1993) Lipid peroxidation. Eur J Clin Nutr., 47: 759–764.

Engelhart, M.J., Geerlings, M.I., Ruitenberg, A., et al. (2002) Dietary intake of antioxidants and risk of Alzheimer disease. JAMA, 287: 3223–3229.

Ferrari, C.K.B. (1998) Oxidação lipídica em alimentos e sistemas biológicos: mecanismos gerais e conseqüências nutricionais e patológicas. Rev Nutr., 11: 3–14.

Ferrari, C.K.B. (1999) Oxidação de gorduras em alimentos: produção de substâncias tóxicas na dieta do homem. Rev Instit Hig Med Soc., 3: 22–26.

Ferrari, C.K.B. (2000) Free radicals, lipid peroxidation and antioxidants in apoptosis: implications in cancer, cardiovascular and neurological diseases. Biologia, 55: 581–590.

Ferrari, C.K.B. (2001) Oxidative stress pathophysiology: Searching for an effective antioxidant protection. Int Med J., 8: 175–184.

Ferrari, C.K.B. (2004) Functional foods, herbs, and nutraceuticals: Towards biochemical mechanisms of healthy aging. Biogerontol., 5: 275–289.

Ferrari, C.K.B, and Torres, E.A.F.S. (2002) Novos compostos com propriedades anticarcinogênicas. Rev Bras Cancerol., 48: 375–382.

Ferrari, C.K.B. and Torres, E.A.F.S. (2003) Biochemical pharmacology of functional foods and prevention of chronic diseases of aging. Biomed Pharmacother., 57: 251–260.

Fontaine, K.R., Redden, D.T., Wang, C., Westfall, A.O. and Allison, D.B. (2003) Years of life lost due to obesity. JAMA, 289: 187–193.

Ford, E.S. and Mokdad, A.H. (2001) Fruit and vegetable consumption and diabetes mellitus incidence among U.S. adults. Prev Med., 32: 33–39.

Fox, C., Ramsoomair, D. and Carter, C. (2001) Magnesium: its proven and potential clinical significance. South Med J., 94: 1195–1201.

Franceschi, C. and Bonafé, M. (2003) Centenarians as a model for healthy aging. Biochem Soc Transact., 31: 457–461.

Fraser, G.E. (1999) Nut consumption, lipids, and risk of coronary event. Clin Cardiol., 22(7suppl.): III11–15.

Freedman, J.E., Parker, C., Li, L., et al. (2001) Select flavonoids and whole juice from purple grapes inhibit platelet function and enhance nitric oxide release. Circulation, 103: 2792–2798.

Garrett, R.H. and Grisham, C.M. (1995) Biochemistry. Saunders College Publ., Orlando, 1995.

Grievink, L., Smith, H.A., Ocké, M.C., van't Veer, P. and Kromhout, D. (1998) Dietary intake of the antioxidant (pro)-vitamins, respiratory symptoms and pulmonary function: the Morgen Study. Thorax, 53: 166–171.

Gutteridge, J.M. (1995). Lipid peroxidation and antioxidants as biomarkers of tissue damage. Clin Chem., 41: 1819–1828.

Hagen, T.M., Ingersoll, R.T., Wehr, C.M., et al. (1998) Acetyl-L-carnitine fed to old rats partially restored mitochondrial function and ambulatory activity. Proc Natl Acad Sci., 95: 9562–9566.

Halliwell, B. (1994). Free radicals and antioxidants: a personal view. Nutr Rev., 52: 253–265.

Hamilton, C.A., Brosnan, J., McIntyre, M., Graham, D and Dominiczak, A.F. (2001) Superoxide excess in hypertension and aging. Hypertension, 37: 529–534.

He, K., Rimm, E.B., Merchant, A., et al. (2002) Fish consumption and risk of stroke in men. JAMA, 288: 3130–3136.

Heber, D. (2004) Vegetables, fruits and phytoestrogens in the prevention of diseases. J Postgrad Med., 50: 145–149.

Heys, S.D., Schofield, A.C. and Wahle, K.W. (2004) Immunonutrition in clinical practice: what is the current evidence? Nutr Hosp., 19: 325–332.

Horrocks, L.A. and Farooqui, A.A. (2004) Docosahexaenoic acid in the diet: its importance in maintenance and restoration of neural embrane function. Prostagl Leukot Essent Fatty Acid., 70: 361–372.

Hu, F.B., Stampfer, M.J., Manson, J.E., et al. (1998) Frequent nut consumption and risk of coronary heart disease in women: prospective cohort study. BMJ, 317: 1341–1345.

Hughes, D.A. (2000) Dietary antioxidants and human immune function. Nutr Bul., 25: 35–41.

Ishii, N., Senoo-Matsuda, N., Miyake, K., et al. (2004) Coenzyme Q10 prolong C. elegans lifespan by lowering oxidative stress. Mech Ageing Dev., 125: 41–46.

Kamat, J.P. and Devasagayam, T.P.A. (2000) Oxidative damage to mitochondria in normal and cancer tissues, and its modulation. Toxicology, 155: 73–82.

Kelloff, G.J., Crowell, J.A., Steele, V.E., et al. (2000) Progress in cancer chemoprevention: Development of diet-derived chemopreventive agents. J Nutr., 130: 467S–471S.

Kelso, G.F., Porteous, C.M., Coulter, C.V., et al. (2001) Selective targeting of a redox-active ubiquinone to mitochondria within cells. J Biol Chem., 276: 4588–4596.

Kim, G.W., Kondo, T., Noshita, N. and Chan, P.H. (2002) Manganse superoxide dismutase deficiency exacerbates cerebral infarction after focal cerebral ischemia/reperfusion in mice: implications for the production and role of superoxide radicals. Stroke, 33: 809–815.

Knekt, P., Järvinen, R., Reunanen, A. and Maatela, J. (1996) Flavonoid intake and coronary mortality in Finland: a cohort study. BMJ, 312: 478–481.

Ku, H.H. and Sohal, R.S. (1993) Comparison of mitochondrial pro-oxidant generation and antioxidant defenses between rat and pigeon: possible basis for variation in longevity and metabolic potential. Mech Aging Dev., 72: 67–76.

Larsson, N.-G. and Luft, R. (1999) Revolution in mitochondrial medicine. FEBS Lett., 455: 199–202.

Lebovitz, R.M., Zhang, H., Vogel, H., et al. (1996) Neurodegeneration, myocardial injury, and perinatal death in mitochondrial superoxide dismutase-deficient mice. Proc Natl Acad Sci., 93: 9782–9787.

Lenaz, G. (1998) Role of mitochondria in oxidative stress and ageing. Biochim Biophys Acta, 1366: 53–67.

Levander, O.A. (2000) The selenium-coxsackievirus connection: chronicle of a collaboration. J Nutr., 130: 485S–488S.

Linnane, A.W., Marzuki, S., Osawa, T. and Tanaka, M. (1989) Mitochondrial DNA mutations as an important contribution to ageing and degenerative diseases. The Lancet, I: 642–645.

Lipkin, M. (1999) Preclinical and early human studies of calcium and colon cancer prevention. Ann NY Acad Sci., 889: 120–127.

Liu, J., Head, E., Gharib, A.M., et al. (2002) Memory loss in old rats is associated with brain mitochondrial decay and RNA/DNA oxidation: partial reversed by feeding acetyl-L-carnitine and/or R-α-lipoic acid. Proc Natl Acad Sci., 99: 2356–2361.

Lukaski, H.C. (2004) Vitamin and mineral status: effects on physical performance. Nutriton, 20: 632–644.

Lyle, B.J., Mares–Perlman, J.A., Klein, B.E.K., et al. (1999) Serum carotenoids and tocopherols and incidence of age-related nuclear cataract. Am J Clin Nutr., 69: 272–277.

Möller, P., Wallin, H. and Knudsen, L.E. (1996) Oxidative stress associated with exercise, psychological stress and life-style factors. Chem-Biol Inter., 102: 17–36.

Mahoney, D.J., Parise, G. and Tarnopolsky, M.A. (2002) Nutritional and exercise-based therapies in the treatment of mitochondrial disease. Curr Opin Clin Nutr Metab Care, 5: 619–629.

Matthews, R.T., Yang, L., Browne, S., Baik, M. and Beal, F. (1998) Coenzyme Q10 adminstration increases brain mitochondrial concentrations and exerts neuroprotective effects. Proc Natl Acad Sci., 95: 8892–8897.

McKeown, N.M., Meigs, J.B., Liu, S., Wilson, P.W.F. and Jacques P.F. (2002) Whole-grain intake is favorably associated with metabolic risk factors for type 2 diabetes and cardiovascular disease in the Framingham Offspring Study. Am J Clin Nutr., 76: 390–398.

Mecocci, P., Polidori, M.C., Troiano, L., et al. (2000) Plasma antioxidants and longevity: a study on healthy centenarians. Free Rad Biol Med., 28: 1243–1248.

Michels, K.B. and Ekbom, A. (2004) Caloric restriction and incidence of breast cancer. JAMA, 291: 1226–1230.

Morley, J.E. (2004) Nutrition and aging. Nutritional assessment outline. In: Preventing ADL decline in nursing homes. Process improvement manual. Pages 1–5, NCHCQF, Saint Louis University, School of Medicine, St. Louis, USA.

Morley, J.E. (2004) The top 10 hot topics in aging. J Gerontol., 59: 24–33.

Nève, J. (2002) Selenium as a 'nutraceutical': how to conciliate physiological and supra-nutritional effects for an essential trace element. Curr Opin Nutr Metab Care, 5: 659–663.

Namura, S., Nagata, I., Takami, S. et al. (2001) Ebselen reduces cytochrome c release from mitochondria and subsequent DNA fragmentation after transient focal cerebral ischemia in mice. Stroke, 32: 1906–1911.

Nappo, F., De Rosa, N., Marfella, R. et al. (1999) Impairment of endothelial functions by acute hyperhomocysteinemia and reversal by antioxidant vitamins. JAMA, 281: 2113–2118.

Padma, V.V. and Devi, C.S. (2002) Effect of fish oil on mitochondrial respiration in isoproterenol induced myocardial infarction in rats. Indian J Exp Biol., 40: 268–272.

Pahor, M. and Applegate, W.B. (1997) Recent advances: geriatric medicine. BMJ, 315: 1071–1074.

Paolisso, G., Barbieri, M., Rizzo, M.R. (2001) Low insulin resistance and preserved beta-cell function contribute to human longevity but are not associated with TH-INS genes. Exp Gerontol., 37: 149–156.

Pawelec, G., Barnett, Y., Forsey, R., et al. (2002) T cells and aging, January 2002 update. Front Biosci., 7: d1056–d1183.

Pedrosa, L.F.C. and Cozzolino, S.M.F. (1999) Alterações metabólicas e funcionais do cobre em diabetes mellitus. Rev Nutr., 12: 213–224.

Pepe, S., Tsuchiya, N., Lakatta, E.G. and Hansford, R.G. (1999) PUFA and aging modulate cardiac mitochondrial membrane lipid composition and Ca^{2+} aactivation of PDH. Am J Physiol., 45: H149–H158.

Perkins, A.J., Hendrie, H.C., Callahan, C.M., et al. (1999) Association of antioxidants with memory in a multiethnic elderly sample using the Third National Health and Nutrition Examination Survey. Am J Epidemiol., 150: 37–44.

Perls, T. and Terry, D (2003) Understanding the determinants of exceptional longevity. Ann Intern Med., 139: 445–449.

Pioro, E.P. (2000) Antioxidant therapy in ALS. ALS Motor Neuron Dis., 1 (Suppl 4): 5–15.

Rafique, R., Shapira, A.H. and Coper, J.M. (2004) Mitochondrial respiratory chain dysfunction in ageing; influence of vitamin E deficiency. Free Radic Res., 38: 157–165.

Ramalingaswami, V. (1992) Una vitamina y dos elementos minerals, claves de la salud. Foro Mund Salud, 13: 220–229.

Rattan, S.I.S. (2003a) Biology of aging and possibilities of gerontomodulation. Proc Indian Nat Sci Acad., B69: 157–164.

Rattan, S.I.S. (ed.) (2003b) Modulating aging and longevity. Biology of aging and its modulation v5. Kluwer Academic Publishers, Dordrecht, The Netherlands.

Richard, M.-J. and Roussel, A.-M. (1999) Micronutrients and ageing: intakes and requirements. Proc Nutr Soc., 58: 573–578.

Rosenfeldt, F.L., Pepe, S., Linnane, A., et al. (2002) Coenzyme Q10 protects the aging heart against oxidative stress. Studies in rats, human tissues, and patients. Ann NY Acad Sci., 959: 355–359.

Rossi, L., Lippe, G., Marchese, E., et al. (1998) Decrease of cytochrome c oxidase protein in heart of mitochondria of copper-deficient rats. Biometals, 11: 207–212.

Sato, Y. (2000) Diabetes and life-styles: role of physical exercise for primary prevention. Brit J Nutr., 84(suppl.2): S187–S190.

Schrauzer, G.N. (2000) Anticarcinogenic effects of selenium. Cel Mol Life Sci., 57: 1864–1873.

Sheard, N.F. (1998) Fish consumption and risk of sudden cardiac death. Nutr Rev., 56: 177–179.

Smith, M.A., Petot, G.J. and Perry, G. (1999) Diet and oxidative stress: a novel synthesis of epidemiological data no Alzheimer's disease. J Alzheim Dis., 1: 203–206.

Sohmiya, M., Tanaka, M., Suzuki, Y., Tanino, Y., Okamoto, K. and Yamamoto, Y. (2005) An increase of oxidized coenzyme Q-10 in the plasma of sporadic ALS patients. J Neurol Sci., 228: 49–53.

Somer, E. (2003) Nutrition basics. *In*: Nutrition for women. 2nd ed., Pages 14–39, OWL Books, New York.

Suh, J.H., Shigeno, E.T., Morrow, J.D., et al. (2001) Oxidative stress in the aging rat heart is reversed by dietary supplementation with (R)-α-lipoic acid. Faseb J., 15: 700–706.

Tabak, C., Arts, I.C.W., Smith, H.A., Heederik, D. and Kromhout, D. (2001) Chronic obstructive pulmonary disease and intake of catechins, flavonols, and flavones. Am J Respir Crit Care Med., 164: 61–64.

Tanguy, S., Toufektsian, M.-C., Besse, S., Ducros, V., Leiris, J. de and Boucher, F. (2003) Dietary selenium intake affects cardiac susceptibility to ischemia/reperfusion in male senescent rats. Age Ageing, 32: 273–278.

Thaler, D.E., Hope, R.A. and Longmore, J.M. (1999) Oxford handbook of clinical medicine. Oxford University Press, New York, 1999.

Tong, G.M. and Rude, R.K. (2005) Magnesium deficiency and critical illness. J Intens Care Med., 20: 3–17.

Tornero, D., Ceña, V. and Jordán, J. (2002) La mitocondria como diana farmacológica en los procesos neurodegenerativos. Offarm, 21: 98–102.

Trichopoulou, A. and Vasilopoulou, E. (2000) Mediterranean diet and longevity. Brit J Nutr., 84 (Suppl 2): 205–209.

Trichopoulou, A., Orfanos, P., Norat, T., et al. (2005) Modified Mediterranean diet and survival: EPIC-elderly prospective cohort study. BMJ, 330: 991.

Viña, J., Sastre, J., Pallardó, F.V. and Bonás, C. (2004) Posibles mecanismos por los que las mujeres viven más que los varones. Rev Esp Geriatr Gerontol., 39: 381–384.

Virmani, A., Gaetani, F., Imam, S., Binienda, Z. and Ali, S. (2003) Possible mechanism for the neuroprotective effects of L-carnitine on methamphetamine-evoked neurotoxicity. Ann NY Acad Sci., 993: 197–207.

Walter, P.B., Knutson, M.D., Paler-Martinez, A., et al. (2002) Iron deficiency and iron excess damage mitochondria and mitochondrial DNA in rats. Proc Natl Acad Sci., 99: 2264–2269.

WHO. (1985) Energy and protein requirements: report of a joint FAO/WHO/UNU expert consultation. WHO Tech Rep Ser, n.724, Geneva.

Wolf, C. and Tanner, M. (2002) Obesity. West J Med., 176: 23–28.

Yoshikawa, T.T. (1997) Aging and infectious diseases: past, present, and future. J Infect Dis., 176: 1053–1057.

Zimmermann, M. (2001) Micronutrients in health and disease. Georg Thieme Verlag, Stuttgart.

Zyczkowska, J., Klich-Raczka, A., Mossakowska, M., Gasowski, J., Wieczorowska-Tobis, K. and Grodzicki, T. (2004) Blood pressure in centenarians in Poland. J Hum Hypertens., 18: 713–716.

CHAPTER 17

DIETARY FATS AND AGE-RELATED DISEASES

KAUSTUV BHATTACHARYA AND SURESH I.S. RATTAN*
**Laboratory of Cellular Ageing, Department of Molecular Biology, University of Aarhus, Denmark.*
(Emails: papai_kb@yahoo.com, Rattan@mb.au.dk)

Abstract: Balanced diet, which includes fats and oils, is one of the important factors for attaining and maintaining a healthy life. Numerous clinical studies have shown the detrimental effects of trans- and saturated-fats in the origin and progression of various age-related diseases, such as coronary heart disease, diabetes, cancer and neurodegenerative diseases. This article reviews the role of dietary lipids in various age-related diseases, and discusses the appropriate dietary fat requirements for the prevention of such diseases

Keywords: Polyunsaturated fatty acids, saturated fats, diseases, antioxidants

1. INTRODUCTION

Fats and oils, as a specific component of diet, provide essential fatty acids and facilitate the delivery of various other nutrients that are vitally important for normal physiological functions. As structural units, fats and lipids are the integral parts of the cellular and organellar membranes, and of the nerve sheathing. Normal physical and mental growth, development and maturation depend on the optimal availability of dietary fats. Additionally, body fat or adipose tissue helps to protect vital organs from injuries and shocks, and provides a source of energy during prolonged exercise. Fats have the highest caloric density among foodstuffs (9 kcal/g), and are also the carriers of vitamins such as A, D, E and K. Vegetable oils are important sources of natural antioxidants, such as tocopherols, tocotrienols and carotenoids. Dietary lipids also play an important role in the immune function by modulating eicosanoid production (Formo, 1979; Lands, 1986; Robert, 1990).

Oils and fats are consumed for caloric reasons and also for their non-caloric functions such as flavour, palatability, appearance, consistency and texture. Intake of oils and fats is primarily through cooking oils, baked products, margarines and

S. Rattan and M. Kassem (eds.), Prevention and Treatment of Age-related Diseases, 335–356.

spreads, various fried products, chocolate and sugar confectionery, dairy products and desserts, salad oils, mayonnaise and other dressings. Fats are also consumed when meat, poultry or fish are eaten. All these sources make up a complex matrix of various visible and invisible oils and fats that end up in our body.

The content and composition of dietary fat, especially the carbon chain length, degree of saturation, positioning of the double bonds and cis and trans configuration of the unsaturated fatty acids and region-specific distribution of the fatty acids in the triacylglycerols have significant contribution to human health. Good health is dependent not only on the quantity but also on the quality of the fat. Several diseases such as hypercholesterolemia and related cardiovascular disorders, type 2 diabetes, inflammation, certain types of cancer, renal diseases and Alzheimer's disease are directly or indirectly related to dietary fats. Very often such diseases are associated with excessive and improper intake of dietary fats or deficiency of essential fatty acids. Excessive amounts of free radicals generated from oxidised oils are also related to the origin of various diseases. This chapter discusses the effects of different types of dietary fats on the origin and progression of various age-related diseases.

2. TYPES AND SOURCES OF DIETARY FATS

Fats and oils of animal and plant origin consist almost exclusively of the simple lipid class triacylglycerols (often termed "triglycerides"). They consist of a glycerol moiety with each hydroxyl group esterified to a fatty acid. Triacylglycerols are synthesised by enzyme systems, which determine that a centre of asymmetry is created about carbon-2 of the glycerol backbone, so they exist in enantiomeric forms, i.e. with different fatty acids in each position. The positions of the fatty acids in the glycerol backbone are denoted by sn-1 or sn-3, the two terminal positions and sn-2, the middle position. (The abbreviation 'sn' stands for 'stereospecific numbering'). Generally, in case of vegetable oils, unsaturated fatty acids are situated in the sn-2 position while SFA's occupy the sn-1 and sn-3 positions (Hunter, 1992). The naturally occurring fatty acids are mainly straight-chain compounds containing an even number of carbon atoms. Fatty acids can be divided into the following three groups: (i) saturated; (ii) monounsaturated and polyunsaturated; and (iii) branched-chain. Unsaturated fatty acids may contain one or more double or triple bonds and can be classified as monounsaturated, polyunsaturated, and acetylenic fatty acids.

The distribution of the fatty acids in triacylglycerols can be rearranged or 'structured' and if desired, new fatty acids can also be introduced through a process called interesterification. The rationale behind the development of structured lipids is based on the effects of dietary fatty acids and the importance of their relative position (sn-1 or sn-3 and sn-2) in triacylglycerol molecules. Triacylglycerols can be tailored to contain appropriate proportions of n-3, n-6, n-9 and SFAs which are beneficial in lowering serum LDL cholesterol and triacylglycerol levels, preventing thrombosis, enhancing immune system, reducing the risk of cancer and improving nitrogen balance (Akoh, 2002).

The nomenclature omega-9, omega-6 and omega-3 fatty acids are related to the position of the first unsaturation in the fatty acid chain relative to the methyl end. Position of the double bond can also be denoted in the form (n-x), where n is the chain-length of the fatty acid and x is the number of carbon atoms from the double bond in the terminal region of the molecule. In case of linoleic acid, it lies at the sixth carbon and as regards linolenic acid it lies at the third carbon atom from the methyl end of the molecule. Thus linoleic acid is termed omega-6 (or n-6) and alpha-linolenic acid is called omega-3 (n-3) fatty acid.

2.1 Saturated fatty acids (SFA)

Saturated alkanoic acids have the general formula R-COOH where R represents straight-chain hydrocarbons having the formula C_nH_{2n+1} or $CH_3(CH_2)_nCOOH$. SFA range from short-chain volatile liquids to waxy solids. Common saturated fatty acids are lauric, (C_{12}), myristic (C_{14}), palmitic (C_{16}) and stearic (C_{18}). Milk fats are characterised with C_4 to C_{10} fatty acids while C_{12} to C_{24} occur in fats and oils. Higher members up to C_{38} are found in waxes. SFA are present in appreciable amounts (50–90% of total fatty acids) in milk fat, coconut oil, palm oil and palm kernel oil.

2.2 Monounsaturated fatty acids (MUFA)

Monounsaturated fatty acids contain one double bond which is present mostly at the ninth carbon atom from the methyl end. They are referred to as omega-9 or n-9 fatty acids. Though more than 100 monounsaturated fatty acids are known, oleic acid (cis-9-octadecaenoic acid) is the most widely distributed of all fatty acids. It acts as the precursor of biosynthesis of omega-9 families of fatty acids. Petroselinic, vaccenic and erucic acids are some examples of other commonly found MUFA. Two most common sources of oleic acid are olive oil and rapeseed oil. However, genetic mutation and selective breeding have developed ‘high-oleic’ version of commodity oils such as sunflower, safflower, peanut, soybean and canola oil. These ‘high-oleic’ oils typically contain more than 70% oleic acid and are commercially available for various food applications (Kristott, 2003).

2.3 Polyunsaturated fatty acids (PUFA)

PUFA contain more than one carbon-carbon unsaturation. There are two major PUFA families: one based on linoleic acid (delta-9,12-18:2 omega-6) and the other on alpha-linolenic acid (delta-9,12,15-18:3 omega-3). The importance of PUFAs in human health and nutrition was postulated first in the 1920s. Linoleic acid and alpha-linolenic acid were termed essential fatty acids (EFA) since these cannot be synthesised in vivo by animals, including humans. Therefore EFA must be consumed from plant-derived dietary sources. Once consumed, both linoleic

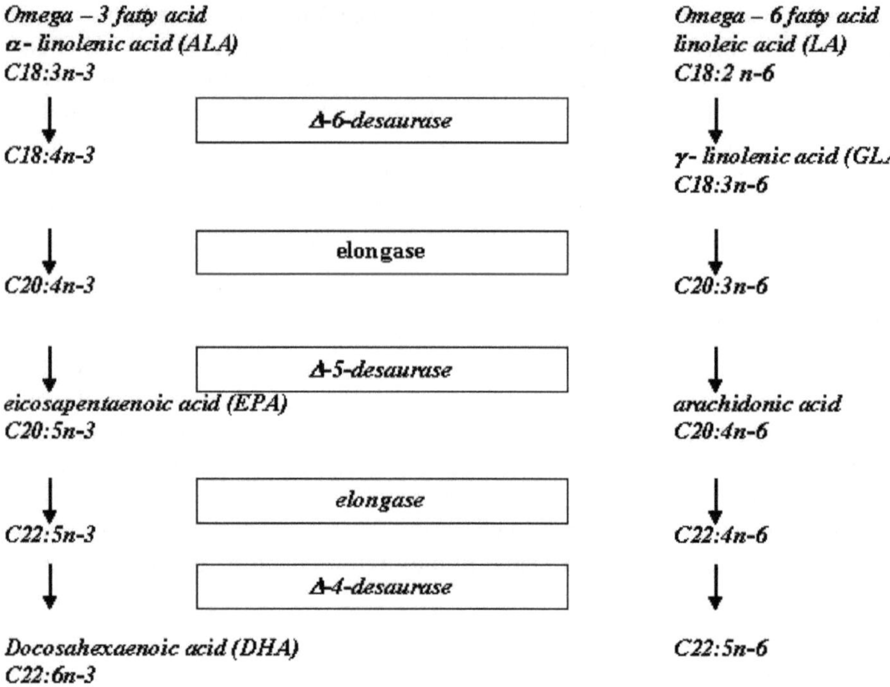

Figure 1. Metabolic pathways of conversion of linoleic and linolenic acid (Adapted from Simopoulos, 1999)

and alpha-linolenic acid are converted to other long chain omega-6 and omega-3 fatty acids by metabolic pathways in mammals through enzymatic catalysis (see Figure 1).

These changes require chain-elongation and desaturation. The most important omega-6 metabolite is arachidonic acid (AA, 20:4) and the most important omega-3 metabolites are eicosapentaenoic acid (EPA, 20:5) and docosahexaenoic acid (DHA, 22:6). EPA and DHA are the most bioavailable forms of omega-3 for humans. Linoleic acid is the major fatty acid in vegetable oils such as soybean, sunflower, safflower, peanut and corn. Vegetable oils such as flax, blackcurrant, rape, perilla and chia contain moderate to high amounts of alpha-linolenic acid. Soybeans, navy beans and walnuts are also sources of alpha-linolenic acid. It is also present in phytoplankton, zooplankton and many marine species.

2.4 Trans fatty acids

Trans fatty acids are those fatty acids that contain double-bond geometry in the trans (E) configuration, i.e. the hydrogen atoms are placed on the opposite sides of the double bond (Hunter, 1992). They naturally occur in small amounts (<1%) in unmodified vegetable oils and fats. The majority of trans fatty acids in our diets come from partially hydrogenated oils. Hydrogenation is a chemical reaction in which hydrogen is added to the ethylenic linkages (double bonds) of unsaturated fatty acids (Hastert, 1996). Small amounts of trans fatty acids occur naturally in milk, butter and tallow as a result of biohydrogenation in ruminants. Blends of hydrogenated and non-hydrogenated oils and fats have been used to produce base stocks for margarine, frying oils and a variety of general purpose fats where solid and stable fats are required. Hydrogenated fats have been given the generic name "vanaspati" in India, and are used for numerous edible applications. The most abundant of the trans fatty acids in partially hydrogenated oils is elaidic acid, the trans isomer of natural cis-oleic acid.

3. FATS AND AGE-RELATED DISEASES

The detrimental effects of improper dietary fats are not observed overnight but the damages undergo a slow, yet certain cumulative pattern and surface at later stages of life. Thus, very often the root causes of various diseases during old age stem from the dietary habits at the young age. The effects of dietary fats on the major age-related diseases are discussed in the following sections.

3.1 Cardiovascular diseases

Early epidemiological observations suggested an association between dietary fat and cardiovascular diseases (Keys et al., 1957; Keys et al., 1959). One of the major risk factors for cardiovascular disease is hypercholesterolemia. Coronary heart disease (CHD) is caused by atherosclerosis, a process characterized by endothelial dysfunction, in connection with cholesterol deposition macrophages and smooth muscle cells in the arterial walls and various other factors. The risk of CHD increases proportionally with serum levels of total and low density lipoprotein (LDL) cholesterol and decreases with increase in high density lipoprotein (HDL) cholesterol (Martin et al., 1986; Castelli et al., 1986). The increased ratio of total cholesterol to HDL is associated with a rise in risk for all-cause mortality in men aged 65 years and above. When considered alone, an elevated level of HDL seems to be protective against mortality from all causes in men aged 65–74 years but this effect diminishes over the age of 75 years (Chyou et al., 2000).

3.1.1 Effect of saturated fatty acids on cholesterol

Saturated fatty acids are reported to be cholesterol-raising but not all acids in this class show the same effect (Mensink et al., 2002). Alkanoic acids can be divided into three major classes: (i) fatty acids having less than 12 carbon atoms; (ii) fatty acids with 12, 14 or 16 carbons atoms; and (iii) the 18 carbon homologue, stearic acid. It has been suggested that the first group slightly reduces LDL cholesterol relative to palmitic acid but raises it when compared to oleic acid (Cater et al., 1997). From the second group, lauric acid has been reported (Denke and Grundy 1992) to increase plasma total cholesterol and LDL cholesterol concentrations compared to oleic acid but to a lower extent relative to palmitic acid while effects on HDL cholesterol were not observed. However, an increase of total cholesterol due to an increase of HDL cholesterol as compared to palmitic acid has also been reported (Temme et al., 1996). Myristic acid has an increasing effect on both LDL cholesterol and HDL cholesterol and hence on total cholesterol concentration relative to palmitic acid (Zock et al., 1997). Despite being hypercholesterolemic compared to stearic acid (Mensink et al., 2002), palmitic acid has not been labeled in all cases as a cholesterol-elevating saturated fatty acid (Ng et al., 1992; Choudhury et al., 1995). This holds true when dietary cholesterol intake is less than 300 mg/day and 6–7% of daily energy comes from linoleic acid. Stearic acid had been shown not to elevate plasma total cholesterol concentration (Keys et al., 1965; Grande et al., 1970). In fact, later studies revealed that stearic acid has a neutral effect on plasma lipoproteins similar to that of cis-monounsaturated oleic acid (Bonanome and Grundy, 1988). Overall, it can be concluded that, saturated fatty acids such as lauric, myristic and palmitic acids raise the levels of both total and LDL cholesterol.

3.1.2 Effect of trans fatty acids on cholesterol

Trans monounsaturated fatty acids, raise LDL cholesterol concentrations (Katan et al., 1995) and decrease HDL cholesterol concentrations (Mensink et al., 2002) in contrast to intake of cis-monounsaturated fatty acids. Investigations on whether TFAs from ruminant sources differ from those resulting from partial hydrogenation with respect to CHD have shown that below an intake level of 2.5 g/day, there were no differences in effects on CHD between the two sources of TFAs but that at total intake levels of above 3 g/day industrial TFAs cause bigger risk of CHD (Weggemans et al., 2004).

3.1.3 Effect of PUFA on CHD

In the recent years, the beneficial cardiac health effects of PUFA, especially omega-3 fatty acids have attracted considerable scientific and public interest. The present consensus is that the cardio protective effects of EPA and DHA at the low dosage used in recent secondary prevention trials mainly results from an effect on the ischemic myocardium and probably not from an effect on blood lipids and hemostasis. On the other hand, dietary α-linolenic acid, the precursor of EPA and DHA may be protective through mechanisms other than the myocardial (anti-arrhythmic) ones (De Lorgeril and Salen, 2004a). Epidemiological studies and dietary trials in humans suggest that α-linolenic acid is a major cardio-protective nutrient (De Lorgeril et al., 2004b).

One of the studies showing the effect of alpha-linolenic acid on heart was the Multiple Risk Factor Intervention Trial (MRFIT). It involved 12,000 men aged between 35 and 57 years who had high risk of heart diseases. It was found that risk of death from CHD was lowest in subjects with highest intakes of alpha-linolenic acid (Dolecek, 1992). The Lyon Diet Heart Study had shown the effect of alpha-linolenic acid on people who had survived one heart attack. Participants in the test group had an increased intake of alpha-linolenic acid by 68% and had lower blood cholesterol and triglyceride levels. In fact, alpha-linolenic acid rich diets were associated with a 70% reduction in coronary problems and cardiac deaths (De Lorgeril et al., 1999). Other investigations indicate that dietary alpha-linolenic acid reduces inflammatory and lipid cardiovascular risk factors in hypercholesterolemic men and women, possibly by favourably changing vascular inflammation and endothelial dysfunction (Zhao, 2004). These authors have also reported that high-PUFA diets (diets rich in linoleic acid and alpha-linolenic acid) decrease serum total cholesterol, LDL cholesterol and triglycerides.

Fish oils, rich in long-chain omega-3 fatty acids, have been found to reduce plasma triacylglycerols of hyperlipidemic subjects, especially in patients with elevated triglyceride concentrations (Harris, 1989). In the case of normocholesterolemic subjects, long-chain PUFA from fish oils do not induce any changes in plasma LDL cholesterol or HDL cholesterol concentrations but have a lowering effect on plasma triacylglycerols and the concentration of cholesterol in very low density lipoprotein (VLDL) (Harris et al., 1983).

3.2 Effect of PUFA on cardiac mitochondrial membranes

Biological membranes are made of complex matrices of lipids, proteins, lipid-proteins complexes, glycolipids and glycoproteins. With aging, both the hormonal status and lipid component of a membrane change and the remodelling of myocardial cell membrane is a major occurrence. Age-related mitochondrial changes include increase of membrane rigidity, cholesterol, phosphatidylcholine, omega-6 fatty acids, and decrease in omega-3 fatty acids and cardiolipin (Pepe, 2005). Studies have shown how specific age-related changes of phospholipids and fatty acid composition in the cardiac mitochondrial membranes can influence vital mitochondrial processes and the heart's adaptive response to stress and survival. The various constitutive changes that occur in heart cells with increased age reduce the cellular capacity to tolerate and adapt to ischemic stress.

The primary PUFA in myocardial membranes are omega-6 linoleic and arachidonic acid and omega-3 DHA. With abundant use of linoleic acid rich oils such as, soybean, sunflower and corn, and low consumption of fish in the western world, there is a much higher intake of linoleic acid and a very low amount of omega-3 fatty acids. Such vast excess of omega-6 fatty acids compete with the omega-3 fatty acids and utilise the delta-5 and delta-6 desaturase enzymes to a greater extent for subsequent conversion into higher homologues of the omega-6 series. Delta-5 and delta-6 desaturase enzymes are crucial for the conversion of linoleic acid to arachidonic acid and conversion of α-linolenic acid into EPA and DHA. The activity of microsomal delta-6 desaturase is less than that of delta-5, making it the rate limiting step involved in two stages of DHA production (Cho et al., 1999). Thus, α-linolenic acid cannot be converted into adequate levels of EPA and DHA (Sprecher et al., 1995). In a situation where there is excess linoleic acid and insufficient omega-3 fatty acids, there occurs a reduction of omega-3 fatty acids in cell membranes. Such deficiency of PUFA in cell membranes is further augmented during senescence.

It has been reported that there is an age-related increase in sarcolemmal and michondrial membrane content of aracidonic acid and reduction in DHA in the heart (Pepe, 2005). The decline in the content of cardiac cell membrane omega-3 fatty acids with aging may result in increased vulnerability to Ca^{2+} overload induced by high work stress, ischemia and reperfusion or oxidative stress itself. However, these age-related qualitative changes in membrane fatty acid composition can be reversed and rectified through dietary manipulation. Studies with young and senescent rats indicate that diets enriched with omega-3 fatty acids can prevent the age-related decline in omega-3 fatty acids in cardiac mitochondrial membranes (Pepe et al., 1999). Reports (Pepe et al., 1999; Demaisson et al., 1994) show that higher ratio of omega-3: omega-6 fatty acids in cardiac mitochondrial membranes displayed greater capacity to recover contractile functions after ischemia and reperfusion compared to that with low ratio of omega-3: omega-6 fatty acids.

Membrane ion permeability is also associated with the PUFA present in the membrane phospholipids. A common aspect of cardiac ischemia and reperfusion during advanced age is increased vulnerability to the perturbation of Ca^{2+}-management systems resulting in highly elevated intracellular Ca^{2+} that precipitates systolic and diastolic contractile dysfunctions (Hano et al., 1995). A higher omega-3/omega-6 ratio in the membrane phospholipids modifies the relative activity of Ca^{2+}-Mg^{2+} -ATPase, Ca^{2+} uptake into sarcoplasmic reticulum, voltage dependence of inactivation of Na^{+} current, and Na^{+} -Ca^{2+} exchanger activity (Phillipson and Ward, 1985; Swanson et al., 1989; Taffet et al., 1993; Leifert et al., 2000).

It is suggested that immunosenescence through increased inflammatory cytokines play important roles in promoting cardiac infections and heart failure (Watson et al., 2005). It is suggested that cytokine polarization due to aging directly dysregulates fibroblasts, leading to altered cardiac structure and dysfunction (Watson et al., 2005). Elderly people with heart diseases have high cytokine levels in the T-helper 2 cells due to suppressed resistance to cardiotrophic pathogens. It is also suggested that reduction of T-helper 2 cells and increase of T-helper 1 cytokines by supplementation with omega-3 fatty acids might provide a way to treat and prevent excessive inflammatory cytokines and their detrimental effects on the heart (Watson et al., 2005).

The recommended intake of omega-3 fatty acids for primary prevention of CHD could be 2–3 g/day of fish oil. This will occur if there is a regular consumption of 200–300 g fish and shellfish per week (Connor and Connor, 1997). The present knowledge of omega-3 fatty acids justifies that physicians in the context of secondary prevention of CHD suggest their patients to increase their consumption of these fatty acids. Apart from advising them to adequately adapt their diet, the systematic prescription of capsules containing oils enriched in α-linolenic acid, EPA and DHA should become a common practice.

4. IMMUNE RESPONSE AND INFLAMMATORY DISEASES

The immune system provides us protection from pathogens but the immunologic vitality has been shown to diminish with age. Immune cells such as lymphocytes contain high amounts of PUFA in their membrane phospholipids. Numerous studies have shown that diets high in fat content suppress immune function (Wu et al., 1999). This is more pronounced, at least in animal studies, when the fat belongs mainly to the omega-6 family of PUFA (Boissonneult and Hayek, 1992). However, the effect of dietary fats is related much more to its quality i.e. the degree of its saturation and unsaturation.

Effects of hydrogenated fats on immunity of human subjects with moderate hypercholesterolemia have been studied. Though trans fatty acids have not been reported to directly affect cellular immunity, they increase the production of inflammatory cytokines (such as interleukin-1beta) known to be associated with atherosclerosis (Han et al., 2002). Investigations (Mozaffarian et al., 2004) on the correlation between the trans fatty acids content and inflammatory marker concentrations in the red blood cell membranes of 86 patients with established heart failure suggest that trans fatty acids are strongly associated with systemic inflammation in patients with cardiac problems.

The role of eicosanoids in immune regulation is well documented. However, excessive omega-6 eicosanoid signalling has been associated with numerous inflammatory/immune vascular disorders, thrombic heart attacks and cardiac arrhythmic events, arthritis, asthma, cancer proliferation and various other chronic illnesses in aging adults (Lands, 2004). Dietary lipids are capable of influencing the fatty acid composition of membrane phospholipids. Such alterations are largely responsible for changes in immune function, through either influences on membrane-bound enzyme activity or the availability of fatty acid precursors of immune-modelling eicosanoids (Boissonneult and Hayek, 1992). Among the different fatty acid types, omega-3 and omega-6 fatty acids are most capable of influencing eicosanoids production. The omega-6 fatty acids are precursors to the 1- and 2-series prostaglandins and leukotrienes of 3- and 4-series while omega-3 fatty acids are the precursors to the 3-series prostaglandins and leukotrienes of 5-series. Leukotriene B_4 is known to enhance natural killer cell activity compared to less potent leukotriene B_5. It is also a powerful inducer of inflammation and leukocyte chemotaxis and adherence (Simopoulos, 1999).

Intake of omega-3 fatty acids either as α-linolenic acid or as EPA or DHA results in the accumulation of these fatty acids into the membrane lipids of the tissues, including cells of the immune system such lymphocytes and phagocytes (Conroy et al., 1986; Marshall and Johnston, 1983; Bankey et al., 1989). In fact, ingestion of EPA partially replaces the omega-6 fatty acids (particularly AA, 20:4) in the cell membranes of platelets, erythrocytes, neutrophils, monocytes and liver cells (Simopoulos, 1999). Intake of EPA and DHA decreases production of prostaglandin E_2 metabolites; reduces formation of leukotriene B_4; lowers the concentrations of throboxane A_2, which is a powerful platelet aggregator and vasoconstrictor;

and increases the concentrations of leukotriene B_5 (Simopoulos, 1999). Selective inhibition of inflammatory responses without inhibiting T- and B-cell functions by DHA supplementation has also been reported (Kelly, 2001).

Production of proinflammatory eicosanoids through metabolic pathways of fatty acids modulates the course of inflammatory diseases such as arthritis and psoriasis. Dietary supplements ranging 1–8 g per day of omega-3 PUFA have been reportedly beneficial in the treatment of inflammatory bowel disease, eczema, psoriasis and rheumatoid arthritis. In addition, experimental studies in rats with experimental ulcerative colitis, induced by intrarectal injection of trinitrobenzene sulphonic acid, have documented that treatment with long-chain omega-3 PUFA reduces mucosal damage as assessed by biochemical and histological markers of inflammation (Gil, 2002).

Psoriasis is one of the most common inflammatory diseases of the skin which can happen at all ages. The epidermis and scale chamber fluid of psoriatic lesions contain several lipoxygenase compounds such as leukotriene B_4, leukotriene C_4, leukotriene D_4, 12-HETE (hydroxy fatty acids) and 15-HETE (Fogh, 1990). Modulation of eicosanoid and lipoxygenase production, through dietary lipids provides a therapeutic treatment. A number of trials have demonstrated the anti-inflammatory effects of fish oils. Various studies (Ziboh et al., 1986) demonstrate the mild to moderate improvement of psoriatic patients from fish oil supplements (11–14g EPA/day for 8 weeks). The improvement in clinical response was associated with the incorporation of EPA and DHA present in the fish oils into the epidermal tissues. Successful reduction of itching, scaling and erythema from 8 week supplementation with fish capsules (1.8g EPA/day) has also been reported (Bittiner et al., 1988). Following similar trials with 3.6g EPA ethyl-ester/day for 3–6 months, significant improvement of scaling and erythema in their patients was reported (Terano et al., 1989). Reduction in neutrophil production of leukotriene B_4 was observed from one month after start of the study along with marked increase of leukotriene B_5 and 5-HETE. These reports demonstrate the potency of omega-3 fatty acids in prevention and treatment of inflammatory skin disorders like psoriasis.

Rheumatoid arthritis is a chronic inflammatory disease of the joints which trouble a large number of the elderly population. It is characterised by inflammation of the synovium and infiltration of the joint by neutrophils, macrophages and T lymphocytes and subsequent erosion of articular cartilage and bone (Boissonneult and Hayek, 1992). Eicosanoids derived from the metabolic pathways of omega-6 fatty acids, arachidonic acid, and the cytokines interleukin-1beta and tumour necrosis factor-alpha are related with the symptoms of inflammatory joint disease, as well as the cartilage degradation seen in established rheumatoid arthritis (James et al., 2003). The presence of leukotriene B_4 and 5-HETE in the synovial fluid from patients with rheumatoid arthritis (Fogh, 1990) suggest that restricting the production of these eicosanoids can probably slow down the inflammatory processes associated with rheumatoid arthritis. The use of omega-3 fatty acids as a part of dietary treatment of rheumatoid arthritis had been investigated by various researchers. Rheumatoid arthritis patients who consumed a supplementation of 1.8g of EPA/day showed fewer clinical symptoms of their disease after 12 weeks (Kremer et al., 1985). Similar improvement of symptoms of rheumatoid arthritis in patients supplemented with omega-3 fatty acids have been observed by others (Sperling et al., 1987; Magaro et al., 1988).

5. CARCINOGENESIS

The correlation between dietary fats and cancer has been investigated through epidemiological and experimental studies in several organs such as the skin, liver, colon, pancreas and mammary gland. Most of the experimental studies concerning dietary fats have been on the rat model system, and have been followed up till complete carcinogenesis induced by polyaromatic hydrocarbons (PAH) or ultra-violet (UV) light. However, non-conforming results, even from similar studies have also been reported.

The incidence of skin cancer has been undergoing a steady increase in recent years. Skin cancer is most common among the elderly, but is now also more frequently found in younger people (Tarstedt et al., 2005). Early studies have shown the effect of fatty acids on the initiation and promotion of skin carcinogenesis. Daily application of lauric acid (20:0) and oleic acid (18:1) on mouse skin after a single administration of 7,12-dimethylbenz[a]anthracene (DMBA) showed cancer promoting activity. Stearic acid (18:0) and palmitic acid (16:0) however showed no effect. In case of DMBA-initiated carcinogenesis, high fat diets slightly inhibited initiation but enhanced the promotion (Birt et al., 1989). Such enhancing effect has been attributed mostly to the increased consumption of calories. While reports show that high fat diets increased UV induced skin carcinogenesis in rats, others found no such effects (Black et al., 1983). On the contrary, they concluded that diets containing saturated fatty acids inhibited tumorigenesis.

Occurrence of colon cancer in the industrialized countries has risen since the early 1970's and it is estimated that more than one-third of such cases are diet related (Roynette et al., 2004). Though a number of correlational and case control epidemiological studies have established a positive association between dietary fats and development of colon cancer many prospective epidemiological studies have concluded otherwise (Glauert, 1992). However, interpretations of such studies are complicated by the total energy intake which has been correlated to colon cancer in various correlational and case control studies (Kolonel, 1987; Lyon et al., 1987). Studies on the effect of different dietary fatty acids show that the promotional phase of colon carcinogenesis (induced by multiple injections of azoxymethane) is affected more by PUFA compared to saturated fatty acids (Sakaguchi et al., 1984). The initiation of colon carcinogenesis, however, is affected more by increasing the level of saturated fats and not by the amount of PUFA (Reddy and Sugie, 1998). It has been hypothesized that dietary fats increase the concentration of metabolites with carcinogenic or promoting activity in fecal stream (Glauert, 1992).

Several researchers have reported the effect of dietary fats on initiation and promotion of carcinogenesis in liver. Increase in fat content of the diet enhanced the development of artificially induced tumors in rat livers (Reddy and Sugie, 1998). The enhancement of hepatocarcinogenesis by dietary fats is primarily due to the effect on initiation of carcinogenesis and polyunsaturated fats have greater effect then saturated fats (Glauert, 1992).

Pancreatic carcinogenesis in humans has been connected with dietary fats (Baldwin and Parker, 1986) and detailed investigations have been carried out in animal models with rats and hamsters. Since the tumors are derived primarily from ductal cells in both hamsters and humans, the hamster model may be considered to be more pertinent to human pancreatic cancer. Higher dietary fat intake increases the incidence of pancreatic carcinogenesis in hamsters (Birt et al., 1989) with polyunsaturated fats having greater enhancing effect, compared to saturated fats. Conversely, one study (Birt et al., 1990) showed that intake of a saturated fat (beef tallow) promoted pancreatic carcinogenesis more than that by polyunsaturated fat (corn oil) in hamster model.

Prostate cancer is the second major cause of cancer related death in men in the US (Pienta and Esper, 1993). Epidemiological studies have demonstrated that men with higher dietary intake of omega-3 fatty acids have a lower incidence of prostate cancer. Moreover, omega-6 and omega-3 fatty acids have respectively displayed promotional and inhibitory effects in prostate cancer cell lines as studied by Pandalai et al. (Pandalai et al., 1996). Their results revealed that EPA inhibits prostate cell growth at high concentration.

Most of the polyunsaturated oils such as corn and safflower, used in various carcinogenic studies are rich in LA i.e. omega-6 fatty acids (typically 55% for corn and 75% for safflower) and have a very low content of omega-3 fatty acids (typically 0–1% for both). Fish oils, rich in long chain omega-3 PUFA however have beneficial effects. Substitution of corn oil with oils rich in omega-3 fatty acids (such as fish oils) generally has inhibitory effects on chemically induced carcinogenesis (O'Connor et al., 1989). Various researchers have observed similar effects in the colon, mammary glands and the pancreas of their animal subjects. Radiation therapy and chemotherapy drugs such as doxorubicin, epirubicin, tamoxifen etc. show higher efficacy when omega-3 fatty acids are included in the diet (Hardman, 2004). Data from 24 European countries indicate that a high ratio of omega-6/omega-3 fatty acids in diet has greater risk for colon cancer (Caygill and Hill, 1995). The mechanisms of action of omega-3 fatty acids on colon carcinogenesis is proposed to be that n-3 PUFAs are able to influence colon carcinogenesis by altering enzyme expression and/or activity and, therefore, the concentrations of end-products or by modulating the levels of available precursors for biosynthetic pathways (Roynette et al., 2004). Other probable mechanisms for the effect of omega-3 fatty acids against carcinogenesis include modulation of eicosanoid production and inflammation, angiogenesis, proliferation, susceptibility for apoptosis, and estrogen signalling (Hardman, 2004).

6. DIABETES

One of the silent killers of modern times is diabetes mellitus. Hyperglycemia and dyslipidemia (a condition marked by abnormal concentrations of lipids or lipoproteins in the blood) are two main abnormalities associated with both insulin-dependent diabetes mellitus (IDDM, type 1) and non-insulin-dependent diabetes mellitus (NIDDM, type 2). Diabetes mellitus is characterised by hyperglycemia in presence of insulin resistance, hypertriglyceridemia, increased VLDL, altered lipogenesis and accelerated lipolysis (Bhathena, 1992). High intake of dietary fats has been correlated with development of insulin resistance in both animals and humans with different types of fats having different effects on insulin action. Saturated fats and trans fats cause insulin resistance while monounsaturated and polyunsaturated fats improve it (Rivellese and Lilli, 2003).

Recent evidence from epidemiological studies show that risk factors for type 2 diabetes is connected to high trans fatty acid and low ratio of unsaturated: saturated fat intake (Parillo and Riccardi, 2004). Increased levels of palmitic acid and palmitoleic (16:1n-7) and reduced levels of linoleic acid have been linked with insulin resistance and consequent complications (Vessby, 2000). Animal studies using primates reveal that similar to saturated fatty acids, trans fatty acids affect the insulin receptors by reducing their numbers and increasing their affinity (Barnard et al., 1990). Markedly higher proportions of saturated fats and decreased PUFA have been observed in the phospholipids of red blood cells of both IDDM and NIDDM subjects (van Doormaal et al., 1984; Prisco et al., 1989). A study involving more than 84,000 women aged between 34–59 years was conducted to examine the relations between dietary fat intakes and risk of type 2 diabetes in USA (Salmeron et al., 2001). None of the subjects had any cardiovascular problems, cancer or diabetes at start. From the data collected over a period of 14 years, it was concluded that total fats and saturated and monounsaturated fatty acids do not increase the risk of type 2 diabetes in women, but trans fatty acids enhance, whereas PUFAs reduce the risk. Trans fatty acids are incorporated into cell membrane phospholipids causing decrease in membrane fluidity and binding of insulin to its receptor, leading to impaired insulin action, insulin resistance and hyperinsulinemia (Simopoulos, 1999).

Most of the studies concerning human diabetic subjects have used linoleic acid rich vegetable oils. Linoleic acid has a protective effect on diabetic retinopathy (Howard-Williams et al., 1985). However, some have reported increased insulin resistance in liver and muscle in diabetic rats from saturated fatty acids and linoleic acid rich diet (Storlien et al., 1987). γ-linolenic acid (GLA), an omega-6 metabolite, has been reported to have many beneficial effects in both NIDDM and IDDM such as prevention and treatment of distal diabetic polyneuropathy (Jamal and Carmichael, 1990). Feeding animal subjects an essential fatty acid deficient diet which lowers the concentration of AA, decreased the incidence of spontaneous diabetes. It has been hypothesised that AA or its eicosanoid metabolites may be responsible for the inflammatory conditions of autoimmune diabetes in the experimental rat model system (Lefkowith et al., 1990), which is similar to human IDDM.

Hyperinsulinemia and insulin resistance are inversely linked with the content of C20 and C22 fatty acids in the phospholipids of muscle cell membranes (Borkman et al., 1993). A reduction of EPA in the livers of diabetic patients was also observed (Singer et al., 1980). In another study, a higher EPA content was reported in the liver triglycerides of diabetic subjects without hyperlipoproteinemia (Singer et al., 1984). Chronic deficiency of EPA may lead to complications of diabetes such as retinopathy, peripheral neuropathy and nephropathy (Sinclair, 1962). Dietary fish oils have various beneficial effects on diabetic subjects, for example, an augmentation of 20- and 22-carbon PUFAs leads to increase in membrane fluidity, the number of insulin receptors and insulin action (Harris, 1996).

People suffering from diabetes mellitus have an increased cardiovascular morbidity and mortality. The most consistent beneficial effect of long chain PUFAs is the reduction of triglyceride levels in serum. There is also considerable evidence that fish oils lower cholesterol/phospholipids ratio and cholesterol/HDL ratio which is considered to be a measure of atherogenic index (Bhathena, 1992). Fish oil also increases lipoprotein lipase activity in NIDDM but has no effect in IDDM (Kasim et al., 1988; Bagdade et al., 1990). Dietary omega-3 fatty acids reportedly reduce blood viscosity (Rillaerts et al., 1989), lower blood pressure (Kasim et al., 1988) and increase neutrophil in diabetic subjects (Schmidt et al., 1989).

Despite the physiological benefits on diabetic subjects, unrestricted or unmonitored use of omega-3 fatty acids is not recommended. Omega-3 fatty acids have detrimental effects on carbohydrate metabolism and inversely affect glycemic control even though insulin sensitivity is improved. Plus, the positive effects on lipid metabolism cannot be sustained by prolonged use of fish oil and are reversed when fish oil supplementation is discontinued (Bhathena, 1992). Another concern for excessive use of omega-3 fatty acids is their susceptibility to oxidation. This aspect of PUFA is discussed in later sections.

7. ALZHEIMER'S DISEASE (AD)

AD is the most common dementing illness of the aged and is characterised by global impairment of cognitive functions. Though the environmental risk factors for AD have not been identified with certainty, a number of dietary elements have been reported to be associated with the development of dementia. Evidence shows that oxidative stress, homocysteine-related vitamins, dietary fats and alcohol play a role in the pathogenesis of AD (Luchsinger and Mayeux, 2004). It has been postulated that AD may be promoted by insulin resistance, excess free radicals, inflammatory metabolites, homocysteine and oestrogen deficiency (Berrino, 2002). Vascular risk factors such as type 2 diabetes, hypertension, high dietary fat intake, high cholesterol, and obesity are also suspected of increasing the risk of both vascular and AD (Haan and Wallace, 2004).

Based on epidemiological risk factors, it has been suggested that dietary lipids may be the principal risk factors for the development of late-onset AD (Cooper, 2003). The nature of saturation and unsaturation of fatty acids are crucial in determining the effect on AD. The Mediterranean diet comprising mostly of oleic acid rich olive oil appear to provide high protection against cognitive decline as observed for the aged population in Southern Italy (Solfrizzi et al., 2003). Omega-3 fatty acids offer some protection against AD while saturated and omega-6 fatty acids increase the risk (Cooper, 2003).

DHA is the principal fatty acid of neurological and retinal membranes and it makes up more than 30% of the structural lipid of the neuron (Kyle et al., 1999). Reduced blood levels of omega-3 fatty acids have been related to many neuropsychiatric disorders such as attention deficit (hyperactivity) disorder, AD, schizophrenia and depression (Young and Conquer, 2005). Innvestigations have been made on the protective relationship between fish consumption and intake of different types of omega-3 fatty acids against AD (Morris et al., 2003). A total of 815 subjects, aged 65 to 94 years, who were initially unaffected by AD completed a dietary questionnaire on average 2.3 years before clinical evaluation of incident disease. It was concluded that participants who consumed fish once per week or more had 60% less risk of AD compared with those who rarely or never ate fish. Total intake of omega-3 PUFA was associated with reduced risk of AD, as was intake of DHA. EPA was not associated with AD. Other clinical studies with DHA have also shown to bring improvement in senile dementia (Yazawa, 2004).

Some investigations have shown reduced levels of AA and DHA in phospholipids fractions such as phosphatidylcholine (PC) and phosphatidylethanolamine (PE) from various parts of the brain (frontal grey frontal white, hippocampus, pons) of patients suffering from AD (Söderburg et al., 1991; Prasad et al., 1998). However, senescence itself has no influence on the fatty acid composition of PC and/or PE in these areas of the brain (Söderburg et al., 1991). The plasma fatty acid analysis of various phospholipids fractions of patients suffering from AD (mean age 82.7 yrs) and other forms of cognitive impairment (but nondemented) (mean age 83.3 yrs) and dementia (mean age 79.4 yrs) show lower levels of EPA, DHA, total omega-3 fatty acids and omega-3/omega-6 ratio in plasma phospholipids, PC and PE (Conquer et al., 2000). No other differences in the fatty acid composition of the different phospholipids fractions were noted in this study.

Kyle et al. (1999) have investigated the correlation between circulating DHA of 1188 elderly American subjects (mean age 75 yrs) and AD diagnosis and scores on the Minimental State Exam (MMSE). The serum PC was used as the biomarker. Their data present low levels of circulating PC-DHA as a risk factor for low scores on the MMSE and development of AD in the elderly patients. Due to the declining activity of the delta-6-desaturase enzyme it is difficult for the elderly to maintain a healthy level of serum DHA. Thus it is very important for the elderly to have DHA supplementation or eat adequate amounts of fish. A relatively small pilot study with 10 elderly subjects (average age 83 yrs) suffering from senile dementia of cerebrovascular disorders has been performed (Terano et al., 1999). They administered DHA (0.72g daily for 1 year) to the subjects and evaluated the effect on dementia using psychometric tests such as MMSE and Hasegawa's Dementia rating scale. Their findings show that DHA supplementation improved the dementia scores in the elderly suffering from moderately severe dementia from thrombotic cerebrovascular disorders.

Thus, one may conclude that, dietary intake of omega-3 fatty acids and weekly consumption of fish may reduce the risk of incident AD. However, there cannot be any compromise with the oxidative quality of the omega-3 supplements and the freshness of the fish, as discussed below.

8. OXIDATION OF LIPIDS AND USE OF ANTIOXIDANTS

Though by and large, PUFAs have numerous beneficial effects on human health, their susceptibility to autoxidation is a serious concern associated with all forms of their intake. Exposure to air, heat and light causes the unsaturated moieties of the fatty acids to undergo a spontaneous free radical-initiated chemical reaction called autoxidation. It proceeds in three steps, initiation, propagation and termination. Autoxidation is commonly characterised by an induction period during which very little change occurs in lipids. After the end of the induction period, oxidative deterioration of the lipids occurs much more quickly. It is well known that the greater the number of unsaturated sites, the greater is the tendency of oxidation. For example, if the rate of oxidation for oleic acid (18:1) is 1, then the relative rates of oxidation for linoleic acid and alpha-linolenic acid are 12 and 25 respectively.

Autoxidation of PUFAs generates hydroperoxides as primary oxidation products and further oxidation leads to cyclic peroxides as secondary oxidation products. Monocyclic peroxides, bicyclic endoperoxides, serial cyclic peroxides, and a new class of endoperoxides (dioxolane-isoprostane peroxides) have been identified from the oxidation of arachidonate (Yin and Porter, 2005). These oxidation products are a potential source of free radicals which may cause damaging effects in vivo. The excess free radicals may react with proteins, DNA and other molecules and these reactions represent pathways whereby cancer, CHD and a host other disorders can develop. Thus it is vitally important that oils and fats are protected from oxidation.

Addition of antioxidants to oils and fats prevent oxidation by extending the induction period. However, use of antioxidants after the end of this period is generally ineffective because by that time, the oil or fat has developed considerable degree of rancidity. Storage of oils and fats in closed containers and in cool, dark places away from heat sources also prolongs the induction period.

Antioxidants can be both synthetic and natural. The major synthetic antioxidants which are widely used in various food products are t-butylhydroquinone, butylated hydroxy toluene, butylated hydroxy anisole and propyl gallate. However, possible harmful side effects of the synthetic antioxidants have created a demand for natural antioxidants. Various herb extracts, spices, teas, oilseeds and oils, cereals, legumes, fruits and vegetables contain minor components that act as natural antioxidants. The different types of natural antioxidants investigated include: (i) tocopherols and tocotrienols; (ii) phenolic acids (carnosic acid and rosmarinic acid) found mainly in the Lamiaceae family of herbs; (iii) flavonoids (e.g. quercetin, kaemferol, luteolin, morin, myricetin) from plant sources; and (iv) catechins or phenols (carnosol, rosmanol, epirosmanol) from tea and Labiatae family of herbs. Several beneficial properties have been attributed to these dietary compounds, including anti-inflammatory and anticarcinogenic effects (Galati and O'Brien, 2004). Though these natural phenolics/flavonoids are generally regarded safe, controlled clinical trials to show efficacy and potential for toxicity of many of these natural antioxidants are still required. These natural compounds are generally lipophilic and dietary lipids can act as the carrier of such active ingredients which would provide multiple benefits.

9. CONCLUSIONS AND FUTURE PERSPECTIVES

It is evident that dietary fats have significant contribution to our well being during all stages of life. It is never too late to initiate and benefit from the nutritive effects of dietary lipids. Modern life style is making us increasingly dependant on bulk-prepared foods. When designed with healthy and top quality oils, the innumerable varieties of prepared foods available can play an important role in the diet schemes. However, dietary fats have to be consumed with prudence and in moderate amounts. Nordic Nutrition Recommendations suggests a limitation of the intake of saturated plus trans fatty acids to about 10% of the total energy intake (E%), and of the total fat intake to 30E%. It is also recommended that cis-MUFA should provide 10–15E% and PUFA 5–10E% including approximately 1E% from omega-3 fatty acids.

Direct consumption of EPA and DHA for vegetarian people is almost non-existent due to absence of fish in their diets and alpha-linolenic acid from plant sources is the primary omega-3 fatty acid in their diet. Although alpha-linolenic acid is transformed to EPA and DHA, consistent quantification seems to be a debatable issue. Though DHA from algal sources is now available in encapsulated forms, future research should concentrate on incorporating such long chain PUFAs into the seed oils. Such development might even require genetic modification. Scientific studies from both academic and industrial areas will continue to discover newer benefits of specific lipids and at the same time caution us about some. Simultaneously, conscious effort has to be made to translate scientific findings into a language understood by consumers who need to feel confident and comfortable about what they eat.

REFERENCES

Akoh, C.C. (2002) Structured lipids. In: Food Lipids–Chemistry, Nutrition and Biotechnology. (Ed.: Akoh, CC), Marcel Dekker, Inc. New York, USA, 877–908.

Bagdade, J.D., et al. (1990) Effects of omega-3 fish oils on plasma lipids, lipoprotein composition, and postheparin lipoprotein lipase in women with IDDM. Diabetes, 39: 426–431.

Baldwin, S., Parker, R.S. (1986) The effect of dietary fat and selenium on the development preneoplastic lesions in rat lever. Nutr Cancer, 8: 273.

Bankey, E.M., et al. (1989) Modulation of Kupffer cell membrane phospholipid function by n-3 polyunsaturated fatty acids. J Lipid Res., 30: 1703–1710.

Barnard, J.D., et al. (1990) Dietary trans fatty acids modulate erythrocyte membrane fatty acyl composition and insulin binding in monkeys. J Nutr. Biochem., 1: 190–195.

Berrino, F. (2002) Western diet and Alzheimer's disease. Epidemiol. Prev., 26: 107–115.

Bhathena, S.J. (1992) Fatty acids and diabetes. In: Fatty Acids in Foods and Their Health Implications. (Ed.: Chow CK), Marcel Dekker Inc., New York, USA, 823–855.

Birt, D.F., White, L.T., Choi, B. and Pelling, J.C. (1989) Dietary fats effects on the initiation and promotion of two-stage skin tumorigenesis in the SENCAR mouse. Cancer Res, 49: 4170.

Birt, D.F., et al. (1990) Comparison of the effects of dietary beef tallow and corn oil on pancreatic carcinogenesis in the hamster model. Carcinogenesis, 11: 745.

Bittiner, S.B., Tucker, W.F., Cartwright, I. and Bleehen, S.S. (1988) A double-blind, randomised placebo-controlled trial of fish oil psoriasis. Lancet 1(8582): 378–380.

Black, H.S., Lenger, W., Phelps, A.W. and Thornby, J.I. (1983) Influence of dietary lipid upon ultraviolet carcinogenesis. Nutr. Cancer, 5: 59.

Boissonneult, G.A., Hayek, M.G. (1992) Dietary fat, immunity, and inflammatory disease. In: Fatty Acids in Foods and Their Health Implications. (Ed.: Chow CK), Marcel Dekker Inc., New York, USA, 707–734.

Bonanome, A., Grundy, S.M. (1988) Effect of dietary stearic acid on plasma cholesterol and lipoprotein levels. N. Engl. J. Med., 318: 1244–1248.

Borkman, M., et al. (1993) The relation between insulin sensitivity and the fatty acid composition of skeletal-muscle phospholipids. N Eng J Med., 328: 238–244.

Castelli, W.P., et al. (1986) Incidence of coronary heart disease and lipoprotein cholesterol levels: The Framingham Study. J Am Med Assoc., 256: 2835–8.

Cater, N.B., Heller, H.J., Denke, M.A. (1997) Comparison of the effects of the medium-chain triglycerols, palm oil, and high oleic sunflower oil on plasma triacylglycerol fatty acids and lipid and lipoprotein concentration in humans. Am. J. Clin. Nutr., 65: 41–45.

Caygill, C., Hill, M. (1995) Fish n-3 fatty acids and human colorectal and breast cancer mortality. Eur J Cancer Prev, 4: 329–332.

Cho, H.P., Nakamura, M., Clarke, S.D. (1999) Cloning, expression and fatty acid regulation of the mammalian Delta-6 desaturase. J. Biol. Chem., 274: 471–477.

Choudhury, N., Tan, L., Truswell, A.S. (1995) Comparison of palm olein and olive oil: Effects on plasma lipids and vitamin E in young adults. Am. J. Clin. Nutr., 61: 1043–51.

Chyou, P., Elaine, D., Eaker, D. (2000) Serum cholesterol concentrations and all-cause mortality in older people. Age and Ageing, 29: 69–74.

Connor, S.L., Connor, W.E. (1997) Are fish oils beneficial in the prevention and treatment of coronary artery disease? Am J Clin Nutr, 66: 1020–31.

Conquer, J.A., et al. (2000) Fatty acid analysis of blood plasma of patients with Alzheimer's disease, other types of dementia, and cognitive impairment. Lipids, 35: 1305–12.

Conroy, D.M., et al. (1986) The effects of dietary oils on the production of n-3 and n-6 metabolites of leukocyte 5-lipoxygenase in five rat strains. Biochim Biophysics Acta, 861: 457–462.

Cooper, J.L. (2003) Dietary lipids in the aetiology of Alzheimer's disease: implication for therapy. Drugs Aging, 20: 399–418.

De Lorgeril, M., Salen, P. (2004b) Alpha-linolenic acid and coronary heart disease. Nutr Metab Cardiovasc. Dis., 14: 162–9.

De Lorgeril, M., Salen, P. (2004a) Use and misuse of dietary fatty acids for the prevention and treatment of coronary heart disease. Reprod Nutr Dev., 44: 283–288.

De Lorgeril, M., et al. (1999) Mediterranean diet, traditional risk factors, and the rate of cardiovascular complicationa after myocardial infarction: Final report of the Lyon Diet Heart Study. Circulation, 99: 779–785.

Demaisson, L., Sergiel, J.P., Moreau, D. and Grynberg, A. (1994) Influence of the phospholipid n-3/n-6 PUFA ratio on mitochondrial oxidative metabolism before and after myocardial ischemia. Biochim Biophys. Acta, 1366: 69–78.

Denke, M.A., Grundy, S.M. (1992) Comparison of effects of lauric acid and palmitic acid on plasma lipids and lipoproteins. Am. J. Clin. Nutr., 56: 895–98.

Dolecek, T.A. (1992) Epidemiological evidence of relationships between dietary polyunsaturated fatty acids and mortality in the Multiple Risk Factor Intervention Trial. Proc Soc Exp Biol. Med., 200: 177–182.

Fogh, K. (1990) Lipoxygenase products of arachidonic acid in psoriasis, atopic dermatitis, and experimental arthritis. Dan. Med. Bull., 37: 289–308.

Formo, M.W. (1979) Fats in the diet. In: Bailey's Industrial Oil & Fat Products. (Ed. Hui), 4th edition, 1, New York, John Wiley & Sons, 233–270.

Galati, G., O'Brien, P.J. (2004) Potential toxicity of flavonoids and other dietary phenolics: significance for their chemopreventive and anticancer properties. Free Radic Biol Med., 37: 287–303.

Gil, A. (2002) Polyunsaturated fatty acids and inflammatory diseases. Biomed Pharmacother, 56: 388–96.

Glauert, P.H. (1992) Dietary fatty acids and cancer. In: Fatty Acids in Foods and Their Health Implications. (Ed.: Chow CK), Marcel Dekker Inc., New York, USA, 753–768.

Grande, F., Andersson, J.T., Keys, A. (1970) Comparison of effects of palmitic and stearic acids in the diet on serum cholesterol in man. Am. J. Clin. Nutr., 23. 1184–93.

Haan, M.N., Wallace, R. (2004) Can dementia be prevented? Brain aging in a population-based context, Annu Rev Public Health, 25: 1–24.

Han, S.N., et al. (2002) Effect of hydrogenated and saturated, relative to polyunsaturated, fat on immune and inflammatory responses of adults with moderate hypercholesterolemia. J. Lipid Res., 43: 445–452.

Hano, O., et al. (1995) Reduced threshold for myocardial cell calcium intolerance in the rat heart with aging. Am. J. Physiol., 269: 1607–1612.

Hardman, W.E. (2004) n-3 fatty acids and cancer therapy. J Nutr., 134: 3427–3430.

Harris, W.S., Connor, W.E., McMurry, M.P. (1983) The comparative reductions of the plasma lipids and lipoproteins by dietary polyunsaturated fats: Salmon oil versus vegetable oils. Metabolism, 32: 179–84.

Harris, W.S. (1989) Fish oils and plasma lipids and lipoprotein metabolism in humans: A critical review. J. Lipid Res., 30: 785–807.

Harris, W.S. (1996) Do omega-3 fatty acids worsen glycemic control in NIDDM? ISSFAL Newsletter, 3: 6–9.

Hastert, R.C. (1996) Hydrogenation. In: Bailey's industrial oil & fat products. (Ed.: Hui) Vol 3, John Wiley & Sons, New York, USA, 212–300.

Howard-Williams, J., et al. (1985) Polyunsaturated fatty acids and diabetic retinopathy. Br. J. Opthalmol., 69: 15.

Hunter, J.E. (1992) Safety and health effects of isomeric fatty acids. In: Fatty Acids in Foods and Their Health Implications. (Ed.: Chow CK), Marcel Dekker, Inc. New York, USA, 857–868.

Jamal, G.A., Carmichael, H. (1990) The effect of gamma-linolenic acid on human diabetic peripheral neuropathy: a double-blind placebo-controlled trial. Diabetic Med., 7: 319–323.

James, M.J., Proudman, S.M., Cleland, L.G. (2003) Dietary n-3 fats as adjunctive therapy in a prototypic inflammatory disease: issues and obstacles for use in rheumatoid arthritis, Prostaglandins Leukot Essent Fatty Acids, 68: 399–405.

Kasim, S.E., et al. (1988) Effects of omega-3 fish oils on lipid metabolism, glycemic control and blood pressure in type II diabetic patients. J Clin. Endocrinol. Metab., 67: 1–5.

Katan, M.B., Zock, P.L., Mensink, R.P. (1995) Trans fatty acids and their effects on lipoproteins in humans. Annu. Rev. Nutr., 15: 473–93.

Kelly, D.S. (2001) Modulation of human immune and inflammatory responses by dietary fatty acids. Nutrition, 17: 669–673.

Keys, A., Andersen, J.T., Grande, F. (1957) Prediction of serum cholesterol responses of man to change in fats in the diet, Lancet 2: 959.

Keys, A., Andersen, J.T. and Grande F. (1959) 'Serum cholesterol in man: dietary fat and intrinsic responsiveness', Circulation, 19: 201.

Keys, A., Andersson, J.T., Grande, F. (1965) Serum cholesterol response to changes in the diet IV. Particular saturated fatty acids in the diet. Metabolism, 14: 776–87.

Kolonel, L.N. (1987) Fat and colon cancer: how firm is the epidemiological evidence? Am. J. Clin. Nutr. 45: 336.

Kremer, J.M., et al. (1985) Effects of manipulation of dietary fatty acids on clinical manifestations of rheumatoid arthritis, Lancet 1(8422): 184–187.
Kristott, J. (2003) High-oleic oils – how good they are for frying? Lipid Technology, 15: 29–2.
Kyle, D.J., Schaefer, E., Patton, G., Beiser, A. (1999) Low serum docosahexaenoic acid is a significant risk factor for Alzheimer's dementia. Lipids, 34: 245.
Lands, W.E.M. (ed) (1986) Fish and Human Health, Academic Press, Inc., Orlando, Florida.
Lands, W.E.M. (2004) Essential fatty acid metabolism to self-healing agents. In: Healthful Lipids. (Eds.: Akoh CC and Lai O), AOCS Press, Champaign, Illinois.
Lefkowith, J., et al. (1990) Prevention of diabetes in the BB rat by essential fatty acid deficiency. Relationship between physiological and biochemical changes. J Exp Med., 171: 729–743.
Leifert, W.R., Jahangiri, A. Saint, D.A., McMurchie, E.J. (2000) Effects of dietary n-3 fatty acids on contractility, Na(+) and K(+) currents in a rat cardiomyocyte model of arrhythmia. J. Nutr. Biochem., 11: 382–392.
Luchsinger, J.A., Mayeux, R. (2004) Dietary factors and Alzheimer's disease. Lancet Neurol., 3: 579–587.
Lyon, J.L., et al. (1987) Energy intake: its relationship to colon cancer risk, J. Natl. Cancer Inst. 78: 853.
Magaro, M., et al. (1988) Influence of diet with different lipid composition on neutrohil chemiluminescence and disease activity in patients with rheumatoid arthritis. Ann Rhem. Dis., 47: 793–796.
Marshall, L.A., Johnston, P.V. (1983) The effect of dietary essential fatty acid in the rat on fatty acid profiles of immunocompetent cell populations. Lipids, 23: 623–625.
Martin, M.J., et al. (1986) Serum cholesterol, blood pressure, and mortality: implications from a cohort 361,662 men. Lancet. 2: 933–6.
Mensink, R.P., Plat, J., Temme EHM. (2002) Dietary fats and coronary heart disease. In.: Food Lipids-Chemistry, Nutrition and Biotechnology. (Ed.: Akoh, CC.), Marcel Dekker Inc., New York, USA, 603–36.
Morris, M.C., et al. (2003) Consumption of fish and n-3 fatty acids and risk of incident Alzheimer disease. Arch Neurol., 60: 940–946.
Mozaffarian, D., et al. (2004) Trans fatty acids and systemic inflammation in heart failure. Am J Clin Nutr, 80: 1521–1525.
Ng, T.K., et al. (1992) Dietary palmitic acid and oleic acids exert similar effects on serum cholesterol and lipoprotein profiles in normocholesterolemic men and women. J. Am. Coll. Nutr., 11: 383–90.
Norell, S.E., et al. (1986) Diet and pancreatic cancer: a case-control study. Am. J Epidemiol., 124: 894
O'Connor, T.P., et al. (1989) Effect of dietary omega-3 and omega-6 fatty acids, on development of azaserine-induced prenoplastic lesions in rat pancreas. J Natl. Cancer Inst., 81: 858.
Pandalai, P.K., et al. (1996) The effects of omega-3 and omega-6 fatty acids on in Vitro prostate cancer growth. Anticancer Res, 16: 815–820.
Parillo, M., Riccardi, G. (2004) Diet composition and the risk of type 2 diabetes: epidemiological and clinical evidence. Br J Nutr., 92: 7–19.
Pepe, S., Tsuchiya, N., Lakatta, E., Hansford, R. (1999) PUFA and aging modulate cardiac mitochondrial membrane lipid composition and Ca^{2+} activation of PDH. Am. J. Phsiol., 276: 149–158.
Pepe, S. (2005) Effect of dietary polyunsaturated fatty acids on age-related changes in cardiac mitochondrial membranes. Experimental Gerontology., 40: 369–376.
Phillipson, R., Ward, R. (1985) Effects of fatty acids on Na^{+}/Ca^{2+} exchange and calcium permeability of cardiac sarcoplasmic reticulum vesicles. J. Biol. Chem., 260: 9666–9671.
Pienta, K.J., Esper, P.S. (1993) Risk factors for prostate cancer. Annals Int Med, 118: 793–803.
Prasad, M.R., Lovell, M.A., Yatin, M., Dhillon, H., Markesbery, W.R. (1998) Regional membrane phospholipid alterations in Alzheimer's disease. Neurochem. Res., 23: 81–88.
Prisco, D., et al. (1989) Altered membrane fatty acid composition and increased thromboxane A_2 generation in platelets from patients with diabetes. Prostaglandins, Leukotrienes, Essent. Fatty Acids., 35: 15–23.
Reddy, B.S., Sugie, S. (1998) Effect of different levels of omega-3 and omega-6 fatty acids on azoxymethane-induced colon carcinogenesis in F344 rats. J Natl. Cancer Inst, 77: 815.
Rillaerts, E.G., Engelmann, G.J., Van Camp, K.M., De Leeuw, I. (1989) Effect of omega-3 fatty acids in diet type I diabetic subjects on lipid values and hemorheological parameters. Diabetes, 38: 1412–1416.

Rivellese, A.A., Lilli, S. (2003) Quality of dietary fatty acids, insulin sensitivity and type 2 diabetes. Biomed Pharmacother., 57: 84–87.

Robert, L.S. (1990) Impact of dietary fat on human health. In: Omega-3 fatty Acids in Health and Diseases. (Eds.: Lees, SR., Karel, M.) Marcel Dekker Inc, New York, 1–38.

Roynette, C.E., et al. (2004) n-3 polyunsaturated fatty acids and colon cancer prevention. Clinical Nutr., 23: 139–151.

Söderburg, M., Edlund, C., Kristensson, K., Dallner, G. (1991) Fatty acid composition of brain phospholipids in aging and Alzheimer's disease. Lipids, 26: 421–428.

Sakaguchi, M., et al. (1984) Effect of dietary unsaturated and saturated fatty acids on azoxymethane-induced colon carcinogenesis in rats. Cancer Res, 44: 1472.

Salmeron, J., et al. (2001) Dietary fat intake and risk of type 2 diabetes in women. Am J Clin Nutr., 73: 1019–26.

Schmidt, E.B., et al. (1989) The effect of n-3 polyunsaturated fatty acids on lipids, haemostasis, neutrophil and monocyte chemotaxis in insulin-dependent diabetes mellitus. J Intern. Med. Suppl., 225: 201–206.

Simopoulos, A.P. (1999) Essential fatty acids in health and chronic diseases. Am J Clin Nutr., 70(suppl): 560–569.

Sinclair, H.M. (1962) Essential fatty acids. In: Clinical Nutrition. 2nd ed. (Ed.: N. Jolliffe), Harper, New York, USA, 206–215.

Singer, P., Honigman, G., Schiliack, V. (1980) Decrease in eicosapentaenoic acid in fatty liver of diabetic subjects. Prostaglandins Med., 5: 183–200.

Singer, P., Honigman, G., Schiliack, V. (1984) Negative correlation of eicosapentaenoic acid and lipid accumulation in hepatocytes of diabetes. Biomed. Biochem. Acta., 43: 438–442.

Solfrizzi, V., Panza, F., Capruso, A. (2003) The role of diet in cognitive decline. J Neural Transm., 110: 95–110.

Sperling, R.I., et al. (1987) Effects of dietary supplementation with marine fish oil on leukocyte lipid mediator generation and function in rheumatoid arthritis. Arthritis Rheum, 30: 988–997.

Sprecher, H., Lutharia, D.L., Mohammed, B.S., Baykousheva, S.P. (1995) Re-evaluation of the pathways for the biosynthesis of polyunsaturated fatty acids. J. Lipid Res, 36: 2471–2477.

Storlien, L.H., et al. (1987) Fish oil prevents insulin resistance induced by high-fat feeding in rats. Science, 237: 885.

Swanson, J., Likesh, B., Kinsella, J. (1989) Ca^{2+}/Mg^{2+}ATPase of mouse cardiac sarcoplasmic reticulum is affected by membrane n-6 and n-3 polyunsaturated fatty acid content. J. Nutr., 119: 364–372.

Taffet, G., et al. (1993) The calcium uptake of the rat heart sarcoplasmic reticulum is altered by dietary lipid. J. Membr. Biol., 131: 35–42.

Tarstedt, M., Larko, O., Molin, L., Wennberg, A.M. (2005) Increasing number of skin cancer cases-also among the younger. Lakartidningen, 102: 1972–5.

Temme, E.H.M., Mensink, R.P., Hornstra, G. (1996) Comparison of the effects of diets enriched in lauric, palmitic, or oleic acids on serum lipids and lipoproteins in healthy women and men. Am. J. Clin. Nutr., 63: 897–903.

Terano, T., et al. (1989) The effect of highly purified eicosapentaenoic acid in patients with psoriasis. Adv. Prostaglandin Thromboxane Leukotriene Res., 17: 880–885.

Terano, T., et al. (1999) Docosahexaenoic acid supplementation improves the moderately severe dementia from thrombotic cerebrovascular diseases. Lipids, 34: 345.

van Doormaal, J.J., et al. (1984) The plasma and erythrocyte fatty acid composition of poorly controlled, insulin-dependent (type I) diabetic patients and the effect of improved metabolic control. Clin. Chim. Acta., 144: 203.

Vessby, B. (2000) Dietary fat and insulin resistance. Br J Nutr., 83 Suppl 1: 91–96.

Watson, R.R., Zibadi, S., Vazquez, R., Larson, D. (2005) Nutritional regulation of immunosenescence for heart health. J Nutr Biochem., 16: 85–87.

Weggemans, R.M., Rudrum, M., Trautwein, E.A. (2004) Intake of ruminant versus industrial trans fatty acids and rise of coronary heart disease. Eur J Lipid Sci Technol., 6: 390–7.

Wu, D., et al., (1999) Effect of dietary supplementation with black currant seed oil on the immune response of healthy elderly subjects. Am J Clin Nutr, 70: 536–43.

Yazawa, K. (2004) Importance of 'health foods', EPA and DHA for preventive medicine. Rinsho Byori., 52: 249–253.

Yin, H., Porter, N.A. (2005) New insights regarding the autoxidation of polyunsaturated fatty acids. Antioxid Redox Signal, 7: 170–184.

Young, G., Conquer, J. (2005) Omega-3 fatty acids and neuropsychiatric disorders. Reprod Nutr Dev., 45: 1–28.

Zhao, G. (2004) Dietary alpha-linolenic acid reduces inflammatory and lipid cardiovascular risk factors in hypercholesterolemic men and women. Nutr., 134: 2991–7.

Ziboh, V.A., et al. (1986) Effects of dietary supplementation of fish oil on neutrophil and epidermal fatty acids. Modulation of clinical course of psoriatic subjects. Arch. Dermatol., 122: 1277–1282.

Zock, P.L., et al. (1997) Butter, Margarine and Serum Lipoproteins. Atherosclerosis, 131: 7–16.

INDEX

MIX
Papier aus verantwortungsvollen Quellen
Paper from responsible sources
FSC® C105338

If you have any concerns about our products,
you can contact us on
ProductSafety@springernature.com

In case Publisher is established outside the EU,
the EU authorized representative is:
Springer Nature Customer Service Center GmbH
Europaplatz 3, 69115 Heidelberg, Germany

Printed by Libri Plureos GmbH
in Hamburg, Germany